T0094307

Verilog® HDL

Digital Design and Modeling

Verilog® HDL

Digital Design and Modeling

Joseph Cavanagh
Santa Clara University

CRC Press
Taylor & Francis Group
Boca Raton   London   New York

CRC Press is an imprint of the
Taylor & Francis Group, an informa business

CRC Press
Taylor & Francis Group
6000 Broken Sound Parkway NW, Suite 300
Boca Raton, FL 33487-2742

© 2007 by Taylor & Francis Group, LLC
CRC Press is an imprint of Taylor & Francis Group, an Informa business

No claim to original U.S. Government works
Printed in the United States of America on acid-free paper
10 9 8 7 6 5 4 3

International Standard Book Number-10: 1-4200-5154-7 (Hardcover)
International Standard Book Number-13: 978-1-4200-5154-4 (Hardcover)

This book contains information obtained from authentic and highly regarded sources. Reprinted material is quoted
with permission, and sources are indicated. A wide variety of references are listed. Reasonable efforts have been made to
publish reliable data and information, but the author and the publisher cannot assume responsibility for the validity of
all materials or for the consequences of their use.

No part of this book may be reprinted, reproduced, transmitted, or utilized in any form by any electronic, mechanical, or
other means, now known or hereafter invented, including photocopying, microfilming, and recording, or in any information storage or retrieval system, without written permission from the publishers.

For permission to photocopy or use material electronically from this work, please access www.copyright.com (http://
www.copyright.com/) or contact the Copyright Clearance Center, Inc. (CCC) 222 Rosewood Drive, Danvers, MA 01923,
978-750-8400. CCC is a not-for-profit organization that provides licenses and registration for a variety of users. For orga-
nizations that have been granted a photocopy license by the CCC, a separate system of payment has been arranged.

Trademark Notice: Product or corporate names may be trademarks or registered trademarks, and are used only for
identification and explanation without intent to infringe.

Library of Congress Cataloging-in-Publication Data

Verilog HDL : digital design and modeling / Joseph Cavanagh.
    p. cm.
   Includes bibliographical references and index.
   ISBN 1-4200-5154-7 (alk. paper)
   1. Digital electronics. 2. Logic circuits--Computer-aided design. 3. Verilog (Computer hardware
description language) I. Title.

TK7868.D5C395 2007
621.39'2--dc22
                                         2006052725

Visit the Taylor & Francis Web site at
http://www.taylorandfrancis.com

and the CRC Press Web site at
http://www.crcpress.com

By the same author:

DIGITAL COMPUTER ARITHMETIC: Design and Implementation

SEQUENTIAL LOGIC: Analysis and Synthesis

To my children,
Brad, Janice, and Valerie

CONTENTS

Appendix B   Verilog Project Procedure 771

Appendix C   Answers to Select Problems 773

Index 891

PREFACE

There are two dominant hardware description languages (HDLs): Verilog HDL and Very High Speed Integrated Circuit (VHSIC) HDL (VHDL). Both are Institute of Electrical and Electronics Engineers (IEEE) standards: Verilog IEEE standard 1364-1995 and VHDL IEEE standard 1076-1993. Of the two hardware description languages, Verilog HDL is the most widely used. The Verilog language provides a means to model a digital system at many levels of abstraction from a logic gate to a complex digital system to a mainframe computer.

The purpose of this book is to present the complete Verilog language together with a wide variety of examples so that the reader can gain a firm foundation in the design of digital systems using Verilog HDL. The different modeling constructs supported by Verilog are described in detail. Numerous examples are designed in each chapter, including both combinational and sequential logic.

The examples include counters of different moduli, half adders, full adders, a carry lookahead adder, array multipliers, the Booth multiply algorithm, different types of Moore and Mealy machines, including sequence detectors, a Hamming code error detection and correction circuit, a binary-coded decimal (BCD) adder/subtractor, arithmetic and logic units (ALUs), and the complete design of a pipelined reduced instruction set computer (RISC) processor. Also included are synchronous sequential machines and asynchronous sequential machines, including pulse-mode asynchronous sequential machines.

Emphasis is placed on the detailed design of various Verilog projects. The projects include the design module, the test bench module, the outputs obtained from the simulator, and the waveforms obtained from the simulator that illustrate the complete functional operation of the design. Where applicable, a detailed review of the theory of the topic is presented together with the logic design principles. This includes state diagrams, Karnaugh maps, equations, and the logic diagram.

The book is intended to be tutorial, and as such, is comprehensive and self contained. All designs are carried through to completion — nothing is left unfinished or partially designed. Each chapter includes numerous problems of varying complexity to be designed by the reader.

Chapter 1 provides a short history of HDLs and introduces Verilog HDL. Different modeling constructs are presented as well as different ways to indicate the active level (or assertion level) of a signal.

Chapter 2 presents an overview of Verilog HDL and discusses the different design methodologies used in designing a project. The chapter is intended to introduce the reader to the basic concepts of Verilog modeling techniques, including dataflow modeling, behavioral modeling, and structural modeling. Examples are presented to illustrate the different modeling techniques. There is also a section that incorporates more than one modeling construct into a mixed-design model. Later chapters present these modeling constructs in more detail. The concept of ports and modules is introduced in conjunction with the use of test benches for module design verification.

Chapter 3 presents the language elements used in Verilog. These consist of comments, identifiers, keywords, data types, parameters, and a set of values that determine the logic state of a net. Comments are placed in the Verilog code to explain the function of a line of code or a block of code. Identifiers are names given to an object or variable so that they can be referenced elsewhere in the module. Verilog provides a list of predefined keywords that are used to define the language constructs. There are two predefined data types: nets and registers. Nets connect logical elements; registers provide storage elements. Compiler directives are used to induce changes during the compilation of a Verilog module.

Chapter 4 covers the expressions used in Verilog. Expressions consist of operands and operators, which are the basis of the language. Operands can be any of the following data types: constant, parameter, net, register, bit-select, part-select, or a memory element. Verilog contains a large set of operators that perform various operations on different types of data. The following categories of operators are available in Verilog: arithmetic, logical, relational, equality, bitwise, reduction, shift, conditional, concatenation, and replication.

Chapter 5 introduces gate-level modeling using built-in primitive gates. Verilog has a profuse set of built-in primitive gates that are used to model nets, including **and**, **nand**, **or**, **nor**, **xor**, **xnor**, and **not**, among others. This chapter presents a design methodology that is characterized by a low level of abstraction, in which the logic hardware is described in terms of gates. This is similar to designing logic by drawing logic gate symbols. Gate delays are also introduced in this chapter. All gates have a propagation delay, which is the time necessary for a signal to propagate from the input terminals, through the internal circuitry, to the output terminal. Examples of iterative networks and a priority encoder are presented as design examples using built-in primitives.

Chapter 6 covers user-defined primitives (UDPs), which are primitive logic functions that are designed according to user specifications. These primitive functions are usually at a higher level of abstraction than the built-in primitives. They are independent primitives and do not instantiate other primitives or modules. They can be used in the design of both combinational and sequential logic circuits. Sequential primitives include level-sensitive and edge-sensitive circuits. Several design examples are included in this chapter, including a binary-to-Gray code converter, a full adder designed from two half adders, multiplexers, a level-sensitive latch, an edge-sensitive flip-flop, a modulo-8 counter, and a Moore finite-state synchronous sequential machine, among others.

Chapter 7 presents dataflow modeling, which is the first of three primary modeling constructs. Dataflow modeling is at a higher level of abstraction than either built-in primitives or UDPs. Dataflow modeling uses the continuous assignment statement to design combinational logic without employing logic gates and interconnecting wires. Design examples presented in this chapter include the use of reduction operators, an octal-to-binary encoder, a multiplexer design using the conditional operator, a 4-bit adder, a high-speed carry lookahead adder, asynchronous sequential machines, and a Moore pulse-mode asynchronous sequential machine. All examples include test benches to test the design modules for correct functional operation, outputs from simulation, and waveforms.

Chapter 8 covers the concepts of behavioral modeling, which describe the behavior of a digital system and is not concerned with the direct implementation of logic gates, but rather the architecture of the system. Behavioral modeling represents a higher level of abstraction than previous modeling methods. A Verilog module that is designed using behavioral modeling contains no internal structural details; it simply defines the behavior of the hardware in an abstract, algorithmic description. Verilog contains two structured procedure statements or behaviors: **initial** and **always**. This chapter introduces procedural assignments and different delay techniques. Procedural assignments include blocking and nonblocking assignments. Conditional statements, which alter the flow within a behavior based upon certain conditions, are addressed. An alternative to conditional statements is the **case** statement, which is a multiple-way conditional branch. Looping statements are also presented. Many complete design examples are illustrated, which include a carry lookahead adder, an add-shift unit, an odd parity generator, a parallel-in, serial-out shift register, counters that count in different sequences, ALUs, various Moore and Mealy synchronous sequential machines, and asynchronous sequential machines.

Chapter 9 covers the third main modeling technique, structural modeling. Structural modeling consists of instantiating one or more of the following objects into a design module: built-in primitives, UDPs, or other design modules. This chapter presents several complete design examples using structural modeling constructs. The examples include a Gray-to-binary code converter, a BCD-to-decimal decoder, a modulo-10 counter, an adder-subtractor, an adder and high-speed shifter, an array multiplier, Moore and Mealy synchronous and asynchronous sequential machines, and a Moore pulse-mode asynchronous sequential machine. As in other chapters, the examples contain the design module, the test bench module, the outputs, and the waveforms.

Chapter 10 presents tasks and functions, which are similar to procedures or subroutines found in other programming languages. These constructs allow a behavioral module to be partitioned into smaller segments. Tasks and functions permit modules to execute common code segments that are written once and then called when required. Examples are given for both tasks and functions.

Chapter 11 presents several design examples utilizing the modeling methods covered in previous chapters. The designs are usually more complex than those previously given. As in other chapters, the designs are complete and include the design module, the test bench module, the outputs, and the waveforms. The examples include a Johnson counter, a counter shifter module, a universal shift register, a Hamming code error detection and correction circuit with accompanying theory, the Booth multiply algorithm with the Booth method described in detail, various Moore and Mealy sequential machines, including a Mealy one-hot machine, a BCD adder/subtractor, and the complete design of a pipelined RISC processor.

Appendix A presents a brief discussion on event handling using the event queue. Operations that occur in a Verilog module are typically handled by an event queue. Appendix B presents a procedure to implement a Verilog project. Appendix C contains the solutions to selected problems in each chapter.

The material presented in this book represents more than two decades of computer equipment design by the author. The book is not intended as a text on logic design, although this subject is reviewed where applicable. It is assumed that the reader has an adequate background in combinational and sequential logic design. The book presents the Verilog HDL with numerous design examples to help the reader thoroughly understand this popular HDL.

This book is designed for practicing electrical engineers, computer engineers, and computer scientists; for graduate students in electrical engineering, computer engineering, and computer science; and for senior-level undergraduate students.

A special thanks to Dr. Ivan Pesic, CEO of Silvaco International, for allowing use of the SILOS Simulation Environment software for the examples in this book. SILOS is an intuitive, easy to use, yet powerful Verilog HDL simulator for logic verification.

I would like to express my appreciation and thanks to the following people who gave generously of their time and expertise to review the manuscript and submit comments: Professor Daniel W. Lewis, Chair, Department of Computer Engineering, Santa Clara University who supported me in all my endeavors; Dr. Geri Lamble; Steve Midford, who reviewed the entire manuscript and offered many helpful suggestions and comments; and Ron Lewerenz. Thanks also to Nora Konopka and the staff at Taylor & Francis for their support.

Joseph Cavanagh

1

Introduction

This book covers the design of combinational and sequential logic using the Verilog hardware description language (HDL). An HDL is a method of designing digital hardware by means of software. A considerable saving of time can be realized when designing systems using an HDL. This offers a competitive advantage by reducing the time-to-market for a system. Another advantage is that the design can be simulated and tested for correct functional operation before implementing the system in hardware. Any errors that are found during simulation can be corrected before committing the design to expensive hardware implementation.

1.1 History of HDL

HDLs became popular in the 1980s and were used to describe large digital systems using a textual format rather than a schematic format such as logic diagrams. With the advent of application-specific integrated circuits (ASICs), field programmable gate arrays (FPGAs), and complex programmable logic devices (CPLDs), computer-aided engineering techniques became necessary. This allowed the engineers to use a programming language to design the logic of the system. Using this technique, test benches could be designed to simulate the entire system and obtain binary outputs and waveforms. Many of these languages were proprietary and not placed in the public domain. Verilog HDL is one of two primary hardware description languages. The other main HDL is the Very High Speed Integrated Circuit (VHSIC) hardware description language (VHDL). VHDL was developed for the United States Department of Defense and was created jointly by IBM, Texas Instruments, and Intermetrics. Although VHDL has not achieved the widespread acceptance of Verilog, both are IEEE standards.

HDLs allow the designer to easily and quickly express architectural concepts in a precise notation without the aid of logic diagrams. All HDLs express the same fundamental concepts, but in slightly different notations.

1.2 Verilog HDL

The Verilog hardware description language is the state-of-the-art method for designing digital and computer systems. Verilog HDL is a C-like language — with some Pascal syntax — used to model a digital system at many levels of abstraction from a logic gate to a complex digital system to a mainframe computer. The combination of C and Pascal syntax makes Verilog easy to learn. The completed design is then simulated to verify correct functional operation. Verilog HDL is the most widely used HDL in the industry.

The Verilog HDL is able to describe both combinational and sequential logic, including level-sensitive and edge-triggered storage devices. Verilog provides a clear relationship between the language syntax and the physical hardware.

The Verilog simulator used in this book is easy to learn and use, yet powerful enough for any application. It is a logic simulator — called SILOS — developed by Silvaco International for use in the design and verification of digital systems. The SILOS simulation environment is a method to quickly prototype and debug any ASIC, FPGA, or CPLD design. It is an intuitive environment that displays every variable and port from a module to a logic gate. SILOS allows single stepping through the Verilog source code, as well as drag-and-drop ability from the source code to a data analyzer for waveform generation and analysis.

Verilog HDL supports a top-down design approach of hierarchical decomposition as well as a bottom-up approach. In a top-down design method, the top-level block is defined, then each sub-block that is used to build the top-level is defined. These second-level blocks are then further subdivided until the lowest level is defined. In a bottom-up method, the building blocks (modules) are first defined. These modules are then used to build larger modules, which are then instantiated into a structural module. Verilog can be used to model algorithms, Boolean equations, and individual logic gates. Simulation occurs at different levels. The low-level modules are first designed and tested for correct functional operation by the simulator using a test bench. These modules are then instantiated into the top-level (structural) module, which is then simulated by means of a test bench.

1.2.1 IEEE Standard

Verilog HDL was developed by Phillip Moorby in 1984 as a proprietary HDL for Gateway Design Automation. Gateway was later acquired by Cadence Design Systems, which placed the language in the public domain in 1990. The Open Verilog International was then formed to promote the Verilog HDL language. In 1995, Verilog was made an IEEE standard HDL (IEEE Standard 1364-1995) and is described in the Verilog Hardware Description Language Reference Manual.

1.2.2 Features

Logic primitives such as AND, OR, NAND, and NOR gates are part of the Verilog language. These are built-in primitives that can be instantiated into a module. The designer also has the option of creating a user-defined primitive (UDP), which can then be instantiated into a module in the same way as a built-in primitive. UDPs can be any logic function such as a multiplexer, decoder, encoder, or flip-flop.

Different types of delays can be introduced into a logic circuit including: interstatement, intrastatement, inertial, and transport delays. These will be defined later in the appropriate sections.

Designs can be modeled in three different modeling constructs: (1) dataflow, (2) behavioral, and (3) structural. Module design can also be done in a mixed-design style, which incorporates the above constructs as well as built-in and user-defined primitives. Structural modeling can be described for any number of module instantiations.

Verilog does not impose a limit to the size of the system; therefore, SILOS can be used to design any size system. Verilog can be used not only to design all the modules of a system, but also to design the test bench that is used for simulation.

Verilog also has available bitwise logic functions such as bitwise AND (&) and bitwise OR (|). High-level programming language constructs such as multiway branching (case statements), conditional statements, and loops are also available.

1.3 Assertion Levels

There are different ways to indicate the active level (or assertion) of a signal. Table 1.1 lists various methods used by companies and textbooks. This book will use the +A and −A method. The AND function can be represented three ways, as shown in Figure 1.1, using an AND gate, a NAND gate, and a NOR gate. Although only two inputs are shown, both AND and OR circuits can have three or more inputs. The plus (+) and minus (−) symbols that are placed to the left of the variables indicate a high or low voltage level, respectively. This indicates the asserted (or active) voltage level for the variables; that is, the *logical 1* (or true) state, in contrast to the *logical 0* (or false) state.

Table 1.1 Assertion Levels

Active high assertion +A	A	A(H)	A	A	A
Active low assertion −A	¬A	A(L)	*A	$\overline{A}$	A'

Thus, a signal can be asserted either plus or minus, depending upon the active condition of the signal at that point. For example, Figure 1.1(a) specifies that the AND function will be realized when both input *A* and input *B* are at their more positive potential, thus generating an output at its more positive potential. The word *positive* as used here does not necessarily mean a positive voltage level, but merely the more positive of two voltage levels. Therefore, the output of the AND gate of Figure 1.1(a) can be written as +(*A* & *B*).

To illustrate that a plus level does not necessarily mean a positive voltage level, consider two logic families: transistor-transistor logic (TTL) and emitter-coupled logic (ECL). The TTL family uses a +5 volt power supply. A plus level is approximately +3.5 volts and above; a minus level is approximately +0.2 volts. The ECL family uses a −5.2 volt power supply. A plus level is approximately −0.95 volts; a minus level is approximately −1.7 volts. Although −0.95 volts is a negative voltage, it is the more positive of the two ECL voltages.

The logic symbol of Figure 1.1(b) is a NAND gate in which inputs *A* and *B* must both be at their more positive potential for the output to be at its more negative potential. A small circle (or wedge symbol for IEEE std 91-1984 logic functions) at the input or output of a logic gate indicates a more negative potential. The output of the NAND gate can be written as −(*A* & *B*).

Figure 1.1(c) illustrates a NOR gate used for the AND function. In this case, inputs *A* and *B* must be active (or asserted) at their more negative potential in order for the output to be at its more positive potential. The output can be written as +(*A* & *B*). A variable can be active (or asserted) at a high and a low level at the same time, as shown in Figure 1.2.

Figure 1.1(d) shows a NAND gate used for the OR function. Either input *A* or *B* (or both) must be at its more negative potential to assert the output at its more positive potential. The output can be written as +(*A* | *B*), where the symbol (|) indicates the OR operation in Verilog.

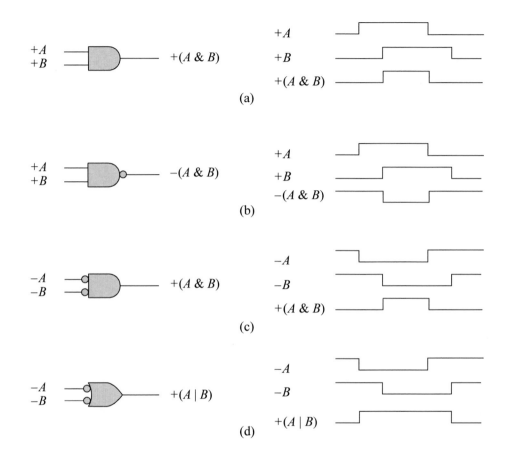

Figure 1.1   Logic symbols and waveforms for AND, NAND, NOR, and NAND (negative-input OR).

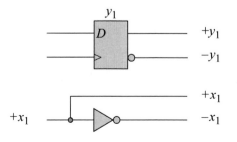

Figure 1.2   Signals can be active high and low at the same time.

2

Overview

This chapter provides a brief introduction to the design methodologies and modeling constructs of the Verilog hardware description language (HDL). Modules and ports will be presented. Modules are the basic units that describe the design of the Verilog hardware. Ports allow the modules to communicate with the external environment; that is, other modules and input/output signals. Different methods will be presented for designing test benches. Test benches are used to apply input vectors to the module in order to test the functional operation of the module in a simulation environment.

Three module constructs will be described together with applications of each type of module. The different modules are: dataflow modeling, behavioral modeling, and structural modeling. Examples will be shown for each type of modeling. There will also be an introduction to mixed-design modeling, which uses a combination of the three main modeling techniques.

2.1 Design Methodologies

There are two main types of design methodologies: *top-down* design and *bottom-up* design. Figure 2.1 shows the layout for a top-down design. In a top-down design, the top-level block is identified, then the blocks in the next lower level are defined. This process continues until all levels in the structure have been defined. The bottom level of the structure contains blocks that cannot be further divided and can be considered as leaf cells in the tree structure. Figure 2.2 shows the layout for a bottom-up design. In a bottom-up design, the lowest level — containing the leaf cells — is defined first.

These become the building blocks with which to design the next higher level. The blocks from each level become the building blocks for the next higher level. This process continues until the top level is reached.

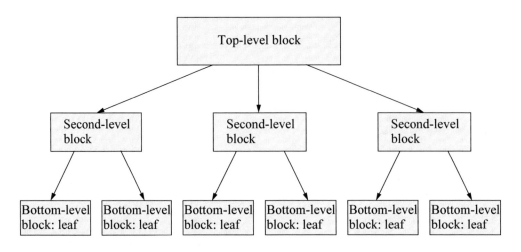

Figure 2.1   Top-down design methodology.

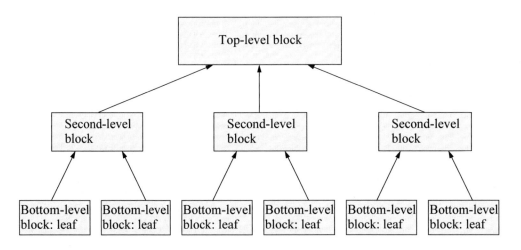

Figure 2.2   Bottom-up design methodology.

The system architect defines the specifications for the machine in the top-level block; for example, a pipelined reduced instruction set computer (RISC). The second-level blocks would consist of the different units in the computer such as instruction cache, instruction unit, decode unit, execution unit, register file, and data cache. The bottom level consists of the hardware in each of the six units.

For example, the instruction cache contains a program counter and an increment-er. The decode unit contains hardware to decode the instruction into its constituent parts such as operation code, source address, and destination address. The execution unit contains the arithmetic and logic unit and associated registers and multiplexers.

2.2 Modulo-16 Synchronous Counter

The top-down approach will now be illustrated for a modulo-16 synchronous counter using D flip-flops. The counting sequence is: $y_3 y_2 y_1 y_0$ = 0000, 0001, 0010, 0011, 0100, 0101, 0110, 0111, 1000, 1001, 1010, 1011, 1100, 1101, 1110, 1111, 0000, . . . , where y_i is the name of the flip-flop. The low-order flip-flop is y_0. Figure 2.3 shows the tree structure for top-down design.

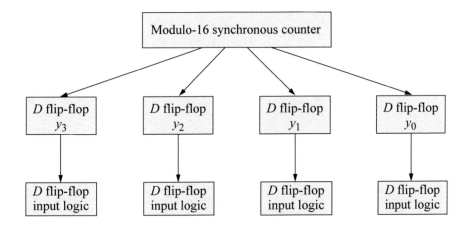

Figure 2.3   Top-down design for a modulo-16 synchronous counter.

The D flip-flop input logic blocks will now be expanded to show more detail. These blocks represent the logic for the input equations of the D flip-flops. The equations will be obtained using Karnaugh maps in the traditional manner. Using the counting sequence shown above, the Karnaugh maps are illustrated in Figure 2.4. The equations for the D flip-flops are shown in Equation 2.1. The logic diagram, obtained from the D flip-flop input equations, is shown in Figure 2.5.

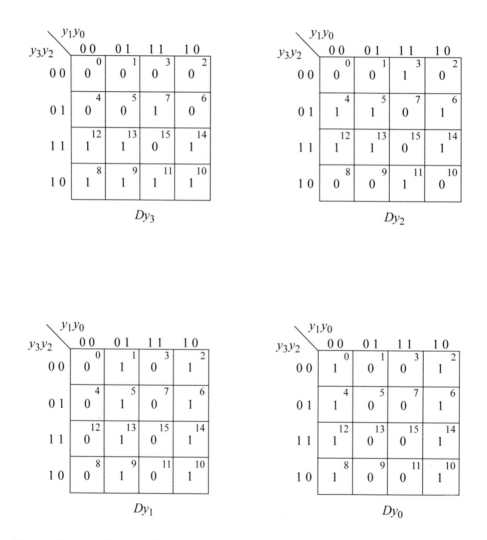

Figure 2.4   Karnaugh maps for the modulo-16 synchronous counter.

The reconfigured top-down design is illustrated in Figure 2.6 showing more detail for the bottom-level leaf nodes. The leaf nodes represent the input logic for the D flip-flops. The input logic is obtained by designing the appropriate gates as separate modules and then instantiating the modules into the structural module that represents the modulo-16 synchronous counter. Modules for the 2-input, 3-input, and 4-input AND gates are labeled AND2, AND3, and AND4, respectively. Modules for the 3-input and 4-input OR gates are labeled OR3 and OR4, respectively. There is one exclusive-OR gate module labeled XOR2.

$$Dy_3 = y_3y_2' + y_3y_1' + y_3y_0' + y_3'y_2y_1y_0$$

$$Dy_2 = y_2y_1' + y_2y_0' + y_2'y_1y_0$$

$$Dy_1 = y_1'y_0 + y_1y_0'$$

$$Dy_0 = y_0' \tag{2.1}$$

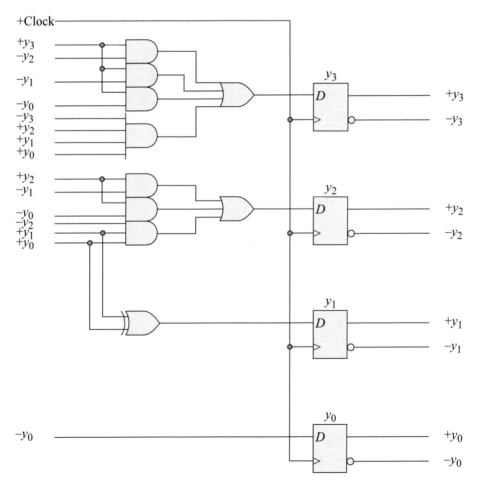

Figure 2.5   Logic diagram for the modulo-16 synchronous counter.

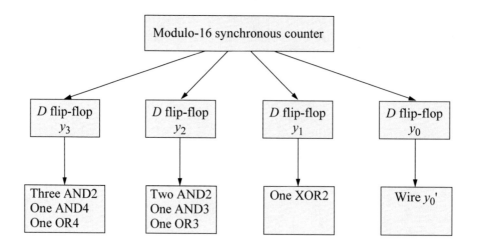

Figure 2.6   Reconfigured top-down design for the modulo-16 synchronous counter showing expanded detail for the bottom-level leaf nodes.

In the top-down design methodology, the functionality of the counter is defined in the top-level block as a modulo-16 synchronous counter. In the next level, the storage elements are established as D flip-flops. The bottom level defines the input logic for the D flip-flops. No further levels are required for this design, because the leaf nodes define the logic gates, which are the lowest level. In a bottom-up methodology, the flow is in the opposite direction by combining small building blocks into larger building blocks.

2.3 Four-Bit Ripple Adder

Another example of top-down methodology is shown in Figure 2.7. This is a ripple adder, which is a relatively low-speed adder because there is no carry-lookahead feature. The tree structure consists of four levels, each successive level providing additional detail of the 4-bit adder.

The 4-bit adder consists of four full adders connected serially in which the carry-out of stage$_i$ is carry-in to stage$_{i+1}$. A full adder adds the two operand bits — augend and addend — plus the carry-in from the previous lower-order stage and produces two outputs: sum and carry. Each full adder is designed using two half adders and one OR gate. A half adder adds the two operand bits only, and produces two outputs: sum and carry. Each half adder is designed using one exclusive-OR gate and one AND gate.

The truth tables for a half adder and full adder are shown in Table 2.1 and Table 2.2, respectively. The corresponding equations are listed in Equation 2.2 and Equation 2.3. The logic diagram for the half adder is shown in Figure 2.8 and for the full adder in Figure 2.9.

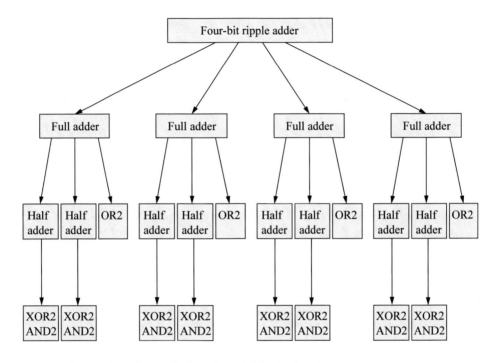

Figure 2.7   Top-down design for a 4-bit ripple adder.

**Table 2.1 Truth Table
for a Half Adder**

a	b	sum	cout
0	0	0	0
0	1	1	0
1	0	1	0
1	1	0	1

**Table 2.2 Truth Table for
a Full Adder**

a	b	cin	sum	cout
0	0	0	0	0
0	0	1	1	0
0	1	0	1	0
0	1	1	0	1
1	0	0	1	0
1	0	1	0	1
1	1	0	0	1
1	1	1	1	1

$$sum = a'b + ab'$$
$$= a \oplus b$$
$$cout = ab \qquad\qquad (2.2)$$

$$sum = a'b'cin + a'bcin' + ab'cin' + abcin$$
$$= cin\,(a \oplus b)' + cin'\,(a \oplus b)$$
$$= a \oplus b \oplus cin$$
$$cout = a'bcin + ab'cin + abcin' + abcin$$
$$= cin\,(a \oplus b) + ab \qquad\qquad (2.3)$$

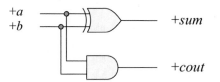

Figure 2.8   Logic diagram for a half adder.

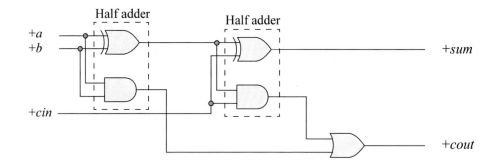

Figure 2.9   Logic diagram for a full adder.

2.4 Modules and Ports

A *module* is the basic unit of design in Verilog. It describes the functional operation of some logical entity and can be a stand-alone module or a collection of modules that are instantiated into a structural module. *Instantiation* means to use one or more lower-level modules in the construction of a higher-level structural module. A module can be a logic gate, an adder, a multiplexer, a counter, or some other logical function.

A module consists of declarative text which specifies the function of the module using Verilog constructs; that is, a Verilog module is a software representation of the physical hardware structure and behavior. The declaration of a module is indicated by the keyword **module** and is always terminated by the keyword **endmodule**.

Verilog has predefined logical elements called *primitives*. These built-in primitives are structural elements that can be instantiated into a larger design to form a more complex structure. Examples are: AND, OR, XOR, and NOT. Built-in primitives are discussed in more detail in Chapter 5.

Modules contain *ports* which allow communication with the external environment or other modules. For example, the full adder of Figure 2.9 has input ports a, b, and cin and output ports sum and $cout$. The general structure and syntax of a module is shown in Figure 2.10. An AND gate can be defined as shown in the module of Figure 2.11, where the input ports are x_1 and x_2 and the output port is z_1.

```
module <module name> (port list);
   declarations
      reg, wire, parameter,
      input, output, . . .
      . . .
   <module internals>
      statements
      initial, always, module instantiation, . . .
      . . .
endmodule
```

Figure 2.10   General structure of a Verilog module.

A Verilog module defines the information that describes the relationship between the inputs and outputs of a logic circuit. A structural module will have one or more instantiations of other modules or logic primitives. In Figure 2.11, the first line is a comment, indicated by (//). In the second line, *and2* is the module name; this is followed by left and right parentheses containing the module ports, which is followed by a semicolon. The inputs and outputs are defined by the keywords **input** and **output**. The ports are declared as **wire** in this dataflow module. Dataflow modeling is covered in detail in Chapter 7. The keyword **assign** describes the behavior of the circuit. Output z_1 is assigned the value of x_1 ANDed (&) with x_2. This continuous assignment statement is discussed in detail in Chapter 7.

```
//dataflow and gate with two inputs
module and2 (x1, x2, z1);

input x1, x2;
output z1;

wire x1, x2;
wire z1;

assign z1 = x1 & x2;

endmodule
```

Figure 2.11   Verilog module for an AND gate with two inputs.

2.4.1 Designing a Test Bench for Simulation

This section describes the techniques for writing test benches in Verilog HDL. When a Verilog module is finished, it must be tested to ensure that it operates according to the machine specifications. The functionality of the module can be tested by applying stimulus to the inputs and checking the outputs. The test bench will display the inputs and outputs in a radix (binary, octal, hexadecimal, or decimal) as well as the waveforms. It is good practice to keep the design module and test bench module separate.

The test bench contains an instantiation of the unit under test and Verilog code to generate input stimulus and to monitor and display the response to the stimulus. Figure 2.12 shows a simple test bench to test the 2-input AND gate of Figure 2.11. Line 1 is a comment indicating that the module is a test bench for a 2-input AND gate. Line 2 contains the keyword **module** followed by the module name, which includes *tb* indicating a test bench module. The name of the module and the name of the module under test are the same for ease of cross-referencing.

Line 4 specifies that the inputs are **reg** type variables; that is, they contain their values until they are assigned new values. Outputs are assigned as type **wire** in test benches. Output nets are driven by the output ports of the module under test. Line 7 contains an **initial** statement, which executes only once. Verilog provides a means to monitor a signal when its value changes. This is accomplished by the **$monitor** task. The **$monitor** continuously monitors the values of the variables indicated in the parameter list that is enclosed in parentheses. It will display the value of the variables whenever a variable changes state. The quoted string within the task is printed and specifies that the variables are to be shown in binary (%b). The **$monitor** is invoked only once. Line 11 is a second **initial** statement that allows the procedural code between the **begin** . . . **end** block statements to be executed only once.

```
 1 //and2 test bench
   module and2_tb;

   reg x1, x2;
 5 wire z1;

   //display variables
   initial
   $monitor ("x1 = %b, x2 = %b, z1 = %b", x1, x2, z1);

10 //apply input vectors
   initial
   begin
       #0      x1 = 1'b0;
               x2 = 1'b0;

15
       #10     x1 = 1'b0;
               x2 = 1'b1;

       #10     x1 = 1'b1;
20             x2 = 1'b0;

       #10     x1 = 1'b1;
               x2 = 1'b1;

25     #10     $stop;

   end

   //instantiate the module into the test bench
   and2 inst1 (
30     .x1(x1),
       .x2(x2),
       .z1(z1)
       );

   endmodule
```

Figure 2.12   Test bench for the 2-input AND gate of Figure 2.11.

Lines 13 and 14 specify that at time 0 (#0), inputs x_1 and x_2 are assigned values of 0, where 1 is the width of the value (one bit), ' is a separator, b indicates binary, and 0 is the value. Line 16 specifies that 10 time units later, the inputs change to: $x_1 = 0$ and $x_2 = 1$. This process continues until all possible values of two variables have been applied to the inputs. Simulation stops at 10 time units after the last input vector has been applied (**$stop**). The total time for simulation is 40 time units — the sum of all the time units.

Line 29 begins the instantiation of the module into the test bench. The name of the instantiation must be the same as the module under test, in this case, *and2*. This is followed by an instance name (*inst1*) followed by a left parenthesis. The $.x_1$ variable in line 30 refers to a port in the module that corresponds to a port (x_1) in the test bench. All the ports in the module under test must be listed. The keyword **endmodule** is the last line in the test bench.

The binary outputs for this test bench are shown in Figure 2.13 and the waveforms in Figure 2.14. The output can be presented in binary (b or B), in octal (o or O), in hexadecimal (h or H), or in decimal (d or D). The waveforms show the values of the input and output variables at the specified simulation times. Notice that the inputs change value every 10 time units and that the duration of simulation is 40 time units. The time units can be specified for any duration.

The Verilog syntax will be covered in greater detail in subsequent chapters. It is important at this point to concentrate on how the module under test is simulated and instantiated into the test bench.

```
x1 = 0, x2 = 0, z1 = 0
x1 = 0, x2 = 1, z1 = 0
x1 = 1, x2 = 0, z1 = 0
x1 = 1, x2 = 1, z1 = 1
```

Figure 2.13   Binary outputs for the test bench of Figure 2.12 for a 2-input AND gate.

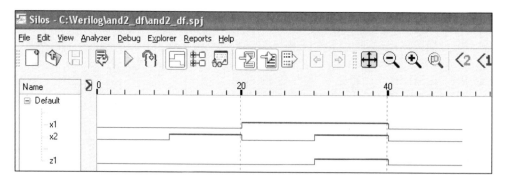

Figure 2.14   Waveforms for the test bench of Figure 2.12 for a 2-input AND gate.

Several different methods to generate test benches will be shown in subsequent sections. Each design in the book will be tested for correct operation by means of a test bench. Test benches provide clock pulses that are used to control the operation of a synchronous sequential machine. An **initial** statement is an ideal method to generate a waveform at discrete intervals of time for a clock pulse. The Verilog code in Figure 2.15 illustrates the necessary statements to generate clock pulses that have a duty cycle of 20%. The clock pulse waveform is shown in Figure 2.16.

```verilog
//generate clock pulses of 20% duty cycle
module clk_gen (clk);
output clk;
reg clk;

initial
begin
    #0      clk = 0;
    #5      clk = 1;
    #5      clk = 0;
    #20     clk = 1;
    #5      clk = 0;
    #20     clk = 1;
    #5      clk = 0;
    #10     $stop;
end
endmodule
```

Figure 2.15   Verilog code to generate clock pulses with a 20% duty cycle.

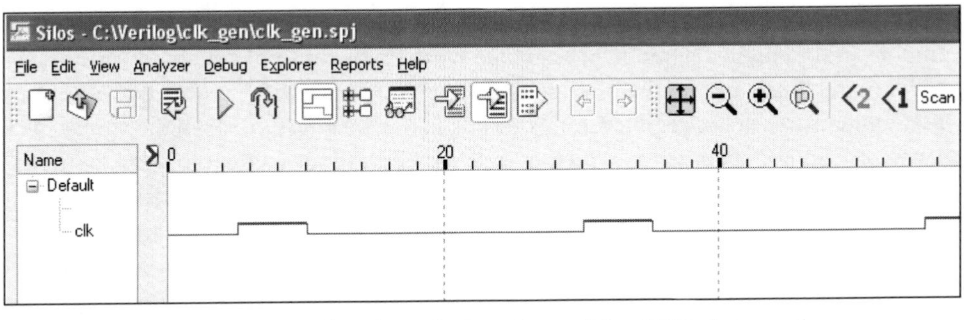

Figure 2.16   Waveform showing clock pulses with a 20% duty cycle.

In Figure 2.15, the clock begins at time 0 with a value of 0. Five time units later, the clock is assigned a value of 1; five time units later (at time 10), the clock goes low (0), 20 time units later (at time 30), the clock goes high (1); five time units later (at time 35), the clock is again assigned a value of 0. It is important to note that the times specified in the code are accumulative; that is, when the clock goes low five time units after the second pulse, this occurs at time 35. Simulation stops at time 70, 10 time units after the last pulse goes low.

Before introducing dataflow, behavioral, structural, and mixed modeling techniques, some definitions will be presented. These definitions fall into the general category of procedural constructs.

2.4.2 Construct Definitions

Continuous assignment   The continuous assignment statement is used to describe combinational logic where the output of the circuit is evaluated whenever an input changes; that is, the value of the right-hand side expression is *continuously assigned* to the left-hand side net. Continuous assignments are similar to Boolean algebra, which is a systematic treatment of logical operations. Continuous assignments can be used only for nets, not for register variables. A continuous assignment, as shown below, establishes a relationship between a right-hand side expression and a left-hand side net. A continuous assignment occurs outside of an **initial** or an **always** statement. The syntax for a continuous assignment statement is:

$$\textbf{assign} \ \ [\text{delay}] \ \text{lhs\_net} = \text{rhs\_expression}$$

Whenever the value of a variable in the right-hand side expression changes, the right-hand side is evaluated and the value is assigned to the left-hand side after any specified delay. That is, the right-hand side creates a value that is assigned to the target variable. The delay specifies the time duration between the time a variable on the right-hand side changes and the time the new value is assigned to the left-hand side.

Procedural continuous assignment   This is a procedural version of the continuous assignment statement and is made within a behavioral construct (**initial** or **always**) and creates a dynamic binding to a variable. This assignment overrides existing assignments to a net or register; for example, using **assign** ... **deassign** or **force** ... **release** procedural statements, which allows for continuous assignments to be made to registers for a specific period of time. A continuous assignment, used for combinational logic, is a static binding because its effect is retained for the duration of simulation. The dynamic binding of a procedural continuous assignment means that the assignment can be changed during execution of a behavioral procedure.

Procedure   As in any programming language, a procedure is simply a sequence of operations that produce a result.

Procedural statement   A procedural statement is a synonym for instruction.

Procedural assignment   A procedural assignment statement represents a logic function that is derived from the right-hand side of an assignment statement and assigned to the left-hand side target. A procedural assignment statement can occur only within an **initial** or an **always** statement. The value assigned to a variable remains unchanged until another procedural assignment changes the value. There are two types of procedural assignment statements: blocking and nonblocking.

Blocking statement   Blocking assignments are used only when describing combinational logic in a procedural block. Blocking assignment statements are executed in the order in which they are listed in a procedural block. In a blocking assignment statement, the assignment to the left-hand side variable takes place before the following statement in the sequential block is executed. The symbol that represents a blocking statement is (=).

Nonblocking statement   Nonblocking assignments allow scheduling of assignments without blocking the following statements in a procedural block. Nonblocking assignments are used to describe only sequential elements in a procedural assignment. The simulator processes all blocking assignments first. Once the assignments have been made, all the nonblocking assignments are evaluated and placed on the event queue (the event queue is presented in Chapter 11).

All variables are evaluated and assigned in the current simulation time. Once all the variables in the blocking assignments have been evaluated, the variables using nonblocking assignments are evaluated and placed on the event queue. Nonblocking assignment statements are used to synchronize assignment statements so that they appear to happen simultaneously. They allow scheduling of assignments without blocking execution of the statements that follow in a sequential block. The symbol that represents a nonblocking statement is (<=).

2.5   Introduction to Dataflow Modeling

Dataflow modeling is used to design combinational logic only. It allows designers to implement a logical function at a higher level of abstraction than gate level modeling using built-in primitives. The fundamental method of designing in dataflow is the continuous assignment statement **assign**, an example of which is shown below.

$$\textbf{assign}\ \ sum = a \wedge b \wedge cin$$

For example, the 2-input AND gate shown in the Verilog code of Figure 2.11 uses the continuous assignment statement to assign a value to z_1 whenever x_1 or x_2 changes value (**assign** $z_1 = x_1$ & x_2;). Since no delay is specified, the default value is zero delay.

2.5.1 Two-Input Exclusive-OR Gate

This section presents the design of a 2-input exclusive-OR gate using dataflow modeling. The circuit is shown in Figure 2.17 and the Verilog code is in Figure 2.18. Note that the symbol for the exclusive-OR function is the caret.

Figure 2.17   Two-input exclusive-OR gate to be designed using dataflow modeling.

```
//dataflow 2-input exclusive-or gate
module xor2 (x1, x2, z1);

input x1, x2;
output z1;

wire x1, x2;
wire z1;

assign z1 = x1 ^ x2;

endmodule
```

Figure 2.18   Verilog code for the 2-input exclusive-OR gate of Figure 2.17.

The test bench for the exclusive-OR gate will be different than the one shown in Figure 2.12 for the 2-input AND gate. Several new modeling constructs are shown in the exclusive-OR test bench of Figure 2.19. Since there are two inputs to the exclusive-OR gate, all four combinations of two variables must be applied to the circuit. This is accomplished by a **for**-loop statement, which is similar in construction to a **for** loop in the C programming language.

Following the keyword **begin** is the name of the block: *apply_stimulus*. In this block, a 3-bit **reg** variable is declared called *invect*. This guarantees that all combinations of the two inputs will be tested by the **for** loop, which applies input vectors of $x_1x_2 = 00, 01, 10, 11$ to the circuit. The **for** loop stops when the pattern 100 is detected by the test segment (*invect* < 4). If only a 2-bit vector were applied, then the expression (*invect* < 4) would always be true and the loop would never terminate. The increment segment of the **for** loop does not support an increment designated as *invect*++; therefore, the long notation must be used: *invect* = *invect* + 1.

The target of the first assignment within the **for** loop ($\{x_1, x_2\} = invect \, [2:0]$) represents a concatenated target. The concatenation of inputs x_1 and x_2 is performed by positioning them within braces: $\{x_1, x_2\}$. A vector of three bits ([2:0]) is then assigned to the inputs. This will apply inputs of 00, 01, 10, 11, and stop when the vector is 100.

The **initial** statement also contains a system task (**$display**) which prints the argument values — within the quotation marks — in binary. The concatenated variables x_1 and x_2 are listed first; therefore, their values are obtained from the first argument to the right of the quotation marks: $\{x_1, x_2\}$. The value for the second variable z_1 is obtained from the second argument to the right of the quotation marks. The delay time (#10) in the system task specifies that the task is to be executed after 10 time units; that is, the delay between the application of a vector and the response of the module. This delay represents the propagation delay of the logic. The simulation results are shown in binary format in Figure 2.20 and the waveforms in Figure 2.21.

```
//2-input exclusive-or gate test bench
module xor2_tb;

reg x1, x2;
wire z1;

//apply input vectors
initial
begin: apply_stimulus
   reg [2:0] invect;
   for (invect = 0; invect < 4; invect = invect + 1)
      begin
          {x1, x2} = invect [2:0];
          #10 $display ("{x1x2} = %b, z1 = %b",
                         {x1, x2}, z1);
      end
end

//continued on next page
```

Figure 2.19   Test bench for the 2-input exclusive-OR gate module of Figure 2.18.

```
//instantiate the module into the test bench
xor2 inst1 (
    .x1(x1),
    .x2(x2),
    .z1(z1)
    );
endmodule
```

Figure 2.19   (Continued)

```
{x1x2} = 00, z1 = 0
{x1x2} = 01, z1 = 1
{x1x2} = 10, z1 = 1
{x1x2} = 11, z1 = 0
```

Figure 2.20   Binary outputs for the test bench of Figure 2.19.

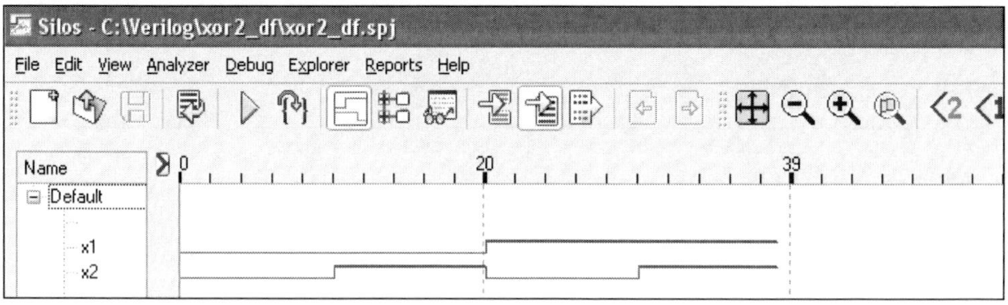

Figure 2.21   Waveforms for the 2-input exclusive-OR gate test bench of Figure 2.19.

2.5.2 Four 2-Input AND Gates with Delay

As a final example in the introduction to dataflow modeling, a circuit will be designed to illustrate a technique to introduce delays in logic gates. All delays in Verilog are specified in terms of time units. The delay value is indicated after the keyword **assign**. For example, a continuous assignment with delay is written as

$$\textbf{assign } \#5\ z_1 = x_1\ \&\ x_2;$$

The #5 indicates five time units. Whenever x_1 or x_2 changes value, the right-hand expression is evaluated and the value is assigned to the left-hand variable z_1 after five time units. The actual unit of time is specified by the `timescale compiler directive. For example, the statement

`timescale 10ns / 100ps

indicates that one time unit is specified as 10 nanoseconds (ns) with a precision of 100 picoseconds (ps). This property is called *inertial* delay.

Figure 2.22 illustrates a logic circuit with four 2-input AND gates. Two scalars — x_1 and x_2 — are specified as the inputs; the output of the circuit is a vector of four bits: z_1, where $z_1 = z_1[0], z_1[1], z_1[2],$ and $z_1[3]$. Figure 2.23 shows the Verilog code for Figure 2.22 using a time unit of 10 ns with a precision of 1 ns.

When input x_1 or x_2 changes value, the new value of the right-hand statement is assigned to the left-side after a delay of 20 ns. The test bench of Figure 2.24 lists six sets of vectors to be assigned to the inputs. The resulting waveforms are shown in Figure 2.25. Note that when x_1 or x_2 changes state, the output of the circuit is delayed by 20 ns.

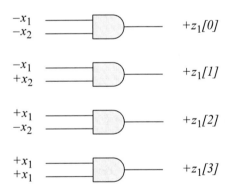

Figure 2.22   Four 2-input AND gates with scalar inputs and a vector output.

```
//dataflow with delay
`timescale 10ns / 1ns
module four_and_delay (x1, x2, z1);

input x1, x2;
output [3:0] z1;              //continued on next page
```

Figure 2.23   Verilog code for the logic circuit of Figure 2.22.

```
assign #2 z1[0] = ~x1 & ~x2;
assign #2 z1[1] = ~x1 & x2;
assign #2 z1[2] = x1 & ~x2;
assign #2 z1[3] = x1 & x2;

endmodule
```

Figure 2.23   (Continued)

```
//four_and_delay test bench          #5 x1 = 1'b0;
module four_and_delay_tb;               x2 = 1'b1;

reg x1, x2;                          #5 x1 = 1'b0;
wire [3:0] z1;                          x2 = 1'b1;

initial                              #5 x1 = 1'b0;
$monitor ("x1 x2=%b, z1=%b",            x2 = 1'b0;
        {x1, x2}, z1);
                                     #5 $stop;
//apply input vectors                end
initial
begin                                //instantiate the module
   #0 x1 = 1'b0;                     //into the test bench
      x2 = 1'b0;                     four_and_delay inst1 (
                                        .x1(x1),
   #5 x1 = 1'b1;                        .x2(x2),
      x2 = 1'b0;                        .z1(z1)
                                        );
   #5 x1 = 1'b1;
      x2 = 1'b1;                     endmodule
```

Figure 2.24   Test bench for the Verilog code of Figure 2.23 for the circuit of Figure 2.22.

2.6 Introduction to Behavioral Modeling

Describing a module in *behavioral* modeling is an abstraction of the functional operation of the design. It does not describe the implementation of the design at the gate level. The outputs of the module are characterized by their relationship to the inputs. The behavior of the design is described using procedural constructs. These constructs are the **initial** statement and the **always** statement.

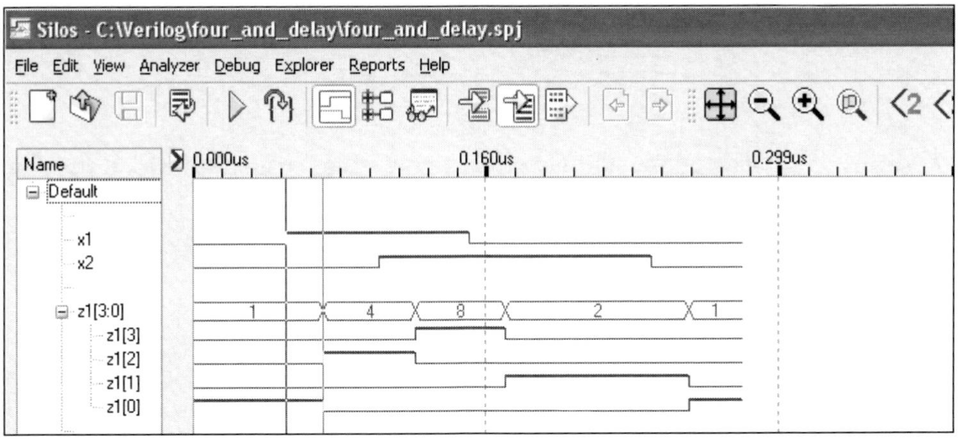

Figure 2.25   Waveforms for the four AND gates with delay using the test bench of Figure 2.24.

The **initial** statement is executed only once during a simulation — beginning at time 0 — and then suspends forever. The **always** statement also begins at time 0 and executes the statements in the **always** block repeatedly in a looping manner. Both statements use only register data types. Objects of the **reg** data types resemble the behavior of a hardware storage device because the data retains its value until a new value is assigned.

2.6.1 Three-Input OR Gate

Figure 2.26 shows a 3-input OR gate, which will be designed using behavioral modeling. The behavioral module is shown in Figure 2.27 using an **always** statement. The expression within the parentheses is called an *event control* or *sensitivity list*. Whenever a variable in the event control list changes value, the statements in the **begin** . . . **end** block will be executed; that is, if either x_1 or x_2 or x_3 changes value, the following statement will be executed:

$$z_1 = x_1 \mid x_2 \mid x_3;$$

where the symbol (|) signifies the logical OR operation.

If only a single statement appears after the **always** statement, then the keywords **begin** and **end** are not required. It is often useful, however, to include the **begin** and **end** keywords because additional statements may have to be added later. The **always** statement has a sequential block (**begin** . . . **end**) associated with an event control. The statements within a **begin** . . . **end** block execute sequentially and execution suspends when the last statement has been executed. When the sequential block completes execution, the **always** statement checks for another change of variables in the event control list.

Figure 2.26   Three-input OR gate to be implemented using behavioral style modeling.

The test bench for the module is shown in Figure 2.28. As stated previously, the inputs for a test bench are of type **reg** because they retain their value until changed, and the outputs are of type **wire**. To ensure that all eight combinations of the inputs are tested, a register vector called *invect* is specified as four bits: 3 through 0, where bit 0 is the low-order bit. This guarantees that a vector of $x_1 x_2 x_3 = 111$ will be applied to the OR gate inputs. The inputs are applied in sequence, $x_1 x_2 x_3 = 000$ through 111. When an input vector of $x_1 x_2 x_3 = 1000$ is reached, the test in the **for** loop returns a value of false and the simulator exits the statements in the **begin** . . . **end** sequence. The binary outputs of the simulator are shown in Figure 2.29 listing the output value for z_1 for all combinations of inputs. The waveforms are shown in Figure 2.30.

```
//behavioral 3-input or gate
module or3 (x1, x2, x3, z1);

input x1, x2, x3;
output z1;

wire x1, x2, x3;
reg z1;

always @ (x1 or x2 or x3)
begin
   z1 = x1 | x2 | x3;
end

endmodule
```

Figure 2.27   Behavioral module for the 3-input OR gate of Figure 2.26.

The statements within the sequential block are called *blocking* statements because execution of the following statement is blocked until the current statement completes execution; that is, the statements within a sequential block execute sequentially. An optional delay may be associated with a procedural assignment. There are two types of delays: interstatement and intrastatement.

```verilog
//or3 test bench
module or3_tb;

reg x1, x2, x3;
wire z1;

//apply input vectors
initial
begin: apply_stimulus
   reg [3:0] invect;
   for (invect = 0; invect < 8; invect = invect + 1)
      begin
         {x1, x2, x3} = invect [3:0];
         #10 $display ("{x1x2x3} = %b, z1 = %b",
                   {x1, x2, x3}, z1);
      end
end

//instantiate the module into the test bench
or3 inst1 (
   .x1(x1),
   .x2(x2),
   .x3(x3),
   .z1(z1)
   );

endmodule
```

Figure 2.28 Test bench for the 3-input OR gate module of Figure 2.27.

```
{x1x2x3} = 000, z1 = 0
{x1x2x3} = 001, z1 = 1
{x1x2x3} = 010, z1 = 1
{x1x2x3} = 011, z1 = 1
{x1x2x3} = 100, z1 = 1
{x1x2x3} = 101, z1 = 1
{x1x2x3} = 110, z1 = 1
{x1x2x3} = 111, z1 = 1
```

Figure 2.29 Binary outputs for the test bench of Figure 2.28.

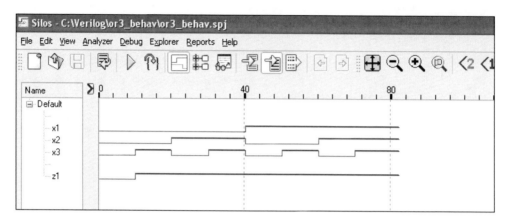

Figure 2.30   Waveforms for the test bench of Figure 2.28 for the 3-input OR gate
module of Figure 2.26.

An *interstatement* delay is the delay by which a statement's execution is delayed;
that is, it is the delay between statements. The following is an example of an interstatement delay:

$$z_1 = x_1 \mid x_2$$
$$\#5\ z_2 = x_1\ \&\ x_2$$

When the first statement has completed execution, a delay of five time units will
be taken before executing the second statement. An *intrastatement* delay is a delay on
the right-hand side of the statement and indicates that the right-hand side is to be evaluated, wait the specified number of time units, and then assign the value to the left-
hand side. This can be used to simulate logic gate delays. The following is an example
of an intrastatement delay:

$$z_1 = \#5\ x_1\ \&\ x_2$$

The statement evaluates the logical function x_1 AND x_2, waits five time units, then assigns the result to z_1. If no delay is specified in a procedural assignment, then zero delay is the default delay and the assignment occurs instantaneously.

Example 2.1   Following is an example of generating waveforms using intrastatement delays in a behavioral module. There are two inputs x_1 and x_2 that will be assigned values based on a timescale of 10 ns / 1 ns. The behavioral module is shown in Figure 2.31 and the waveforms in Figure 2.32.

```
//behavioral intrasegment delay example
`timescale 10ns / 1ns
module intra_stmt_dly_delay (x1,x2);

output x1, x2;
reg x1, x2;

initial
begin
   x1 = #0    1'b0;
   x2 = #0    1'b0;

   x1 = #1    1'b1;
   x2 = #0.5  1'b1;

   x1 = #1    1'b0;
   x2 = #2    1'b0;

   x1 = #1    1'b1;
   x2 = #2    1'b1;

   x1 = #2    1'b0;
   x2 = #1    1'b0;
end
endmodule
```

Figure 2.31   Behavioral module to generate waveforms using intrastatement delays.

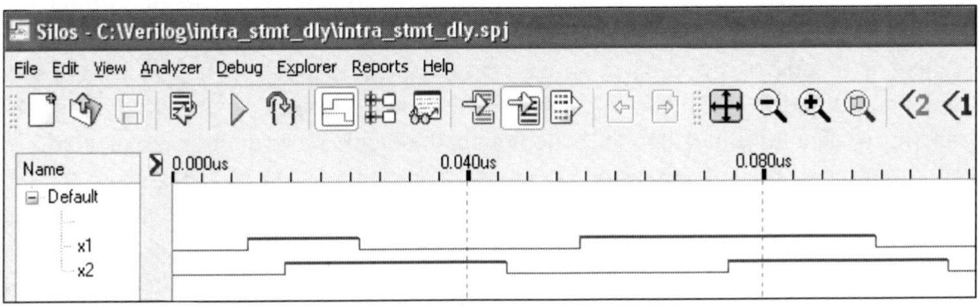

Figure 2.32   Waveforms showing intrastatement delays for Figure 2.31.

2.6.2 Four-Bit Adder

The top-down structural architecture of a 4-bit ripple adder was shown in Section 2.3. A similar 4-bit adder will now be designed using behavioral modeling to illustrate the differences between the two modeling techniques. Figure 2.33 depicts the block diagram of the adder, but does not indicate the internal logic. Figure 2.34 shows the behavioral module for the 4-bit adder. Note that only the *behavior* of the adder is specified, not the internal logic elements as in the structural architecture. The behavior is specified as:

$$sum = a + b + cin$$

The behavioral module does not indicate the specifics as to how to design the adder; the Verilog code simply specifies the functionality of the module. Also, the *sum* variable is specified as a 5-bit vector to include the carry-out of the adder. The augend *a[3:0]* and addend *b[3:0]* remain as 4-bit vectors with the carry-in (*cin*) as a scalar.

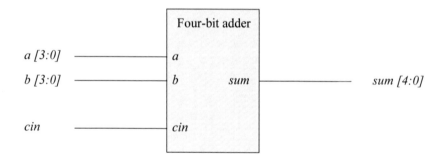

Figure 2.33   Block diagram of a 4-bit adder.

Inputs are declared as type **wire** and outputs as type **reg** in behavioral modeling. Operands *a* and *b* are specified as 4-bit vectors: *a[3]*, . . . , *a[0]* and *b[3]*, . . . , *b[0]*, where *a[0]* and *b[0]* are the low-order bits of *a* and *b*, respectively. The **always** statement continuously checks for a change of value to the operands and the carry-in. If *a* or *b* or *cin* change value, then the statement in the **begin** . . . **end** block is executed. The right-hand side of the expression is evaluated and then assigned to *sum*.

The test bench is shown in Figure 2.35, in which six sets of input vectors are applied to the operands. The binary outputs obtained from the test bench are shown in Figure 2.36 and the corresponding waveforms in Figure 2.37. The values of operands *a*, *b*, and *sum* are given in hexadecimal. By double-clicking the *sum[4:0]* entry in the SILOS Data Analyzer, the individual bits of the sum can be displayed.

```
//behavioral 4-bit adder           wire cin;
module adder_4_behav (a, b,        reg [4:0] sum;
         cin, sum);
                                   always @ (a or b or cin)
input [3:0] a, b;                  begin
input cin;                            sum = a + b + cin;
                                   end
output [4:0] sum;
                                   endmodule
wire [3:0] a, b;
```

Figure 2.34   Behavioral module for a 4-bit adder.

```
//behavioral 4-bit adder test      #10    a = 4'b1001;
//bench                                   b = 4'b0111;
module adder_4_behav_tb;                  cin = 1'b1;

reg [3:0] a, b;                    #10    a = 4'b1101;
reg cin;                                  b = 4'b0111;
                                          cin = 1'b1;
wire [4:0] sum;
                                   #10    a = 4'b1111;
//display variables                       b = 4'b0110;
initial                                   cin = 1'b1;
$monitor ("a b cin = %b_%b_%b,
sum = %b", a, b, cin, sum);        #10    $stop;
                                   end
//apply input vectors
initial                            //instantiate the module into
begin                              //the test bench
   #0    a = 4'b0011;              adder_4_behav inst1 (
         b = 4'b0100;                 .a(a),
         cin = 1'b0;                  .b(b),
                                      .cin(cin),
   #10   a = 4'b1100;                 .sum(sum)
         b = 4'b0011;                 );
         cin = 1'b0;
                                   endmodule
   #10   a = 4'b0111;
         b = 4'b0110;
         cin = 1'b1;
```

Figure 2.35   Test bench for the 4-bit adder module of Figure 2.34.

```
a b cin = 0011_0100_0, sum = 00111
a b cin = 1100_0011_0, sum = 01111
a b cin = 0111_0110_1, sum = 01110
a b cin = 1001_0111_1, sum = 10001
a b cin = 1101_0111_1, sum = 10101
a b cin = 1111_0110_1, sum = 10110
```

Figure 2.36   Binary outputs for the 4-bit adder obtained from the test bench of Figure 2.35.

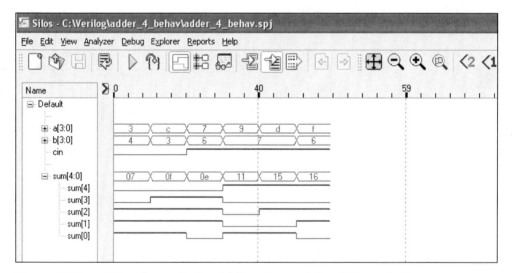

Figure 2.37   Waveforms for the 4-bit adder module of Figure 2.34.

2.6.3 Modulo-16 Synchronous Counter

Figure 2.5 showed the gate-level design of a modulo-16 synchronous counter. The same counter will now be designed using behavioral modeling. This is a much simpler approach because the counter is not designed using Karnaugh maps and equations to provide discrete logic gates and flip-flops. Also, the storage elements are not specified. Verilog is simply given the counting sequence; the counter is then designed by Verilog according to the specifications. Figure 2.38 shows the symbol for the modulo-16 counter.

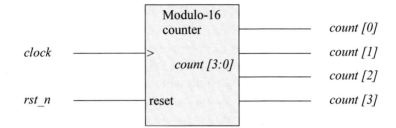

Figure 2.38   Symbol for a modulo-16 counter.

The Verilog module is shown in Figure 2.39 using nonblocking statements in the procedural block. There are three ports: *clk*, *rst_n*, and *count*, which is defined as a 4-bit vector, where *count [0]* is the low-order bit. The **always** statement contains two events in the sensitivity list: *posedge clk* and *negedge rst_n*. When the system clock transitions from a low level to a high level, the counter is incremented. The % symbol indicates a modulus or remainder such that the counter increments from 0 to 15, then begins again at a count of 0. The variable *rst_n* is the conventional way to indicate an active-low reset signal. If the reset signal is at a low voltage level, then the counter is reset to a value of *count* = 0000.

```
//behavioral modulo-16 counter
module ctr_mod_16 (clk, rst_n, count);

input clk, rst_n;
output [3:0] count;

wire clk, rst_n;
reg [3:0] count;

//define counting sequence
always @ (posedge clk or negedge rst_n)
begin
   if (rst_n == 0)
      count <= 4'b0000;
   else
      count <= (count + 1) % 16;
end
endmodule
```

Figure 2.39   Verilog code for a modulo-16 synchronous counter.

The test bench is shown in Figure 2.40. When a change occurs on *count*, the **&monitor** task will display the current count as a binary value. The reset signal is conditioned to an active-low level for five time units initially, then made inactive five time units later. The clock signal is initially set to 0, then alternates between positive and negative levels indefinitely. This is accomplished by the keyword **forever**, which assigns the clock the value of the complemented clock every 10 time units. The duration of simulation is defined to be 320 time units, which is sufficient duration to generate a counting sequence of 0000, ... , 1111, 0000 to show the modulo-16 counting sequence.

The binary outputs are shown in Figure 2.41 as obtained from the **$monitor** task in the test bench. The waveforms are shown in Figure 2.42 and indicate the same counting sequence, but in hexadecimal notation. By clicking the (+) sign to left of the *count [3:0]* signal name or by double-clicking the *count [3:0]* signal name, the individual bits of the counter are obtained.

```
//modulo-16 counter test bench          //define length of
module ctr_mod_16_tb;                    //simulation
                                         initial
reg clk, rst_n;                          begin
wire [3:0] count;                           #320 $stop;
                                         end
initial
$monitor ("count=%b", count);            //instantiate the module
                                         //into the test bench
//define reset                           ctr_mod_16 inst1 (
initial                                     .clk(clk),
begin                                       .rst_n(rst_n),
   #0 rst_n = 1'b0;                         .count(count)
   #5 rst_n = 1'b1;                         );
end                                      endmodule

//define clock
initial
begin
   #0 clk = 1'b0;
   forever
      #10clk = ~clk;
end
```

Figure 2.40   Test bench for the modulo-16 counter of Figure 2.39.

```
count = 0000                    count = 1001
count = 0001                    count = 1010
count = 0010                    count = 1011
count = 0011                    count = 1100
count = 0100                    count = 1101
count = 0101                    count = 1110
count = 0110                    count = 1111
count = 0111                    count = 0000
count = 1000
```

Figure 2.41   Binary outputs for the modulo-16 counter of Figure 2.39 obtained from the test bench of Figure 2.40.

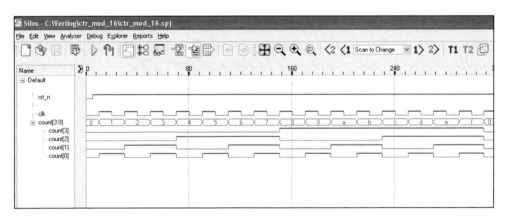

Figure 2.42   Waveforms for the modulo-16 counter of Figure 2.39 obtained from the test bench of Figure 2.40.

2.7 Introduction to Structural Modeling

Structural modeling is a top-level design that is synthesized by instantiating lower-level logic modules into a larger structural module. Each of the lower-level modules (submodules) must have previously been compiled and tested for correct functional operation. A structural module can be described using built-in gate primitives, user-defined primitives, or module instances in any combination. Interconnections between instances are specified using nets.

2.7.1 Sum-of-Products Implementation

This section will implement the sum-of-products expression of Equation 2.4 using structural modeling. The structural module will instantiate the following four sub-modules: *and2*, *and3*, *and4*, and *or3* as shown in Figure 2.43. The ports of the sub-modules are indicated by small black squares and correspond to the port names in the respective modules. The module name is shown outside the dashed line and the instance name associated with the module is shown within the dashed line. The structural module is called *sop*.

$$z_1 = x_1 x_2 + x_2 x_3' x_4 + x_1' x_2' x_3 x_4 \qquad (2.4)$$

The modules for *and2*, *and3*, *and4*, and *or3* are shown in Figure 2.44. The structural module is shown in Figure 2.45, which instantiates the four sub-modules. The instantiation modules and the instance names directly correspond to those shown in Figure 2.43. For example, in instance 2 of the structural module, port $.x_1$ of the *and3* module connects to x_2 ($+x_2$) of the structural module; port $.x_2$ connects to $\sim x_3$ ($-x_3$), port $.x_3$ connects to x_4 ($+x_4$); and port $.z_1$ connects to *net2*. This represents the port connections and internal nets of the structural module. The structural module test bench is shown in Figure 2.46.

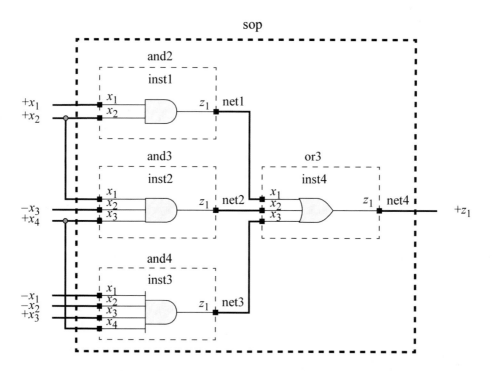

Figure 2.43   Structural module for the sum-of-products equation of Equation 2.4.

```
//dataflow 2-input and gate        //dataflow 4-input and gate
module and2 (x1, x2, z1);          module and4 (x1, x2, x3, x4,
                                                     z1);
input x1, x2;
output z1;                         input x1, x2, x3, x4;
                                   output z1;
wire x1, x2;
wire z1;                           wire x1, x2, x3, x4;
                                   wire z1;
assign z1 = x1 & x2;
                                   assign z1 = x1 & x2 & x3 & x4;
endmodule                          endmodule
```

```
//dataflow 3-input and gate        //behavioral 3-input or gate
module and3 (x1, x2, x3, z1);      module or3 (x1, x2, x3, z1);

input x1, x2, x3;                  input x1, x2, x3;
output z1;                         output z1;

wire x1, x2, x3;                   wire x1, x2, x3;
wire z1;                           reg z1;

assign z1 = x1 & x2 & x3;          always @ (x1 or x2 or x3)
                                   begin
endmodule                            z1 = x1 | x2 | x3;
                                   end
                                   endmodule
```

Figure 2.44   Modules for *and2*, *and3*, *and4*, and *or3* that will be instantiated into the sum-of-products structural module of Figure 2.45.

The binary values for output z_1 as obtained from the test bench are shown in Figure 2.47 for all combinations of the input variables. Output z_1 is a logical 1 for the conditions that satisfy Equation 2.4 and can be verified by the Karnaugh map of Figure 2.48. The resulting waveforms are shown in Figure 2.49.

```
//structural sum of products       //define internal nets
module sop (x1, x2, x3, x4,        wire net1, net2, net3, net4;
           z1);
                                   wire z1;
input x1, x2, x3, x4;
output z1;                         assign z1 = net4;
wire x1, x2, x3, x4;               //continued on next page
```

Figure 2.45   Structural module for the sum-of-products equation of Equation 2.4.

```
//instantiate the gate modules        and4 inst3 (
//into the structural module              .x1(~x1),
and2 inst1 (                              .x2(~x2),
   .x1(x1),                               .x3(x3),
   .x2(x2),                               .x4(x4),
   .z1(net1)                              .z1(net3)
   );                                     );


and3 inst2 (                          or3 inst4 (
   .x1(x2),                              .x1(net1),
   .x2(~x3),                             .x2(net2),
   .x3(x4),                              .x3(net3),
   .z1(net2)                             .z1(net4)
   );                                    );

                                      endmodule
```

Figure 2.45   (Continued)

```
//structural sum of products test bench
module sop_tb;


reg x1, x2, x3, x4;
wire z1;


//apply input vectors
initial
begin: apply_stimulus
   reg [4:0] invect;
   for (invect = 0; invect < 16; invect = invect + 1)
      begin
         {x1, x2, x3, x4} = invect [4:0];
         #10 $display ("{x1x2x3x4} = %b, z1 = %b",
                        {x1, x2, x3, x4}, z1);
      end
end
//continued on next page
```

Figure 2.46   Test bench for the structural module of Figure 2.45.

```
//instantiate the module into the test bench
sop inst1 (
    .x1(x1),
    .x2(x2),
    .x3(x3),
    .x4(x4),
    .z1(z1)
    );
endmodule
```

Figure 2.46   (Continued)

```
{x1x2x3x4} = 0000, z1 = 0        {x1x2x3x4} = 1000, z1 = 0
{x1x2x3x4} = 0001, z1 = 0        {x1x2x3x4} = 1001, z1 = 0
{x1x2x3x4} = 0010, z1 = 0        {x1x2x3x4} = 1010, z1 = 0
{x1x2x3x4} = 0011, z1 = 1        {x1x2x3x4} = 1011, z1 = 0
{x1x2x3x4} = 0100, z1 = 0        {x1x2x3x4} = 1100, z1 = 1
{x1x2x3x4} = 0101, z1 = 1        {x1x2x3x4} = 1101, z1 = 1
{x1x2x3x4} = 0110, z1 = 0        {x1x2x3x4} = 1110, z1 = 1
{x1x2x3x4} = 0111, z1 = 0        {x1x2x3x4} = 1111, z1 = 1
```

Figure 2.47   Binary outputs for the test bench of Figure 2.46.

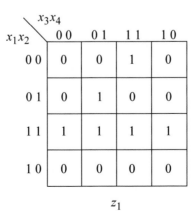

Figure 2.48   Karnaugh map for the sum-of-products expression of Equation 2.4.

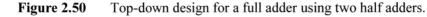

Figure 2.49   Waveforms for Figure 2.45 using the test bench of Figure 2.46.

2.7.2 Full Adder

A full adder will now be synthesized with two half adders and an OR gate using structural modeling. Recall that a full adder can be designed using a top-down approach as shown in Figure 2.50. The equations for a half adder and a full adder are shown in Equation 2.5 and 2.6, respectively, and the corresponding gate level designs in Figure 2.51 and Figure 2.52.

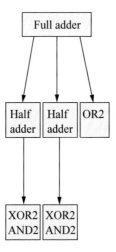

Figure 2.50   Top-down design for a full adder using two half adders.

$$sum = a \oplus b$$
$$cout = ab \tag{2.5}$$

$$sum = a \oplus b \oplus cin$$
$$cout = cin(a \oplus b) + ab \tag{2.6}$$

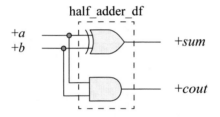

Figure 2.51   Logic diagram for a half adder to be instantiated into a full adder structural design.

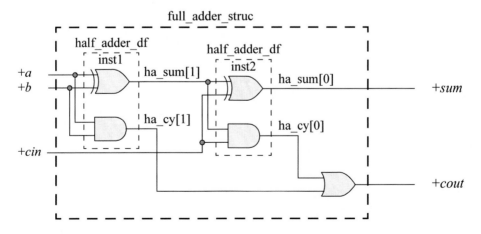

Figure 2.52   Logic diagram for a full adder using two half adders instantiated into a structural module.

The half adder module is named *half_adder_df*, which is then instantiated two times into the structural module *full_adder_struc*. In the full adder module, internal wires are specified for the sum and carry-out of the two half adders. The sum from the two half adders is a 2-bit vector with bits named *ha_sum[1]* and *ha_sum[0]*, where bit 0 is the low-order bit. The carry-out from the two half adders is a 2-bit vector with bits named *ha_cy[1]* and *ha_cy[0]*, where bit 0 is the low-order bit.

The Verilog code for the half adder is shown in Figure 2.53, where the symbol (^) is the exclusive-OR function and (&) is the AND function. As stated previously, all modules that are to be instantiated into a structural module should be tested for correct functional operation before they are instantiated. The test bench for the half adder is shown in Figure 2.54 in which all combinations of the two operands are applied to the inputs.

The **$monitor** system task continuously monitors the values of the variables specified in the parameter list. When any of the variables change value, all of the variables are displayed. The test bench displays the variables in the binary radix (%b). The resulting binary outputs are shown in Figure 2.55. The **$stop** task provides a stop during simulation. System tasks are covered in detail in Chapter 10. The waveforms shown in Figure 2.56 display the binary variables in a graphical representation.

The full adder structural module is shown in Figure 2.57. In instantiation *inst1*, input port *.a* of the half adder connects to input *a* (+*a*) of the full adder structural module and port *.b* connects to input *b* (+*b*). The output port *.sum* connects to net *ha_sum[1]* in the structural module, which is an output of the half adder and declared as a **wire**. The carry-out *cout* of the half adder connects to net *ha_cy[1]* in the structural module.

In instantiation *inst2*, input port *.a* of the half adder connects to net *ha_sum[1]* and port *.b* connects to input *cin* (+*cin*) of the full adder module. The *sum* and *cout* ports of the half adder connect to nets *ha_sum[0]* and *ha_cy[0]* of the full adder, respectively. The two continuous assignment statements assign output *sum* of the full adder the value of *ha_sum[0]* and *cout* the value of the logical OR of *ha_cy[0]* and *ha_cy[1]*.

```
//dataflow half_adder
module half_adder_df (a, b, sum, cout);

input a, b;              //list which are input
output sum, cout;        //list which are output

wire a, b, sum, cout;    //all are wire

assign sum = a ^ b;
assign cout = a & b;

endmodule
```

Figure 2.53   Verilog code for the half adder of Figure 2.51.

```
//dataflow half_adder test bench
module half_adder_df_tb;

reg a, b;               //inputs are reg for test bench
wire sum, cout;         //outputs are wire for test bench

initial
$monitor ("ab = %b, sum = %b, cout = %b",
         {a, b}, sum, cout);

initial
begin
   #0  a = 1'b0;
       b = 1'b0;

   #10a = 1'b0;
       b = 1'b1;

   #10a = 1'b1;
       b = 1'b0;

   #10a = 1'b1;
       b = 1'b1;

   #10    $stop;
end

//instantiate the dataflow module into the test bench
half_adder_df inst1 (
   .a(a),
   .b(b),
   .sum(sum),
   .cout(cout)
   );
endmodule
```

Figure 2.54   Test bench for the half adder module of Figure 2.53.

```
ab = 00, sum = 0, cout = 0
ab = 01, sum = 1, cout = 0
ab = 10, sum = 1, cout = 0
ab = 11, sum = 0, cout = 1
```

Figure 2.55   Binary outputs obtained from the test bench of Figure 2.54 for the half adder module of Figure 2.53.

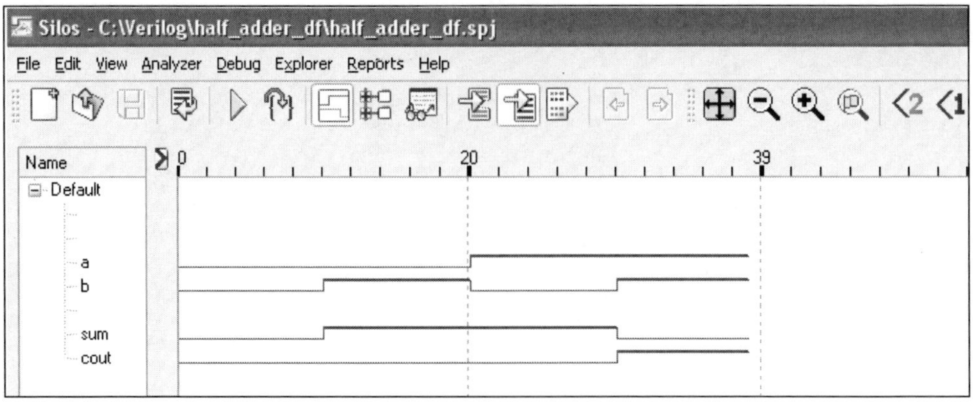

Figure 2.56   Waveforms for the half adder of Figure 2.53.

```
//structural full adder
module full_adder_struc (a, b, cin, sum, cout);

input a, b, cin;
output sum, cout;

wire [1:0] ha_sum, ha_cy;

//instantiate the half adder
half_adder_df inst1 (
   .a(a),
   .b(b),
   .sum(ha_sum[1]),
   .cout(ha_cy[1])
   );

half_adder_df inst2 (
   .a(ha_sum[1]),
   .b(cin),
   .sum(ha_sum[0]),
   .cout(ha_cy[0])
   );

assign sum = ha_sum[0];
assign cout = ha_cy[0] | ha_cy[1];

endmodule
```

Figure 2.57   Structural module for the full adder.

The test bench for the full adder structural module is shown in Figure 2.58 in which all possible combinations of three variables are applied to the inputs. The binary outputs are shown in Figure 2.59 and the waveforms in Figure 2.60.

```verilog
//structural full adder test          #10a = 1'b1;
//bench                                   b = 1'b0;
module full_adder_struc_tb;               cin = 1'b1;

//inputs are reg for tb                #10a = 1'b1;
reg a, b, cin;                            b = 1'b1;
                                          cin = 1'b0;
//outputs are wire for tb
wire sum, cout;                        #10a = 1'b1;
                                          b = 1'b1;
initial                                   cin = 1'b1;
$monitor ("ab = %b, cin = %b,
          sum = %b, cout = %b",        #10 $stop;
          {a, b}, cin, sum,            end
          cout);
                                       //instantiate the module
initial                                //into the test bench
begin                                  full_adder_struc inst1 (
#0 a = 1'b0;                           .a(a),
   b = 1'b0;                           .b(b),
   cin = 1'b0;                         .cin(cin),
                                       .sum(sum),
#10a = 1'b0;                           .cout(cout)
   b = 1'b0;                           );
   cin = 1'b1;
                                       endmodule
#10a = 1'b0;
   b = 1'b1;
   cin = 1'b0;

#10a = 1'b0;
   b = 1'b1;
   cin = 1'b1;

#10a = 1'b1;
   b = 1'b0;
   cin = 1'b0;
```

Figure 2.58   Test bench for the full adder module of Figure 2.57.

```
ab = 00, cin = 0, sum = 0, cout = 0
ab = 00, cin = 1, sum = 1, cout = 0
ab = 01, cin = 0, sum = 1, cout = 0
ab = 01, cin = 1, sum = 0, cout = 1
ab = 10, cin = 0, sum = 1, cout = 0
ab = 10, cin = 1, sum = 0, cout = 1
ab = 11, cin = 0, sum = 0, cout = 1
ab = 11, cin = 1, sum = 1, cout = 1
```

Figure 2.59   Binary outputs for the test bench of Figure 2.58.

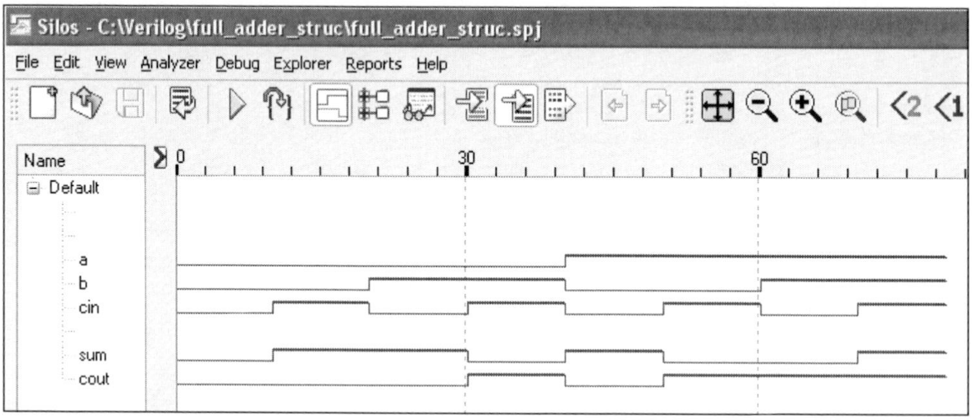

Figure 2.60   Waveforms for the test bench of Figure 2.58 for the structural full adder of Figure 2.57.

2.7.3 Four-Bit Ripple Adder

The 4-bit ripple adder of Section 2.3 will now be designed using structural modeling. The full adder of Section 2.7.2 will be instantiated four times into a structural module to produce a 4-bit ripple adder. Recall that a full adder adds the two operand bits plus the carry-out from the previous lower-order stage. Figure 2.61 shows the logic symbol for a full adder. In a 4-bit ripple adder, the carry-out from stage$_i$ becomes the carry-in to stage$_{i+1}$. The 4-bit ripple adder will be designed using 4-bit vectors for the two operands — augend and addend — and a 4-bit vector for the sum.

The full adder will be verified for correct operation by means of a test bench, then instantiated into the structural module, which will also be verified for correct functionality. The binary outputs and waveforms will be obtained for several combinations of the augend and addend.

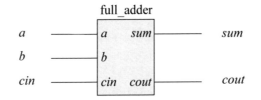

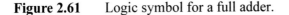

Figure 2.61   Logic symbol for a full adder.

Figure 2.62 shows the Verilog code for the full adder. The module is called *full_adder* and has five ports: *a*, *b*, *cin*, *sum*, and *cout*. Figure 2.63 shows the module for the test bench. The test bench provides all combinations of the three bits assigned to augend *a*, addend *b*, and the carry-in *cin*. The binary outputs and waveforms are shown in Figure 2.64 and Figure 2.65, respectively.

```
//dataflow full adder
module full_adder (a, b, cin, sum, cout);

//list inputs and outputs
input a, b, cin;
output sum, cout;

//define wires
wire a, b, cin;
wire sum, cout;

//continuous assign
assign sum = (a ^ b) ^ cin;
assign cout = cin & (a ^ b) | (a & b);

endmodule
```

Figure 2.62   Verilog code for a full adder.

When the full adder has been verified for correct functional operation, it is then instantiated into the 4-bit ripple adder structural model. When the operation of the ripple adder has been verified, it can then be used as a 4-bit-slice module to design a 16-bit or larger adder. For a 16-bit adder, the 4-bit slice is instantiated four times into a structural module to produce a 16-bit adder. Because Verilog supports a mixture of modeling styles within a module, the 4-bit adder and 16-bit adder can be designed using built-in primitives and continuous assignment statements.

In a later chapter, the design of a carry-lookahead adder will be presented. A carry-lookahead adder negates the drawback of the slow carry propagation delay in a ripple adder in which the carry is propagated serially between stages. The carry-lookahead method adds additional logic in each adder stage such that the carry-in to any stage is a function only of the two operand bits of that stage and the low-order carry-in to the adder. The carry-in to a stage is no longer the result of the carry-out from the previous lower-order stage.

```verilog
//full adder test bench
module full_adder_tb;

reg a, b, cin;
wire sum, cout;

//apply input vectors
initial
begin: apply_stimulus
   reg [3:0] invect;
   for (invect = 0; invect < 8; invect = invect + 1)
      begin
         {a, b, cin} = invect [3:0];
         #10 $display ("{abcin} = %b, sum = %b, cout = %b",
               {a, b, cin}, sum, cout);
      end
end

//instantiate the module into the test bench
full_adder inst1 (
   .a(a),
   .b(b),
   .cin(cin),
   .sum(sum),
   .cout(cout)
   );

endmodule
```

Figure 2.63   Test bench for the full adder module of Figure 2.62.

```
{abcin} = 000, sum = 0, cout = 0
{abcin} = 001, sum = 1, cout = 0
{abcin} = 010, sum = 1, cout = 0
{abcin} = 011, sum = 0, cout = 1
{abcin} = 100, sum = 1, cout = 0
{abcin} = 101, sum = 0, cout = 1
{abcin} = 110, sum = 0, cout = 1
{abcin} = 111, sum = 1, cout = 1
```

Figure 2.64   Binary outputs for the full adder of Figure 2.62 obtained from the test bench of Figure 2.63.

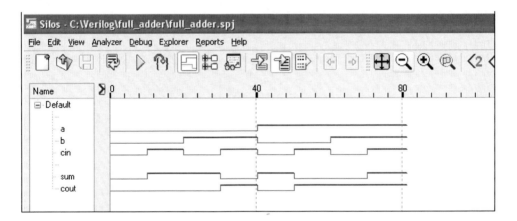

Figure 2.65   Waveforms for the full adder of Figure 2.62 obtained from the test bench of Figure 2.63.

The logic diagram for the 4-bit ripple adder is shown in Figure 2.66. The module for the 4-bit ripple adder is shown in Figure 2.67 using four instantiations of the Verilog code of Figure 2.62. The name of the module is *ripple_adder_4*; the ports are *a*, *b*, *cin*, *sum*, and *cout*. The keywords **input** and **output** tell Verilog which ports are used for external communication. Input ports are defaulted to **wire** in Verilog; however, the keyword **wire** is listed for all ports for completeness. The carry-out (*cout*) of the 4-bit adder is assigned the value of the internal carry *c[3]*.

The full adder is instantiated four times into the structural module of the ripple adder. In all instantiations, the name of the full adder (*full_adder*) is used. This is followed by the instance name; for example, *inst1* and *inst2*. The form *.a(a[0])* indicates a connection between port *a* in the instance module (*full_adder*) to the expression *a[0]* in the test bench. This method is *instantiation by name* and eliminates the need to know the order of the ports in the instantiated module. This is particularly useful when a large number of ports must be instantiated. The other method is *instantiation by position*, which is prone to errors when many ports are to be instantiated in the same order as indicated in the instance module.

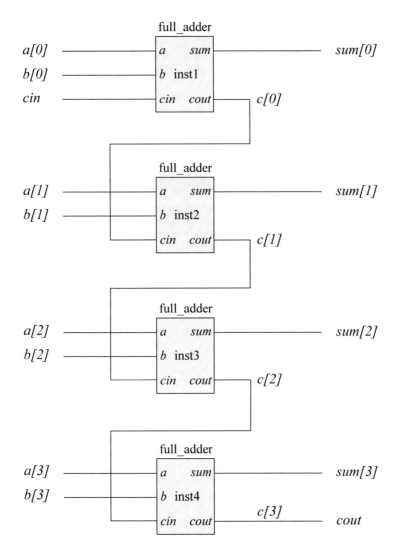

Figure 2.66   Logic diagram for a 4-bit ripple adder.

In instantiation *inst2* of the *ripple_adder_4* module, note that port *.a* of the full adder connects to *a[1]* of the ripple adder module. This is also shown in the logic diagram of Figure 2.66. In a similar manner, in *inst3*, the *.a* port of the instance module connects to *a[2]* of the ripple adder module. The carry-out (*.cout*) of *inst3* connects to *c[2]* of the structural module. The final carry-out (*cout*) of the full adder in *inst4* connects to the carry-out of the 4-bit ripple adder.

The test bench for the 4-bit ripple adder module is shown in Figure 2.68. The name of the module under test is normally used as the name of the test bench module (with *_tb* appended). This allows for ease of cross-referencing. All Verilog modules are saved with the .v extension. Inputs for test benches are declared as type **reg** because they retain their values during simulation. Outputs are declared as **wire** for test benches.

As explained previously, Verilog provides a mechanism to monitor a variable when it changes state. This facility is provided by the **$monitor** task. All parameters within the quotation marks are displayed in the radix indicated. The parameter values are obtained from the variables to the right of the quotation marks, in the same positional relationship. The **$monitor** continuously monitors the values of the variables specified in the parameter list (within the quotation marks) and displays all the parameters in the list whenever any variable changes value.

```
//dataflow 4-bit ripple adder
module ripple_adder_4 (a, b, cin, sum, cout);

input [3:0]a, b;
input cin;

output [3:0] sum;
output cout;

wire [3:0]a, b;
wire cin;
wire [3:0] sum;
wire cout;
wire [3:0]c;//define internal carries

assign cout = c[3];
//instantiate the full adder
full_adder inst1 (
    .a(a[0]),
    .b(b[0]),
    .cin(cin),
    .sum(sum[0]),
    .cout(c[0])
    );

//continued on next page
```

Figure 2.67   Structural module for a 4-bit ripple adder.

```
full_adder inst2 (
    .a(a[1]),
    .b(b[1]),
    .cin(c[0]),
    .sum(sum[1]),
    .cout(c[1])
    );

full_adder inst3 (
    .a(a[2]),
    .b(b[2]),
    .cin(c[1]),
    .sum(sum[2]),
    .cout(c[2])
    );

full_adder inst4 (
    .a(a[3]),
    .b(b[3]),
    .cin(c[2]),
    .sum(sum[3]),
    .cout(c[3])
    );

endmodule
```

Figure 2.67   (Continued)

Following the **$monitor** task is a section of code that applies a sequence of input vectors to the module under test. Six sets of input vectors are shown. The first set of vectors is applied at time 0, where the augend a is assigned a value of 3. The statement $a = 4'b0011$ specifies that the field for the value of a is four bits wide, the radix is binary, and the value is 3. The addend b is assigned a value of 4; the carry-in c is 0. The result of the first set of vectors is: $sum = 7$, $cout = 0$ as shown in the binary outputs of Figure 2.69. In the last set of input vectors, $a = 15$, $b = 6$, and $cin = 1$. This generates the following results: $sum = 6$, $cout = 1$, where $cout$ has a weight of 2^4.

Simulation stops 10 time units after the last set of inputs has been applied. The unit under test is instantiated by name, not by position. Thus, port $.a$ in the ripple adder module connects to a of the test bench. The waveforms are shown in Figure 2.70.

```
//4-bit ripple adder test          #10    a = 4'b1001;
//bench                                   b = 4'b0111;
module ripple_adder_4_tb;                 cin = 1'b1;

reg [3:0]a, b;                     #10    a = 4'b1101;
reg cin;                                  b = 4'b0111;
                                          cin = 1'b1;
wire [3:0] sum;
wire cout;                         #10    a = 4'b1111;
                                          b = 4'b0110;
//display variables                       cin = 1'b1;
initial
$monitor ("a b cin = %b_%b_%b,     #10    $stop;
        sum = %b, cout = %b",      end
      a, b, cin, sum, cout);
                                   //instantiate the module into
//apply input vectors              //the test bench
initial                            ripple_adder_4 inst1 (
begin                                 .a(a),
    #0    a = 4'b0011;                 .b(b),
          b = 4'b0100;                 .cin(cin),
          cin = 1'b0;                  .sum(sum),
                                       .cout(cout)
    #10   a = 4'b1100;                 );
          b = 4'b0011;
          cin  = 1'b0            endmodule

    #10   a = 4'b0111;
          b = 4'b0110;
          cin = 1'b1;
```

Figure 2.68   Test bench module for the 4-bit ripple adder of Figure 2.67.

```
a b cin = 0011_0100_0, sum = 0111, cout = 0
a b cin = 1100_0011_0, sum = 1111, cout = 0
a b cin = 0111_0110_1, sum = 1110, cout = 0
a b cin = 1001_0111_1, sum = 0001, cout = 1
a b cin = 1101_0111_1, sum = 0101, cout = 1
a b cin = 1111_0110_1, sum = 0110, cout = 1
```

Figure 2.69   Binary outputs obtained from the adder test bench of Figure 2.68.

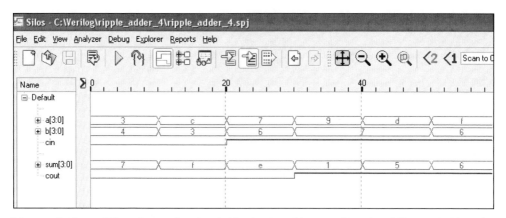

Figure 2.70   Waveforms for the 4-bit ripple adder test bench of Figure 2.68. The values for the variables are shown in hexadecimal.

2.8 Introduction to Mixed-Design Modeling

Mixed-design modeling incorporates different modeling styles in the same module. This includes gate and module instantiations, as well as continuous assignments and behavioral constructs. For example, a full adder can be designed that incorporates built-in primitives, dataflow, and behavioral modeling. When using behavioral modeling, variables within **always** and **initial** statements must be a register data type

2.8.1 Full Adder

In Section 2.7.2, a full adder was designed using structural modeling. The same adder will now be designed using mixed-design modeling. The equations for a full adder are restated below for convenience, where *a* and *b* are the augend and addend, respectively, *cin* is the carry-in to the adder, *sum* is the result, and *cout* is the carry-out.

$$sum = (a \oplus b) \oplus cin$$
$$cout = cin\,(a \oplus b) + ab$$

The full adder of Figure 2.52 is redrawn in Figure 2.71. The Verilog module for the full adder is shown in Figure 2.72. The variable *cout* is used in an **always** statement; therefore, it must be defined as type **reg**. There is an internal net declared as *net1*. Gate instantiation using a built-in primitive is used for the exclusive-OR function that generates *net1*. Behavioral modeling uses the **always** statement to define *cout* and dataflow modeling uses continuous assignment to define the *sum*.

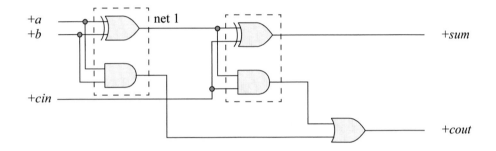

Figure 2.71   Full adder used in mixed-design modeling.

```
//mixed-design full adder
module full_adder_mixed (a, b, cin, sum, cout);

//list inputs and outputs
input a, b, cin;
output sum, cout;

//define reg and wires
reg cout;
wire a, b, cin;
wire sum;
wire net1;

//built-in primitive
xor (net1, a, b);

//continued on next page
```

Figure 2.72   Verilog module for the full adder of Figure 2.71 using mixed-design modeling.

```
//behavioral
always @ (a or b or cin)
begin
   cout = cin & (a ^ b) | (a & b);
end

//dataflow
assign sum = net1 ^ cin;

endmodule
```

Figure 2.72   (Continued)

The test bench is shown in Figure 2.73 in a slightly different version. The input vectors for each time unit are listed on the same line — a semicolon separates the individual bits. The complete set of vectors generates all combinations of the three inputs. The binary outputs are listed in Figure 2.74 and the data analyzer waveforms are shown in Figure 2.75. The binary outputs and waveforms are identical to those of the full adder of Section 2.7.2.

```
//mixed-design full adder test bench
module full_adder_mixed_tb;

reg a, b, cin;
wire sum, cout;

//display variables
initial
$monitor ("a b cin = %b %b %b, sum = %b, cout = %b",
      a, b, cin, sum, cout);

//apply input vectors
initial
begin
   #0    a = 1'b0; b = 1'b0; cin = 1'b0;
   #10   a = 1'b0; b = 1'b0; cin = 1'b1;
   #10   a = 1'b0; b = 1'b1; cin = 1'b0;
   #10   a = 1'b0; b = 1'b1; cin = 1'b1;
   #10   a = 1'b1; b = 1'b0; cin = 1'b0;
   #10   a = 1'b1; b = 1'b0; cin = 1'b1;
//continued next page
```

Figure 2.73   Test bench for the mixed-design full adder of Figure 2.72.

```
    #10    a = 1'b1; b = 1'b1; cin = 1'b0;
    #10    a = 1'b1; b = 1'b1; cin = 1'b1;
    #10    $stop;
end

//instantiate the module into the test bench
full_adder_mixed inst1 (
    .a(a),
    .b(b),
    .cin(cin),
    .sum(sum),
    .cout(cout)
    );
endmodule
```

Figure 2.73   (Continued)

```
a b cin = 0 0 0, sum = 0, cout = 0
a b cin = 0 0 1, sum = 1, cout = 0
a b cin = 0 1 0, sum = 1, cout = 0
a b cin = 0 1 1, sum = 0, cout = 1
a b cin = 1 0 0, sum = 1, cout = 0
a b cin = 1 0 1, sum = 0, cout = 1
a b cin = 1 1 0, sum = 0, cout = 1
a b cin = 1 1 1, sum = 1, cout = 1
```

Figure 2.74   Binary outputs for the mixed-design full adder of Figure 2.72.

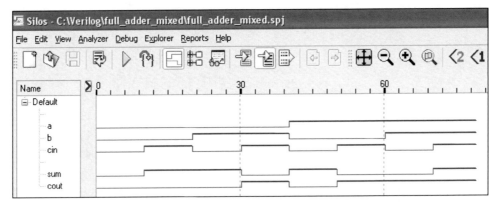

Figure 2.75   Waveforms for the mixed-design full adder of Figure 2.72.

2.9 Problems

2.1   Design a 4-input AND gate using dataflow modeling. Design a test bench and obtain the binary outputs and waveforms.

2.2   Design a 4-input AND gate using behavioral modeling. Design a test bench and obtain the binary outputs and waveforms.

2.3   Design a 2-input NAND gate using dataflow modeling. Design a test bench and obtain the binary outputs and waveforms.

2.4   Design a 2-input NAND gate using behavioral modeling. Design a test bench and obtain the binary outputs and waveforms.

2.5   Design a 2-input NOR gate using dataflow modeling. Design a test bench and obtain the binary outputs and waveforms.

2.6   Design a 2-input NOR gate using behavioral modeling. Design a test bench and obtain the binary outputs and waveforms.

2.7   Design a 3-input exclusive-NOR gate using behavioral modeling. Design a test bench and obtain the binary outputs and waveforms.

2.8   Implement the logic diagram shown below using dataflow modeling. Design a test bench and obtain the binary outputs and waveforms.

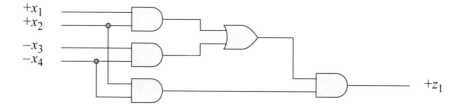

2.9   Implement the logic diagram shown in the previous problem using structural modeling.   Design a test bench and obtain the binary outputs and waveforms.

2.10   Implement the logic diagram shown below using dataflow modeling. Design a test bench and obtain the binary outputs and waveforms.

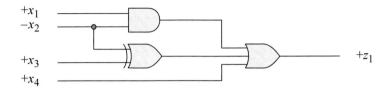

2.11   Implement the logic diagram shown below using dataflow modeling. Design a test bench and obtain the binary outputs and waveforms.

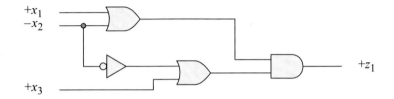

2.12   Given the Karnaugh map shown below, obtain the minimum sum-of-products expression and the minimum product-of-sums expression. Then implement both expressions using behavioral modeling. Design two test benches and compare the binary outputs and waveforms.

$x_1 x_2$ \ $x_3 x_4$	0 0	0 1	1 1	1 0
0 0	0	0	1	0
0 1	0	1	1	0
1 1	0	1	1	0
1 0	1	1	1	1

z_1

2.13   Obtain the minimized product-of-sums expression for function z_1 represented by the Karnaugh map shown below, then implement the equation using dataflow modeling.

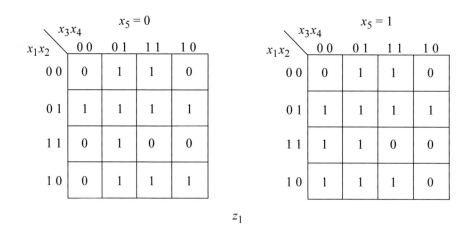

$$z_1$$

2.14   Minimize the following equation, then implement the result using structural modeling:

$$z_1 = x_1'x_3'x_4' + x_1'x_3x_4' + x_1x_3'x_4' + x_2x_3x_4 + x_1x_3x_4'$$

2.15   Minimize the following equation, then implement the result using structural modeling:

$$z_1 = x_1'x_2x_3'x_4' + x_3'x_4 + x_1x_2x_3'x_4' + x_3'x_4 + x_1x_2'x_3'x_4$$

2.16   Obtain the equation for a logic circuit that generates an output when the 4-bit unsigned binary number $z_1 = x_1x_2x_3x_4$ is greater than 5, but less than 10, where x_4 is the low-order bit. The equation is to be in a minimal *sum-of-product* form. Then implement the equation using behavioral modeling.

2.17   Plot the expression shown below on a Karnaugh map using x_4 as a map-entered variable and obtain the minimized equation in a sum-of-products form. Then implement the equation using structural modeling.

$$z_1 = x_1'x_2'x_3'(x_4) + x_1'x_2x_3'(x_4) + x_1x_2x_3'(x_4) + x_1x_2'x_3(x_4) + x_1x_2'x_3(x_4')$$

2.18   Given the Karnaugh map shown below, obtain the minimized expression for z_1. Then implement the result using structural modeling.

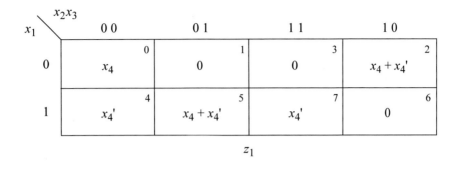

z_1

2.19   Obtain the minimized product-of-sums expression for the function z_1 represented by the Karnaugh map shown below. Then implement the result using dataflow, behavioral, structural, and mixed-design modeling.

$x_5 = 0$

x_1x_2 \ x_3x_4	0 0	0 1	1 1	1 0
0 0	0 [0]	0 [2]	0 [6]	0 [4]
0 1	0 [8]	0 [10]	0 [14]	0 [12]
1 1	0 [24]	1 [26]	1 [30]	0 [28]
1 0	1 [16]	1 [18]	1 [22]	0 [20]

$x_5 = 1$

x_1x_2 \ x_3x_4	0 0	0 1	1 1	1 0
0 0	1 [1]	1 [3]	1 [7]	1 [5]
0 1	0 [9]	1 [11]	1 [15]	0 [13]
1 1	0 [25]	1 [27]	1 [31]	0 [29]
1 0	1 [17]	1 [19]	1 [23]	1 [21]

z_1

3

Language Elements

Language elements are the constituent parts of the Verilog language. They consist of comments, identifiers, keywords, data types, parameters, and a set of values which determines the logic value of a net. Also included in this chapter are compiler directives and an introduction to system tasks and functions, which are covered in more detail in Chapter 10.

3.1 Comments

Comments can be inserted into a Verilog module to explain the function of a particular block of code or a line of code. There are two types of comments: single line and multiple lines. A single-line comment is indicated by a double forward slash (//) and may be placed on a separate line or at the end of a line of code, as shown below.

```
//This is a single-line comment on a dedicated line
assign z_1 = x_1 | x_2   //This is a comment on a line of code
```

A single-line comment usually explains the function of the following block of code. A comment on a line of code explains the function of that particular line of code. All characters that follow the forward slashes are ignored by the compiler.

A multiple-line comment begins with a forward slash followed by an asterisk (/*) and ends with an asterisk followed by a forward slash (*/), as shown below. Multiple-line comments cannot be nested. All characters within a multiple-line comment are ignored by the compiler.

```
/*This is a multiple-line comment.
   More comments go here.
   More comments. */
```

3.2 Identifiers

An identifier is a name given to an object or variable so that it can be referenced elsewhere in the design. Identifier names are used for modules, registers, ports, wires, or module instance names. Also, **begin** . . . **end** blocks may include an identifier. An identifier consists of a sequence of characters that can be letters, digits, $, or underscore (_). The first character of an identifier must be a letter or an underscore. The $ character is reserved for system tasks. Identifiers are case sensitive. For example, *Clock* and *clock* are different identifiers. An identifier refers to a unique object in the module in which it is defined.

When a module is instantiated into another module, the module instance name is considered to be an identifier. An identifier can contain up to 1024 characters; however, the first character cannot be a digit. Examples of identifiers are shown below, where **input**, **output**, and **reg** are keywords.

```
input a, b, cin;      //a, b, and cin are identifiers
output sum, cout;     //sum and cout are identifiers
reg z1;               //z1 is an identifier
```

Escaped identifiers   *Escaped identifiers* begin with a backslash (\) and end with a white space (space, tab, or new line) and provide a means to include any printable ASCII character in an identifier. The backslash and whitespace are not part of the identifier. An escaped identifier that contains a keyword is different than the keyword. For example, \assign is different than the keyword **assign**. Examples of escaped identifiers are shown below.

```
\2005
\~$~
\*****
```

3.3 Keywords

Verilog HDL reserves a list of special predefined, nonescaped identifiers called *keywords* which are used to define the language constructs. Only lower case keywords are used for reserved words. A list of keywords is shown by category in Table 3.1.

Table 3.1 Verilog HDL Keywords

Category	Keywords		
Bidirectional Gates	rtran	rtranif0	rtranif1
	tran	tranif0	tranif1
Charge Storage Strengths	large	medium	small
CMOS Gates	cmos	rcmos	
Combinational Logic Gates	and	buf	nand
	nor	not	or
	xnor	xor	
Continuous Assignment	assign		
Data Types	integer	real	realtime
	reg	scalared	time
	tri	tri0	tri1
	triand	trior	trireg
	vectored	wand	wire
	wor		
Module Declaration	endmodule	module	
MOS Gates	nmos	pmos	rnmos
	rpmos		
Multiway Branching	case	casex	casez
	default	endcase	
Named Event	event		
Parameters	defparam	parameter	specparam
Port Declaration	inout	input	output
Procedural Constructs	always	initial	
Procedural Continuous Assignment	assign	deassign	force
	release		
Procedural Flow Control	begin	disable	else
	end	for	forever
	fork	if	join
	repeat	wait	while
Pull Gates	pulldown	pullup	

Table 3.1 Verilog HDL Keywords

Category	Keywords		
Signal Strengths	highz0	highz1	pull0
	pull1	strong0	strong1
	supply0	supply1	weak0
	weak1		
Specify Block	endspecify	specify	
Tasks and Functions	endfunction	endtask	function
	task		
Three-State Gates	bufif0	bufif1	notif0
	notif1		
Timing Control	edge	negedge	posedge
User-Defined Primitives	endprimitive	endtable	primitive
	table		

3.3.1 Bidirectional Gates

tran, tranif0, tranif1, rtran, rtranif0, rtranif1   These are bidirectional primitive gates. The signals on either side of the gates can be specified as inputs or outputs; that is, either signal can be the driver. The inputs and outputs are classified as scalar signals. The **tran** gate (switch) acts as a buffer between the two signals. One terminal can be declared as **input** or **inout**, the other declared as **output** or **inout**. The **tran** primitive is instantiated as shown below, where the instance name *inst1* is optional.

tran   inst1 (inout1, inout2);

The primitives **tranif0** and **tranif1** also have two bidirectional terminals plus a control input. The **tranif0** gate connects the two signals only if the control input is a logical 0; otherwise, the output of the gate is a high impedance. The **tranif1** gate transmits only if the control input is a logical 1; otherwise, the output of the gate is a high impedance. That is, the output logic value is the same as the input logic value or a high impedance. The **tranif0** and **tranif1** primitives are instantiated as shown below, where the control input is listed last.

tranif0   inst1 (inout1, inout2, control);
tranif1   inst1 (inout1, inout2, control);

There are also three bidirectional gates — called resistive gates — that operate the same as the previous gates, but have a higher source-to-drain impedance. This reduces the signal strength. The resistive gates are declared with the same keywords, but prefixed with an **r** — for example, **rtran**, **rtranif0**, and **rtranif1**. The resistive gates have the same syntax as the non-resistive gates. Figure 3.1 shows the logic diagrams for the six types of bidirectional gates.

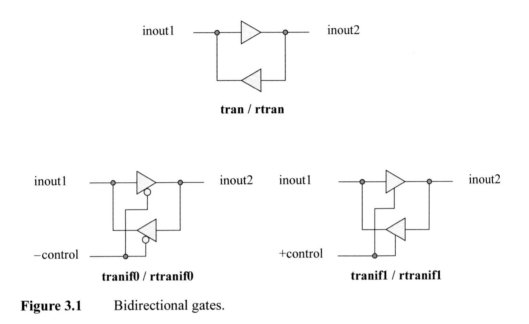

Figure 3.1   Bidirectional gates.

3.3.2 Charge Storage Strengths

small, medium, large   There are three charge storage strengths that relate to the capacitance of a net: **small**, **medium**, and **large**. A **trireg** net, which models the charge stored on a net, can have a charge strength associated with the net when the drivers are in a high impedance state. The amount of charge stored determines the size of the capacitance associated with the net. A driver for a **trireg** net will generate three states: logic 0, logic 1, or **x**. If all drivers on the net generate a high impedance output, then the net retains the last value driven onto the net. The default strength is **medium**.

Optional delay times can also be specified for a **trireg** net, indicating the rise delay, the fall delay, and the charge decay time. The charge decays when all the drivers of the net are in a high impedance state. If no delay is specified, then the charge on the **trireg** net does not decay. A **trireg** net can be forced to a high impedance state by a **force** procedural continuous assignment.

3.3.3 CMOS Gates

cmos   The **cmos** gate can be modeled with an **nmos** and a **pmos** device to implement a **cmos** transmission gate. There are two enable inputs: one for the n-channel device and one for the p-channel device. A **cmos** gate transmits data from input to output when its two control signals are in complementary states; that is, the n-channel control is a logic 1 and the p-channel control is a logic 0. If the states of the control signals are reversed, then the output is in a high impedance state. These control inputs allow both transistors to be either on (transmit) or off (high impedance). A **cmos** gate can be instantiated as shown below, where the instance *inst1* name is optional.

> **cmos**   inst1 (output, data_input, n_enable, p_enable);

rcmos   The **rcmos** gate is a high resistive version of the **cmos** gate.

3.3.4 Combinational Logic Gates

These are built-in primitive gates used to describe a net and have one or more scalar inputs, but only one scalar output. The output signal is listed first, followed by the inputs in any order. The outputs are declared as **wire;** the inputs can be declared as either **wire** or **reg**. The gates represent a combinational logic function and can be instantiated into a module, as follows, where the instance name is optional:

> **gate_type**   inst1 (output, input_1, input_2, . . . , input_n);

Two or more instances of the same type of gate can be specified in the same construct, as follows:

> **gate_type**   inst1 (output_1, input_11, input_12, . . . , input_1n),
>   inst2 (output_2, input_21, input_22, . . . , input_2n),
>
>   .
>   .
>   .
>
>   instm (output_m, input_m1, input_m2, . . . , input_mn);

and   This is a built-in primitive gate that operates according to the truth table shown in Table 3.2 for a 2-input AND gate.

Table 3.2 Truth Table for the Logical AND Built-In Primitive

Inputs x_1 x_2	Output z_1
0 0	0
0 1	0
1 0	0
1 1	1
0 x	0
0 z	0
1 x	x
1 z	x

Inputs x_1 x_2	Output z_1
x 0	0
x 1	x
x x	x
x z	x
z 0	0
z 1	x
z x	x
z z	x

The **x** entry in Table 3.2 represents an unknown logic value. The **z** entry represents a high impedance state. Simulators use the **z** value to indicate when the driver of a net is disabled or not connected. AND gates can be represented by two symbols as shown below for the AND function and the OR function.

AND gate for the AND function   AND gate for the OR function

buf   A **buf** gate is a noninverting primitive with one scalar input and one or more scalar outputs. The output terminals are listed first when instantiated; the input is listed last, as shown below. The instance name is optional.

   buf   inst1 (output, input);   //one output

   buf   inst2 (output_1, output_2, . . . , output_n, input); //multiple outputs

The truth table for a **buf** gate is shown in Table 3.3 for one output.

Table 3.3 Truth Table for a buf Gate

Input	Output
0	0
1	1
x	x
z	x

nand   This is a built-in primitive gate that operates according to the truth table shown in Table 3.4 for a 2-input NAND gate.

Table 3.4 Truth Table for the Logical NAND Built-In Primitive

Inputs x_1 x_2	Output z_1		Inputs x_1 x_2	Output z_1
0 0	1		x 0	1
0 1	1		x 1	x
1 0	1		x x	x
1 1	0		x z	x
0 x	1		z 0	1
0 z	1		z 1	x
1 x	x		z x	x
1 z	x		z z	x

NAND gates can be represented by two symbols as shown below for the AND function and the OR function.

NAND gate for the AND function   NAND gate for the OR function

DeMorgan's theorems are associated with NAND and NOR gates and convert the complement of a sum term or a product term into a corresponding product or sum term, respectively. For every $x_1, x_2 \in B$,

(a) $(x_1 \bullet x_2)' = x_1' + x_2'$ Nand gate
(b) $(x_1 + x_2)' = x_1' \bullet x_2'$ NOR gate

DeMorgan's laws can be generalized for any number of variables.

nor   This is a built-in primitive gate that operates according to the truth table shown in Table 3.5 for a 2-input NOR gate.

Table 3.5 Truth Table for the Logical NOR Built-In Primitive

Inputs x_1 x_2	Output z_1		Inputs x_1 x_2	Output z_1
0 0	1		x 0	x
0 1	0		x 1	0
1 0	0		x x	x
1 1	0		x z	x
0 x	x		z 0	x
0 z	x		z 1	0
1 x	0		z x	x
1 z	0		z z	x

NOR gates can be represented by two symbols as shown below for the OR function and the AND function.

NOR gate for the OR function   NOR gate for the AND function

not   A **not** gate is an inverting built-in primitive with one scalar input and one or more scalar outputs. The output terminals are listed first when instantiated; the input is listed last, as shown below. The instance name is optional.

     not   inst1 (output, input);   //one output
     not   inst2 (output_1, output_2, . . . , output_n, input);   //multiple outputs

The truth table for a **not** gate is shown in Table 3.6 for one output.

**Table 3.6 Truth Table for
the Logical NOT (Inverter)
Built-In Primitive**

Input	Output
0	1
1	0
x	x
z	x

The NOT function can be represented by two symbols as shown below depending on the assertion levels required. The function of the inverters is identical; the low assertion is placed at the input or output for readability with associated logic.

NOT (inverter) function
with low assertion output

NOT (inverter) function
with low assertion input

or   This is a built-in primitive gate that operates according to the truth table shown in Table 3.7 for a 2-input OR gate.

Table 3.7 Truth Table for the Logical OR Built-In Primitive

Inputs x_1 x_2	Output z_1		Inputs x_1 x_2	Output z_1
0 0	0		x 0	x
0 1	1		x 1	1
1 0	1		x x	x
1 1	1		x z	x
0 x	x		z 0	x
0 z	x		z 1	1
1 x	1		z x	x
1 z	1		z z	x

OR gates can be represented by two symbols as shown below for the OR function and the AND function.

OR gate for the OR function   OR gate for the AND function

xnor   This is a built-in primitive gate that operates according to the truth table shown in Table 3.8 for a 2-input exclusive-NOR gate.

Table 3.8 Truth Table for the Logical Exclusive-NOR Built-In Primitive

Inputs $x_1 x_2$	Output z_1		Inputs $x_1 x_2$	Output z_1
0 0	1		x 0	x
0 1	0		x 1	x
1 0	0		x x	x
1 1	1		x z	x
0 x	x		z 0	x
0 z	x		z 1	x
1 x	x		z x	x
1 z	x		z z	x

Exclusive-NOR gates can be represented by the symbol shown below. An exclusive-NOR gate is also called an equality function because the output is a logical 1 whenever the two inputs are equal, as can be seen in the first four rows of Table 3.8.

Exclusive-NOR gate

The equation for the exclusive-NOR gate shown above is

$$z_1 = (x_1 x_2) + (x_1' x_2')$$

xor   This is a built-in primitive gate that operates according to the truth table shown in Table 3.9 for a 2-input exclusive-OR gate. Exclusive-OR gates can be represented by the symbol shown below. The output of an exclusive-OR gate is a logical 1 whenever the two inputs are different, as can be seen in the first four rows of Table 3.9.

Exclusive-OR gate

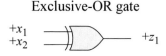

The equation for the exclusive-OR gate shown above is

$$z_1 = (x_1 x_2') + (x_1' x_2)$$

Table 3.9 Truth Table for the Logical Exclusive-OR Built-In Primitive

Inputs $x_1\ x_2$	Output z_1		Inputs $x_1\ x_2$	Output z_1
0 0	0		x 0	x
0 1	1		x 1	x
1 0	1		x x	x
1 1	0		x z	x
0 x	x		z 0	x
0 z	x		z 1	x
1 x	x		z x	x
1 z	x		z z	x

3.3.5 Continuous Assignment

assign   This is a continuous assignment statement used in dataflow modeling to describe combinational logic. Continuous assignments can be applied only to nets. The left-hand side is declared as type **wire** not **reg**. The syntax is as follows:

assign <Optional delay> Left-hand side net = Right-hand side expression;

When a variable on the right-hand side changes value, the right-hand side expression is evaluated and the value is assigned to the left-hand side net after the specified delay. The continuous assignment is used to place a value on a net. An example of using continuous assignment in the design of a 4-bit adder is shown in Figure 3.2.

```
//dataflow for a 4-bit adder        wire [3:0] a, b;
module adder4_df (a, b, cin,         wire cin;
                  sum);              wire [3:0] sum;

input [3:0] a, b;                   //continuous assignment for
input cin;                          //dataflow
output [3:0] sum;                   assign sum = a + b + cin;

                                    endmodule
```

Figure 3.2   Module using continuous assignment for a 4-bit adder.

3.3.6 Data Types

Verilog supports two built-in data types: nets for connectivity between hardware elements and registers for storage. This section provides a brief introduction to data types. Data types are covered in detail in Section 3.5.

integer   An **integer** is a general-purpose register that supports procedural computation. The size of an integer is represented by the word-length of the host machine.

real   Data values specified as **real** are represented in a double precision floating-point format. They can be specified in either decimal notation or in scientific (exponential) notation.

realtime   A **realtime** register is identical to a **real** register, except that the values that it stores are time values in real number format.

reg   Register data types are declared by the keyword **reg** because a register holds a value. The register value is retained in memory until it is changed by a subsequent assignment. A variable of type **reg** closely resembles a hardware register that is synthesized with *D* flip-flops, *JK* flip-flops, or *SR* latches.

scalared   A **scalared** keyword explicitly declares a net whose bits can be selected either individually or as part-select.

time   This stores the simulation time as a 64-bit unsigned quantity.

tri   The **tri** keyword specifies a net with multiple drivers. It has the same function as **wire**, but describes a three-state net.

tri0   This keyword models a net with a pulldown resistor to ground. A **tri0** net has a value of 0 if the net is not being driven.

tri1   This keyword models a net with a pullup resistor to the power supply. A **tri1** net has a value of 1 if the net is not being driven.

triand   The **triand** specifies a three-state net with multiple drivers. It models a **wand** hardware implementation. If the output of any driver is 0, then the value of the net is 0. It has the same syntax and functionality as the **wand** net.

trior   The **trior** specifies a three-state net with multiple drivers. It models a **wor** hardware implementation. If the output of any driver is 1, then the value of the net is 1. It has the same syntax and functionality as the **wor** net.

trireg   This net specifies a register that stores a value and is used to model the charge stored on a net.

vectored   The **vectored** keyword explicitly declares a net whose bits cannot be selected either individually or as part-select; that is, the net is an indivisible entity in which the entire net must be referenced.

wand   This represents a wired-AND net in which the value of the net is 0 if the output of any driver is 0. The circuit is implemented using open-collector logic.

wire   This represents a physical connection between hardware elements. The connection can be a wire or a group of wires, both of which are called a net. Nets are 1-bit scalar values unless declared otherwise. The output of a logic gate is declared as **wire** and represents a net with a single driver. The **wire** net is identical in syntax to a **tri** net.

wor   This represents a wired-OR net in which the value of the net is 1 if the output of any driver is 1. The circuit is implemented using emitter-coupled logic.

3.3.7 Module Declaration

A Verilog module contains descriptive information that describes the functional operation of the circuit. The keywords **module** and **endmodule** are mandatory and act as delimiters for a module.

endmodule   The **endmodule** keyword is the last statement in a module and signifies the end of the module definition.

module   This is the first statement in a Verilog module and defines the beginning of the module declaration. It does not necessarily have to be the first line; the first line in a module is usually a comment indicating the function of the module. A module is a software representation that specifies the structure and behavior of a hardware unit. The module functional operation is defined between the delimiters using various declarations, statements, and instantiations.

3.3.8 MOS Switches

There are four MOS switches (primitives) defined in Verilog. They are declared by the keywords **nmos**, **pmos**, **rnmos**, and **rpmos** and represent n-channel and p-channel devices. They propagate a signal in one direction only, as determined by a control input. Each primitive models a MOS transistor, but the **rnmos** and **rpmos** gates have a higher resistance compared to the **nmos** and **pmos** transistors. The resistive switches have the same syntax as the non-resistive switches, but are prefixed by an **r**. The switches are instantiated as shown below for **nmos** and **pmos** primitives with optional instance names.

> **nmos**   inst1 (output, input, control);
> **pmos**   inst2 (output, input, control);

3.3.9 Multiway Branching

When there are many paths from which to chose, nested **if-else if** statements can be cumbersome. A convenient way to achieve multiway branching is by means of a **case** statement. The keywords **case**, **endcase**, and **default** are used in a **case** statement, as shown below.

```
case (case_expression)
    alternative_1 : statement_1;
    alternative_2 : statement_2;
            .

            .

            .

    alternative_n : statement_n;
    default : default_statement;
endcase
```

case   The case_expression is evaluated, then compared to each alternative in the order listed — alternative_1 through alternative_n. The **case** statement compares the case_expression with each alternative on a bit-by-bit basis. The first alternative that matches the case_expression will cause the corresponding statement of that alternative to be executed. The statement can be a single statement or a block of statements bounded by the keywords **begin** and **end**. If none of the alternatives match, then the **default** statement is executed.

casex   The **casex** statement is a variation of the **case** statement that allows "don't care" values to be considered. In the **case** statement, values of **x** and **z** in the case_expression and the alternatives are treated as unknown and high impedance, respectively. The **casex** statement, however, treats all **x** and **z** values as "don't cares."

casez   The **casez** statement is a variation of the **case** statement that treats both **x** and **z** as "don't cares." Another type of explicit "don't care" is the symbol **?**. This is useful in a decode unit when only the operation code is to be decoded, not the remaining fields. The case_expression would contain the symbol **?** in all fields that were to be treated as "don't cares."

default   In the event that there is no match, the **default** statement is executed.

endcase   This keyword terminates the **case** statement.

A typical application for the **case** statement is to model a multiplexer, as shown in the behavioral program of Figure 3.3 for a 4:1 multiplexer, where the select inputs are *sel[1:0]*, the data inputs are *data[3:0]*, and the output is *out*.

```
//4:1 multiplexer using a          always @ (sel or data)
//case statement                   begin
module mux_4_1_case (sel,          case (sel)
          data, out);                  (0) : out = data[0];
                                        (1) : out = data[1];
input [1:0] sel;                        (2) : out = data[2];
input [3:0] data;                       (3) : out = data[3];
output out;                          endcase
                                   end
reg out;
                                   endmodule
```

Figure 3.3   Multiplexer design using a **case** statement.

3.3.10 Named Event

An *event* is a change in value of a net or register. A *regular event* control is specified by the @ symbol and is executed whenever there is a positive or negative transition on a signal. This is referred to as *edge-sensitive* control.

An OR event control allows for a change on multiple signals — the signals are ORed together. If one or more signals change, then the following statement or block of statements is executed. The list of signals that are listed in the OR expression is called a *sensitivity* list.

The **wait** keyword allows for *level-sensitive* control. This method waits for a condition to become true, then executes the following statement or block of statements.

3.3.11 Parameters

Constants can be defined in a module by means of the keyword **parameter**. A constant cannot be changed during simulation; however, the value of a constant can be changed during compilation.

defparam   One or more **defparam** statements can be used in a module to permit parameters to be changed during compilation. An example is shown in Figure 3.4. Figure 3.4(a) shows the Verilog code that uses the **parameter** keyword to define a constant named x_1. Figure 3.4(b) shows how the **defparam** keyword changes the value of x_1 to 4 and 8. Figure 3.4(c) shows the simulation output.

```
//example of defparam              //define top level module
module def_param1;                 //for defparam1
                                   module top_level;
parameter x1 = 0;
                                   defparam value1.x1 = 4;
initial                            defparam value2.x1 = 8;
$display ("value=%d", x1);
                                   def_param1 value1 ( );
endmodule                          def_param1 value2 ( );
            (a)
                                   endmodule
                                              (b)
```

```
value = 8
value = 4
value = 0
            (c)
```

Figure 3.4   Verilog code to illustrate the use of the **defparam** keyword: (a) **parameter** keyword, (b) **defparam** keyword to change the value, and (c) the outputs.

parameter   A constant is declared using the keyword **parameter**, which assigns a value to the constant. Parameters cannot be used as variables. Parameters are used frequently to specify delays and ranges in declarations. A syntax for a parameter declaration is as follows:

> **parameter**   constant_name_1 = constant_value_1,
>   constant_name_2 = constant_value_2,
>
>   .
>
>   .
>
>   .
>
>   constant_name_n = constant_value_n,

Examples of parameter declarations are shown below.

```
parameter    bus_width = 32;                      //integer
parameter    cache_size = 1024;                   //integer
parameter    initialize_counter = 1000_0011;  //register
parameter    out_port_id = 8;                     //integer
parameter    real_value = 6.72;                   //real
parameter    width = 8, depth = 32;               //integers
parameter    byte = 8, word = byte * 4;           //integers
```

Note that a constant expression can be used in declaring the value of a constant.

specparam   The specify parameter **specparam** is used to declare parameters within a **specify** block. Specify parameters are local to the **specify** block in which they are declared. They are typically used to specify timing considerations between the inputs and outputs of a module. Any constants declared by the **parameter** keyword are not visible within the **specify** block.

3.3.12 Port Declaration

Ports in a module provide connections between the module internal elements and the module external environments. A port is declared as **input**, **output**, or **inout**. When a module instantiates other modules, certain rules must be followed. These are defined in the port descriptions below. Module ports can be declared as either scalar or vector, where a scalar is a single value and a vector is a one-dimensional array.

inout   Within the module, **inout** ports must be declared as type net; externally, they must connect to a net.

input   Within the module, input ports must be declared as type **wire** (net); external-ly, they can connect to a variable of type **reg** or type **wire**. However, since the default for inputs within the module is **wire** they do not have to be explicitly declared as **wire**, unless desired.

output   Within the module, outputs can be declared as type **reg** or as type **wire**, the default being **wire**. Externally, they must always connect to net (**wire**), not to a type **reg**. If an **inout** or **output** port was defaulted to **wire** and it is to be used in an **initial** or an **always** statement, then the port must be redeclared as a type **reg**.

3.3.13 Procedural Constructs

always   This is a behavioral construct that begins at time 0 and executes the state-ments in the associated block continuously in a looping fashion. The **always** construct never exits the corresponding block. There can be more than one **always** statement in a behavioral module and, together with the **initial** statement, is one of the basic con-structs for representing concurrency.

initial   An **initial** block begins at time zero, executes the statements in the block only once, then suspends forever. Multiple **initial** blocks in the same module all execute concurrently at time zero. If there is only one behavioral statement in the **initial** block, then the statements do not have to be grouped. If multiple behavioral statements are contained in the **initial** block, then the statements are enclosed within **begin** . . . **end** delimiters. Data types within an **initial** block must be declared as registers. The **ini-tial** blocks are typically used for module initialization and monitoring.

3.3.14 Procedural Continuous Assignment

Procedural continuous assignment statements are continuous assignments made with-in a behavior; that is, within an **initial** or an **always** statement. They allow values to be continuously assigned to nets or registers. The assignments override other assign-ments to a net or register. A procedural continuous assignment is different than a con-tinuous assignment — a continuous assignment occurs external to an **initial** or an **always** statement

assign . . . deassign   The **assign** keyword is used to override all procedural as-signments to a register or a concatenation of registers; it is not used with nets. The re-sult of a procedural continuous assignment remains in effect until another procedural assignment is executed or until a **deassign** statement is executed. The **deassign** pro-cedural statement terminates the continuous assignment to a register.

force . . . release   The **force** and **release** procedural statements are similar in concept to the **assign** and **deassign** statements, with the exception that **force** and **release** can be applied to both registers and nets. The **force** and **release** statements are primarily used in test benches.

When **force** is applied to a net, the state of the net is changed to the forced value. The net returns to its previous state when **release** is applied, unless the driver for the net changes value during simulation, in which case, the net assumes the new value. When **force** is applied to a register, the state of the register is changed to the forced value; the register retains the forced value after **release** is applied, but can be changed by a subsequent procedural assignment.

3.3.15 Procedural Flow Control

Procedural flow control statements modify the flow in a behavior by selecting branch options, repeating certain activities, selecting a parallel activity, or terminating an activity. The activity can occur in sequential blocks or in parallel blocks.

begin . . . end   The **begin** . . . **end** keywords are used to group multiple statements into sequential blocks. The statements in a sequential block execute in sequence; that is, a statement does not execute until the preceding statement has executed, except for nonblocking statements. An optional name, called a block identifier, can be given to a **begin** . . . **end** block, as shown below. The name is preceded by a colon.

        begin [: optional block identifier]
            block of sequential procedural statements
        end

If there is only one procedural statement in the block, then the **begin** . . . **end** keywords may be omitted.

disable   The **disable** statement terminates a named block of procedural statements or a task and transfers control to the statement immediately following the block or task. The **disable** statement can also be used to exit a loop.

for   The keyword **for** is used to specify a loop. The **for** loop repeats the execution of a procedural statement or a block of procedural statements a specified number of times. The **for** loop is used when there is a specified beginning and end to the loop. The format and function of a **for** loop is similar to the **for** loop used in the C programming language. The parentheses following the keyword **for** contain three expressions separated by semicolons, as shown below.

        for (register initialization; test condition; update register control variable)
            procedural statement or block of procedural statements

forever   The **forever** loop statement executes the procedural statements continuously. The loop is primarily used for timing control constructs, such as clock pulse generation. The **forever** procedural statement must be contained within an **initial** or an **always** block. In order to exit the loop, the **disable** statement may be used to prematurely terminate the procedural statements. An **always** statement executes at the beginning of simulation; the **forever** statement executes only when it is encountered in a procedural block.

fork . . . join   These keywords specify parallel blocks in which the statements execute concurrently. In contrast, a sequential block has the delimiters **begin . . . end** in which the statements execute sequentially. Delay times within a parallel block are relative to the time at which the block was entered. All statements in a parallel block must complete execution before control is passed to the statements following the block.

if . . . else   These keywords are used as conditional statements to alter the flow of activity through a behavioral module. They permit a choice of alternative paths based upon a Boolean value obtained from a condition. The syntax is shown below.

> **if** (condition)
>   {procedural statement 1}
> **else**
>   {procedural statement 2}

If the result of the *condition* is true, then procedural statement 1 is executed; otherwise, procedural statement 2 is executed. The procedural statement following the **if** and **else** statements can be a single procedural statement or a block of procedural statements. Two uses for the **if . . . else** statement are to model a multiplexer or decode an instruction register operation code to select alternative paths depending on the instruction. The **if** statement can be nested to provide several alternative paths to execute procedural statements. The syntax for nested **if** statements is shown below.

> **if** (condition 1)
>   {procedural statement 1}
> **else if** (condition 2)
>   {procedural statement 2}
> **else if** (condition 3)
>   {procedural statement 3}
> **else**
>   {procedural statement 4)

repeat   The **repeat** keyword is used to execute a loop a fixed number of times as specified by a constant contained within parentheses following the **repeat** keyword.

The loop can be a single statement or a block of statements contained within **begin . . . end** keywords. The syntax is shown below.

> **repeat** (expression)
>> statement or block of statements

When the activity flow reaches the **repeat** construct, the expression in parentheses is evaluated to determine the number of times that the loop is to be executed. The *expression* can be a constant, a variable, or a signal value. If the expression evaluates to **x** or **z**, then the value is treated as 0 and the loop is not executed. An example of the **repeat** keyword is shown in Figure 3.5, where a variable *count* is incremented eight times. The outputs are shown in Figure 3.6. Execution of the loop can be terminated by a **disable** statement before the loop has executed the specified number of times.

The **repeat** keyword can also be used as a *repeat event control* to delay the assignment of the right-hand expression to the left-hand target. This is a form of intra-statement delay. The syntax is shown below.

> **repeat** (expression) @ (event_expression)

```verilog
//example of the repeat keyword
module repeat_example;

integer count;

initial
begin
        count = 0;
        repeat (8)
        begin
                $display ("count = %d", count);
                count = count + 1;
        end
end
endmodule
```

Figure 3.5   Example of the **repeat** keyword for loop control.

count =	0	count =	4
count =	1	count =	5
count =	2	count =	6
count =	3	count =	7

Figure 3.6   Outputs for the Verilog code of Figure 3.5.

An example using the **repeat** construct is shown below in which the sum of *reg1* plus *reg2* is assigned to *reg3* after the occurrence of three consecutive positive edges of *clk*. The expression is (3) and the event_expression is (**posedge** clk). The operation is as follows:

- Evaluate the sum of *reg1* and *reg2*
- Wait three positive edges of the clock
- Assign the result to *reg3*

      reg3 = **repeat** (3) @ (**posedge** clk) reg1 + reg2;

The above statement is equivalent to the following statements, which use a temporary register to store the sum of *reg1* and *reg2* before assigning the sum to *reg3*:

```
      begin
          temp = reg1 + reg2;
          @ (posedge clk);
          @ (posedge clk);
          @ (posedge clk);
          reg3 = temp;
      end
```

wait   The **wait** keyword is a level-sensitive control that waits for an expression to become true before a statement or a block of statements is executed. The syntax is shown below.

      wait (expression) statement;

The value of the expression is monitored continuously. If the expression is true (a logical 1) when the **wait** statement is detected, then execution is not suspended. If the expression is false (a logical 0), then execution is suspended until the expression becomes true. An example of the **wait** construct is shown below.

      wait (data_ready) data_reg = data_in;

The **wait** construct provides a means to synchronize two concurrent processes. Therefore, a means must be provided to ensure that the value for the expression will never remain false; otherwise, the statement following the conditional expression would never be executed. The **wait** construct provides an ideal method to synchronize communication between two units that operate over an asynchronous interface.

while   The **while** statement executes a statement or a block of statements while an expression is true. The syntax is shown below.

      while (expression) statement

The expression is evaluated and a Boolean value, either true (a logical 1) or false (a logical 0) is returned. If the expression is true, then the procedural statement or block of statements is executed. The **while** loop executes until the expression becomes false, at which time the loop is exited and the next sequential statement is executed. If the expression is false when the loop is entered, then the procedural statement is not executed. If the value returned is **x** or **z**, then the value is treated as false.

An example illustrating the use of the **while** construct is shown in Figure 3.7 with the outputs shown in Figure 3.8. This is similar to the **repeat** example, but with a higher count. The example demonstrates a Verilog module that counts in unit increments until a count of 15 is reached, then exits the loop.

3.3.16 Pull Gates

These are MOS pull gates with no inputs and one output. The syntax is shown below.

   pull_gate [optional_instance_name) (output);

pulldown   The **pulldown** gate places a value of logic 0 on the output with a default strength of **pull**. The symbol for a **pulldown** gate is shown in Figure 3.9(a).

pullup   The **pullup** gate places a value of logic 1 on the output with a default strength of **pull**. The symbol for a **pullup** gate is shown in Figure 3.9(b).

```
//illustrates the use of the while statement
module while_example;

integer count;
initial
begin
      count = 0;
      while (count < 16)
      begin
            $display ("count = %d", count);
            count = count + 1;
      end
end
endmodule
```

Figure 3.7   Example of the **while** keyword for loop control.

count = 0	count = 4	count = 8	count = 12
count = 1	count = 5	count = 9	count = 13
count = 2	count = 6	count = 10	count = 14
count = 3	count = 7	count = 11	count = 15

Figure 3.8   Outputs for the Verilog code of Figure 3.7.

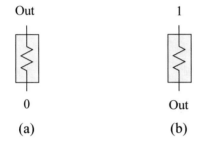

Figure 3.9   Symbols for **pull** gates: (a) **pulldown** gate and (b) **pullup** gate.

3.3.17 Signal Strengths

Signal strengths are useful when multiple drivers contend for the same net. Drive strengths can be specified for the outputs of primitive gates, continuous assignments, or net declaration assignments. The logic values in Verilog are 0, 1, **x**, and **z**. Associated with these values are the logic strengths listed in Table 3.10. The default strengths are **strong0** and **strong1**.

Examples of strength declarations for primitive gates are shown below.

> **nor (pull1, strong0)** instance_name (output, input_1, input_2, input_3);
>
> **nand (pull1, strong0)** instance_name (output, input_1, input_2);

Examples of strength declarations for continuous assignments are shown below.

> **assign (pull1, weak0)** op_code reg = instr_reg [15:8];
>
> **assign (weak1, pull0)** z1 = x1 ^ x2;

Table 3.10 Strength Levels for Logic Values

Strength level	Increasing strength
supply1	
strong1	
pull1	
weak1	
highz1	
highz0	
weak0	
pull0	
strong0	
supply0	

3.3.18 Specify Block

A *module path delay* is the delay between an **input** or **inout** and an **output** or **inout**. Module path delays are assigned within a specify block which is delimited by the keywords **specify** and **endspecify**. A specify block can be used to specify delays across a module; that is, from input to output. The keyword **specparam** is used to declare a parameter contained within the specify block.

A specify block is a separate block in a module and is not contained within any other block of the module such as an **initial** or an **always** block. The syntax is shown below.

```
specify
      timing specifications and timing checks
      define specparam constants
endspecify
```

An example showing the use of the **specify** block is shown in Figure 3.10 for a 2-input NOR gate. The propagation delay from a low to a high voltage level and from a high to a low voltage level is given as *tplh* and *tphl*, respectively. Delays for gates can be specified in a *minimum:typical:maximum* form as shown in the **specify** block. The `timescale directive specifies the time units and time precision for delays. The notation (x1 => z1) represents the input port and the output port associated with the delay. Propagation delays and timing are discussed in more detail in Chapter 5.

```
//example of specify block with delays
module specify_block (x1, x2, z1);

input x1, x2;
output z1;

nor (z1, x1, x2);

specify
     specparam
            tplh = 0.55 : 0.90 : 1.20,    //min : typ : max
            tphl = 0.50 : 0.70 : 1.55;

          (x1 => z1) = tplh, tphl;
          (x2 => z1) = tplh, tphl;
endspecify

endmodule
```

Figure 3.10   Example of a **specify** block.

3.3.19 Tasks and Functions

The symbol **$** preceding an identifier indicates a system task or a system function. Tasks and functions are similar to procedures (or subroutines) in a programming language in which a commonly used functionality can be invoked more than once. A task returns zero or more values. A function returns only one value and must have at least one input. Both tasks and functions are local to the module in which they are invoked. A task can invoke other tasks and functions; a function can invoke other functions but not other tasks. Tasks and functions are covered in more detail in Chapter 10. The syntax for tasks and functions is shown in the following:

```
task task_name;                    function [range] function_name;
    declarations                       input declaration(s)
    procedural statements              other declarations
endtask                                procedural statements
                                   endfunction
```

3.3.20 Three-State Gates

Three-state gates model three-state drivers and have one scalar data input, one or more scalar outputs, and one control input. The outputs are listed first followed by the input. The control input is listed last in instantiation. The syntax is shown below.

> **gate_type** inst1 (output_1, output_2, . . . , output_n, input, control);

A delay can also be assigned to the gates. The logic diagrams for the **bufif0**, **bufif1**, **notif0**, and **notif1** gates are shown in Figure 3.11.

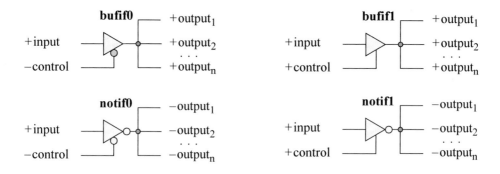

Figure 3.11   Three-state gates: **bufif0**, **bufif1**, **notif0**, and **notif1**.

The output will be in a high impedance state if the control input is inactive. For a **bufif0** gate, if the control input is at a low voltage level, then the input is transferred to the output; if the control input is at a high voltage level, then the output is in a high impedance state. For a **bufif1** gate, the output is in a high impedance state if the control input is at a low voltage level; otherwise, the input is transferred to the output.

For a **notif0** gate, if the control input is at a low voltage level, then the output is the inverted value of the input; if the control input is a high voltage level, then the output is in a high impedance state. Similarly, the **notif1** gate produces an inverted output or a high impedance output if the control input is a high or low voltage level, respectively. Truth tables for the four three-state gates are shown in Table 3.11, Table 3.12, Table 3.13, and Table 3.14.

3.3.21 Timing Control

Activity in a module can be controlled by the edges of a signal; for example, a negative transition or a positive transition. If the system clock is used in an event control statement, then the negative edge or positive edge of the clock is specified. These edges are typically used to trigger flip-flops and other edge-triggered devices. There are three keywords that specify edges: **edge**, **negedge**, and **posedge**.

edge This edge-control specifier is used to further refine the timing of an event to achieve greater precision. For example, the edges 01, 10, 0x, x1, 1x, and x0 can be used with the keyword **edge**. An example is shown below which specifies the setup and hold time for a flip-flop using the system task **\$setuphold**. Six time units is the setup time for data to be stable at the flip-flop data input before the positive transition of the clock $(0 \rightarrow 1)$. Three time units is the hold time for data to remain valid after the positive transition of the clock.

> **\$setuphold** (data, **edge** 01 clock, 6, 3);

Table 3.11 Truth Table for bufif0		
Input	Control	Output
0	0	0
0	1	z
0	x	0/z
0	z	0/z
1	0	1
1	1	z
1	x	1/z
1	z	1/z
x	0	x
x	1	z
x	x	x
x	z	x
z	0	x
z	1	z
z	x	x
z	z	x

Table 3.12 Truth Table for bufif1		
Input	Control	Output
0	0	z
0	1	0
0	x	0/z
0	z	0/z
1	0	z
1	1	1
1	x	1/z
1	z	1/z
x	0	z
x	1	x
x	x	x
x	z	x
z	0	z
z	1	x
z	x	x
z	z	x

Table 3.13 Truth Table for notif0

Input	Control	Output
0	0	1
0	1	z
0	x	1/z
0	z	1/z
1	0	0
1	1	z
1	x	0/z
1	z	0/z
x	0	x
x	1	z
x	x	x
x	z	x
z	0	x
z	1	z
z	x	x
z	z	x

Table 3.14 Truth Table for notif1

Input	Control	Output
0	0	z
0	1	1
0	x	1/z
0	z	1/z
1	z	z
1	1	0
1	x	0/z
1	z	0/z
x	0	z
x	1	x
x	x	x
x	z	x
z	0	z
z	1	x
z	x	x
z	z	x

The entries 0/z and 1/z in the tables indicate that the output could be 0, 1, or z depending on the strength of the data or control line.

negedge   This keyword is an edge-triggered event control that waits for a high to low transition on a wire or reg type variable before executing a statement or a block of statements. An example is a procedural statement that is executed when a clock pulse produces a negative transition, as shown below.

> **always** @ (**negedge** clk)
>   procedural statement

The following transitions are examples of negative edge transitions:

$1 \rightarrow x$	Transition from a logic 1 to an unknown value
$1 \rightarrow z$	Transition from a logic 1 to a high impedance
$1 \rightarrow 0$	Transition from a logic 1 to a logic 0
$x \rightarrow 0$	Transition from an unknown value to a logic 0
$z \rightarrow 0$	Transition from a high impedance to a logic 0

posedge   This keyword is an edge-triggered event control that waits for a low to high transition on a wire or reg type variable before executing a statement or a block of statements. An example is a procedural statement that is executed when a clock pulse produces a positive transition, as shown below.

> **always** @ (**posedge** clk)
> procedural statement

The following transitions are examples of positive edge transitions:

$0 \rightarrow x$ Transition from a logic 0 to an unknown value
$0 \rightarrow z$ Transition from a logic 0 to a high impedance
$0 \rightarrow 1$ Transition from a logic 0 to a logic 1
$x \rightarrow 1$ Transition from an unknown value to a logic 1
$z \rightarrow 1$ Transition from a high impedance to a logic 1

3.3.22 User-Defined Primitives

Primitives can be added to the set of built-in primitives supplied by Verilog. These are called *user-defined primitives* (UDPs). The user can create primitives for combinational or sequential modules. A UDP is created by means of a table that defines the entire functionality of the primitive. UDPs can have one or more scalar inputs, but only one scalar output. A UDP begins with the keyword **primitive** and ends with the keyword **endprimitive**. The syntax for a UDP is shown below.

> **primitive** udp_name (output, input_1, input_2, . . . , input_n);
>   output declaration
>   input declarations
>   **table**
>   define the functionality of the primitive
>   **endtable**
> **endprimitive**

An example of defining a UDP is shown in Figure 3.12 for a 2-input OR gate. UDPs are covered in detail in Chapter 6. Note the comment line above the table entries in Figure 3.12. This line provides column headings for the table.

```
//used-defined primitive for a 2-input OR gate
primitive udp_or2 (out, a, b);        //list output first

//declarations
output out;                           //must be output (not reg)
                                      //for comb logic
input a, b;

//state table definition
      table
      //inputs are in same order as input list
      //    a     b     :      out;  comment is for readability
            0     0     :      0;
            0     1     :      1;
            1     0     :      1;
            1     1     :      1;
      endtable

endprimitive
```

Figure 3.12   A UDP for a 2-input OR gate.

3.4 Value Set

The output of a logic element drives the inputs of other logic elements in a net. Verilog expresses the output states in terms of four predefined values, as shown in Table 3.15.

Table 3.15 Predefined 4-Value Logic Set

Value Level	Definition
0	Logic 0; false condition
1	Logic 1; true condition
x	Unknown logic value
z	High impedance condition; floating state

The values 0 and 1 represent low and high levels, respectively, of the logic family being used. The value **x** indicates ambiguity when the simulator cannot determine whether the logic value is a 0 or a 1. The value **z** indicates a high impedance condition in which the driver of a net is disabled or the driver is disconnected. Three-state drivers are normally used to connect circuits to a bus that is driven by multiple three-state drivers. When the value **z** is present at the input of a logic gate, it is interpreted as an **x** value. The values **x** and **z** are not case sensitive. The 4-valued logic set also applies to registers on a bit-by-bit basis.

3.5 Data Types

Verilog defines two data types: nets and registers. These predefined data types are used to connect logical elements and to provide storage. A net is a physical wire or group of wires connecting hardware elements in a module or between modules. The value of a net is determined by the logic that drives the net such as a continuous assignment. A register represents a storage element that retains its value until a new value is entered. It is assigned a value by means of an **initial** statement or an **always** statement. This section presents some of the more common data types.

3.5.1 Net Data Types

An example of net data types is shown in Figure 3.13, where five internal nets are defined: *net1*, *net2*, *net3*, *net4*, and *net5*. The value of *net1* is determined by the inputs to the *and1* gate represented by the term $x_1 x_2'$, where x_2 is active low; the value of *net2* is determined by the inputs to the *and2* gate represented by the term $x_1' x_2$, where x_1 is active low; the value of *net3* is determined by the input to the inverter represented by the term x_3, where x_3 is active low. Nets are declared as **wire**.

   Output z_1 is connected to *net4*, which is driven by the output of the *or1* gate represented by the term *net1* + *net2*. Output z_2 is connected to *net5*, which is driven by the output of the *or2* gate represented by the term *net2* + *net3*. The equations for z_1 and z_2 are shown in Equation 3.1. The Verilog module for Figure 3.13 is shown in Figure 3.14 using dataflow modeling with built-in primitives to illustrate the use of the net data type. The test bench, binary outputs, and waveforms are shown in Figure 3.15, Figure 3.16, and Figure 3.17, respectively.

tri   The **tri** data type has the same syntax and functionality as **wire**, but is listed separately to indicate a three-state driver. The keyword **wire** indicates a connection between logic elements using a single driver; the keyword **tri** indicates a connection between logic elements using multiple drivers.

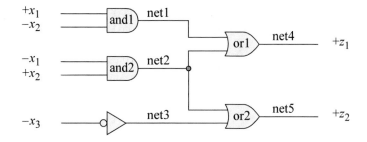

Figure 3.13   A logic diagram showing single-wire nets and one multiple-wire net, where the connectivity of the logic gates is determined by nets.

$$z_1 = x_1 x_2' + x_1' x_2$$

$$z_2 = x_1' x_2 + x_3 \tag{3.1}$$

```
//module showing use of wire          and inst1 (net1, x1, ~x2);
//connecting logic primitives         and inst2 (net2, ~x1, x2);
                                      not inst3 (net3, ~x3);
module log_diag_eqn4 (x1, x2,          or inst4 (net4, net1, net2);
            x3, z1, z2);               or inst5 (net5, net2, net3);

input x1, x2, x3;                      assign z1 = net4;
output z1, z2;                         assign z2 = net5;

wire x1, x2, x3;                      endmodule
wire z1, z2;

//define internal nets as wire
wire net1,net2,net3,net4,net5;

//instantiate the built-in
//primitives
```

Figure 3.14   Module showing the use of wires to connect logic primitives.

```
//logic diagram test bench          #10  x1=1'b1; x2=1'b0; x3=1'b0;
module log_diag_eqn4_tb;            #10  x1=1'b1; x2=1'b0; x3=1'b1;
                                    #10  x1=1'b1; x2=1'b1; x3=1'b0;
reg x1, x2, x3;                     #10  x1=1'b1; x2=1'b1; x3=1'b1;
wire z1, z2;                        #10  $stop;
                                    end
//display variables
initial                             //instantiate the module into
$monitor ("x1x2x3 = %b,            //the test bench
    z1 = %b, z2 = %b",             log_diag_eqn4 inst1 (
    {x1, x2, x3}, z1, z2);              .x1(x1),
                                        .x2(x2),
//apply stimulus                        .x3(x3),
initial                                 .z1(z1),
begin                                   .z2(z2)
#0   x1=1'b0; x2=1'b0; x3=1'b0;         );
#10  x1=1'b0; x2=1'b0; x3=1'b1;
#10  x1=1'b0; x2=1'b1; x3=1'b0;    endmodule
#10  x1=1'b0; x2=1'b1; x3=1'b1;
```

Figure 3.15   Test bench for the module of Figure 3.14.

```
x1x2x3 = 000, z1 = 0, z2 = 0       x1x2x3 = 100, z1 = 1, z2 = 0
x1x2x3 = 001, z1 = 0, z2 = 1       x1x2x3 = 101, z1 = 1, z2 = 1
x1x2x3 = 010, z1 = 1, z2 = 1       x1x2x3 = 110, z1 = 0, z2 = 0
x1x2x3 = 011, z1 = 1, z2 = 1       x1x2x3 = 111, z1 = 0, z2 = 1
```

Figure 3.16   Binary outputs for the module of Figure 3.14.

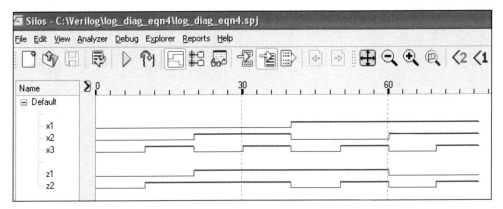

Figure 3.17   Waveforms for the test bench of Figure 3.15.

wand   The **wand** data type signifies a net with multiple drivers. It models a *wired AND* circuit using open collector technology. If any of the drivers has a logic 0 output, then the value of the net is 0. A wired AND circuit is shown in Figure 3.18 using transistor-transistor logic (TTL) gates, where a logic 0 output is approximately at ground potential.

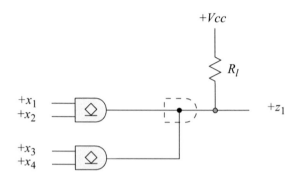

Figure 3.18   Wired AND function using two open collector AND gates.

A wired AND circuit is a collection of two or more open collector AND gates whose outputs are wired together forming the AND function without the use of a physical AND gate. The equation for output z_1 is shown in Equation 3.2.

$$z_1 = (x_1 x_2)(x_3 x_4) \tag{3.2}$$

triand   This net has multiple drivers and models the wired AND circuit as a three-state circuit.

wor   The **wor** data type signifies a net with multiple drivers. It models a *wired OR* circuit using open collector technology. A wired OR circuit is shown in Figure 3.19(a) using TTL gates and in Figure 3.19(b) using emitter-coupled logic (ECL) gates. In Figure 3.19(a), if any of the drivers has a logic 0 output, then the value of the net is 0. In Figure 3.19(b), if any of the drivers has a logic 1 output, then the value of the net is 1.

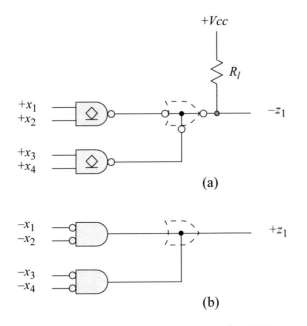

Figure 3.19   Wired OR circuits: (a) using TTL gates and (b) using ECL gates.

Figure 3.19(a) represents a TTL wired OR configuration using open collector gates. The circuit can be considered as an AND-OR-INVERT function with an active low output. In TTL, a low voltage level is approximately at ground potential. The equation is shown in Equation 3.3. The same circuit can also be considered as a wired AND function represented by Equation 3.4. Figure 3.19(b) represents a wired OR configuration using ECL gates. In this case, the inputs are active low. A high voltage level is approximately at ground potential. The equation is shown in Equation 3.5.

$$z_1 = x_1 x_2 + x_3 x_4 \tag{3.3}$$

$$
\begin{aligned}
z_1 &= (x_1 x_2 + x_3 x_4)' \\
&= (x_1' + x_2')(x_3' + x_4')
\end{aligned}
\tag{3.4}
$$

$$z_1 = x_1 x_2 + x_3 x_4 \tag{3.5}$$

trior   This net has multiple drivers and models the wired OR circuit as a three-state circuit.

trireg   This net is similar to a register. It stores a value and is used to model a capacitive charge stored on the net when its drivers are disabled. Nets of this type are in one of two states: capacitive state or driving state. In the capacitive state, the strengths are **small**, **medium**, and **large**, with the default strength being **medium**. When the three-state drivers are disabled, the net retains the last logic value that was present before the drivers were disabled (0, 1, or **x**). In the driving state, a 0, 1, or **x** from the driver outputs is stored on the net.

3.5.2 Register Data Types

A register data type represents a variable that can retain a value. Verilog registers are similar in function to hardware registers, but are conceptually different. Hardware registers are synthesized with storage elements such as *D* flip-flops, *JK* flip-flops, and *SR* latches. Verilog registers are an abstract representation of hardware registers and are declared as **reg**.

The default size of a register is one bit; however, a larger width can be specified in the declaration. The general syntax to declare a width of more than one bit is as follows:

reg [msb:lsb] register_name

To declare a byte width in the *little endian* notation, a *data_register* would be declared as follows:

reg [7:0] data_register

where bit 7 is the high-order bit and bit 0 is the low-order bit. In any width declaration, the left-hand bit is the high-order, or most significant bit; the right-hand bit is the low-order, or least significant bit. Verilog also supports the *big endian* notation, as shown below for the same *data_register*.

reg [0:7] data_register

As before, the left-hand bit is the high-order bit and the right-hand bit is the low-order bit. Single bit registers are designated as *scalar*; multiple bit registers are designated as *vector*.

Registers can also be formed by the concatenation of bits. For example, for the *data_register* shown above, {*data_register [5:4], data_register [3:2]*} defines a 4-bit register composed of the middle four bits of the 8-bit *data_register*. Alternatively, the same 4-bit register can be specified as *data_register [5:2]*, which represents an abbreviated method for assigning partial bits from a larger register.

Memories   Memories can be represented in Verilog by an array of registers and are declared using a **reg** data type as follows:

Number of bits per register   Number of registers
↓   ↓
reg [msb:lsb] memory_name [first address:last address];

A 32-word register with one byte per word would be declared as follows:

$$\text{reg } [7:0] \text{ memory\_name } [0:31];$$

An array can have only two dimensions. Memories must be declared as **reg** data types, not as **wire** data types. A register can be assigned a value using one statement, as shown below.

$$\text{reg } [15:0] \text{ buff\_reg};$$
$$\text{buff\_reg} = 16'\text{h7ab5};$$

Register *buff_reg* is assigned the value $0111\ 1010\ 1011\ 0101_2$. Values can be stored in memories by assigning a value to each word individually, as shown below for an instruction cache of eight registers with eight bits per register.

$$\text{reg } [7:0] \text{ instr\_cache } [0:7];$$

instr_cache [0] =	8'h08;
instr_cache [1] =	8'h09;
instr_cache [2] =	8'h0a;
instr_cache [3] =	8'h0b;
instr_cache [4] =	8'h0c;
instr_cache [5] =	8'h0d;
instr_cache [6] =	8'h0e;
instr_cache [7] =	8'h0f;

Alternatively, memories can be initialized by means of one of the following system tasks:

$readmemb	for binary data
$readmemh	for hexadecimal data

A text file is prepared for the specified memory in either binary or hexadecimal format. The system task reads the file and loads the contents into memory. An example of loading the *instr_cache* memory described above with binary data is shown in Example 3.1.

Example 3.1   A block diagram of the instruction cache is shown in Figure 3.20 with a 3-bit program counter *pc [2:0]* as an input vector to address the instruction cache and an 8-bit *ic_data_out [7:0]* bus as an output vector. Each address in the cache is considered to be a cache line containing a block of data, in this case one byte. The contents of the cache represent the instructions to execute a computer program. The instructions are decoded to generate the operation code, source operand address, and destination operand address.

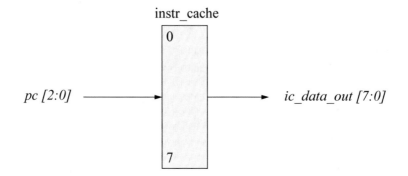

Figure 3.20   Block diagram of the instruction cache for Example 3.1.

The file shown in Table 3.16 is created and saved as *icache.instr* (the addresses are shown for reference). The file is saved as a separate file in the project folder without the **.v** extension.

Table 3.16 *Icache.instr* File to Be Loaded into the Memory *instr_cache*

Address	Data	Address	Data
Word 0	0000_1000	Word 4	0000_1100
Word 1	0000_1001	Word 5	0000_1101
Word 2	0000_1010	Word 6	0000_1110
Word 3	0000_1011	Word 7	0000_1111

The contents of *icache.instr* are loaded into the memory *instr_cache* beginning at location 0, by the following two statements:

reg [7:0] instr_cache [0:7];

$readmemb ("icache.instr", instr_cache);

The Verilog code for the above procedure is shown in Figure 3.21 using binary data for the file called *icache.instr*. The test bench, the contents of the *icache.instr* file, the binary output, and the waveforms are shown in Figure 3.22, Figure 3.23, Figure 3.24, and Figure 3.25, respectively.

There are two ports in Figure 3.18: input port *pc* and output port *ic_data_out*, which is declared as **reg** because it operates as a storage element. The instruction cache is defined as an array of eight 8-bit registers by the following statement:

reg [7:0] instr_cache [0:7];

An **initial** procedural construct is used to load data from the *icache.instr* file into the instruction cache *instr_cache* by means of the system task **$readmemb**. The **initial** statement executes only once to initialize the instruction cache. An **always** procedural construct is then used to read the contents of the instruction cache based on the value of the program counter; that is, *ic_data_out* receives the contents of the instruction cache at the address specified by the program counter. The program counter is an event control used in the **always** statement — when the program counter changes, the statement in the **begin** . . . **end** block is executed.

```
//procedure for loading memory with
//binary data from file icache.instr
module mem_load (pc, ic_data_out);

//list inputs and outputs
input pc;
output ic_data_out;

//list wire and reg
wire [2:0] pc;              //a program counter to address 8 words
reg [7:0] ic_data_out;

//define memory size
//instr_cache is an array of eight 8-bit regs
reg [7:0] instr_cache [0:7];

//continued on next page
```

Figure 3.21   Verilog module to illustrate the use of **$readmemb** to load an instruction cache.

```
//define memory contents
//load instr_cache from file icache.instr
initial
begin
    $readmemb ("icache.instr", instr_cache);
end

//use a program counter to access the instr_cache
always @ (pc)
begin
    ic_data_out = instr_cache [pc];
end

endmodule
```

Figure 3.21   (Continued)

The test bench of Figure 3.22 has a 3-bit input vector *pc* declared as **reg** because it retains its value and an 8-bit output vector *ic_data_out* declared as **wire**. All eight combinations of the program counter are listed in 10-time-unit increments. Then the contents of the instruction cache are displayed by means of a **for** loop using **integer** *i* as a control variable.

```
//mem_load test bench
module mem_load_tb;

integer i;//used to display contents

reg [2:0] pc;
wire [7:0] ic_data_out;

//assign values to the program counter
initial
begin
    #0     pc = 3'b000;
    #10    pc = 3'b001;
    #10    pc = 3'b010;
    #10    pc = 3'b011;
    #10    pc = 3'b100;
    #10    pc = 3'b101;

//continued on next page
```

Figure 3.22   Test bench for the module of Figure 3.21.

```
      #10    pc = 3'b110;
      #10    pc = 3'b111;
      #15    $stop;
end

//display the contents of the instruction cache
initial
begin
   for (i=0; i<8; i=i+1)
   begin
      #10 $display ("address %h = %b", i, ic_data_out);
   end
   #150 $stop;
end

//instantiate the module into the test bench
mem_load inst1 (
   .pc(pc),
   .ic_data_out(ic_data_out)
   );

endmodule
```

Figure 3.22   (Continued)

```
0000_1000
0000_1001
0000_1010
0000_1011
0000_1100
0000_1101
0000_1110
0000_1111
```

Figure 3.23   Instruction cache file *icache.instr* saved as a separate file. It is saved in the project folder without the .v extension.

```
address 00000000 = 00001000      address 00000004 = 00001100
address 00000001 = 00001001      address 00000005 = 00001101
address 00000002 = 00001010      address 00000006 = 00001110
address 00000003 = 00001011      address 00000007 = 00001111
```

Figure 3.24   Outputs obtained from the test bench of Figure 3.22.

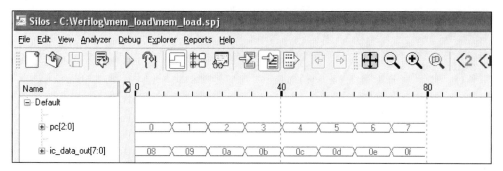

Figure 3.25   Waveforms for the test bench of Figure 3.22.

Integer registers   An integer register is a general-purpose register that is used for computation and manipulating data. Integers are declared by the keyword **integer**. Integers have a wordlength of at least 32 bits and are specified in 2s complement representation, where the high-order bit indicates the sign of the number. For example, the integers $+24_{10}$ and -24_{10} are written in 2s complement representation as:

$$0000\_0000\_0000\_0000\_0000\_0000\_0001\_1000_2 \ (+24)$$

$$1111\_1111\_1111\_1111\_1111\_1111\_1110\_1000_2 \ (-24)$$

The sign of a positive number in any radix is 0 and the sign of a negative number in any radix is $r-1$, where r is the radix. Integer registers store signed numbers and are used to model high-level behavior, providing a result in 2s complement representation. Registers that are declared as type **reg** store unsigned values; registers declared as type **integer** store signed values. Integer registers are declared in the following form:

integer integer_1, integer_2, . . . , integer_n [msb:lsb];

The notation *[msb:lsb]* specifies an optional range; however, a bit-select or part-select is not allowed. Thus, for an integer declared as *instr*, the bit-select *instr[5]* and the part-select *instr[15:8]* are not allowed.

If it is necessary to select a particular bit or string of bits, then the **integer** register can be assigned to a **reg** register. For example,

```
reg [31:0] mach_instr;
integer instr;
    .
    .
    .
mach_instr = instr;
```

Once the assignment has been made, then *mach_instr[5]* and *mach_instr[15:8]* can be obtained.

Real registers   Real constants and register data types are declared by the keyword **real**. Real data objects are stored in double precision floating-point representation of 64 bits as shown below.

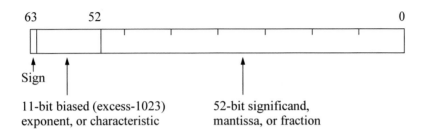

The bias makes all exponents unsigned internally; that is, a bias of 1023 is added to the signed exponent to create an unsigned exponent. This makes exponent comparison hardware more efficient. Real numbers can be specified in decimal notation or in scientific notation; for example, 0.000003 in decimal notation or 3.0×10^{-6} in scientific notation. When a real number is assigned to an integer, the real number is rounded to the nearest integer.

Time registers   A time register is used to store simulation time. It is declared by the keyword **time** as follows:

$$\textbf{time} \quad time_1, time_2, \dots, time_n \, [msb:lsb];$$

Time variables are stored as unsigned values of at least 64 bits. The system function **$time** obtains the current simulation time. The keyword **realtime** stores time values in the real number format.

3.6 Compiler Directives

The syntax for compiler directives is `<keyword>, where the symbol ( `) is a back-quote or grave accent mark. A list of compiler directives as defined in the IEEE standard 1364-1995 is shown in Table 3.17.

Table 3.17 Compiler Directives

`celldefine	`include
`default_nettype	`nounconnected_drive
`define	`resetall
`else	`timescale
`endcelldefine	`unconnected_drive
`endif	`undefine
`ifdef	

This section defines some of the more common compiler directives. When a compiler directive is compiled, it remains in effect for the duration of the compilation until changed by another compiler directive.

`define**    The **`define directive is used to create text substitution similar to the #define construct in the C programming language. The Verilog compiler substitutes the defined macro text whenever it encounters a **`define** directive. The syntax for a **`define** directive is

`define   text_macro macro_text

The directive, the text_macro, and the macro_text are separated by white space. Also, there is no semicolon terminating the line. An example of the directive is shown below.

```
`define bus_size 16            //define a bus of 16 bits
         .
         .
//addr_in bus is 16 bits wide, 15:0
reg [`bus_size-1:0] addr_in;
```

`undefine   This directive removes the definition of a previously defined macro. An example is shown below.

```
`define bus_size 16       //define a bus of 16 bits
         .
         .
//addr_in bus is 16 bits wide, 15:0
reg [`bus_size-1:0] addr_in;
         .
         .
`undefine bus_size         //definition of bus_size is removed
```

`ifdef, `else, `endif**    These compiler directives are used for conditional compilation as illustrated in the example below. The **`ifdef directive is followed by a text macro name. If the text macro has been previously defined by a **`define** directive, then the first statement (or block of code) is included in the compilation; otherwise, the second statement (or block of code) is compiled. The **`else** compiler directive is optional.

```
        `define TEXT_MACRO_NAME
            .
            .
            .
        `ifdef TEXT_MACRO_NAME
            statement_1;
            statement_2;
                .
                .
                .
            statemnet_n
        `else
            alternative_statement_1;
            alternative_statement_2;
                .
                .
                .
            alternative_statement_n;
        `endif
```

```
        `ifdef TEXT_MACRO_NAME
            statement_1;
            statement_2;
                .
                .
                .
            statemnet_n
        `endif
```

`include**    The **`include directive allows a Verilog file to be included inline in another file during compilation. During compilation, the line containing the **`include** directive is replaced with the contents of the source file. This is similar to the #include construct in the C programming language. For example, the Verilog code **`include** filename.v is replaced by the contents of filename.v.

`resetall   This directive resets all compiler directives to their default values; for example, the default net type is reset to **wire**.

`timescale**    In a Verilog module, all delays are specified in terms of time units. This is accomplished by the **`timescale compiler directive, which specifies the time units and time precision to be used in the module, and has the following syntax:

`timescale   time_unit / time_precision

The *time_unit* and *time_precision* values are specified in integers of 1, 10, and 100. The units and precision associated with the directive are listed in Table 3.18. For example,

`timescale 10ns / 10ps

specifies a time unit of 10 nanoseconds with a precision of 10 picoseconds.

Table 3.18 Verilog Timescale Units and Precision

Time units	Definition
s	Seconds
ms	Milliseconds
μs	Microseconds
ns	Nanoseconds
ps	Picoseconds
fs	Femtoseconds

3.7 Problems

3.1   Write the following decimal numbers as 16-bit binary numbers in 2s complement representation:

(a) 2170_{10}
(b) -858_{10}
(c) 32767_{10}
(d) -32768_{10}
(e) -1_{10}

The binary weight assigned to the 4-bit segments is:

$2^{12} = 4096$	$2^8 = 256$	$2^4 = 16$	$2^0 = 1$

3.2 Obtain the 2s complement of the following binary numbers:

(a) $0010\_0111\_0110\_1011_2$
(b) $1111\_1111\_1111\_1000_2$
(c) $0101\_0101\_0101\_0101_2$

3.3 Indicate the error, if any, in the following identifiers:

(a) out_reg
(b) $in_reg
(c) addr_$reg
(d) 2augend

3.4 Obtain the module, test bench, binary outputs, and waveforms for the equations shown below. Use built-in primitives and declare any internal wires.

$$z_1 = (x_1 \oplus x_2)x_3'$$
$$z_2 = (x_1 \oplus x_2)' \oplus x_3$$

3.5 After the following program has executed, determine the contents of reg_out:

```
//test of a register
module out_reg (reg_out);


output reg_out;


reg [31:0] reg_out;
initial
begin
      reg_out = 31'bx;
      reg_out = 16'bz;
end


initial
begin
      #10    $display ("reg_out = %h", reg_out);
end
endmodule
```

3.6   Using dataflow modeling, implement the following truth table:

z_1	x_1 x_2 x_3 x_4	z_1	x_1 x_2 x_3 x_4
0	0 0 0 0	1	1 0 0 0
1	0 0 0 1	0	1 0 0 1
0	0 0 1 0	0	1 0 1 0
1	0 0 1 1	0	1 0 1 1
0	0 1 0 0	1	1 1 0 0
0	0 1 0 1	0	1 1 0 1
0	0 1 1 0	1	1 1 1 0
1	0 1 1 1	1	1 1 1 1

3.7   (a)   Specify the number of registers and the number of bits per register in the following statement:

reg [0:7] name [0:4];

(b)   What does the following declaration specify?

reg a [1:7];

(c)   What do the following declarations represent (be specific)?

reg [1:n] reg_a;
reg reg_a [1:n];

3.8   Which *one* of the following statements is true?

(a)   Vectors and arrays are the same.
(b)   Every element in an array can be assigned using one statement.
(c)   A RAM can be modeled by declaring an array of registers.

3.9   Assume that floating-point numbers are represented in a 12-bit format as shown below. The scale factor has an implied base of 2 and a 5-bit, excess-15 exponent. The 6-bit mantissa is normalized as in the IEEE format, with an implied 1 to the left of the binary point. Represent the number +19 in this format.

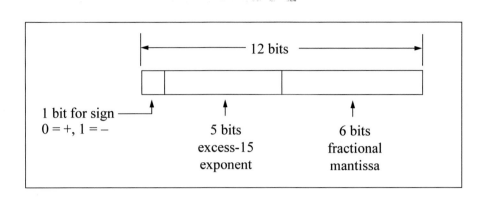

3.10   Load a 16-byte data cache called *dcache* with the hexadecimal characters 80, C0, E0, F0, F8, FC, FE, FF, 7F, 3F, 1F, 0F, 07, 03, 01, 00. Then design a test bench and obtain the binary outputs and waveforms. The data cache has a vector input called *dc_addr[3:0]* and a vector output called *dc_data_out[7:0]*. Generate a hexadecimal file called *dcache.data* that will be loaded into the data cache.

4

Expressions

Expressions consist of operands and operators, which are the basis of Verilog HDL. The result of a right-hand side expression can be assigned to a left-hand side net variable or register variable. The value of an expression is determined from the combined operations on the operands. An expression can consist of a single operand or two or more operands in conjunction with one or more operators. The result of an expression is represented by one or more bits. Examples of expressions are as follows:

```
assign  z1 = x1 & x2 & x3;
assign  z1 = x1 | x2 | x3;
assign cout = (a & cin) | (b & cin) | (a & b);
```

4.1 Operands

Operands can be any of the data types listed in Table 4.1.

Table 4.1 Operands

Operands	Comments
Constant	Signed or unsigned
Parameter	Similar to a constant
Continued on next page	

Table 4.1 Operands

Operands	Comments
Net	Scalar or vector
Register	Scalar or vector
Bit-select	One bit from a vector
Part-select	Contiguous bits of a vector
Memory element	One word of a memory
Function call	System or user function

4.1.1 Constant

Constants can be signed or unsigned. A decimal integer is treated as a signed number. An integer that is specified by a base is interpreted as an unsigned number. Examples of both types are shown in Table 4.2.

Table 4.2 Signed and Unsigned Constants

Constant	Comments
127	Signed decimal: Value = 8-bit binary vector: 0111_1111
-1	Signed decimal: Value = 8-bit binary vector: 1111_1111
-128	Signed decimal: Value = 8-bit binary vector: 1000_0000
4'b1110	Binary base: Value = unsigned decimal 14
8'b0011_1010	Binary base: Value = unsigned decimal 58
16'h1A3C	Hexadecimal base: Value = unsigned decimal 6716
16'hBCDE	Hexadecimal base: Value = unsigned decimal 48,350
9'o536	Octal base: Value = unsigned decimal 350
-22	Signed decimal: Value = 8-bit binary vector: 1110_1010
-9'o352	Octal base: Value = 8-bit binary vector: 1110_1010 = unsigned decimal 234

The last two entries in Table 4.2 both evaluate to the same bit cofiguration, but represent different decimal values. The number -22_{10} is a signed decimal value; the number -9'o352 is treated as an unsigned number with a decimal value of 234_{10}.

4.1.2 Parameter

A parameter is similar to a constant and is declared by the keyword **parameter**. Parameter statements assign values to constants; the values cannot be changed during simulation. Examples of parameters are shown in Table 4.3.

Table 4.3 Examples of Parameters

Examples	Comments
parameter width = 8	Defines a bus width of 8 bits
parameter width = 16, depth = 512	Defines a memory with two bytes per word and 512 words
parameter out_port = 8	Defines an output port with an address of 8

Parameters are useful in defining the width of a bus. For example, the adder shown in Figure 4.1 contains two 8-bit vector inputs *a* and *b* and one scalar input *cin*. There is also one 9-bit vector output *sum* comprised of an 8-bit result and a scalar carry-out.

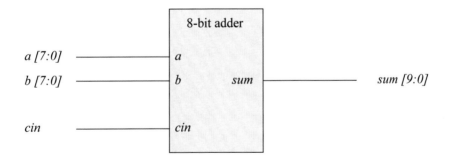

Figure 4.1   Eight-bit adder to illustrate the use of a **parameter** statement.

The Verilog code for the adder is shown in Figure 4.2. Note the use of the **parameter** statement which defines a bus width of eight bits. Wherever *width* appears in the code, it is replaced by the value eight. The test bench is shown in Figure 4.3, the outputs in Figure 4.4, and the data analyzer waveforms in Figure 4.5.

```
//example of using a parameter
module param1 (a, b, cin, sum);


parameter width = 8;


input [width-1:0] a, b;        //a and b are 8 bits (7:0)
input cin;                     //cin is a scalar
output [width:0] sum;          //sum is 9 bits (8:0)
                               //to include cout
//inputs default to wire
reg [width:0] sum;


always @ (a or b or cin)
begin
     sum = a + b + cin;
end


endmodule
```

Figure 4.2   Module for the 8-bit adder of Figure 4.1 illustrating the use of the **parameter** statement.

```
//param1 test bench
module param1_tb;


parameter width = 8;


reg [width-1:0] a, b;
reg cin;


wire [width:0] sum;


//display variables
initial
$monitor ("a b cin = %b_%b_%b, sum = %b", a, b, cin, sum);


//continued on next page
```

Figure 4.3   Test bench for the module of Figure 4.2.

```
//apply input vectors
initial
begin
     #0    a = 8'b0000_0011;
           b = 8'b0000_0100;
           cin = 1'b0;

     #5    a = 8'b0000_1100;
           b = 8'b0000_0011;
           cin = 1'b0;

     #5    a = 8'b0000_0111;
           b = 8'b0000_0110;
           cin = 1'b1;

     #5    a = 8'b0001_1001;        //25 (19h)
           b = 8'b0010_0111;        //39 (27h)
           cin = 1'b1;              //1.  sum = 65 (41h)

     #5    a = 8'b0111_1101;        //125 (7dh)
           b = 8'b0110_0111;        //103 (67h)
           cin = 1'b1;              //1.  sum = 229 (e5h)

     #5    a = 8'b1000_1111;        //143 (8fh)
           b = 8'b1100_0110;        //198 (c6h)
           cin = 1'b1;              //1.  sum = 342 (156h)

     #5    $stop;

end

//instantiate the module into the test bench
param1 inst1 (
     .a(a),
     .b(b),
     .cin(cin),
     .sum(sum)
     );

endmodule
```

Figure 4.3   (Continued)

```
a b cin = 00000011_00000100_0, sum = 000000111
a b cin = 00001100_00000011_0, sum = 000001111
a b cin = 00000111_00000110_1, sum = 000001110
a b cin = 00011001_00100111_1, sum = 001000001
a b cin = 01111101_01100111_1, sum = 011100101
a b cin = 10001111_11000110_1, sum = 101010110
```

Figure 4.4   Outputs obtained from the test bench of Figure 4.3.

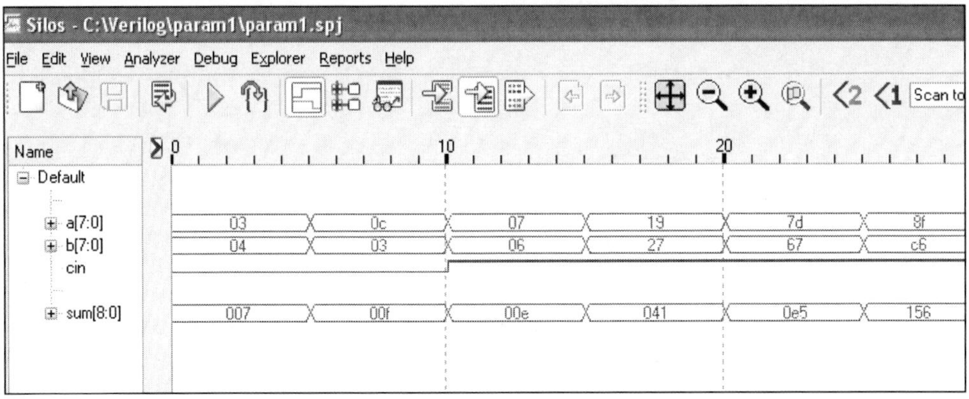

Figure 4.5   Data analyzer waveforms for the test bench of Figure 4.4.

4.1.3 Net

Nets were discussed previously in Section 3.5.1, but will be reviewed briefly in this section. Nets can be scalar (one bit) or vector (multiple bits) and are used to connect hardware elements. The value assigned to a net is interpreted as an unsigned value, as shown in the following statements:

> **wire** [7:0] bus_in, bus_out;
> .
> .
> .
> **assign** bus_in = −59;   //1100_0101 = 197 unsigned
> **assign** bus_out = 16'he0;   //1110_0000 = −32 signed; 224 unsigned

4.1.4 Register

Registers were presented in Section 3.5.2, but will be reviewed briefly in this section. Registers represent storage elements and are declared by the keywords **reg**, **integer**, **time**, **real**, or **realtime**. Registers can contain scalar or vector values. A value in a register declared by the keyword **reg** is interpreted as an unsigned number that stores logic values in flip-flops or latches, whereas a value in a register declared by the keyword **integer** stores a quantity that is treated as a signed number in 2s complement representation. Examples of **reg** and **integer** type registers are shown in the following code segment:

```
reg  [15:0] bus_out;
integer  [15:0] accumulator;
     .
     .
     .

initial
begin
    bus_out = −20;        //−20 is 1111_1111_1110_1100 signed
                          //which is 65,516 unsigned
    accumulator = −57     //−57 is 1111_1111_1100_0111 signed
end
```

4.1.5 Bit-Select

A bit-select extracts a single bit or element from a net or register vector to be used in an operation. For example, to determine the sign of a 16-bit operand, bit 15 is tested to determine whether the bit is a 0 or a 1; that is, a positive or negative number in 2s complement representation. A bit-select statement has the following format:

identifier [bit-select_expression]

The selected bit can be specified by either a constant or by an expression, as shown in the following statements.

```
wire  [15:0] bus_out;    //16-bit bus, where bus_out [15] is the high-order bit

bus_out [7]              //selects bit 7 of bus_out

bus_out [x1 + 3]         //selects a bit of bus_out that is dependent on the
                         //evaluation of the expression x1 + 3
```

4.1.6 Part-Select

A part-select examines a contiguous sequence of bits in a vector and is formatted as shown below. The identifier is followed by square brackets enclosing the range of bits — the high-order bit (msb) on the left of the colon and the low-order bit (lsb) on the right of the colon.

identifier [msb_constant_expression : lsb_constant_expression]

One use for part-select is to extract the operation code from an instruction vector, usually the high-order bits. An example of part-select is as follows:

reg [7:0] opcode;
reg [31:0] instr;   //32-bit instruction
opcode = instr [31:24]; //the leftmost byte of the instruction

If the range is declared in the *little endian* notation, then the part-select must be in the same notation; the same is true for the *big endian* notation. Thus, if a vector is declared as **reg** *[31:0] instr*, then the larger number in the part-select must be to the left of the colon, as shown above.

4.1.7 Memory Element

A memory element refers to a word in memory, which may be one or more bits. Memories are referenced by words only — there is no bit-select and no part-select. The individual bits in a memory cannot be accessed directly; the addressed word must first be assigned to a word register and then the individual bits can be accessed. For example, assume a 2048-word memory called *icache* with 32 bits per word as shown below. Assume also a 32-bit register called *instr_reg* that will be used to access particular bits of a memory word.

reg [31:0] icache [0:2047];
reg [31:0] instr_reg;

The *icache* and *instr_reg* are units in a pipelined reduced-instruction-set computer (RISC). The *icache* memory contains instructions that are accessed individually. The addressed instruction is placed in the *instr_reg*, then sent to a *decode* unit, where the individual fields of the instruction are stored. The instruction consists of four fields: *op_code*, *addr_mode*, *dst_addr*, and *src_addr*. This is a very simple example and is used to illustrate the method of selecting bits from memory. A block diagram of the three units is shown in Figure 4.6.

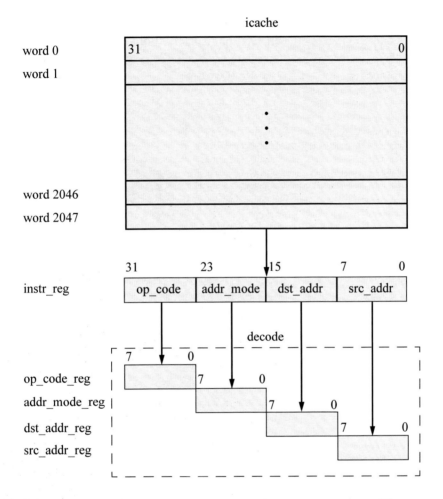

Figure 4.6   Block diagram for a partial RISC processor to illustrate a method to select individual memory bits.

To select the word at address 127 (the 128th word) of the *icache* and load the *op_code* of that word into the *op_code_reg* register, the following statements are executed:

    instr_reg = icache [127];

    op_code_reg = instr_reg [31:24];

4.2 Operators

Verilog HDL contains a profuse set of operators that perform various operations on different types of data to yield results on nets and registers. Some operators are similar to those used in the C programming language. Table 4.4 lists the categories of operators in order of precedence, from highest to lowest.

Table 4.4 Verilog HDL Operators and Symbols

Operator type	Operator Symbol	Operation	Number of Operands
Arithmetic	+	Add	Two or one
	-	Subtract	Two or one
	*	Multiply	Two
	/	Divide	Two
	%	Modulus	Two
Logical	&&	Logical AND	Two
	||	Logical OR	Two
	!	Logical negation	One
Relational	>	Greater than	Two
	<	Less than	Two
	>=	Greater than or equal	Two
	<=	Less than or equal	Two
Equality	==	Logical equality	Two
	!=	Logical inequality	Two
	===	Case equality	Two
	!==	Case inequality	Two
Bitwise	&	AND	Two
	|	OR	Two
	~	Negation	One
	^	Exclusive-OR	Two
	^~ or ~^	Exclusive-NOR	Two
Reduction	&	AND	One
	~&	NAND	One
	|	OR	One
	~|	NOR	One
	^	Exclusive-OR	One
	~^ or ^~	Exclusive-NOR	One
Shift	<<	Left shift	One
	>>	Right shift	One
Conditional	? :	Conditional	Three
Concatenation	{ }	Concatenation	Two or more
Replication	{{ }}	Replication	Two or more

4.2.1 Arithmetic

Arithmetic operations are performed on one (unary) operand or two (binary) operands in the following radices: binary, octal, decimal, or hexadecimal. The result of an arithmetic operation is interpreted as an unsigned value or as a signed value in 2s complement representation on both scalar and vector nets and registers.

The operands shown in Table 4.5 are used for the operations of addition, subtraction, multiplication, and division.

Table 4.5 Operands Used for Arithmetic Operations

	Addition		Subtraction		Multiplication		Division
	Augend		Minuend		Multiplicand		Dividend
+)	Addend	–)	Subtrahend	×)	Multiplier	÷)	Divisor
	Sum		Difference		Product		Quotient, Remainder

The unary + and – operators change the sign of the operand and have higher precedence than the binary + and – operators. Examples of unary operators are shown below.

$$+45 (\text{Positive } 45_{10})$$
$$-72 (\text{Negative } 72_{10})$$

Unary operators treat net and register operands as unsigned values, and treat real and integer operands as signed values.

The binary add operator performs unsigned and signed addition on two operands. Register and net operands are treated as unsigned operands; thus, a value of

$$1111\_1111\_1111\_1111_2$$

stored in a register has a value of $65{,}535_{10}$ unsigned, not -1_{10} signed. Real and integer operands are treated as signed operands; thus, a value of

$$1111\_1110\_1010\_0111_2$$

stored in an integer register has a value of -345_{10} signed, not $65{,}191_{10}$ unsigned. The width of the result of an arithmetic operation is determined by the width of the largest operand. The Verilog code of Figure 4.7 shows examples of addition, subtraction, multiplication, division, and modulus using the **parameter** and **case** constructs.

The operation code is three bits in order to accommodate the five arithmetic operations that are defined by the **parameter** keyword. The **case** statement is used to branch to the appropriate arithmetic operation depending on the operation code.

```
//demonstrate arithmetic operations
module arith_ops1 (a, b, opcode, rslt);

input [3:0] a, b;
input [2:0] opcode;
output [7:0] rslt;

reg [7:0] rslt;

parameter   addop = 3'b000,
            subop = 3'b001,
            mulop = 3'b010,
            divop = 3'b011,
            modop = 3'b100;

always @ (a or b or opcode)
begin
    case (opcode)
        addop: rslt = a + b;
        subop: rslt = a - b;
        mulop: rslt = a * b;
        divop: rslt = a / b;
        modop: rslt = a % b;
        default: rslt = 8'bxxxxxxxx;
    endcase
end
endmodule
```

Figure 4.7   Verilog code illustrating the operations of addition, subtraction, multiplication, division, and modulus.

Figure 4.8 illustrates the test bench for the Verilog code of Figure 4.7. Five sets of vectors are applied to the inputs, one set for each of the five arithmetic operations. The modulus operator produces the remainder of a divide operation; thus 0111/0011 yields a remainder of 00000001 as seen in the output listing of Figure 4.9. The result of a multiply operation is always $2n$ bits, where n is the number of bits of each operand. Therefore, the *rslt* identifier is declared as an 8-bit variable. The waveforms are shown in Figure 4.10.

```
//arithmetic operations test //bench
module arith_ops1_tb;

reg [3:0] a, b;
reg [2:0] opcode;
wire [7:0] rslt ;              //continued on next page
```

Figure 4.8   Test bench for the module of Figure 4.7.

```
initial
$monitor ("a = %b, b = %b,
    opcode = %b, rslt = %b",
    a , b, opcode, rslt);

initial
begin
    #0    a = 4'b0011;
          b = 4'b0111;
          opcode = 3'b000;

    #5    a = 4'b1111;
          b = 4'b1111;
          opcode = 3'b001;

    #5    a = 4'b1110;
          b = 4'b1110;
          opcode = 3'b010;
    #5    a = 4'b1000;
          b = 4'b0010;
          opcode = 3'b011;

    #5    a = 4'b0111;
          b = 4'b0011;
          opcode = 3'b100;

    #5    $stop;
end

//instantiate the module into
//the test bench
arith_ops1 inst1 (
    .a(a),
    .b(b),
    .opcode(opcode),
    .rslt(rslt)
    );
endmodule
```

Figure 4.8   (Continued)

```
a = 0011, b = 0111, opcode = 000, rslt = 00001010  //add
a = 1111, b = 1111, opcode = 001, rslt = 00000000  //subtract
a = 1110, b = 1110, opcode = 010, rslt = 11000100  //multiply
a = 1000, b = 0010, opcode = 011, rslt = 00000100  //divide
a = 0111, b = 0011, opcode = 100, rslt = 00000001  //modulus
```

Figure 4.9   Outputs for the test bench of Figure 4.8.

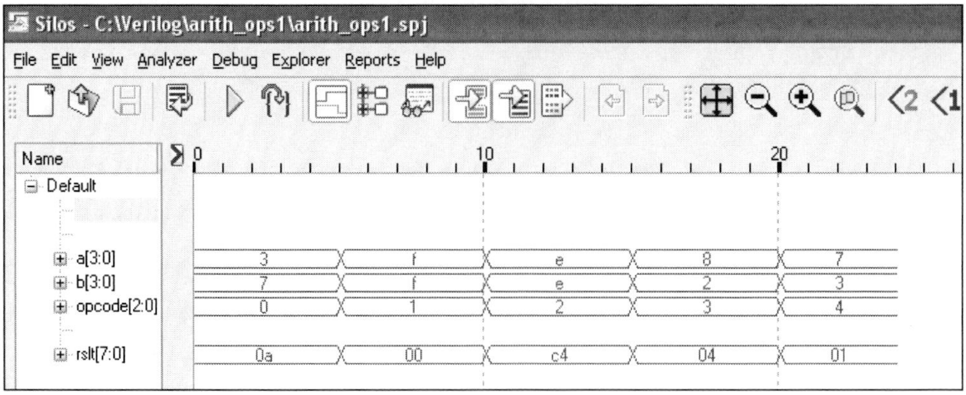

Figure 4.10   Waveforms for the test bench of Figure 4.8.

4.2.2 Logical

There are three logical operators: the binary logical AND operator (&&), the binary logical OR operator (||), and the unary logical negation operator (!). Logical operators evaluate to a logical 1 (true), a logical 0 (false), or an **x** (ambiguous). If a logical operation returns a nonzero value, then it is treated as a logical 1 (true); if a bit in an operand is **x** or **z**, then it is ambiguous and is normally treated as a false condition. Figure 4.11 shows examples of the logical operators using dataflow modeling. Figure 4.12, Figure 4.13, and Figure 4.14 show the test bench, outputs, and waveforms, respectively.

```
//examples of logical operators
module log_ops1 (a, b, z1, z2, z3);

input [3:0] a, b;
output z1, z2, z3;

assign z1 = a && b;
assign z2 = a || b;
assign z3 = !a;

endmodule
```

Figure 4.11   Examples of logical operators.

```
//test bench for logical              #5      a = 4'b0000;
//operators                                   b = 4'b0000;
module log_ops1_tb;

                                      #5      a = 4'b1111;
reg [3:0] a, b;                               b = 4'b1111;
wire z1, z2, z3;

                                      #5      $stop;
initial                           end
$monitor ("z1 = %d, z2 = %d,
    z3 = %d", z1, z2, z3);        //instantiate the module
                                  //into the test bench
//apply input vectors             log_ops1 inst1 (
initial                               .a(a),
begin                                 .b(b),
    #0      a = 4'b0110;                .z1(z1),
            b = 4'b1100;                .z2(z2),
                                        .z3(z3)
    #5      a = 4'b0101;              );
            b = 4'b0000;
                                  endmodule
    #5      a = 4'b1000;
            b = 4'b1001;
```

Figure 4.12   Test bench for the logical operators module.

```
z1 = 1, z2 = 1, z3 = 0      //z1 is logical AND
z1 = 0, z2 = 1, z3 = 0      //z2 is logical OR
z1 = 1, z2 = 1, z3 = 0      //z3 is logical negation
z1 = 0, z2 = 0, z3 = 1
z1 = 1, z2 = 1, z3 = 0
```

Figure 4.13   Outputs for the logical operators obtained from the test bench of Figure 4.12. Output z_1 is the logical AND; output z_2 is the logical OR; output z_3 is the logical negation.

If a vector operand is nonzero, then it treated as a 1 (true). Thus, referring to the module of Figure 4.11 and the test bench of Figure 4.12, where $z_1 = a \,\&\&\, b$ for a vector of $a = 0110$ and $b = 1100$, the value for z_1 is 1 because both a and b are nonzero.

Output z_2 is also equal to 1 for the expression $z_2 = a \,||\, b$. Output z_3 is equal to 0 because a is true.

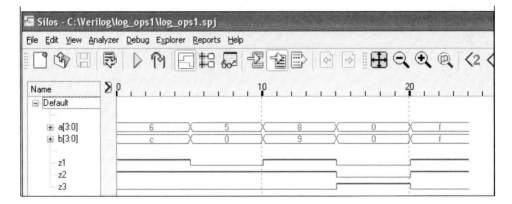

Figure 4.14   Waveforms for the logical operators obtained from the test bench of Figure 4.12. Output z_1 is the logical AND; output z_2 is the logical OR; output z_3 is the logical negation.

Now let $a = 0101$ and $b = 0000$. Thus, $z_1 = a \;\&\&\; b = 1 \;\&\&\; 0 = 0$ because a is true and b is false. Output z_2, however, is equal to 1 because $z_2 = a \,||\, b = 1 \,||\, 0 = 1$. In a similar manner, $z_3 = !a = !1 = 0$, because a is true.

As a final example, let $a = 0000$ and $b = 0000$; that is, both variables are false. Therefore, $z_1 = a \;\&\&\; b = 0 \;\&\&\; 0 = 0$; $z_2 = a \,||\, b = 0 \,||\, 0 = 0$; $z_3 = !a = !0 = 1$. If a bit in either operand is **x**, then the result of a logical operation is **x**. Also, $!\textbf{x}$ is **x**.

4.2.3 Relational

Relational operators compare operands and return a Boolean result, either 1 (true) or 0 (false) indicating the relationship between the two operands. There are four relational operators: greater than ($>$), less than ($<$), greater than or equal ($>=$), and less than or equal ($<=$). These operators function the same as identical operators in the C programming language.

If the relationship is true, then the result is 1; if the relationship is false, then the result is 0. Net or register operands are treated as unsigned values; real or integer operands are treated as signed values. An **x** or **z** in any operand returns a result of **x**. When the operands are of unequal size, the smaller operand is zero-extended to the left.

Figure 4.15 shows examples of relational operators using dataflow modeling, where the identifier *gt* means greater than, *lt* means less than, *gte* means greater than or equal, and *lte* means less than or equal. The test bench, which applies several different values to the two operands, is shown in Figure 4.16. The outputs are shown in Figure 4.17 and the waveforms in Figure 4.18.

```
//examples of relational operators
module relational_ops1 (a, b, gt, lt, gte, lte);
input [3:0] a, b;
output gt, lt, gte, lte;

assign gt = a > b;
assign lt = a < b;
assign gte = a >= b;
assign lte = a <= b;
endmodule
```

Figure 4.15   Verilog module to illustrate the relational operators.

```
//test bench relational ops
module relational_ops1_tb;

reg [3:0] a, b;
wire gt, lt, gte, lte;

initial
$monitor ("a=%b, b=%b, gt=%d,
   lt=%d, gte=%d, lte=%d",
   a, b, gt, lt, gte, lte);

//apply input vectors
initial
begin
      #0    a = 4'b0110;
            b = 4'b1100;
      #5    a = 4'b0101;
            b = 4'b0000;
      #5    a = 4'b1000;
                        b = 4'b1001;
            #5    a = 4'b0000;
                  b = 4'b0000;
            #5    a = 4'b1111;
                  b = 4'b1111;
            #5    $stop;
end

//instantiate the module
relational_ops1 inst1 (
      .a(a),
      .b(b),
      .gt(gt),
      .lt(lt),
      .gte(gte),
      .lte(lte)
      );
endmodule
```

Figure 4.16   Test bench for the relational operators module of Figure 4.15.

```
a=0110, b=1100, gt=0, lt=1, gte=0, lte=1
a=0101, b=0000, gt=1, lt=0, gte=1, lte=0
a=1000, b=1001, gt=0, lt=1, gte=0, lte=1
a=0000, b=0000, gt=0, lt=0, gte=1, lte=1
a=1111, b=1111, gt=0, lt=0, gte=1, lte=1
```

Figure 4.17   Outputs for the test bench of Figure 4.16 for the relational operators.

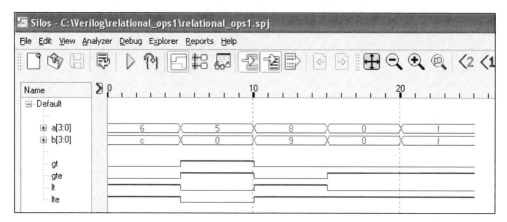

Figure 4.18   Waveforms for the test bench of Figure 4.16 for the relational operators.

4.2.4 Equality

There are four equality operators: logical equality (==), logical inequality (!=), case equality (===), and case inequality (!==).

Logical equality is used in expressions to determine if two values are identical. The result of the comparison is 1 if the two operands are equal, and 0 if they are not equal. The *logical inequality* operator is used to determine if two operands are unequal. A 1 is returned if the operands are unequal; otherwise a 0 is returned. If the result of the comparison is ambiguous for logical equality or logical inequality, then a value of **x** is returned. An **x** or **z** in either operand will return a value of **x**. If the operands are nets or registers, they are treated as unsigned values; real or integer operands are treated as signed values, but are compared as though they were unsigned operands.

The *case equality* operator compares both operands on a bit-by-bit basis, including **x** and **z**. The result is 1 if both operands are identical in the same bit positions, including those bit positions containing an **x** or a **z**. The *case inequality* operator is used to determine if two operands are unequal by comparing them on a bit-by-bit basis, including those bit positions that contain **x** or **z**. Examples of the equality operators are shown in Figure 4.19 using behavioral modeling. The test bench and outputs are shown in Figure 4.20 and Figure 4.21, respectively.

```
//illustrate the use of equality operators
module equality (x1, x2, x3, x4, x5, z1, z2, z3, z4);

input [3:0] x1, x2, x3, x4, x5;
output z1, z2, z3, z4;

wire x1, x2, x3, x4, x5;      //can be omitted.
                              //inputs are wire by default
//z1 is logical equality
//z2 is logical inequality
//z3 is case equality
//z4 is case inequality
reg z1, z2, z3, z4;

always @ (x1 or x2 or x3 or x4 or x5)
begin
      if (x1 == x2)                     //logical equality
            z1 = 1;
      else z1 = 0;
end

always @ (x1 or x2 or x3 or x4 or x5)
begin
      if (x2 != x3)                     //logical inequality
            z2 = 1;
      else z2 = 0;
end

always @ (x1 or x2 or x3 or x4 or x5)
begin
      if (x3 === x4)                    //case equality
            z3 = 1;
      else z3 = 0;
end                                     //continued on next page
```

Figure 4.19   Module to illustrate the use of the equality operators.

```
always @ (x1 or x2 or x3 or x4 or x5)
begin
     if (x4 !== x5)                        //case inequality
          z4 = 1;
     else z4 = 0;
end

endmodule
```

Figure 4.19   (Continued)

```
//equality operators test bench
module equality_tb;

reg [3:0] x1, x2, x3, x4, x5;
wire z1, z2, z3, z4;

initial
$monitor ("x1=%b, x2=%b, x3=%b, x4=%b, x5=%b, z1=%b, z2=%b,
         z3=%b, z4=%b", x1, x2, x3, x4, x5, z1, z2, z3, z4);

//apply input vectors
initial
begin
     #0   x1 = 4'b1000;
          x2 = 4'b1101;
          x3 = 4'b01xz;
          x4 = 4'b01xz;
          x5 = 4'bx1xx;

     #10  x1 = 4'b1011;
          x2 = 4'b1011;
          x3 = 4'bx1xz;
          x4 = 4'bx1xz;
          x5 = 4'b11xx;

//continued on next page
```

Figure 4.20   Test bench for the equality module of Figure 4.19.

```
        #10    x1 = 4'b1100;
               x2 = 4'b0101;
               x3 = 4'bx01z;
               x4 = 4'b11xz;
               x5 = 4'b11xx;
end

//instantiate the module into the test bench
equality inst1 (
        .x1(x1),
        .x2(x2),
        .x3(x3),
        .x4(x4),
        .x5(x5),
        .z1(z1),
        .z2(z2),
        .z3(z3),
        .z4(z4)
        );

endmodule
```

Figure 4.20   (Continued)

```
x1=1000, x2=1101, x3=01xz, x4=01xz, x5=x1xx,
z1=0, z2=1, z3=1, z4=1

x1=1011, x2=1011, x3=x1xz, x4=x1xz, x5=11xx,
z1=1, z2=1, z3=1, z4=1

x1=1100, x2=0101, x3=x01z, x4=11xz, x5=11xx,
z1=0, z2=1, z3=0, z4=1
```

Figure 4.21   Outputs for the test bench of Figure 4.20 for the equality module of Figure 4.19.

Referring to the outputs of Figure 4.21 for the first set of inputs, the logical equality (z_1) of x_1 and x_2 is false because the operands are unequal. The logical inequality

(z_2) of x_2 and x_3 is true. The case equality (z_3) of inputs x_3 and x_4 is 1 because both operands are identical in all bit positions, including the **x** and **z** bits. The case inequality (z_4) of inputs x_4 and x_5 is also 1 because the operands differ in the high-order and low-order bit positions.

4.2.5 Bitwise

The bitwise operators are: AND (&), OR (|), negation (~), exclusive-OR (^), and exclusive-NOR (^~ or ~^). The bitwise operators perform logical operations on the operands on a bit-by-bit basis and produce a vector result. Except for negation, each bit in one operand is associated with the corresponding bit in the other operand. If one operand is shorter, then it is zero-extended to the left to match the length of the longer operand.

The *bitwise AND* operator performs the AND function on two operands on a bit-by-bit basis. The truth table is shown in Table 4.6 for inputs x_1, x_2 and output z_1.

Table 4.6 Truth Table for Bitwise AND

x_1	x_2	z_1		x_1	x_2	z_1
0	0	0		x	0	0
0	1	0		x	1	x
1	0	0		x	x	x
1	1	1		x	z	x
0	x	0		z	0	0
0	z	0		z	1	x
1	x	x		z	x	x
1	z	x		z	z	x

An example of the bitwise AND operator is shown below.

```
         1  0  1  1  0  1  1  0
    &)   1  1  0  1  0  1  0  1
         ──────────────────────
         1  0  0  1  0  1  0  0
```

The *bitwise OR* operator performs the OR function on the two operands on a bit-by-bit basis. The truth table is shown in Table 4.7 for inputs x_1, x_2 and output z_1.

Table 4.7 Truth Table for Bitwise OR

x_1	x_2	z_1		x_1	x_2	z_1
0	0	0		x	0	x
0	1	1		x	1	1
1	0	1		x	x	x
1	1	1		x	z	x
0	x	x		z	0	x
0	z	x		z	1	1
1	x	1		z	x	x
1	z	1		z	z	x

An example of the bitwise OR operator is shown below.

```
    1  0  1  1  0  1  1  0
|)  1  1  0  1  0  1  0  1
    ─────────────────────
    1  1  1  1  0  1  1  1
```

The *bitwise negation* operator performs the negation function on one operand on a bit-by-bit basis. Each bit in the operand is inverted. The truth table is shown in Table 4.8 for input x_1 and output z_1.

Table 4.8 Truth Table for Bitwise Negation

x_1	z_1
0	1
1	0
x	x
z	x

An example of the bitwise negation operator is shown below.

```
~)  1  1  0  1  0  1  0  1
    ─────────────────────
    0  0  1  0  1  0  1  0
```

The *bitwise exclusive-OR* operator performs the exclusive-OR function on two operands on a bit-by-bit basis.   The truth table is shown in Table 4.9 for inputs x_1, x_2 and output z_1.

Table 4.9 Truth Table for Bitwise Exclusive-OR

x_1	x_2	z_1		x_1	x_2	z_1
0	0	0		x	0	x
0	1	1		x	1	x
1	0	1		x	x	x
1	1	0		x	z	x
0	x	x		z	0	x
0	z	x		z	1	x
1	x	x		z	x	x
1	z	x		z	z	x

An example of the bitwise exclusive-OR operator is shown below.

$$
\begin{array}{c}
\quad\; 1\;\; 0\;\; 1\;\; 1\;\; 0\;\; 1\;\; 1\;\; 0 \\
^\wedge)\;\; 1\;\; 1\;\; 0\;\; 1\;\; 0\;\; 1\;\; 0\;\; 1 \\
\hline
\quad\; 0\;\; 1\;\; 1\;\; 0\;\; 0\;\; 0\;\; 1\;\; 1
\end{array}
$$

The *bitwise exclusive-NOR* operator performs the exclusive-NOR function on two operands on a bit-by-bit basis.   The truth table is shown in Table 4.10 for inputs x_1, x_2 and output z_1.

Table 4.10 Truth Table for Bitwise Exclusive-NOR

x_1	x_2	z_1		x_1	x_2	z_1
0	0	1		x	0	x
0	1	0		x	1	x
1	0	0		x	x	x
1	1	1		x	z	x
0	x	x		z	0	x
0	z	x		z	1	x
1	x	x		z	x	x
1	z	x		z	z	x

An example of the bitwise exclusive-NOR operator is shown below.

```
      1  0  1  1  0  1  1  0
^~ )  1  1  0  1  0  1  0  1
      ────────────────────
      1  0  0  1  1  1  0  0
```

Bitwise operators perform operations on operands on a bit-by-bit basis and produce a vector result. This is in contrast to logical operators, which perform operations on operands in such a way that the truth or falsity of the result is determined by the truth or falsity of the operands. That is, the logical AND operator returns a value of 1 (true) only if both operands are nonzero (true); otherwise, it returns a value of 0 (false). If the result is ambiguous, it returns a value of **x**.

The logical OR operator returns a value of 1 (true) if either or both operands are true; otherwise, it returns a value of 0. The logical negation operator returns a value of 1 (true) if the operand has a value of zero and a value of 0 (false) if the operand is nonzero. Figure 4.22 shows a coding example to illustrate the use of the five bitwise operators. The test bench and outputs are in Figure 4.23 and Figure 4.24, respectively.

```
//example of the bitwise operators
module bitwise1 (a, b, and_rslt, or_rslt, neg_rslt,
                 xor_rslt, xnor_rslt);

input [7:0] a, b;
output [7:0] and_rslt, or_rslt, neg_rslt, xor_rslt,
                 xnor_rslt;

wire [7:0] a, b;
reg [7:0] and_rslt, or_rslt, neg_rslt, xor_rslt, xnor_rslt;

always @ (a or b)
begin
   and_rslt = a & b;        //bitwise AND
   or_rslt = a | b;         //bitwise OR
   neg_rslt = ~a;           //bitwise negation
   xor_rslt = a ^ b;        //bitwise exclusive-OR
   xnor_rslt = a ^~ b;      //bitwise exclusive-NOR
end

endmodule
```

Figure 4.22   Module to illustrate the coding for the bitwise operators.

```
//test bench for bitwise1 module
module bitwise1_tb;

reg [7:0] a, b;
wire [7:0] and_rslt, or_rslt, neg_rslt, xor_rslt, xnor_rslt;

$monitor ("a=%b, b=%b, and_rslt=%b, or_rslt=%b, neg_rslt=%b,
          xor_rslt=%b, xnor_rslt=%b",
     a, b, and_rslt, or_rslt, neg_rslt, xor_rslt, xnor_rslt);

initial
$monitor ("a=%b, b=%b, and_rslt=%b, or_rslt=%b, neg_rslt=%b,
          xor_rslt=%b, xnor_rslt=%b",
     a, b, and_rslt, or_rslt, neg_rslt, xor_rslt, xnor_rslt);

//apply input vectors
initial
begin
   #0    a = 8'b1100_0011;
         b = 8'b1001_1001;

   #10   a = 8'b1001_0011;
         b = 8'b1101_1001;

   #10   a = 8'b0000_1111;
         b = 8'b1101_1001;

   #10   a = 8'b0100_1111;
         b = 8'b1101_1001;

   #10   a = 8'b1100_1111;
         b = 8'b1101_1001;
   #10   $stop;
end
//instantiate the module into the test bench
bitwise1 inst1 (
   .a(a),
   .b(b),
   .and_rslt(and_rslt),
   .or_rslt(or_rslt),
   .neg_rslt(neg_rslt),
   .xor_rslt(xor_rslt),
   .xnor_rslt(xnor_rslt)
   );

endmodule
```

Figure 4.23   Test bench for the bitwise module of Figure 4.22.

```
        a = 11000011,                      a = 01001111,
        b = 10011001,                      b = 11011001,

and_rslt = 10000001,              and_rslt = 01001001,
or_rslt  = 11011011,              or_rslt  = 11011111,
neg_rslt = 00111100,              neg_rslt = 10110000,
xor_rslt = 01011010,              xor_rslt = 10010110,
xnor_rslt= 10100101              xnor_rslt= 01101001

        a = 10010011,                      a   = 11001111,
        b = 11011001,                      b   = 11011001,

and_rslt = 10010001,              and_rslt = 11001001,
or_rslt  = 11011011,              or_rslt  = 11011111,
neg_rslt = 01101100,              neg_rslt = 00110000,
xor_rslt = 01001010,              xor_rslt = 00010110,
xnor_rslt= 10110101              xnor_rslt= 11101001

        a = 00001111,
        b = 11011001,

and_rslt = 00001001,
or_rslt  = 11011111,
neg_rslt = 11110000,
xor_rslt = 11010110,
xnor_rslt= 00101001
```

Figure 4.24   Outputs for the bitwise module of Figure 4.22.

4.2.6 Reduction

The reduction operators are: AND (&), NAND (~&), OR (|), NOR (~ |), exclusive-OR (^), and exclusive-NOR (^~ or ~^). Reduction operators are unary operators; that is, they operate on a single vector and produce a single-bit result. If any bit of the operand is **x** or **z**, the result is **x**. The truth tables are the same as those for the bitwise operators. Reduction operators perform their respective operations on a bit-by-bit basis from right to left.

reduction AND   If any bit in the operand is 0, then the result is 0; otherwise, the result is 1. For example, let x_1 be the vector shown below.

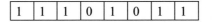

The reduction AND ($\& x_1$) operation is equivalent to the following operation:

$$1 \& 1 \& 1 \& 0 \& 1 \& 0 \& 1 \& 1$$

which returns a result of 1'b0.

reduction NAND   If any bit in the operand is 0, then the result is 1; otherwise, the result is 0. For a vector x_1, the reduction NAND ($\sim\& x_1$) is the inverse of the reduction AND operator.

reduction OR   If any bit in the operand is 1, then the result is 1; otherwise, the result is 0. For example, let x_1 be the vector shown below.

| 1 | 1 | 1 | 0 | 1 | 0 | 1 | 1 |

The reduction OR ($| x_1$) operation is equivalent to the following operation:

$$1 | 1 | 1 | 0 | 1 | 0 | 1 | 1$$

which returns a result of 1'b1.

reduction NOR   If any bit in the operand is 1, then the result is 0; otherwise, the result is 1. For a vector x_1, the reduction NOR ($\sim| x_1$) is the inverse of the reduction OR operator.

reduction exclusive-OR   If there are an even number of 1s in the operand, then the result is 0; otherwise, the result is 1. For example, let x_1 be the vector shown below.

| 1 | 1 | 1 | 0 | 1 | 0 | 1 | 1 |

The reduction exclusive-OR ($^\wedge x_1$) operation is equivalent to the following operation:

$$1 \wedge 1 \wedge 1 \wedge 0 \wedge 1 \wedge 0 \wedge 1 \wedge 1$$

which returns a result of 1'b0. The reduction exclusive-OR operator can be used as an even parity generator.

reduction exclusive-NOR   If there are an odd number of 1s in the operand, then the result is 0; otherwise, the result is 1. For a vector x_1, the reduction exclusive-NOR

$(^\wedge \sim x_1)$ is the inverse of the reduction exclusive-OR operator. The reduction exclusive-NOR operator can be used as an odd parity generator.

Figure 4.25 contains a module that illustrates the coding of the reduction operators. The test bench and outputs are shown in Figure 4.26 and Figure 4.27, respectively.

```
//module to illustrate the use of reduction operators
module reduction (a, and_rslt, nand_rslt, or_rslt, nor_rslt,
                  xor_rslt, xnor_rslt);

input [7:0] a;
output and_rslt, nand_rslt, or_rslt, nor_rslt, xor_rslt,
       xnor_rslt;

wire [7:0] a;
reg and_rslt, nand_rslt, or_rslt, nor_rslt, xor_rslt,
    xnor_rslt;

always @(a)
begin
   and_rslt = &a;        //reduction AND
   nand_rslt = ~&a;      //reduction NAND
   or_rslt = |a;         //reduction OR
   nor_rslt = ~|a;       //reduction NOR
   xor_rslt = ^a;        //reduction exclusive-OR
   xnor_rslt = ^~a;      //reduction exclusive-NOR
end
endmodule
```

Figure 4.25   Module to illustrate the use of the reduction operators.

```
//test bench for reduction module
module reduction_tb;

reg [7:0] a;
wire and_rslt, nand_rslt, or_rslt, nor_rslt, xor_rslt,
     xnor_rslt;

initial
$monitor ("a=%b, and_rslt=%b, nand_rslt=%b, or_rslt=%b,
          nor_rslt=%b, xor_rslt=%b, xnor_rslt=%b",
       a, and_rslt, nand_rslt, or_rslt, nor_rslt, xor_rslt,
          xnor_rslt);
//continued on next page
```

Figure 4.26   Test bench for the reduction operators of Figure 4.25.

```
//apply input vectors
initial
begin
   #0    a = 8'b1100_0011;

   #10   a = 8'b1001_0011;

   #10   a = 8'b0000_1111;

   #10   a = 8'b0100_1111;

   #10   a = 8'b1100_1111;

   #10   $stop;
end

//instantiate the module into the test bench
reduction inst1 (
   .a(a),
   .and_rslt(and_rslt),
   .nand_rslt(nand_rslt),
   .or_rslt(or_rslt),
   .nor_rslt(nor_rslt),
   .xor_rslt(xor_rslt),
   .xnor_rslt(xnor_rslt)
   );

endmodule
```

Figure 4.26   (Continued)

```
a=11000011, and_rslt=0,nand_rslt=1,or_rslt=1,nor_rslt=0,
            xor_rslt=0,xnor_rslt=1
a=10010111, and_rslt=0,nand_rslt=1,or_rslt=1,nor_rslt=0,
            xor_rslt=1,xnor_rslt=0
a=00000000, and_rslt=0,nand_rslt=1,or_rslt=0,nor_rslt=1,
            xor_rslt=0,xnor_rslt=1
a=01001111, and_rslt=0,nand_rslt=1,or_rslt=1,nor_rslt=0,
            xor_rslt=1,xnor_rslt=0
a=11111111, and_rslt=1,nand_rslt=0,or_rslt=1,nor_rslt=0,
            xor_rslt=0,xnor_rslt=1
```

Figure 4.27   Outputs for the reduction operators of Figure 4.25.

4.2.7 Shift

The shift operators shift a single vector operand left or right a specified number of bit positions. These are logical shift operations, not algebraic; that is, as bits are shifted left or right, zeroes fill in the vacated bit positions. The bits shifted out of the operand are lost; they do not rotate to the high-order or low-order bit positions of the shifted operand. If the shift amount evaluates to **x** or **z**, then the result of the operation is **x**. There are two shift operators, as shown below. The value in parentheses is the number of bits that the operand is shifted.

$$\ll \text{(Left-shift amount)}$$
$$\gg \text{(Right-shift amount)}$$

When an operand is shifted left, this is equivalent to a multiply-by-two operation for each bit position shifted. When an operand is shifted right, this is equivalent to a divide-by-two operation for each bit position shifted. The shift operators are useful to model the sequential add-shift multiplication algorithm and the sequential shift-subtract division algorithm. Figure 4.28 shows examples of the shift-left and shift-right operators using behavioral modeling with blocking assignments. The test bench is shown in Figure 4.29 and the outputs in Figure 4.30.

```
//examples of shift operations
module shift (a_reg, b_reg, rslt_a, rslt_b);

input [7:0] a_reg, b_reg;
output [7:0] rslt_a, rslt_b;

wire [7:0] a_reg, b_reg;
reg [7:0] rslt_a, rslt_b;

always @ (a_reg or b_reg)
begin
      rslt_a = a_reg << 3;          //multiply by 8
      rslt_b = b_reg >> 2;          //divide by 4
end

endmodule
```

Figure 4.28   Examples of shift-left and shift-right operators.

```
//shift test bench
module shift_tb;


reg [7:0] a_reg, b_reg;
wire [7:0] rslt_a, rslt_b;


//display variables
initial
$monitor ("a_reg = %b, b_reg = %b, rslt_a = %b, rslt_b = %b",
                a_reg, b_reg, rslt_a, rslt_b);


//apply input vectors
initial
begin
      #0    a_reg = 8'b0000_0010;        //2;    rslt_a = 16
            b_reg = 8'b0000_1000;        //8;    rslt_b = 2

      #10   a_reg = 8'b0000_0110;        //6;    rslt_a = 48
            b_reg = 8'b0001_1000;        //24;   rslt_b = 6

      #10   a_reg = 8'b0000_1111;        //15;   rslt_a = 120
            b_reg = 8'b0011_1000;        //56;   rslt_b = 14

      #10   a_reg = 8'b1110_0000;        //224;  rslt_a = 0
            b_reg = 8'b0000_0011;        //3;    rslt_b = 0

      #10   $stop;
end


//instantiate the module into the test bench
shift inst1 (
      .a_reg(a_reg),
      .b_reg(b_reg),
      .rslt_a(rslt_a),
      .rslt_b(rslt_b)
      );


endmodule
```

Figure 4.29   Test bench for the shift operators of Figure 4.28.

```
a_reg = 00000010, b_reg = 00001000,   //shift a_reg left 3
rslt_a = 00010000, rslt_b = 00000010   //shift b_reg right 2


a_reg = 00000110, b_reg = 00011000,   //shift a_reg left 3
rslt_a = 00110000, rslt_b = 00000110   //shift b_reg right 2


a_reg = 00001111, b_reg = 00111000,   //shift a_reg left 3
rslt_a = 01111000, rslt_b = 00001110   //shift b_reg right 2


a_reg = 11100000, b_reg = 00000011,   //shift a_reg left 3
rslt_a = 00000000, rslt_b = 00000000   //shift b_reg right 2
```

Figure 4.30   Outputs for the test bench of Figure 4.29 showing the results of the shift-left and shift-right operations. Operand *a_reg* is shifted left three bits with the low-order bits filled with zeroes. Operand *b_reg* is shifted right two bits with the high-order bits filled with zeroes.

4.2.8 Conditional

The conditional operator (**?** **:**) has three operands, as shown in the syntax below. The *conditional_expression* is evaluated. If the result is true (1), then the *true_expression* is evaluated; if the result is false (0), then the *false_expression* is evaluated.

conditional_expression **?** true_expression **:** false_expression;

The conditional operator can be used when one of two expressions is to be selected. For example, in the statement below, if x_1 is greater than or equal to x_2, then z_1 is assigned the value of x_3; if x_1 is less than x_2, then z_1 is assigned the value of x_4.

z1 = (x1 >= x2) **?** x3 **:** x4;

If the conditional_expression evaluates to **x**, then the true_expression and the false_expression are evaluated on a bit-by-bit basis according to the truth table shown in Table 4.11.

Table 4.11 Truth Table for a Conditional-Expression of x

True_expression	False_expression	Result
0	0	0
0	1	x
0	x	x
1	0	x
1	1	1
1	x	x
x	0	x
x	1	x
x	x	x

If the operands have different lengths, then the shorter operand is zero-extended on the left. Since the conditional operator selects one of two values, depending on the result of the conditional_expression evaluation, the operator can be used in place of the **if . . . else** construct. The conditional operator is ideally suited to model a 2:1 multiplexer as shown in Figure 4.31. The test bench and outputs are shown in Figure 4.32 and Figure 4.33, respectively. Conditional operators can be nested; that is, each true_expression and false_expression can be a conditional operation. This is useful for modeling a 4:1 multiplexer.

conditional_expression ? (cond_expr1 ? true_expr1 : false_expr1)
                         : (cond_expr2 ? true_expr2 : false_expr2);

```
//dataflow 2:1 mux using conditional operator
module mux_2to1_cond (s0, in0, in1, out);

input s0, in0, in1;
output out;

assign out = s0 ? in1 : in0;
endmodule

//    s0            out
//    1(true)       in1
//    0(false)      in0
```

Figure 4.31   Verilog code to model a 2:1 multiplexer using the conditional operator.

```
//2:1 multiplexer test bench          #10    s0 = 1'b1;
module mux_2to1_cond_tb;                     in0 = 1'b1;
                                             in1 = 1'b0;
reg s0, in0, in1;
wire out;                             #10    s0 = 1'b1;
                                             in0 = 1'b0;
//display variables                          in1 = 1'b1;
initial
$monitor ("s0=%b, in0 in1=%b,        #10    $stop;
         out = %b",                 end
      s0, {in0, in1}, out);
                                     //instantiate the module
//apply stimulus                     //into the test bench
initial                              mux_2to1_cond inst1 (
begin                                      .s0(s0),
      #0     s0 = 1'b0;                     .in0(in0),
             in0 = 1'b0;                    .in1(in1),
             in1 = 1'b0;                    .out(out)
                                           );
      #10    s0 = 1'b0;
             in0 = 1'b1;             endmodule
             in1 = 1'b1;
```

Figure 4.32   Test bench for the 2:1 multiplexer of Figure 4.31.

```
s0 = 0, in0 in1 = 00, out = 0
s0 = 0, in0 in1 = 11, out = 1
s0 = 1, in0 in1 = 10, out = 0
s0 = 1, in0 in1 = 01, out = 1
```

Figure 4.33   Outputs for the test bench of Figure 4.32.

4.2.9 Concatenation

The concatenation operator ({ }) forms a single operand from two or more operands by joining the different operands in sequence separated by commas. The operands to be appended are contained within braces. The size of the operands must be known before concatenation takes place. For example, Figure 4.34 shows the concatenation of scalars and vectors of different sizes. Also, *a_bus [7:4]* and *a_bus [3:0]* are inter-changed. The test bench and outputs are shown in Figure 4.35 and Figure 4.36, respectively.

```verilog
//examples of concatenation
module concat (a, b, c, d, a_bus, z1, z2, z3, z4, z5, z6);

input [1:0] a;
input [2:0] b;
input [3:0] c;
input d;
input [7:0] a_bus;
output [9:0] z1, z2, z3, z4;
output [7:0] z5;
output [11:0] z6;

assign z1 = {a, c};
assign z2 = {b, a};
assign z3 = {c, b, a};
assign z4 = {a, b, c, d};

assign z5 = {a_bus[3:0], a_bus[7:4]};
assign z6 = {b, c, d, 4'b0111};

endmodule
```

Figure 4.34   Verilog module to demonstrate the use of the concatenation operator.

```verilog
//concatenation test bench
module concat_tb;

reg [1:0]a;
reg [2:0]b;
reg [3:0]c;
reg d;
reg [7:0] a_bus;
wire [7:0] z5;
wire [9:0] z1, z2, z3, z4;
wire [11:0] z6;                   //continued on next page
```

Figure 4.35   Test bench for the concatenation module of Figure 4.34.

```
initial
$monitor ("a=%b, b=%b, c=%b, d=%b,
          z1=%b, z2=%b, z3=%b, z4=%b, z5=%b, z6=%b",
          a, b, c, d, z1, z2, z3, z4, z5, z6);


initial
begin
      #0     a = 2'b11;
             b = 3'b001;
             c = 4'b1100;
             d = 1'b1;
             a_bus = 8'b1111_0000;

      #10 $stop;
end

//instantiate the module into the test bench
concat inst1 (
      .a(a),
      .b(b),
      .c(c),
      .d(d),
      .a_bus(a_bus),
      .z1(z1),
      .z2(z2),
      .z3(z3),
      .z4(z4),
      .z5(z5),
      .z6(z6)
      );

endmodule
```

Figure 4.35   (Continued)

The z_1 through z_4 identifiers are declared as 10 bits to accommodate the concatenation of a, b, c, and d. Since the a_bus was defined as a vector [7:0], a_bus [3:0] replaces the high-order bits [7:4] and a_bus [7:4] replaces the low-order bits [3:0]. Note that concatenation applies to vectors of different lengths as well as to explicit vectors. Thus, the concatenation of $z_6 = \{b, c, d, 4\text{'b0111}\}$ results in $z_6 = 001\_1100\_1\_0111$. The underscore character is inserted for readability in both the outputs and the text.

```
z1, z2, z3, and z4 are 10 bits in length.

a = 11, b = 001, c = 1100, d = 1

a_bus = 11110000

z1 = 0000_11_1100           //z1 = {a, c}
z2 = 00000_001_11           //z2 = {b, a}
z3 = 0_1100_001_11          //z3 = {c, b, a}
z4 = 11_001_1100_1          //z4 = {a, b, c, d}
z5 = 0000_1111              //z5 = {a_bus [3:0], a_bus [7:4]}
z6 = 001_1100_1_0111        //z6 = {b, c, d, 4'b0111}
```

Figure 4.36   Outputs for the concatenation test bench of Figure 4.35.

4.2.10 Replication

Replication is a means of performing repetitive concatenation. Replication specifies the number of times to duplicate the expressions within the innermost braces. The syntax is shown below.

{number_ of_ repetitions {expression_1, expression_2, . . . , expression_n}};

A replication example is shown in Figure 4.37 with three input vectors: a [1:0], b [2:0], and c [3:0]. There are two output vectors: z_1 [11:0] and z_2 [21:0] that have a sufficient number of bits to contain the concatenation and replication required by the module. Note also that there is an explicit vector that is to be replicated together with the input vectors. The test bench is shown in Figure 4.38 and the outputs in Figure 4.39. Underscore characters are again inserted for readability.

```
//example of replication
module replication (a, b, c, z1, z2);

input [1:0] a;
input [2:0] b;
input [3:0] c;                        //continued on next page
```

Figure 4.37   Module to illustrate the replication operator.

```
output [11:0] z1;
output [21:0] z2;

assign z1 = {2{a, c}};
assign z2 = {2{b, c, 4'b0111}};

endmodule
```

Figure 4.37   (Continued)

```
//replication test bench
module replication_tb;

reg [1:0] a;
reg [2:0] b;
reg [3:0] c;

wire [11:0] z1;
wire [21:0] z2;

initial
$monitor ("a=%b, b=%b, c=%b, z1=%b, z2=%b",
                  a, b, c, z1, z2);
initial
begin
      #0    a = 2'b11; b = 3'b010; c = 4'b0011;
      #10   $stop;
end

//instantiate the module into the test bench
replication inst1 (
      .a(a),
      .b(b),
      .c(c),
      .z1(z1),
      .z2(z2)
      );

endmodule
```

Figure 4.38   Test bench for the replication module of Figure 4.37.

```
a = 11, b = 010, c = 0011,

z1 = 11_0011_11_0011,                       //z1 = {2{a, c}}


z2 = 010_0011_0111_010_0011_0111     //z2 = {2{b, c, 4'b0111}}
```

Figure 4.39   Outputs for the replication test bench of Figure 4.38.

4.3 Problems

4.1   Design a 4-bit adder using behavioral modeling with a parameter statement and blocking assignment.

4.2   A cache memory contains 16-bit words with 1024 words per cache. The word format is shown below. Select the word at address 511 (the 512th word) and load the contents into the following four 4-bit registers:

> operation code register (*op_code reg*)
> addressing mode register (*addr_mode_reg*)
> destination address register (*dst_addr_reg*)
> source address register (*src_addr_reg*)

	15	11	7	3	0
instr_reg	op_code	addr_mode	dst_addr	src_addr	

4.3   Design a five-function arithmetic and logic unit for addition, subtraction, multiplication, division, and modulus. There will be three 4-bit operands: a, b, and c and one result **reg** variable, which must be of sufficient width to contain the largest result. Design a test bench and obtain the outputs for several values of the inputs. The operations are defined below.

> add operation $= a + b + c$
> subtract operation $= (a + b) - c$
> multiply operation $= a * b * c$
> divide operation $= (a + b) / c$
> modulus operation $= (b + c) \% a$

4.4   Design a four-function arithmetic and logic unit for the three *logical* opera-
        tions: logical AND, logical OR, and logical negation. There will be three 4-bit
        operands: a, b, and c and one result **reg** variable. Generate the test bench and
        obtain the outputs for the following operations:

$$\begin{array}{ll}
\text{AND operation} & = (a \;\&\&\; b) \;\&\&\; c \\
\text{OR operation} & = (a \,||\, c) \,||\, b \\
\text{AND OR operation} & = (b \;\&\&\; c) \,||\, a \\
\text{NOT operation} & = !((b \;\&\&\; c) \,||\, a)
\end{array}$$

4.5   Design a five-function arithmetic and logic unit to show the operation of the
        five bit-wise operators: AND (&), OR (|), negation (~), exclusive-OR (^),
        and exclusive-NOR (^~ or ~^). There will be three 4-bit operands: a, b, and
        c and one result **reg** variable. Generate the test bench and obtain the outputs
        for the following operations:

$$\begin{array}{ll}
\text{AND operation} & = (a \;\&\; b) \;\&\; c \\
\text{OR operation} & = (a \,|\, c) \,|\, b \\
\text{negation operation} & = \sim((b \;\&\; c) \,|\, a) \\
\text{exclusive-OR operation} & = (b \,^\wedge\, c) \,^\wedge\, a \\
\text{exclusive-NOR operation} & = (a \,^\wedge\sim\, c) \,^\wedge\sim\, b
\end{array}$$

4.6   Design an odd parity generator using the exclusive-OR and exclusive-NOR
        operators. There are four data inputs and one output that is a logic 1 when the
        number of 1s in the input vector is even. Use dataflow modeling and test all
        combinations of the inputs. Generate a test bench and obtain the outputs.

4.7   If x = 8'haa, then $^\wedge x$ = 1'b0 _____ or 1'b1 _____.

4.8   Design a Verilog module that adds two 8-bit operands, then shifts the sum left
        four bit positions into a *rslt1* **reg** variable. Then shift the original sum right
        four bit positions into a *rslt2* **reg** variable. Generate a test bench and obtain
        the outputs.

4.9   Design a 4:1 multiplexer using dataflow modeling and the conditional oper-
        ator. Generate a test bench and obtain the outputs.

4.10   Perform concatenation on the following operands: *a [2:0]*, *b [3:0]*, and *c
        [4:0]*. Then replicate the result two times. Generate a test bench and obtain
        the outputs.

5

Gate-Level Modeling

Verilog has a profuse set of built-in primitive gates that are used to model nets. The single output of each gate is declared as type **wire**. The inputs are declared as type **wire** or as type **reg** depending on whether they were generated by a structural or behavioral module. This chapter presents a design methodology that is characterized by a low level of abstraction, where the logic hardware is described in terms of gates. Designing logic at this level is similar to designing logic by drawing gate symbols — there is a close correlation between the logic gate symbols and the Verilog built-in primitive gates. Each predefined primitive is declared by a keyword. Chapter 6 through Chapter 9 continue modeling logic units at progressively higher levels of abstraction.

The multiple-input gates **and, nand, or, nor, xor,** and **xnor** were introduced in Section 3.3.4. The multiple-output gates **buf** and **not** were also covered in Section 3.3.4 in sufficient detail. The pull gates **pullup** and **pulldown** were covered in Section 3.3.16. The three-state gates **bufif0, bufif1, notif0,** and **notif1** were presented in Section 3.3.20 in sufficient detail. Therefore, this chapter will present logic design using the multiple-input gates only.

5.1 Multiple-Input Gates

The multiple-input gates are **and, nand, or, nor, xor,** and **xnor,** which are built-in primitive gates used to describe a net and have one or more scalar inputs, but only one scalar output. The output signal is listed first, followed by the inputs in any order. The outputs are declared as **wire;** the inputs can be declared as either **wire** or **reg**. The gates represent combinational logic functions and can be instantiated into a module, as follows, where the instance name is optional:

> **gate_type** inst1 (output, input_1, input_2, . . . , input_n);

Two or more instances of the same type of gate can be specified in the same construct, as shown below. Note that only the last instantiation has a semicolon terminating the line. All previous lines are terminated by a comma.

> **gate_type** inst1 (output_1, input_11, input_12, . . . , input_1n),
> inst2 (output_2, input_21, input_22, . . . , input_2n),
>
> .
>
> .
>
> .
>
> instm (output_m, input_m1, input_m2, . . . , input_mn);

The best way to learn design methodologies using built-in primitives is by examples. Therefore, several examples will be presented ranging from very simple to moderately complex. When necessary, the theory for the examples will be presented prior to the Verilog design. All examples are carried through to completion at the gate level. Nothing is left unfinished or partially designed.

Example 5.1   A 3-input AND gate and a 3-input OR gate will be designed using built-in primitives. The truth tables for the AND and OR gates are shown in Table 5.1 and Table 5.2, respectively. The module, test bench, outputs, and waveforms are shown in Figure 5.1, Figure 5.2, Figure 5.3, and Figure 5.4, respectively.

Table 5.1 Truth Table for AND Gate

x_1	x_2	x_3	and_out
0	0	0	0
0	0	1	0
0	1	0	0
0	1	1	0
1	0	0	0
1	0	1	0
1	1	0	0
1	1	1	1

Table 5.2 Truth Table for OR Gate

x_1	x_2	x_3	or_out
0	0	0	0
0	0	1	1
0	1	0	1
0	1	1	1
1	0	0	1
1	0	1	1
1	1	0	1
1	1	1	1

```
//gate-level modeling for and/or gates
module and3_or3 (x1, x2, x3, and3_out, or3_out);

input x1, x2, x3;
output and3_out, or3_out;

and (and3_out, x1, x2, x3);
or (or3_out, x1, x2, x3);

endmodule
```

Figure 5.1   Verilog code for a 3-input AND gate and a 3-input OR gate using built-in primitives.

```
//test bench for and3_or3 module
module and3_or3_tb;

reg x1, x2, x3;
wire and3_out, or3_out;

//monitor variables
initial
$monitor ("x1x2x3 = %b, and3_out = %b, or3_out = %b",
                {x1, x2, x3}, and3_out, or3_out);

initial
begin
      #0    x1=1'b0;    x2=1'b0;    x3=1'b0;
      #10   x1=1'b0;    x2=1'b0;    x3=1'b1;
      #10   x1=1'b0;    x2=1'b1;    x3=1'b0;
      #10   x1=1'b0;    x2=1'b1;    x3=1'b1;
      #10   x1=1'b1;    x2=1'b0;    x3=1'b0;
      #10   x1=1'b1;    x2=1'b0;    x3=1'b1;
      #10   x1=1'b1;    x2=1'b1;    x3=1'b0;
      #10   x1=1'b1;    x2=1'b1;    x3=1'b1;
      #10   $stop;
end
//continued on next page
```

Figure 5.2   Test bench for Figure 5.1 for a 3-input AND gate and a 3-input OR gate.

```
//instantiate the module into the test bench
and3_or3 inst1 (
      .x1(x1),
      .x2(x2),
      .x3(x3),
      .and3_out(and3_out),
      .or3_out(or3_out)
      );

endmodule
```

Figure 5.2 (Continued)

```
x1x2x3 = 000, and3_out = 0, or3_out = 0
x1x2x3 = 001, and3_out = 0, or3_out = 1
x1x2x3 = 010, and3_out = 0, or3_out = 1
x1x2x3 = 011, and3_out = 0, or3_out = 1
x1x2x3 = 100, and3_out = 0, or3_out = 1
x1x2x3 = 101, and3_out = 0, or3_out = 1
x1x2x3 = 110, and3_out = 0, or3_out = 1
x1x2x3 = 111, and3_out = 1, or3_out = 1
```

Figure 5.3   Outputs for the test bench of Figure 5.2.

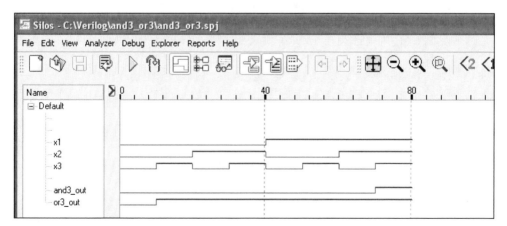

Figure 5.4   Waveforms for the *and3_or3* module of Figure 5.1.

Example 5.2 A 4-input exclusive-OR gate and a 4-input exclusive-NOR gate will be designed using built-in primitives. The truth tables for a 2-input exclusive-OR and a 2-input exclusive-NOR function are shown in Table 5.3 and Table 5.4, respectively. The output of an exclusive-OR gate is a logic 1 if there are an odd number of 1s on the inputs, thus indicating odd parity. The output of an exclusive-NOR gate is a logic 1 if there are an even number of 1s on the inputs, thus indicating even parity. A 2-input exclusive-NOR gate is also called an equality function because the output is a logic 1 if the inputs are the same, either both 0s or both 1s. The module, test bench, outputs, and waveforms are shown in Figure 5.5, Figure 5.6, Figure 5.7, and Figure 5.8, respectively.

Table 5.3 Truth Table for Exclusive-OR Function

x_1	x_2	z_1
0	0	0
0	1	1
1	0	1
1	1	0

Table 5.4 Truth Table for Exclusive-NOR Function

x_1	x_2	z_1
0	0	1
0	1	0
1	0	0
1	1	1

```
//xor/xnor using built-in primitives
module xor_xnor (x1, x2, x3, x4, xor_out, xnor_out);
input x1, x2, x3, x4;
output xor_out, xnor_out;

xor (xor_out, x1, x2, x3, x4);
xnor (xnor_out, x1, x2, x3, x4);
endmodule
```

Figure 5.5 Verilog module illustrating the **xor** and **xnor** built-in primitives.

```
//test bench for xor/xnor module
module xor_xnor_tb;

reg x1, x2, x3, x4;
wire xor_out, xnor_out;        //continued on next page
```

Figure 5.6 Test bench for Figure 5.5.

```
//monitor variables
initial
$monitor ("x1x2x3x4 = %b, xor_out = %b, xnor_out = %b",
                {x1, x2, x3, x4}, xor_out, xnor_out);



initial
begin
      #0     x1=1'b0;    x2=1'b0;    x3=1'b0;    x4=1'b0;
      #10    x1=1'b0;    x2=1'b0;    x3=1'b0;    x4=1'b1;
      #10    x1=1'b0;    x2=1'b0;    x3=1'b1;    x4=1'b0;
      #10    x1=1'b0;    x2=1'b0;    x3=1'b1;    x4=1'b1;
      #10    x1=1'b0;    x2=1'b1;    x3=1'b0;    x4=1'b0;
      #10    x1=1'b0;    x2=1'b1;    x3=1'b0;    x4=1'b1;
      #10    x1=1'b0;    x2=1'b1;    x3=1'b1;    x4=1'b0;
      #10    x1=1'b0;    x2=1'b1;    x3=1'b1;    x4=1'b1;
      #10    x1=1'b1;    x2=1'b0;    x3=1'b0;    x4=1'b0;
      #10    x1=1'b1;    x2=1'b0;    x3=1'b0;    x4=1'b1;
      #10    x1=1'b1;    x2=1'b0;    x3=1'b1;    x4=1'b0;
      #10    x1=1'b1;    x2=1'b0;    x3=1'b1;    x4=1'b1;
      #10    x1=1'b1;    x2=1'b1;    x3=1'b0;    x4=1'b0;
      #10    x1=1'b1;    x2=1'b1;    x3=1'b0;    x4=1'b1;
      #10    x1=1'b1;    x2=1'b1;    x3=1'b1;    x4=1'b0;
      #10    x1=1'b1;    x2=1'b1;    x3=1'b1;    x4=1'b1;

      #10    $stop;

end



//instantiate the module into the test bench
xor_xnor inst1 (
      .x1(x1),
      .x2(x2),
      .x3(x3),
      .x4(x4),
      .xor_out(xor_out),
      .xnor_out(xnor_out)
      );

endmodule
```

Figure 5.6   (Continued)

```
x1x2x3x4 = 0000, xor_out = 0, xnor_out = 1
x1x2x3x4 = 0001, xor_out = 1, xnor_out = 0
x1x2x3x4 = 0010, xor_out = 1, xnor_out = 0
x1x2x3x4 = 0011, xor_out = 0, xnor_out = 1
x1x2x3x4 = 0100, xor_out = 1, xnor_out = 0
x1x2x3x4 = 0101, xor_out = 0, xnor_out = 1
x1x2x3x4 = 0110, xor_out = 0, xnor_out = 1
x1x2x3x4 = 0111, xor_out = 1, xnor_out = 0
x1x2x3x4 = 1000, xor_out = 1, xnor_out = 0
x1x2x3x4 = 1001, xor_out = 0, xnor_out = 1
x1x2x3x4 = 1010, xor_out = 0, xnor_out = 1
x1x2x3x4 = 1011, xor_out = 1, xnor_out = 0
x1x2x3x4 = 1100, xor_out = 0, xnor_out = 1
x1x2x3x4 = 1101, xor_out = 1, xnor_out = 0
x1x2x3x4 = 1110, xor_out = 1, xnor_out = 0
x1x2x3x4 = 1111, xor_out = 0, xnor_out = 1
```

Figure 5.7   Outputs for the test bench of Figure 5.6.

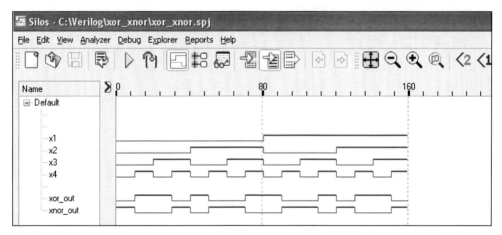

Figure 5.8   Waveforms for the *xor_xnor* module of Figure 5.5.

Example 5.3   The Karnaugh map shown in Figure 5.9 will be used to obtain a minimal expression for output z_1 in a sum-of-products form as shown in Equation 5.1. Figure 5.10 shows the logic diagram indicating the instance names and the internal net names. The inputs are assumed to be available in both high and low assertion. The logic will then be designed using built-in primitives. The module is shown in Figure 5.11 using the built-in primitives **and** and **or** with optional instance names. Output z_1

could also have been generated by a continuous assignment statement by assigning *net6* as the output of the OR gate, as shown below.

        or inst6 (net6, net1, net2, net3, net4, net5);
        assign z1 = net6;

The test bench is shown in Figure 5.12 using input vectors of six bits in order to accommodate vectors $x_1 x_2 x_3 x_4 x_5 = 00000$ through $x_1 x_2 x_3 x_4 x_5 = 11111$. The outputs are listed in Figure 5.13 and can be verified using the Karnaugh map of Figure 5.9.

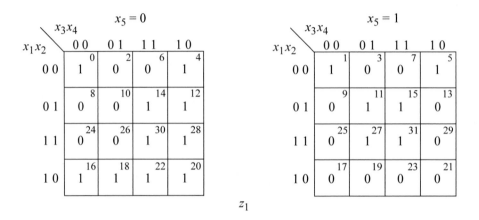

Figure 5.9   Karnaugh map for Example 5.3 used to obtain output z_1 in a minimal sum-of-products form.

$$z_1 = x_2' x_4' x_5' + x_1' x_2' x_4' + x_1 x_2' x_5' + x_2 x_3 x_5' + x_2 x_4 x_5 \qquad (5.1)$$

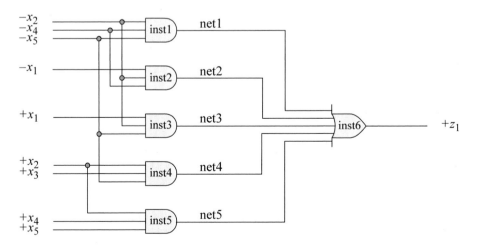

Figure 5.10   Logic diagram for Example 5.3 showing output z_1 in a sum-of-products form.

```
//logic diagram using built-in primitives
module log_eqn_sop7 (x1, x2, x3, x4, x5, z1);

input x1, x2, x3, x4, x5;
output z1;

and    inst1 (net1, ~x2, ~x4, ~x5),
       inst2 (net2, ~x1, ~x2, ~x4),
       inst3 (net3, x1, ~x2, ~x5),
       inst4 (net4, x2, x3, ~x5),
       inst5 (net5, x2, x4,x5);

or inst6 (z1, net1, net2, net3, net4, net5);

endmodule
```

Figure 5.11   Module for the sum-of-products equation of Equation 5.1 that represents the logic diagram of Figure 5.10.

```
//test bench for log_eqn_sop7
module log_eqn_sop7_tb;

reg x1, x2, x3, x4, x5;
wire z1;

//apply input vectors
initial
begin: apply_stimulus
       reg [6:0] invect;   //invect[6] terminates the for loop
       for (invect=0; invect<32; invect=invect+1)
            begin
                 {x1, x2, x3, x4, x5} = invect [6:0];
                 #10 $display ("x1x2x3x4x5 = %b, z1 = %b",
                                 {x1, x2, x3, x4, x5}, z1);
            end
end

//continued on next page
```

Figure 5.12   Test bench for the module of Figure 5.11.

```
log_eqn_sop7 inst1 (
       .x1(x1),
       .x2(x2),
       .x3(x3),
       .x4(x4),
       .x5(x5),
       .z1(z1)
       );

endmodule
```

Figure 5.12   (Continued)

```
x1x2x3x4x5 = 00000, z1 = 1      x1x2x3x4x5 = 10000, z1 = 1
x1x2x3x4x5 = 00001, z1 = 1      x1x2x3x4x5 = 10001, z1 = 0
x1x2x3x4x5 = 00010, z1 = 0      x1x2x3x4x5 = 10010, z1 = 1
x1x2x3x4x5 = 00011, z1 = 0      x1x2x3x4x5 = 10011, z1 = 0
x1x2x3x4x5 = 00100, z1 = 1      x1x2x3x4x5 = 10100, z1 = 1
x1x2x3x4x5 = 00101, z1 = 1      x1x2x3x4x5 = 10101, z1 = 0
x1x2x3x4x5 = 00110, z1 = 0      x1x2x3x4x5 = 10110, z1 = 1
x1x2x3x4x5 = 00111, z1 = 0      x1x2x3x4x5 = 10111, z1 = 0
x1x2x3x4x5 = 01000, z1 = 0      x1x2x3x4x5 = 11000, z1 = 0
x1x2x3x4x5 = 01001, z1 = 0      x1x2x3x4x5 = 11001, z1 = 0
x1x2x3x4x5 = 01010, z1 = 0      x1x2x3x4x5 = 11010, z1 = 0
x1x2x3x4x5 = 01011, z1 = 1      x1x2x3x4x5 = 11011, z1 = 1
x1x2x3x4x5 = 01100, z1 = 1      x1x2x3x4x5 = 11100, z1 = 1
x1x2x3x4x5 = 01101, z1 = 0      x1x2x3x4x5 = 11101, z1 = 0
x1x2x3x4x5 = 01110, z1 = 1      x1x2x3x4x5 = 11110, z1 = 1
x1x2x3x4x5 = 01111, z1 = 1      x1x2x3x4x5 = 11111, z1 = 1
```

Figure 5.13   Outputs for the test bench of Figure 5.12 for the module of Figure 5.11.

Example 5.4   This example repeats Example 5.3, but generates output z_1 in a product-of-sums form. The Karnaugh map of Figure 5.9 is reproduced in Figure 5.14 for convenience. The minimal product-of-sums expression can be obtained by combining the 0s to form sum terms in the same manner as the 1s were combined to form product terms. However, since 0s are being combined, each sum term must equal 0.

Thus, the four 0s in row $x_1 x_2' x_5 = 0000$ combine to yield the sum term $(x_1' + x_2 + x_5')$. In a similar manner, the remaining 0s are combined to yield the product-of-sums expression shown in Equation 5.2. When combining 0s to obtain sum terms, treat a variable value of 1 as false and a variable value of 0 as true. Thus, minterm locations 8, 10, 24, and 26 have variables $x_2' x_3 x_5 = 100$, providing a sum term of $(x_2' + x_3 + x_5)$.

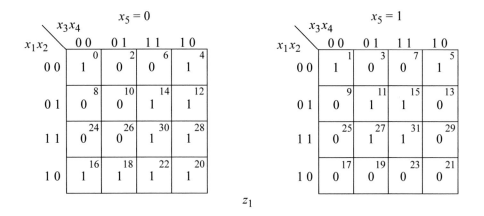

Figure 5.14   Karnaugh map for Example 5.4 which generates output z_1 in a product-of-sums form.

$$z_1 = (x_2' + x_3 + x_5)(x_1 + x_2 + x_4')(x_2' + x_4 + x_5')(x_1' + x_2 + x_5') \quad (5.2)$$

The logic diagram for Example 5.4 is shown in Figure 5.15, which generates output z_1 in a product-of-sums form. The diagram also provides the instance names and the internal net names. The design module is shown in Figure 5.16, the test bench in Figure 5.17, and the outputs in Figure 5.18, which are identical to the outputs of Example 5.3, Figure 5.13.

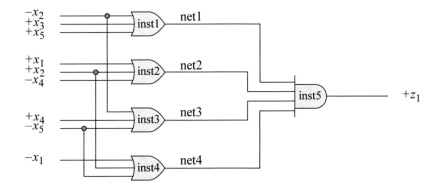

Figure 5.15   Logic diagram for output z_1 in a product-of-sums form.

```
//product of sums using built-in primitives
module log_eqn_pos3 (x1, x2, x3, x4, x5, z1);

input x1, x2, x3, x4, x5;
output z1;

or      inst1 (net1, ~x2, x3, x5),
        inst2 (net2, x1, x2, ~x4),
        inst3 (net3, ~x2, x4, ~x5),
        inst4 (net4, ~x1, x2, ~x5);

and     inst5 (z1, net1, net2, net3, net4);

endmodule
```

Figure 5.16   Module for the product-of-sums logic diagram of Figure 5.15.

```
//test bench for product of sums
module log_eqn_pos3_tb;

reg x1, x2, x3, x4, x5;
wire z1;

//apply input vectors
initial
begin: apply_stimulus
        reg [6:0] invect;
        for (invect=0; invect<32; invect=invect+1)
             begin
                  {x1, x2, x3, x4, x5} = invect [6:0];
                  #10 $display ("x1x2x3x4x5 = %b, z1 = %b",
                                      {x1, x2, x3, x4, x5}, z1);
             end
end

//continued on next page
```

Figure 5.17   Test bench for the product-of-sums module of Figure 5.16.

```
//instantiate the module into the test bench
log_eqn_pos3 inst1 (
        .x1(x1),
        .x2(x2),
        .x3(x3),
        .x4(x4),
        .x5(x5),
        .z1(z1)
        );
endmodule
```

Figure 5.17   (Continued)

```
x1x2x3x4x5 = 00000, z1 = 1      x1x2x3x4x5 = 10000, z1 = 1
x1x2x3x4x5 = 00001, z1 = 1      x1x2x3x4x5 = 10001, z1 = 0
x1x2x3x4x5 = 00010, z1 = 0      x1x2x3x4x5 = 10010, z1 = 1
x1x2x3x4x5 = 00011, z1 = 0      x1x2x3x4x5 = 10011, z1 = 0
x1x2x3x4x5 = 00100, z1 = 1      x1x2x3x4x5 = 10100, z1 = 1
x1x2x3x4x5 = 00101, z1 = 1      x1x2x3x4x5 = 10101, z1 = 0
x1x2x3x4x5 = 00110, z1 = 0      x1x2x3x4x5 = 10110, z1 = 1
x1x2x3x4x5 = 00111, z1 = 0      x1x2x3x4x5 = 10111, z1 = 0
x1x2x3x4x5 = 01000, z1 = 0      x1x2x3x4x5 = 11000, z1 = 0
x1x2x3x4x5 = 01001, z1 = 0      x1x2x3x4x5 = 11001, z1 = 0
x1x2x3x4x5 = 01010, z1 = 0      x1x2x3x4x5 = 11010, z1 = 0
x1x2x3x4x5 = 01011, z1 = 1      x1x2x3x4x5 = 11011, z1 = 1
x1x2x3x4x5 = 01100, z1 = 1      x1x2x3x4x5 = 11100, z1 = 1
x1x2x3x4x5 = 01101, z1 = 0      x1x2x3x4x5 = 11101, z1 = 0
x1x2x3x4x5 = 01110, z1 = 1      x1x2x3x4x5 = 11110, z1 = 1
x1x2x3x4x5 = 01111, z1 = 1      x1x2x3x4x5 = 11111, z1 = 1
```

Figure 5.18   Outputs for the test bench of Figure 5.17 for the product-of-sums module of Figure 5.16.

Example 5.5   This example illustrates the design of a majority circuit using built-in primitives. The output of a majority circuit is a logic 1 if the majority of the inputs is a logic 1; otherwise, the output is a logic 0. Therefore, a majority circuit must have an odd number of inputs in order to have a majority of the inputs be at the same logic level. A 5-input majority circuit will be designed using the Karnaugh map of Figure 5.19, where a 1 entry indicates that the majority of the inputs is a logic 1. Equation 5.3

represents the logic for output z_1 in a sum-of-products form.   The module is shown in Figure 5.20, the test bench in Figure 5.21, and the outputs in Figure 5.22.

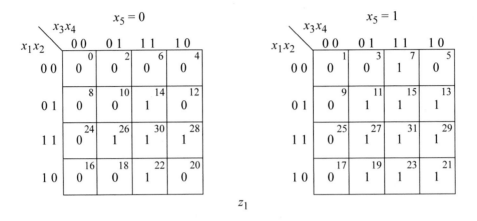

Figure 5.19   Karnaugh map for the majority circuit of Example 5.5.

$$z_1 = x_3 x_4 x_5 + x_2 x_3 x_5 + x_1 x_3 x_5 + x_2 x_4 x_5 + x_1 x_4 x_5$$

$$+ x_1 x_2 x_3 + x_1 x_2 x_4 + x_2 x_3 x_4 + x_1 x_3 x_4 \tag{5.3}$$

```
//5-input majority circuit
module majority (x1, x2, x3, x4, x5, z1);
input x1, x2, x3, x4, x5;
output z1;

and     inst1    (net1, x3, x4, x5),
        inst2    (net2, x2, x3, x5),
        inst3    (net3, x1, x3, x5),
        inst4    (net4, x2, x4, x5),
        inst5    (net5, x1, x4, x5),
        inst6    (net6, x1, x2, x5),
        inst7    (net7, x1, x2, x4),
        inst8    (net8, x2, x3, x4),
        inst9    (net9, x1, x3, x4);
or      inst10 (z1, net1, net2, net3, net4, net5,
                  net6, net7, net8, net9);

endmodule
```

Figure 5.20   Module for the majority circuit of Figure 5.19.

```
//test bench for 5-input majority circuit
module majority_tb;
reg x1, x2, x3, x4, x5;
wire z1;

//apply input vectors
initial
begin: apply_stimulus
      reg [6:0] invect;
      for (invect=0; invect<32; invect=invect+1)
            begin
                  {x1, x2, x3, x4, x5} = invect [6:0];
                  #10 $display ("x1x2x3x4x5 = %b, z1 = %b",
                              {x1, x2, x3, x4, x5}, z1);
            end
end

//instantiate the module into the test bench
majority inst1 (
      .x1(x1),
      .x2(x2),
      .x3(x3),
      .x4(x4),
      .x5(x5),
      .z1(z1)
      );
 endmodule
```

Figure 5.21   Test bench for the majority circuit module of Figure 5.20.

```
x1x2x3x4x5 = 00000, z1 = 0      x1x2x3x4x5 = 01000, z1 = 0
x1x2x3x4x5 = 00001, z1 = 0      x1x2x3x4x5 = 01001, z1 = 0
x1x2x3x4x5 = 00010, z1 = 0      x1x2x3x4x5 = 01010, z1 = 0
x1x2x3x4x5 = 00011, z1 = 0      x1x2x3x4x5 = 01011, z1 = 1
x1x2x3x4x5 = 00100, z1 = 0      x1x2x3x4x5 = 01100, z1 = 0
x1x2x3x4x5 = 00101, z1 = 0      x1x2x3x4x5 = 01101, z1 = 1
x1x2x3x4x5 = 00110, z1 = 0      x1x2x3x4x5 = 01110, z1 = 1
x1x2x3x4x5 = 00111, z1 = 1      x1x2x3x4x5 = 01111, z1 = 1
                                //continued on next page
```

Figure 5.22   Outputs for the majority circuit of Figure 5.20.

```
x1x2x3x4x5 = 10000, z1 = 0          x1x2x3x4x5 = 11000, z1 = 0
x1x2x3x4x5 = 10001, z1 = 0          x1x2x3x4x5 = 11001, z1 = 1
x1x2x3x4x5 = 10010, z1 = 0          x1x2x3x4x5 = 11010, z1 = 1
x1x2x3x4x5 = 10011, z1 = 1          x1x2x3x4x5 = 11011, z1 = 1
x1x2x3x4x5 = 10100, z1 = 0          x1x2x3x4x5 = 11100, z1 = 1
x1x2x3x4x5 = 10101, z1 = 1          x1x2x3x4x5 = 11101, z1 = 1
x1x2x3x4x5 = 10110, z1 = 1          x1x2x3x4x5 = 11110, z1 = 1
x1x2x3x4x5 = 10111, z1 = 1          x1x2x3x4x5 = 11111, z1 = 1
```

Figure 5.22   (Continued)

Example 5.6   The 4:1 multiplexer of Figure 5.23 will be designed using built-in primitives. The design is simpler and takes less code if a continuous assignment statement is used, but this section presents gate-level modeling only. The multiplexer has four data inputs: d_0, d_1, d_2, and d_3, two select inputs: s_0 and s_1, and one *Enable* input. Also, the system function **$time** will be used in the test bench to return the current simulation time. The design module is shown in Figure 5.24, the test bench in Figure 5.25, and the outputs in Figure 5.26.

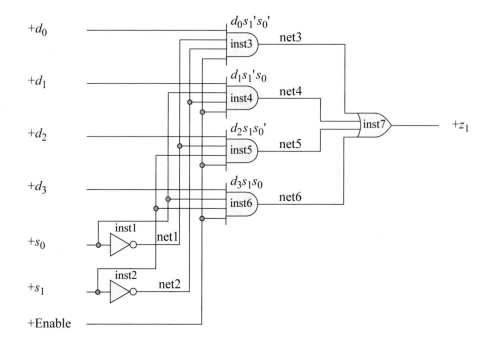

Figure 5.23   Logic diagram of a 4:1 multiplexer to be designed using built-in primitives.

```
//a 4:1 multiplexer using built-in primitives
module mux_4to1 (d, s, enbl, z1);

input [3:0] d;
input [1:0] s;
input enbl;
output z1;

not    inst1 (net1, s[0]),
       inst2 (net2, s[1]);

and    inst3 (net3, d[0], net1, net2, enbl),
       inst4 (net4, d[1], s[0], net2, enbl),
       inst5 (net5, d[2], net1, s[1], enbl),
       inst6 (net6, d[3], s[0], s[1], enbl);

or     inst7 (z1, net3, net4, net5, net6);
endmodule
```

Figure 5.24   Module for a 4:1 multiplexer using built-in primitives.

```
//test bench for 4:1 multiplexer
module mux_4to1_tb;

reg [3:0] d;
reg [1:0] s;
reg enbl;
wire z1;

initial
$monitor ($time,"ns, select:s=%b, inputs:d=%b, output:z1=%b",
                    s, d, z1);

initial
begin
      #0    s[1]=1'b0;   s[0]=1'b0;
            d[3]=1'b1;   d[2]=1'b0;   d[1]=1'b1;   d[0]=1'b0;
            enbl=1'b1;                         //d[0]=0; z1=0
//continued on next page
```

Figure 5.25   Test bench for the 4:1 multiplexer of Figure 5.24.

```
        #10    s[1]=1'b0;  s[1]=1'b0;
               d[3]=1'b1;  d[2]=1'b0;  d[1]=1'b1;  d[0]=1'b1;
               enbl=1'b1;                //d[0]=1; z1=1

        #10    s[1]=1'b0;  s[0]=1'b1;
               d[3]=1'b1;  d[2]=1'b0;  d[1]=1'b1;  d[0]=1'b1;
               enbl=1'b1;                //d[1]=1; z1=1

        #10    s[1]=1'b1;  s[0]=1'b0;
               d[3]=1'b1;  d[2]=1'b0;  d[1]=1'b1;  d[0]=1'b1;
               enbl=1'b1;                //d[2]=0; z1=0

        #10    s[1]=1'b0;  s[0]=1'b1;
               d[3]=1'b1;  d[2]=1'b0;  d[1]=1'b1;  d[0]=1'b1;
               enbl=1'b1;                //d[1]=1; z1=1

        #10    s[1]=1'b1;  s[0]=1'b1;
               d[3]=1'b1;  d[2]=1'b0;  d[1]=1'b1;  d[0]=1'b1;
               enbl=1'b1;                //d[3]=1; z1=1

        #10    s[1]=1'b1;  s[0]=1'b1;
               d[3]=1'b0;  d[2]=1'b0;  d[1]=1'b1;  d[0]=1'b1;
               enbl=1'b1;                //d[3]=0; z1=0

        #10    $stop;

end


//instantiate the module into the test bench
mux_4to1 inst1 (
        .d(d),
        .s(s),
        .z1(z1),
        .enbl(enbl)
        );

endmodule
```

Figure 5.25   (Continued)

```
 0ns,  select:s=00,  inputs:d=1010,  output:z1=0
10ns,  select:s=00,  inputs:d=1011,  output:z1=1
20ns,  select:s=01,  inputs:d=1011,  output:z1=1
30ns,  select:s=10,  inputs:d=1011,  output:z1=0
40ns,  select:s=01,  inputs:d=1011,  output:z1=1
50ns,  select:s=11,  inputs:d=1011,  output:z1=1
60ns,  select:s=11,  inputs:d=0011,  output:z1=0
```

Figure 5.26   Outputs for the 4:1 multiplexer test bench of Figure 5.25.

Example 5.7   This example will design an *SR* latch using NOR logic. The truth table for an *SR* latch is shown in Table 5.5. The logic diagram is shown in Figure 5.27, the module in Figure 5.28, the test bench in Figure 5.29, and the outputs in Figure 5.30. When both *S* and *R* are asserted at a high level, *qbar* is a logic 0 and *q* is a logic 0, which is invalid for a bistable device with complementary outputs.

Table 5.5 Truth Table for an *SR* Latch

Set	Reset	q output	qbar output	Comments
0	0	Previous state	Previous state	
0	1	0	1	Reset
1	1	0	0	Invalid state
1	0	1	0	Set

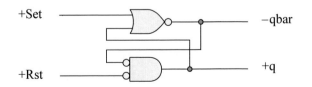

Figure 5.27   Logic diagram for an *SR* latch.

```
//set / reset latch using NOR gates
module latch_nor (set, rst, q, qbar);
input set, rst;
output q, qbar;

nor (qbar, set, q);
nor (q, qbar, rst);

endmodule
```

Figure 5.28   Module for an *SR* latch using NOR logic.

```
//test bench for NOR latch
module latch_nor_tb;

reg set, rst;
wire q, qbar;

initial      //display signals
$monitor ("set = %b, rst = %b, q = %b, qbar = %b",
                    set, rst, q, qbar);

initial      //apply stimulus
begin
        #0     set = 1'b1;        rst = 1'b0;
        #10    set = 1'b0;        rst = 1'b0;
        #10    set = 1'b0;        rst = 1'b1;
        #10    set = 1'b1;        rst = 1'b1;
        #10 $stop;
end

//instantiate the module into the test bench
latch_nor inst1 (
        .set(set),
        .rst(rst),
        .q(q),
        .qbar(qbar)
        );

endmodule
```

Figure 5.29   Test bench for the *SR* latch of Figure 5.28.

```
set = 1, rst = 0, q = 1, qbar = 0
set = 0, rst = 0, q = 1, qbar = 0
set = 0, rst = 1, q = 0, qbar = 1
set = 1, rst = 1, q = 0, qbar = 0
```

Figure 5.30 Outputs for the test bench of Figure 5.29 for the *SR* latch of Figure 5.28.

Example 5.8 A 3-bit binary comparator will be designed using built-in primitives. The operands are 3-bit vectors $a[2:0]$ and $b[2:0]$, as follows:

$$a = a[2]\ a[1]\ a[0]$$
$$b = b[2]\ b[1]\ b[0]$$

where $a[0]$ and $b[0]$ are the low-order bits of a and b, respectively. There are three outputs indicating the relative magnitude of the two operands: $(a < b)$, $(a = b)$, and $(a > b)$. The equations to obtain the outputs are shown in Equation 5.4 and the block diagram is shown in Figure 5.31. The logic diagram is shown in Figure 5.32 as obtained directly from Equation 5.4. The design module, test bench, and outputs are shown in Figure 5.33, Figure 5.34, and Figure 5.35, respectively. To thoroughly test a module, all combinations of the inputs should be applied to the module in the test bench.

$$(a < b) = a[2]'\ b[2] + (a[2] \oplus b[2])'\ a[1]'\ b[1]$$
$$+ (a[2] \oplus b[2])'\ (a[1] \oplus b[1])'\ a[0]'b[0]$$

$$(a = b) = (a[2] \oplus b[2])'\ (a[1] \oplus b[1])'\ (a[0] \oplus b[0])'$$

$$(a > b) = a[2]\ b[2]' + (a[2] \oplus b[2])'\ a[1]\ b[1]'$$
$$+ (a[2] \oplus b[2])'\ (a[1] \oplus b[1])'\ a[0]\ b[0]' \qquad (5.4)$$

Referring to Equation 5.4 for $(a < b)$, the term $a[2]'\ b[2]$ indicates that if the high-order bits of a and b are 0 and 1, respectively, then a must be less than b. If the high-order bits of a and b are equal, then the relative magnitude of a and b depends upon the next lower-order bits $a[1]$ and $b[1]$. This is indicated by the second term of the equation for $(a < b)$.

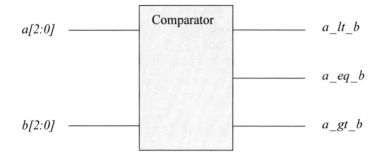

Figure 5.31   Block diagram for the 3-bit binary comparator.

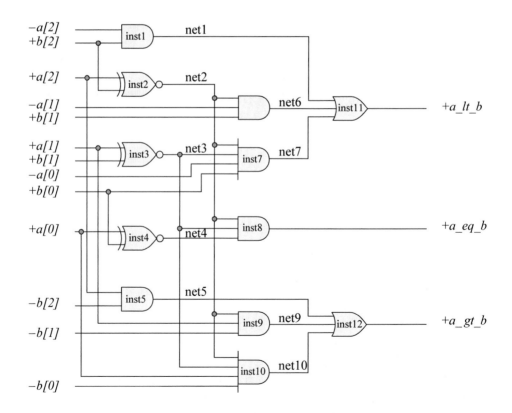

Figure 5.32   Logic diagram for the 3-bit comparator of Example 5.8.

```
//a 3-bit comparator using built-in primitives
module comparator3 (a, b, a_lt_b, a_eq_b, a_gt_b);

input [2:0] a, b;
output a_lt_b, a_eq_b, a_gt_b;

and     inst1        (net1, ~a[2], b[2]);
xnor    inst2        (net2, a[2], b[2]);
xnor    inst3        (net3, a[1], b[1]);
xnor    inst4        (net4, a[0], b[0]);
and     inst5        (net5, a[2], ~b[2]);
and     inst6        (net6, net2, ~a[1], b[1]);
and     inst7        (net7, net2, net3, ~a[0], b[0]);
and     inst9        (net9, net2, a[1], ~b[1]);
and     inst10       (net10, net2, net3, a[0], ~b[0]);

//generate the outputs
or      inst11       (a_lt_b, net1, net6, net7);
and     inst8        (a_eq_b, net2, net3, net4);
or      inst12       (a_gt_b, net5, net9, net10);

endmodule
```

Figure 5.33   Design module for the 3-bit comparator of Figure 5.32.

```
//test bench for 3-bit comparator
module comparator3_tb;

reg [2:0] a, b;
wire a_lt_b, a_eq_b, a_gt_b;

//apply input vectors
initial
begin: apply_stimulus
     reg [7:0] invect;
     for (invect=0; invect<64; invect=invect+1)

//continued on next page
```

Figure 5.34   Test bench for the 3-bit comparator of Figure 5.33.

```
          begin
             {a, b} = invect [7:0];
             #10 $display ("a = %b, b = %b, a_lt_b = %b,
                            a_eq_b = %b, a_gt_b = %b",
                            a, b, a_lt_b, a_eq_b, a_gt_b);
          end
end

//instantiate the module into the test bench
comparator3 inst1 (
   .a(a),
   .b(b),
   .a_lt_b(a_lt_b),
   .a_eq_b(a_eq_b),
   .a_gt_b(a_gt_b)
   );

   endmodule
```

Figure 5.34   (Continued)

```
a = 000, b = 000, a_lt_b = 0, a_eq_b = 1, a_gt_b = 0
a = 000, b = 001, a_lt_b = 1, a_eq_b = 0, a_gt_b = 0
a = 000, b = 010, a_lt_b = 1, a_eq_b = 0, a_gt_b = 0
a = 000, b = 011, a_lt_b = 1, a_eq_b = 0, a_gt_b = 0
a = 000, b = 100, a_lt_b = 1, a_eq_b = 0, a_gt_b = 0
a = 000, b = 101, a_lt_b = 1, a_eq_b = 0, a_gt_b = 0
a = 000, b = 110, a_lt_b = 1, a_eq_b = 0, a_gt_b = 0
a = 000, b = 111, a_lt_b = 1, a_eq_b = 0, a_gt_b = 0
a = 001, b = 000, a_lt_b = 0, a_eq_b = 0, a_gt_b = 1
a = 001, b = 001, a_lt_b = 0, a_eq_b = 1, a_gt_b = 0
a = 001, b = 010, a_lt_b = 1, a_eq_b = 0, a_gt_b = 0
a = 001, b = 011, a_lt_b = 1, a_eq_b = 0, a_gt_b = 0
a = 001, b = 100, a_lt_b = 1, a_eq_b = 0, a_gt_b = 0
a = 001, b = 101, a_lt_b = 1, a_eq_b = 0, a_gt_b = 0
a = 001, b = 110, a_lt_b = 1, a_eq_b = 0, a_gt_b = 0
a = 001, b = 111, a_lt_b = 1, a_eq_b = 0, a_gt_b = 0
a = 010, b = 000, a_lt_b = 0, a_eq_b = 0, a_gt_b = 1

//continued on next page
```

Figure 5.35   Outputs for the 3-bit comparator obtained from the test bench of Figure 5.34.

```
a = 010, b = 001, a_lt_b = 0, a_eq_b = 0, a_gt_b = 1
a = 010, b = 010, a_lt_b = 0, a_eq_b = 1, a_gt_b = 0
a = 010, b = 011, a_lt_b = 1, a_eq_b = 0, a_gt_b = 0
a = 010, b = 100, a_lt_b = 1, a_eq_b = 0, a_gt_b = 0
a = 010, b = 101, a_lt_b = 1, a_eq_b = 0, a_gt_b = 0
a = 010, b = 110, a_lt_b = 1, a_eq_b = 0, a_gt_b = 0
a = 010, b = 111, a_lt_b = 1, a_eq_b = 0, a_gt_b = 0
a = 011, b = 000, a_lt_b = 0, a_eq_b = 0, a_gt_b = 1
a = 011, b = 001, a_lt_b = 0, a_eq_b = 0, a_gt_b = 1
a = 011, b = 010, a_lt_b = 0, a_eq_b = 0, a_gt_b = 1
a = 011, b = 011, a_lt_b = 0, a_eq_b = 1, a_gt_b = 0
a = 011, b = 100, a_lt_b = 1, a_eq_b = 0, a_gt_b = 0
a = 011, b = 101, a_lt_b = 1, a_eq_b = 0, a_gt_b = 0
a = 011, b = 110, a_lt_b = 1, a_eq_b = 0, a_gt_b = 0
a = 011, b = 111, a_lt_b = 1, a_eq_b = 0, a_gt_b = 0
a = 100, b = 000, a_lt_b = 0, a_eq_b = 0, a_gt_b = 1
a = 100, b = 001, a_lt_b = 0, a_eq_b = 0, a_gt_b = 1
a = 100, b = 010, a_lt_b = 0, a_eq_b = 0, a_gt_b = 1
a = 100, b = 011, a_lt_b = 0, a_eq_b = 0, a_gt_b = 1
a = 100, b = 100, a_lt_b = 0, a_eq_b = 1, a_gt_b = 0
a = 100, b = 101, a_lt_b = 1, a_eq_b = 0, a_gt_b = 0
a = 100, b = 110, a_lt_b = 1, a_eq_b = 0, a_gt_b = 0
a = 100, b = 111, a_lt_b = 1, a_eq_b = 0, a_gt_b = 0
a = 101, b = 000, a_lt_b = 0, a_eq_b = 0, a_gt_b = 1
a = 101, b = 001, a_lt_b = 0, a_eq_b = 0, a_gt_b = 1
a = 101, b = 010, a_lt_b = 0, a_eq_b = 0, a_gt_b = 1
a = 101, b = 011, a_lt_b = 0, a_eq_b = 0, a_gt_b = 1
a = 101, b = 100, a_lt_b = 0, a_eq_b = 0, a_gt_b = 1
a = 101, b = 101, a_lt_b = 0, a_eq_b = 1, a_gt_b = 0
a = 101, b = 110, a_lt_b = 1, a_eq_b = 0, a_gt_b = 0
a = 101, b = 111, a_lt_b = 1, a_eq_b = 0, a_gt_b = 0
a = 110, b = 000, a_lt_b = 0, a_eq_b = 0, a_gt_b = 1
a = 110, b = 001, a_lt_b = 0, a_eq_b = 0, a_gt_b = 1
a = 110, b = 010, a_lt_b = 0, a_eq_b = 0, a_gt_b = 1
a = 110, b = 011, a_lt_b = 0, a_eq_b = 0, a_gt_b = 1
a = 110, b = 100, a_lt_b = 0, a_eq_b = 0, a_gt_b = 1
a = 110, b = 101, a_lt_b = 0, a_eq_b = 0, a_gt_b = 1
a = 110, b = 110, a_lt_b = 0, a_eq_b = 1, a_gt_b = 0

//continued on next page
```

Figure 5.35   (Continued)

```
a = 110, b = 111, a_lt_b = 1, a_eq_b = 0, a_gt_b = 0
a = 111, b = 000, a_lt_b = 0, a_eq_b = 0, a_gt_b = 1
a = 111, b = 001, a_lt_b = 0, a_eq_b = 0, a_gt_b = 1
a = 111, b = 010, a_lt_b = 0, a_eq_b = 0, a_gt_b = 1
a = 111, b = 011, a_lt_b = 0, a_eq_b = 0, a_gt_b = 1
a = 111, b = 100, a_lt_b = 0, a_eq_b = 0, a_gt_b = 1
a = 111, b = 101, a_lt_b = 0, a_eq_b = 0, a_gt_b = 1
a = 111, b = 110, a_lt_b = 0, a_eq_b = 0, a_gt_b = 1
a = 111, b = 111, a_lt_b = 0, a_eq_b = 1, a_gt_b = 0
```

Figure 5.35   (Continued)

5.2 Gate Delays

All gates have a propagation delay, which is the time necessary for a signal to propagate from the input terminals (ports), through the internal circuitry of the gate, to the output terminal (port). When no delay is specified, the default delay is zero. The optional propagation delay and optional instance name can be specified in the gate instantiation as follows:

gate_type delay inst1 (output, input_1, input_2, . . . , input_n);

Two or more instances of the same type of gate can be specified in the same construct as shown below with optional delay. Note that only the last instantiation has a semicolon terminating the line. All previous lines are terminated by a comma.

gate_type   delay inst1 (output_1, input_11, input_12, . . . , input_1n),
                   delay inst2 (output_2, input_21, input_22, . . . , input_2n),
                                          .
                                          .
                                          .
                   delay instm (output_m, input_m1, input_m2, . . . ,
                     input_mn);

Three types of delays can be specified: rise delay, fall delay, and turn-off delay. If only one delay is specified, then the delay value applies to all three delays. If two delays are specified, then the delays apply to the rise and fall delays, respectively — the turn-off delay is the minimum of the two delays. If all three delays are specified, then they refer to the rise delay, fall delay, and turn-off delay, respectively. Associated with each delay is a corresponding minimum, typical, and maximum value. The syntax for the three delay values is shown below.

minimum : typical : maximum

The minimum, typical, and maximum values must be constant expressions, as shown below for a 2-input **xor** gate primitive.

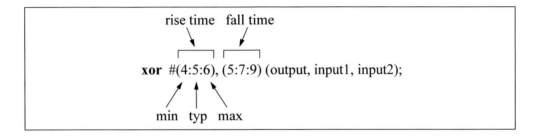

The entries (4:5:6) represent a rise time of 4 time units minimum, 5 time units typical, and 6 time units maximum. The entries (5:7:9) represent a fall time of 5 time units minimum, 7 time units typical, and 9 time units maximum.

Multiple-input gates such as **and**, **or**, **nand**, **nor**, **xor**, and **xnor** cannot have high impedance outputs; therefore, they have only rise time and fall time delays. Three-state gates, however, can have rise time, fall time, and turn-off delays. Examples of primitive gates with different types of delays are shown below.

and   inst1 (output, input1, input2);   //No delay

nand   #5 (output, input1, input2, input3); //Rise time and fall time delays are 5
                                                 //time units

xor   #(4, 6) inst2 (output, input1, input2); //Rise time is 4 time units and fall
                                                   //time is 6 time units. The transition
                                                   //to **x** is 4 time units (the minimum of
                                                   //4 and 6)

bufif1   #(3, 5, 6) (output, data, control);   //Rise delay is 3 time units, fall delay
                                                     //is 5 time units, turn-off delay is 6
                                                     //time units. The transition to **x** is 3
                                                     //time units (the minimum of 3, 5, 6)

Example 5.9   This example illustrates the operation of a logic circuit designed from Equation 5.5. The logic diagram is shown in Figure 5.36 with delay values assigned to the gates. The AND gates will be assigned three time units of delay and the OR gate five time units of delay. Figure 5.37, Figure 5.38, Figure 5.39, and Figure 5.40 show the design module, the test bench, the outputs, and the waveforms, respectively.

$$z_1 = x_1 x_2 + x_3' x_4 \qquad\qquad (5.5)$$

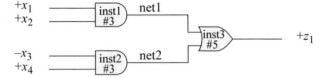

Figure 5.36   Logic diagram to illustrate propagation delays in gates.

```
//a sum-of-products circuit with delays
module log_eqn_sop8 (x1, x2, x3, x4, z1);
input x1, x2, x3, x4;
output z1;

and    #3 inst1 (net1, x1, x2);
and    #3 inst2 (net2, ~x3, x4);
or     #5 inst3 (z1, net1, net2);
endmodule
```

Figure 5.37   Module for Equation 5.5 with propagation delays assigned to the gates.

```
//test bench for sum-of-products with delay
module log_eqn_sop8_tb;
reg x1, x2, x3, x4;
wire z1;

//display variables
initial
$monitor ("x1x2x3x4 = %b, z1 = %b", {x1, x2, x3, x4}, z1);
//continued on next page
```

Figure 5.38   Test bench for the module of Figure 5.37.

```
//apply input vectors
initial
begin

        #0     x1=1'b0;    x2=1'b0;    x3=1'b0;    x4=1'b0;
        #10    x1=1'b1;    x2=1'b1;    x3=1'b1;    x4=1'b0;
        #10    x1=1'b0;    x2=1'b1;    x3=1'b0;    x4=1'b1;
        #10    x1=1'b1;    x2=1'b1;    x3=1'b1;    x4=1'b1;
        #10    x1=1'b0;    x2=1'b0;    x3=1'b0;    x4=1'b0;

        #10    $stop;

end

//instantiate the module into the test bench
log_eqn_sop8 inst1 (
        .x1(x1),
        .x2(x2),
        .x3(x3),
        .x4(x4),
        .z1(z1)
        );

endmodule
```

Figure 5.38   (Continued)

```
x1x2x3x4 = 1110, z1 = 0    //z1 will not change until time 18
x1x2x3x4 = 1110, z1 = 1    //z1 is 1 after 18 time units
x1x2x3x4 = 0101, z1 = 1    //z1 does not change
x1x2x3x4 = 1111, z1 = 1    //z1 does not change
x1x2x3x4 = 0000, z1 = 1    //z1 will not change until time 48
x1x2x3x4 = 0000, z1 = 0    //z1 is 0 at time 48
```

Figure 5.39   Outputs for the test bench of Figure 5.38.

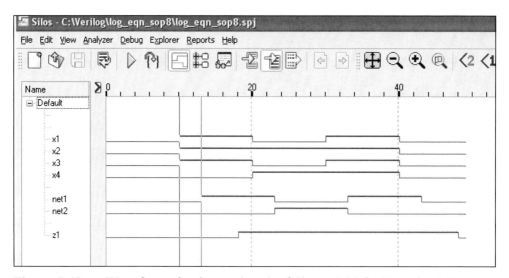

Figure 5.40   Waveforms for the test bench of Figure 5.38 for Equation5.5.

Referring to the logic diagram of Figure 5.36 and the waveforms of Figure 5.40, the output of the AND gate at *inst1* — which is *net1* — is delayed by three time units from the assertion of the two inputs. That is, x_1 and x_2 are asserted at time 10, but the output of the AND gate is not asserted until time 13. When x_3 and x_4 are asserted at a low and high voltage level, respectively, *net2* is asserted three time units later.

In a similar manner, output z_1 is asserted five time units after either *net1* or *net2* is asserted at the OR gate inputs. Thus, when *net1* is asserted at time 13, z_1 is not asserted until time 18. In the same way, when both inputs to the OR gate become inactive at time 43, output z_1 is deasserted at time 48.

Example 5.10   A half adder will be designed in this example with delays for both the exclusive-OR gate and the AND gate. Recall that a half adder is a combinational circuit that performs addition on two binary bits and produces two outputs — a sum bit and a carry-out bit. The half adder does not accommodate a carry-in bit. The truth table for a half adder is shown in Table 5.6 and the equations for the sum and carry-out are shown in Equation 5.6.

Table 5.6 Truth Table for a Half Adder

a	b	sum	carry-out
0	0	0	0
0	1	1	0
1	0	1	0
1	1	0	1

$$sum = a'b + ab'$$
$$= a \oplus b$$

$$carry\text{-}out = ab \qquad\qquad (5.6)$$

The logic diagram for the implementation of a half adder is shown in Figure 5.41 with propagation delays of four time units for the exclusive-OR gate and two time units for the AND gate. The module using built-in primitives, test bench, binary outputs, and waveforms are shown in Figure 5.42, Figure 5.43, Figure 5.44, and Figure 5.45, respectively.

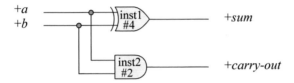

Figure 5.41   Logic diagram for a half adder.

```
//half adder using built-in primitives
module half_adder (a, b, sum, cout);
input a, b;
output sum, cout;

xor   #4 inst1 (sum, a, b);
and   #2 inst2 (cout, a, b);
endmodule
```

Figure 5.42   Module for the half adder of Figure 5.41.

```
//test bench for the half adder
module half_adder_tb;
reg a, b;
wire sum, cout;

initial
$monitor ("ab = %b, sum = %b, cout = %b", {a, b}, sum, cout);
//continued on next page
```

Figure 5.43   Test bench for the half adder module of Figure 5.42.

```
initial      //apply input vectors
begin
     #0    a=1'b0;      b=1'b0;
     #10   a=1'b0;      b=1'b1;
     #10   a=1'b1;      b=1'b0;
     #10   a=1'b1;      b=1'b1;
     #10   $stop;
end
//instantiate the module into the test bench
half_adder inst1 (
     .a(a),
     .b(b),
     .sum(sum),
     .cout(cout)
     );
endmodule
```

Figure 5.43   (Continued)

```
ab = 00, sum = 0, cout = 0    //time unit 0
ab = 01, sum = 0, cout = 0    //time unit 10
ab = 01, sum = 1, cout = 0    //time unit 14
ab = 10, sum = 1, cout = 0    //time unit 20
ab = 11, sum = 1, cout = 0    //time unit 30
ab = 11, sum = 1, cout = 1    //time unit 32
ab = 11, sum = 0, cout = 1    //time unit 34
```

Figure 5.44   Binary outputs for the half adder module of Figure 5.42.

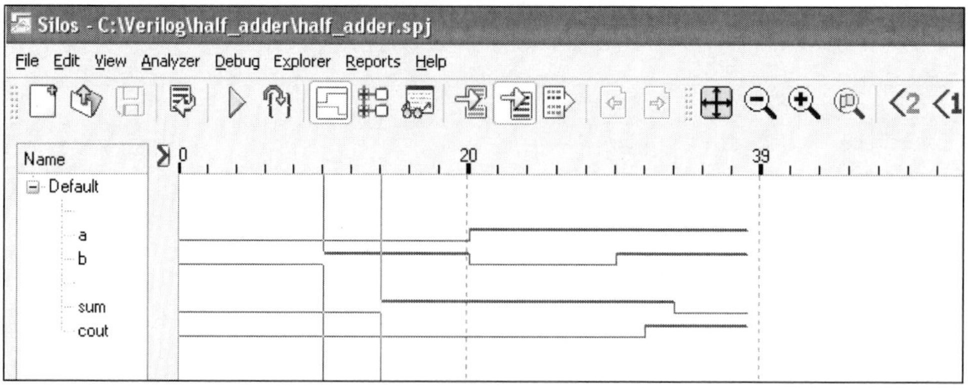

Figure 5.45   Waveforms for the half adder module of Figure 5.42.

Referring to Figure 5.45, when input b changes from 0 to 1 at time 10 with a deasserted, the *sum* output is asserted at time 14 as specified in the design module and *cout* remains inactive. At time 30, input a is already asserted when b becomes asserted causing *sum* to be deasserted at time 34 and *cout* to be asserted at time 32, which is correct for a value of $ab = 11$.

Refer to the binary outputs of Figure 5.44 for the discussion that follows. The second line of the outputs appears to be inaccurate: $ab = 01$, $sum = 0$, $cout = 0$. Likewise, the fifth line appears to be inaccurate: $ab = 11$, $sum = 1$, $cout = 0$. This is because the **$monitor** system task in the test bench displays the variables only when the variables change value, but does not take into account the delays associated with the changes. The exclusive-OR gate has a propagation delay of four time units and the AND gate has a propagation delay of two time units. The waveforms of Figure 5.45, however, accurately reflect the outputs of the half adder.

The waveforms shown in Figure 5.46 add more detail to the waveforms of Figure 5.45 and will be used to explain the apparent inaccuracy of the binary outputs. At #0 time, $ab = 00$, $sum = 0$, $cout = 0$, which is correct. At #10, the test bench changes the inputs to $ab = 01$; however, the outputs remain $sum = 0$ and $cout = 0$ because *cout* does not change until #12 and *sum* does not change until #14. At #14, both *sum* and *cout* are stable. This is reflected in the third line of Figure 5.44, where $ab = 01$, $sum = 1$, and $cout = 0$.

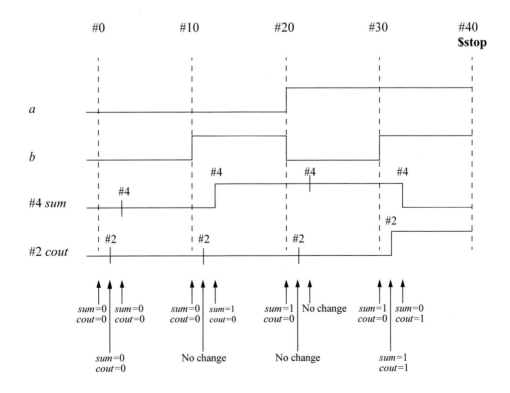

Figure 5.46 Waveforms to explain the apparent inaccuracy of the binary outputs of Figure 5.44.

At #20, the test bench changes the inputs to $ab = 10$, where the outputs remain *sum* = 1, *cout* = 0 because this input combination represents no change to the outputs. At #30, the test bench changes the inputs to $ab = 11$; however, the outputs remain *sum* = 1, *cout* = 0 because *cout* does not change until #32, at which time *sum* is still *sum* = 1 and *cout* = 1. This is indicated in the sixth line of Figure 5.44. At #34, *sum* changes to *sum* = 0, yielding $ab = 11$, *sum* = 0, *cout* = 1, which is correct.

Example 5.11   A full adder will now be designed using built-in primitives with propagation delays assigned to the gates. This will be a high-speed full adder using a sum-of-minterms format. A full adder is a combinational circuit that computes the sum of two operand bits plus a carry-in bit. The carry-in signal represents the carry-out of the previous lower-order bit position. The full adder produces two outputs: a sum bit and a carry-out bit. The truth table for a full adder is shown in Table 5.7.

Table 5.7 Truth Table for a Full Adder

a	b	cin	sum	cout
0	0	0	0	0
0	0	1	1	0
0	1	0	1	0
0	1	1	0	1
1	0	0	1	0
1	0	1	0	1
1	1	0	0	1
1	1	1	1	1

The equations for the sum and carry-out are shown in Equation 5.7 in a canonical sum-of-products notation — also called a sum of minterms or a disjunctive normal form. The carry-out equation can be minimized to a sum-of-products notation while still maintaining two gate delays. The full adder can also be designed using two half adders, but for high-speed addition, the disjunctive normal implementation is preferred. The logic diagram is shown in Figure 5.47. The module, test bench, and waveforms are shown in Figure 5.48, Figure 5.49, and Figure 5.50, respectively.

$$sum = a'b'cin + a'bcin' + ab'cin' + abcin$$

$$carry\text{-}out = a'bcin + ab'cin + abcin' + abcin$$

$$= ab + acin + bcin \qquad (5.7)$$

In the waveforms of Figure 5.50, the delays through the full adder logic are clearly shown. When *cin* is asserted at time 10, *sum* is not asserted until time 15. This is caused by the aggregate delays through the AND gates and the OR gate. At time 30, *abcin* = 011, which generates a *sum* of 0 and a *cout* of 1 at time 35. Finally, at time 70, all three inputs are asserted, providing a *sum* of 1 and a *cout* of 1.

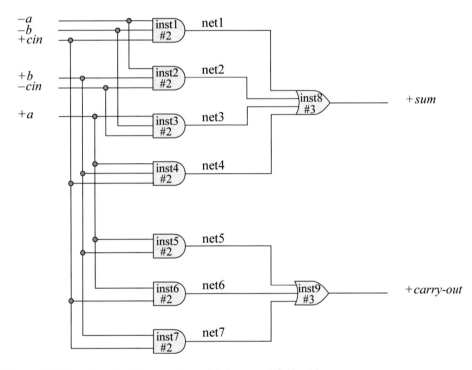

Figure 5.47   Logic diagram for a high-speed full adder.

```
//full adder using built-in primitives with delay
module full_adder_hi_spd (a, b, cin, sum, cout);

input a, b, cin;
output sum, cout;

//continued on next page
```

Figure 5.48   Module for the full adder of Figure 5.47 with delays assigned.

```
and    #2 inst1 (net1, ~a, ~b, cin);
and    #2 inst2 (net2, ~a, b, ~cin);
and    #2 inst3 (net3, a, ~b, ~cin);
and    #2 inst4 (net4, a, b, cin);
and    #2 inst5 (net5, a, b);
and    #2 inst6 (net6, a, cin);
and    #2 inst7 (net7, b, cin);


or     #3 inst8 (sum, net1, net2, net3, net4);
or     #3 inst9 (cout, net5, net6, net7);


endmodule
```

Figure 5.48   (Continued)

```
//test bench for full adder using
//built-in primitives with delays

module full_adder_hi_spd_tb;

reg    a, b, cin;
wire   sum, cout;

initial
$monitor ("abcin = %b, sum = %b, cout = %b",
          {a, b, cin}, sum, cout);

//apply input vectors
initial
begin
        #0     a=1'b0;      b=1'b0;      cin=1'b0;
        #10    a=1'b0;      b=1'b0;      cin=1'b1;
        #10    a=1'b0;      b=1'b1;      cin=1'b0;
        #10    a=1'b0;      b=1'b1;      cin=1'b1;

//continued on next page
```

Figure 5.49   Test bench for the module of Figure 5.48 for the full adder of Figure 5.47.

```
        #10    a=1'b1;      b=1'b0;      cin=1'b0;
        #10    a=1'b1;      b=1'b0;      cin=1'b1;
        #10    a=1'b1;      b=1'b1;      cin=1'b0;
        #10    a=1'b1;      b=1'b1;      cin=1'b1;
        #10    $stop;
end

//instantiate the module into the test bench
full_adder_hi_spd inst1 (
      .a(a),
      .b(b),
      .cin(cin),
      .sum(sum),
      .cout(cout)
      );
endmodule
```

Figure 5.49   (Continued)

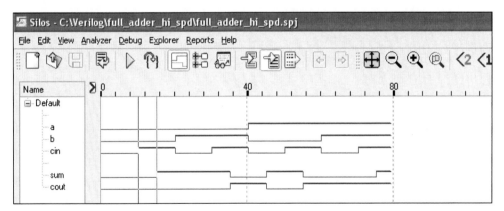

Figure 5.50   Waveforms for the full adder of Figure 5.47 with delays assigned to the logic gates.

5.2.1 Inertial Delay

If the pulse width of an input signal to a built-in primitive gate is narrower than the propagation delay of the gate, then the pulse will not be propagated through the gate. This represents the inertial delay of the gate. The *inertial delay* is the amount of time that the inputs must be stable in order to generate an output; that is, the pulse width must be at least as long as the inertial delay of the gate. If the pulse width is equal to the propagation delay, then the output is indeterminate.

Example 5.12 In this example, a 2-input AND gate will be used to demonstrate inertial delay. The propagation delay of the AND gate is specified to be 10 time units. Two input pulses will be generated for four time units each, both of which will not appear on the output. Figure 5.51 shows the module for a 2-input AND gate. The test bench is shown in Figure 5.52. The system task **$display** and the system function **$time** are used in the test bench to display the various simulation times. This allows correlation between the outputs and the waveforms. The outputs are shown in Figure 5.53.

```
//module to show that a narrow pulse will not propagate
//through a gate with a large propagation delay
module no_glitch (x1, x2, out);

input x1, x2;
output out;

and    #10    (out, x1, x2);

endmodule
```

Figure 5.51   Module to illustrate inertial delay.

```
//test bench for the no_glitch module
module no_glitch_tb;

reg x1, x2;
wire out;

//apply input vectors
initial
begin
   #0   x1=1'b0;   x2=1'b0;
        $display ($time, " x1x2=%b, out=%b", {x1, x2}, out);

   #10 x1 = 1'b1; x2 = 1'b1;
        $display ($time, " x1x2=%b, out=%b", {x1, x2}, out);
//continued on next page
```

Figure 5.52   Test bench to illustrate inertial delay.

```
#10 x1 = 1'b1; x2 = 1'b1;
    $display ($time, " x1x2=%b, out=%b", {x1, x2}, out);

#10 x1 = 1'b0; x2 = 1'b1;
    $display ($time, " x1x2=%b, out=%b", {x1, x2}, out);

#10 x1 = 1'b0; x2 = 1'b1;
    $display ($time, " x1x2=%b, out=%b", {x1, x2}, out);

#3  x1 = 1'b1; x2 = 1'b1;
    $display ($time, " x1x2=%b, out=%b", {x1, x2}, out);

#4  x1 = 1'b0; x2 = 1'b1;
    $display ($time, " x1x2=%b, out=%b", {x1, x2}, out);

#3  x1 = 1'b0; x2 = 1'b1;
    $display ($time, " x1x2=%b, out=%b", {x1, x2}, out);

#10 x1 = 1'b0; x2 = 1'b1;
    $display ($time, " x1x2=%b, out=%b", {x1, x2}, out);

#10 x1 = 1'b1; x2 = 1'b1;
    $display ($time, " x1x2=%b, out=%b", {x1, x2}, out);

#10 x1 = 1'b1; x2 = 1'b1;
    $display ($time, " x1x2=%b, out=%b", {x1, x2}, out);

#3  x1 = 1'b1; x2 = 1'b0;
    $display ($time, " x1x2=%b, out=%b", {x1, x2}, out);

#4  x1 = 1'b1; x2 = 1'b1;
    $display ($time, " x1x2=%b, out=%b", {x1, x2}, out);

#10 x1 = 1'b1; x2 = 1'b1;
    $display ($time, " x1x2=%b, out=%b", {x1, x2}, out);

#10 x1 = 1'b1; x2 = 1'b1;
    $display ($time, " x1x2=%b, out=%b", {x1, x2}, out);

#10 $stop;

end

//continued on next page
```

Figure 5.52 (Continued)

```
//instantiate the module into the test bench
no_glitch inst1 (
    .x1(x1),
    .x2(x2),
    .out(out)
    );

endmodule
```

Figure 5.52   (Continued)

```
0    x1x2=00,  out=0
10   x1x2=11,  out=0
20   x1x2=11,  out=1
30   x1x2=01,  out=1
40   x1x2=01,  out=0
43   x1x2=11,  out=0
47   x1x2=01,  out=0
50   x1x2=01,  out=0
60   x1x2=01,  out=0
70   x1x2=11,  out=0
80   x1x2=11,  out=1
83   x1x2=10,  out=1
87   x1x2=11,  out=1
97   x1x2=11,  out=1
107  x1x2=11,  out=1
```

Figure 5.53   Outputs for the inertial delay module of Figure 5.51.

Figure 5.54 shows the waveforms, which illustrate the inertial delay for the 2-input AND gate. The x_1 pulse is asserted for four time units (43 to 47), which is less than the inertial delay of the AND gate and is, therefore, not propagated. The x_2 pulse is deasserted for four time units (83 to 87), which is less than the inertial delay of the AND gate, and is also not propagated.

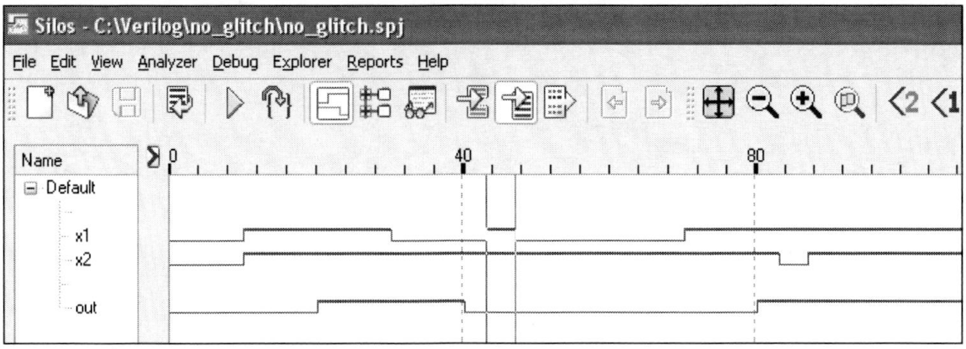

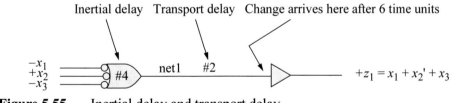

Figure 5.54   Waveforms for the inertial delay module of Figure 5.51.

5.2.2 Transport Delay

Transport delay is the time required for a signal to travel over a net from source to destination. This is a very small delay — approximately 1 ns/foot. However, in some situations it may be important to model the transport delay. Verilog models transport delay as the time delay between when an output changes on a logic primitive (or a continuous assignment that drives a net) and an input changes at a destination gate (or right-hand side of a continuous assignment receiving the net). Figure 5.55 shows a logic diagram that illustrates the difference between inertial delay and transport delay.

Inertial delay   Transport delay   Change arrives here after 6 time units

$$-x_1$$
$$+x_2$$
$$-x_3$$
\#4   net1   \#2 $$+z_1 = x_1 + x_2' + x_3$$

Figure 5.55   Inertial delay and transport delay.

The logic circuit shown in Figure 5.56 will be modeled to illustrate both inertial delay and transport delay. The transport delay of all nets is two time units. The module is shown in Figure 5.57.

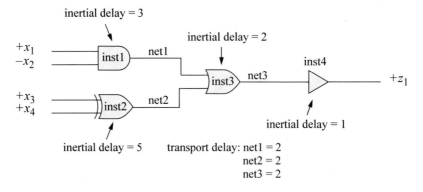

Figure 5.56   Logic circuit used to model inertial delay and transport delay.

```
//module to illustrate inertial delay and
//transport delay using built-in primitives
module inert_trans_dly1 (x1, x2, x3, x4, net1, net2, net3, z1);

input x1, x2, x3, x4;
output net1, net2, net3, z1;

//transport delay
wire #2 net1, net2, net3;

//inertial delay
and #3 inst1 (net1, x1, ~x2);
xor #5 inst2 (net2, x3, x4);
or  #2 inst3 (net3, net1, net2);
buf #1 inst4 (z1, net3);
endmodule
```

Figure 5.57   Module to model inertial delay and transport delay.

5.2.3 Module Path Delay

A *module path delay* is the delay between an event at an input port of a module and the resulting event at an output port. This is shown in Figure 5.58. The syntax that specifies the path is as follows, where the symbol * > specifies the path:

$$\text{source } * > \text{ destination } = \text{ delay value}$$

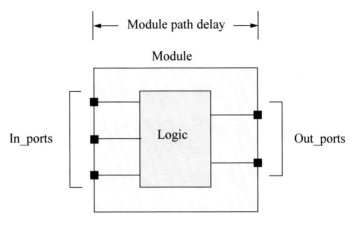

Figure 5.58   Module path delay diagram.

The module path delay can be specified using a **specify** block. The logic diagram of Figure 5.59 will be used to illustrate the module path delay technique. The module, test bench, and waveforms are shown in Figure 5.60, Figure 5.61, and Figure 5.62, respectively.

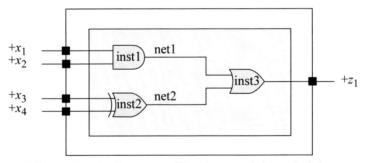

Figure 5.59   Logic diagram to illustrate module path delay.

```
//module path delay
module module_path_dly (x1, x2, x3, x4, z1);

input x1, x2, x3, x4;
output z1;

//continued on next page
```

Figure 5.60   Module to illustrate module path delay.

```
and     inst1 (net1, x1, x2);
xor     inst2 (net2, x3, x4);
or      inst3 (z1, net1, net2);

specify
        (x1, x2 *> z1) = 6;
        (x3, x4 *> z1) = 8;
endspecify

endmodule
```

Figure 5.60   (Continued)

As can be seen in Figure 5.60, the delay through the AND gate and OR gate is six time units; the delay through the exclusive-OR gate and OR gate is eight time units.

```
//test bench for module path delay
module module_path_dly_tb;
reg x1, x2, x3, x4;
wire z1;

initial
$monitor ("x1x2x3x4 = %b, z1 = %b", {x1, x2, x3, x4}, z1);
initial              //apply input vectors
begin
    #0   x1=1'b0;   x2=1'b0;   x3=1'b0;   x4=1'b0;
    #10  x1=1'b1;   x2=1'b1;   x3=1'b0;   x4=1'b0;
    #10  x1=1'b0;   x2=1'b0;   x3=1'b0;   x4=1'b0;
    #10  x1=1'b0;   x2=1'b0;   x3=1'b1;   x4=1'b0;
    #10  $stop;
end
module_path_dly inst1 (        //instantiate the module
  .x1(x1),
  .x2(x2),
  .x3(x3),
  .x4(x4),
  .z1(z1)
  );

endmodule
```

Figure 5.61   Test bench for module path delay.

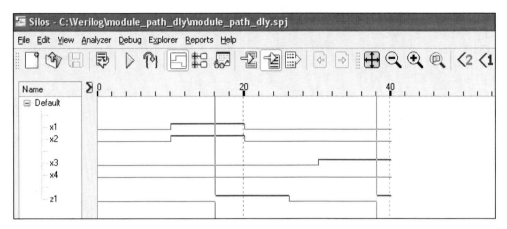

Figure 5.62   Waveforms to illustrate module path delay.

5.3 Additional Design Examples

The previous examples in this chapter were relatively straightforward. This section provides additional examples for gate-level modeling that are more challenging: three for iterative networks and one for a priority encoder.

5.3.1 Iterative Networks

An *iterative network* is a logical structure composed of identical cells. It is a cascade of identical combinational or sequential circuits (cells) in which the first or last cells may be different than the other cells in the network. Since an iterative network consists of identical cells, it is only necessary to design a typical cell, and then to replicate that cell for the entire network.

Example 5.13   In this example, a circuit will be designed to determine if an input vector contains a single 1 bit only. The output will be a logical 1 if there is only a single 1 bit; the output will be a logical 0 if there are no 1 bits or two or more 1 bits. The block diagram for a typical $cell_i$ is shown in Figure 5.63, where input $y_{i-1(1)}$ indicates that a single 1 bit has been detected up to that cell. Input $y_{i-1(0)}$ indicates that all 0s have been detected up to that cell. Output $y_{i(1)}$ indicates that a single 1 bit has been detected up to and including that cell. Output $y_{i(0)}$ indicates that the input vector contained only 0s up to and including that cell. The general equations for $cell_i$ are shown in Equation 5.8.

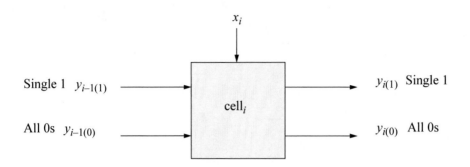

Figure 5.63   General block diagram for a typical cell of a single-bit detector.

$$y_{i(1)} = x_i' y_{i-1(1)} + x_i y_{i-1(0)}$$

$$y_{i(0)} = x_i' y_{i-1(0)} \tag{5.8}$$

The input vector will consist of four bits: $x_1 x_2 x_3 x_4$. The equations for the single-bit detector are shown in Equation 5.9 for each cell. A typical $cell_i$ for this design is shown in Figure 5.64 and the iterative network is shown in Figure 5.65.

Bit 1 cell	$y_{1(1)} = x_1$	One 1
	$y_{1(0)} = x_1'$	No 1s
Bit 2 cell	$y_{2(1)} = y_{1(1)} x_2' + y_{1(0)} x_2$	One 1
	$y_{2(0)} = y_{1(0)} x_2'$	No 1s
Bit 3 cell	$y_{3(1)} = y_{2(1)} x_3' + y_{2(0)} x_3$	One 1
	$y_{3(0)} = y_{2(0)} x_3'$	No 1s
Bit 4 cell	$y_{4(1)} = y_{3(1)} x_4' + y_{3(0)} x_4$	One 1 to assert output z_1
	$y_{4(0)} = y_{3(0)} x_4'$	No 1s (5.9)

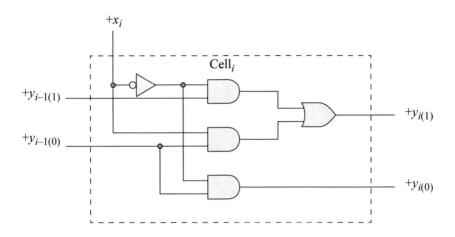

Figure 5.64   Typical cell for the single-bit detection iterative network of Example 5.13.

Referring to Equation 5.9 for the bit 1 cell — which is the leftmost cell — a 1 bit is detected by that cell if the input to the cell is a 1; that is, $x_1 = 1$, as indicated by the term $y_{1(1)} = x_1$. Conversely, if the input to cell$_1$ is a 0, then this is indicated by the term $y_{1(0)} = x_1{}'$. In a similar manner, there will be a single 1 up to and including cell$_2$ if cell$_1$ has an input of $x_1 = 1$ and cell$_2$ has an input of $x_2 = 0$ as indicated by the term $y_{1(1)} x_2{}'$ or if cell$_1$ has an input of $x_1 = 0$ and cell$_2$ has an input of $x_2 = 1$ as indicated by the term $y_{i(0)} x_2$. Therefore, the equation for a single 1 bit detected up to and including cell$_2$ is

$$y_{2(1)} = y_{1(1)}\, x_2{}' + y_{1(0)} x_2$$

The same rationale can be applied to the remaining cells. For example, bit 3 cell will indicate that a single 1 has been detected up to and including cell$_3$ if cell$_2$ has an output of $y_{2(1)}$ and $x_3 = 0$ or if cell$_2$ has an output of $y_{2(0)}$ and $x_3 = 1$. The equation for a single 1 detected up to and including cell$_3$ is

$$y_{3(1)} = y_{2(1)}\, x_3{}' + y_{2(0)} x_3$$

The $y_{i(1)}$ and $y_{i(0)}$ variables will be converted to nets in the logic diagram. For example, the output of cell$_2$ ($y_{2(1)}$) — indicating that the input vector contained a single 1 bit up to and including cell$_2$ — is specified as net 6. The output of cell$_2$ ($y_{2(0)}$) — indicating that the input vector contained only 0s up to and including cell$_2$ — is specified as net 5.

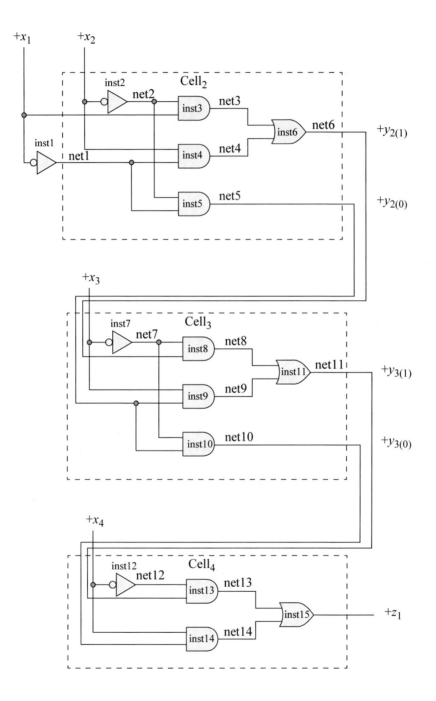

Figure 5.65   Iterative network to detect a single 1 bit in an input vector $x_1x_2x_3x_4$.

Referring to the iterative network of Figure 5.65, the leftmost cell ($cell_1$) is simply an inverter. $Cell_2$ and $cell_3$ are identical, but $cell_4$ is different than the other cells. The output of $cell_4$, indicating that the input vector contained a single 1 bit, is z_1 from instantiation *inst15*. Output z_1 will be a logical 1 if the previous cells detected a single 1 bit and $x_4 = 0$ or if the previous cells detected all 0s and $x_4 = 1$.

The Verilog design module is shown in Figure 5.66 using built-in primitives. The test bench, outputs, and waveforms are shown in Figure 5.67, Figure 5.68, and Figure 5.69, respectively.

```
//single bit detector
module sngl_bit_detect (x1, x2, x3, x4, z1);

input x1, x2, x3, x4;
output z1;


//cell 1 **********************************************
not    inst1    (net1, x1);


//cell 2 **********************************************
not    inst2    (net2, x2);
and    inst3    (net3, net2, x1);
and    inst4    (net4, x2, net1);
and    inst5    (net5, net2, net1);
or     inst6    (net6, net3, net4);


//cell 3 **********************************************
not    inst7    (net7, x3);
and    inst8    (net8, net7, net6);
and    inst9    (net9, x3, net5);
and    inst10   (net10, net7, net5);
or     inst11   (net11, net8, net9);


//cell 4 **********************************************
not    inst12   (net12, x4);
and    inst13   (net13, net12, net11);
and    inst14   (net14, x4, net10);
or     inst15   (z1, net13, net14);

endmodule
```

Figure 5.66   Design module for the iterative network of Figure 5.65.

```
//test bench for single bit detection
module sngl_bit_detect_tb;

reg x1, x2, x3, x4;
wire z1;

initial          //apply input vectors
begin: apply_stimulus
   reg [4:0] invect;
   for (invect=0; invect<16; invect=invect+1)
      begin
          {x1, x2, x3, x4} = invect [4:0];
          #10 $display ("x1x2x3x4 = %b, z1 = %b",
                        {x1, x2, x3, x4}, z1);
      end
end


//instantiate the module into the test bench
sngl_bit_detect inst1 (
   .x1(x1),
   .x2(x2),
   .x3(x3),
   .x4(x4),
   .z1(z1)
   );

endmodule
```

Figure 5.67   Test bench for the iterative network module of Figure 5.66.

```
x1x2x3x4 = 0000, z1 = 0        x1x2x3x4 = 1000, z1 = 1
x1x2x3x4 = 0001, z1 = 1        x1x2x3x4 = 1001, z1 = 0
x1x2x3x4 = 0010, z1 = 1        x1x2x3x4 = 1010, z1 = 0
x1x2x3x4 = 0011, z1 = 0        x1x2x3x4 = 1011, z1 = 0
x1x2x3x4 = 0100, z1 = 1        x1x2x3x4 = 1100, z1 = 0
x1x2x3x4 = 0101, z1 = 0        x1x2x3x4 = 1101, z1 = 0
x1x2x3x4 = 0110, z1 = 0        x1x2x3x4 = 1110, z1 = 0
x1x2x3x4 = 0111, z1 = 0        x1x2x3x4 = 1111, z1 = 0
```

Figure 5.68   Outputs for the iterative network module of Figure 5.66.

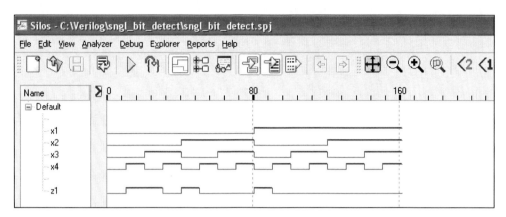

Figure 5.69   Waveforms for the iterative network module of Figure 5.66.

Example 5.14 This repeats Example 5.13 to illustrate an alternative method to design a single-bit detector. In this implementation, all four cells are identical, including the first cell and the fourth cell, as shown in Figure 5.70. The logic diagram is shown in Figure 5.71.

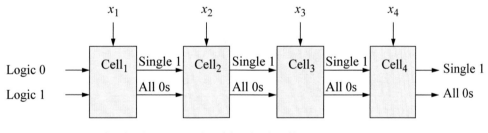

Figure 5.70   Single-detector using identical cells.

The design module is shown in Figure 5.72. The test bench, outputs, and waveforms are shown in Figure 5.73, Figure 5.74, and Figure 5.75, respectively, and are identical to those of Example 5.13.

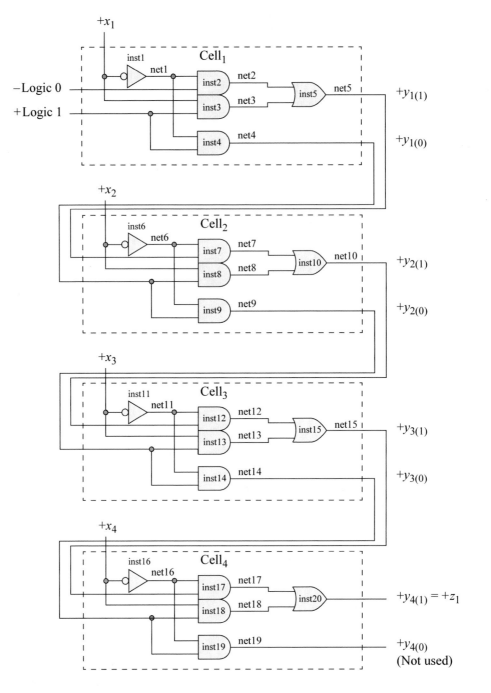

Figure 5.71   Logic diagram for a single-bit detector using four identical cells.

```
//single bit detector
module sngl_bit_detect1 (x1, x2, x3, x4, z1);


input x1, x2, x3, x4;
output z1;


//cell 1 ************************************************
not   inst1    (net1, x1);
and   inst2    (net2, net1, 1'b0);
and   inst3    (net3, x1, 1'b1);
and   inst4    (net4, net1, 1'b1);
or    inst5    (net5, net2, net3);


//cell 2 ************************************************
not   inst6    (net6, x2);
and   inst7    (net7, net6, net5);
and   inst8    (net8, x2, net4);
and   inst9    (net9, net6, net4);
or    inst10   (net10, net7, net8);


//cell 3 ************************************************
not   inst11   (net11, x3);
and   inst12   (net12, net11, net10);
and   inst13   (net13, x3, net9);
and   inst14   (net14, net11, net9);
or    inst15   (net15, net12, net13);


//cell 4 ************************************************
not   inst16   (net16, x4);
and   inst17   (net17, net16, net15);
and   inst18   (net18, x4, net14);
and   inst19   (net19, net16, net14);
or    inst20   (z1, net17, net18);

endmodule
```

Figure 5.72   Design module for the single-bit detection circuit of Example 5.14 using four identical cells.

```
//test bench for single bit detection
module sngl_bit_detect1_tb;

reg x1, x2, x3, x4;
wire z1;


//apply input vectors
initial
begin: apply_stimulus
    reg [4:0] invect;
    for (invect=0; invect<16; invect=invect+1)
        begin
            {x1, x2, x3, x4} = invect [4:0];
            #10 $display ("x1x2x3x4 = %b, z1 = %b",
                            {x1, x2, x3, x4}, z1);
        end
end


//instantiate the module into the test bench
sngl_bit_detect1 inst1 (
    .x1(x1),
    .x2(x2),
    .x3(x3),
    .x4(x4),
    .z1(z1)
    );

endmodule
```

Figure 5.73   Test bench for the single-bit detection module of Figure 5.72.

```
x1x2x3x4 = 0000, z1 = 0      x1x2x3x4 = 1000, z1 = 1
x1x2x3x4 = 0001, z1 = 1      x1x2x3x4 = 1001, z1 = 0
x1x2x3x4 = 0010, z1 = 1      x1x2x3x4 = 1010, z1 = 0
x1x2x3x4 = 0011, z1 = 0      x1x2x3x4 = 1011, z1 = 0
x1x2x3x4 = 0100, z1 = 1      x1x2x3x4 = 1100, z1 = 0
x1x2x3x4 = 0101, z1 = 0      x1x2x3x4 = 1101, z1 = 0
x1x2x3x4 = 0110, z1 = 0      x1x2x3x4 = 1110, z1 = 0
x1x2x3x4 = 0111, z1 = 0      x1x2x3x4 = 1111, z1 = 0
```

Figure 5.74   Outputs for the single-bit detection module of Figure 5.72.

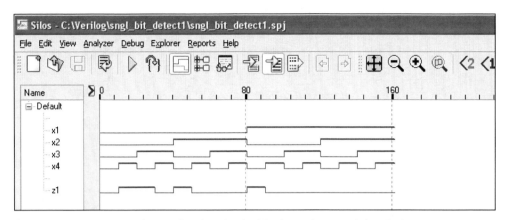

Figure 5.75   Waveforms for the single-bit detection module of Figure 5.72.

Example 5.15   This example demonstrates a third method to design a single-bit detection circuit. In this example, a typical cell will be designed, then instantiated four times into a higher-level module to detect a single bit in a 4-bit input vector $x[1:4]$. Figure 5.76 shows the block diagram of a typical cell and Figure 5.77 shows the internal logic of the cell, which will be instantiated four times into the higher-level circuit of Figure 5.78.

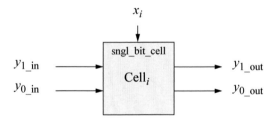

Figure 5.76   Typical cell for a single-bit detection circuit that will be instantiated four times into a higher-level module.

The module for the typical cell is shown in Figure 5.79. The module for the single-bit detection circuit, test bench, outputs, and waveforms are shown in Figure 5.80, Figure 5.81, Figure 5.82, and Figure 5.83, respectively.

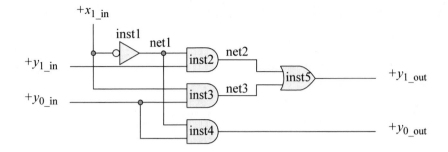

Figure 5.77   Internal logic for a typical cell for the single-bit detection circuit of Example 5.15.

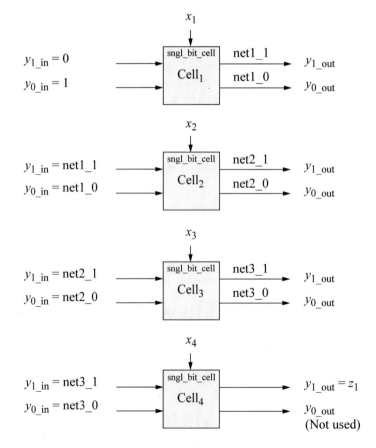

Figure 5.78   Block diagram to detect a single bit in a 4-bit input vector $x[1:4]$.

In Figure 5.78, the input and output lines are defined as follows:

- y_{1_in} is an active-high input line indicating that a single 1 bit was detected up to that cell.

- y_{0_in} is an active-high input line indicating that no 1 bits were detected up to that cell.

- y_{1_out} is an active-high output line indicating that a single 1 bit was detected up to and including that cell.

- y_{0_out} is an active-high output line indicating that no 1 bits were detected up to and including that cell.

```
//typical cell for single-bit detection
module sngl_bit_cell (x1_in, y1_in, y0_in, y1_out, y0_out);

input x1_in, y1_in, y0_in;
output y1_out, y0_out;

not inst1 (net1, x1_in);
and inst2 (net2, net1, y1_in);
and inst3 (net3, x1_in, y0_in);
and inst4 (y0_out, net1, y0_in);
or inst5 (y1_out, net2, net3);

endmodule
```

Figure 5.79   Typical cell that is instantiated four times to detect a single bit in an input vector.

```
//single-bit detection module
//instantiate a typical cell four times
module sngl_bit_detect2 (x1, x2, x3, x4, z1);

input x1, x2, x3, x4;
output z1;

//continued on next page
```

Figure 5.80   Module to detect a single bit in a 4-bit input vector *x[1:4]* in which the typical cell of Figure 5.79 is instantiated four times.

```verilog
//instantiate the single-bit cell modules

//cell 1 *********************************************
sngl_bit_cell  inst1(
   .x1_in(x1),
   .y1_in(1'b0),
   .y0_in(1'b1),
   .y1_out(net1_1),
   .y0_out(net1_0)
   );



//cell 2 *********************************************
sngl_bit_cell  inst2(
   .x1_in(x2),
   .y1_in(net1_1),
   .y0_in(net1_0),
   .y1_out(net2_1),
   .y0_out(net2_0)
   );



//cell 3 *********************************************
sngl_bit_cell  inst3(
   .x1_in(x3),
   .y1_in(net2_1),
   .y0_in(net2_0),
   .y1_out(net3_1),
   .y0_out(net3_0)
   );



//cell 4 *********************************************
sngl_bit_cell  inst4(
   .x1_in(x4),
   .y1_in(net3_1),
   .y0_in(net3_0),
   .y1_out(z1)
   );

endmodule
```

Figure 5.80   (Continued)

```
//test bench for the single-bit detection
//using a typical cell instantiation
module sngl_bit_detect2_tb;

reg x1, x2, x3, x4;
wire z1;


//apply input vectors
initial
begin: apply_stimulus
   reg [4:0] invect;
   for (invect=0; invect<16; invect=invect+1)
      begin
         {x1, x2, x3, x4} = invect [4:0];
         #10 $display ("x1x2x3x4 = %b, z1 = %b",
                        {x1, x2, x3, x4}, z1);
      end
end


//instantiate the module into the test bench
sngl_bit_detect2 inst1 (
   .x1(x1),
   .x2(x2),
   .x3(x3),
   .x4(x4),
   .z1(z1)
   );

endmodule
```

Figure 5.81   Test bench for the single bit detection module of Figure 5.80.

```
x1x2x3x4 = 0000, z1 = 0       x1x2x3x4 = 1000, z1 = 1
x1x2x3x4 = 0001, z1 = 1       x1x2x3x4 = 1001, z1 = 0
x1x2x3x4 = 0010, z1 = 1       x1x2x3x4 = 1010, z1 = 0
x1x2x3x4 = 0011, z1 = 0       x1x2x3x4 = 1011, z1 = 0
x1x2x3x4 = 0100, z1 = 1       x1x2x3x4 = 1100, z1 = 0
x1x2x3x4 = 0101, z1 = 0       x1x2x3x4 = 1101, z1 = 0
x1x2x3x4 = 0110, z1 = 0       x1x2x3x4 = 1110, z1 = 0
x1x2x3x4 = 0111, z1 = 0       x1x2x3x4 = 1111, z1 = 0
```

Figure 5.82   Outputs for the single-bit detector module of Figure 5.80.

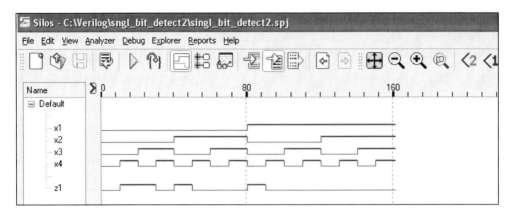

Figure 5.83   Waveforms for the single-bit detector module of Figure 5.80.

5.3.2 Priority Encoder

Unlike a regular encoder, a *priority encoder* can have more than one input asserted simultaneously. Priority encoders are typically used in a time-sharing system. The outputs indicate the input that has the highest priority; that is, if two lines x_i and x_j, where $i > j$, request service simultaneously, then line x_i has priority over line x_j and the outputs will indicate the binary number corresponding to i. In general, there are n input lines and $\log_2 n$ output lines, as shown below.

$$\log_2 n = y$$
$$2^y = n$$

If there are eight input lines, then there will be three output lines encoded in binary to indicate the highest priority input, as shown below.

$$\log_2 8 = y$$
$$2^y = 8$$
$$y = 3$$

A block diagram of an 8-input, 3-output priority encoder is shown in Figure 5.84 which, contains an *enable* input and a *valid request* output. The truth table is shown in Table 5.8. The outputs of the truth table correspond to the highest asserted input, where $z_4 = 2^2$, $z_2 = 2^1$, and $z_1 = 2^0$. The "don't care" entries are indicated by a dash (–). Refer to the third row where x_2 is requesting service. Whether or not inputs x_0 and x_1 are requesting service is immaterial because x_2 has the highest priority. Therefore, dashes are placed in the columns representing x_0 and x_1.

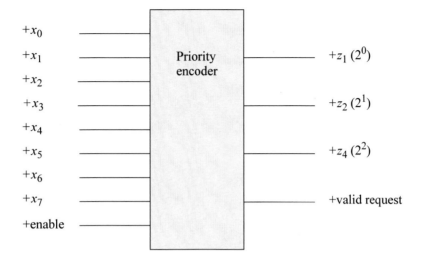

Figure 5.84   Block diagram for an 8-input priority encoder.

Table 5.8 Truth Table for an 8-Input Priority Encoder

x_0	x_1	x_2	x_3	x_4	x_5	x_6	x_7	$z_4 (2^2)$	$z_2 (2^1)$	$z_1 (2^0)$
1	0	0	0	0	0	0	0	0	0	0
–	1	0	0	0	0	0	0	0	0	1
–	–	1	0	0	0	0	0	0	1	0
–	–	–	1	0	0	0	0	0	1	1
–	–	–	–	1	0	0	0	1	0	0
–	–	–	–	–	1	0	0	1	0	1
–	–	–	–	–	–	1	0	1	1	0
–	–	–	–	–	–	–	1	1	1	1

The equations representing the outputs are shown in Equation 5.10 as derived from the truth table. The logic diagram for the 8-input priority encoder is shown in Figure 5.85. The design module, test bench, outputs, and waveforms are shown in Figure 5.86, Figure 5.87, Figure 5.88, and Figure 5.89, respectively. Only a few of the 2^8 = 256 combinations of the inputs will be entered in the test bench.

$$z_1 (2^0) = x_1 x_2' x_3' x_4' x_5' x_6' x_7' + x_3 x_4' x_5' x_6' x_7' + x_5 x_6' x_7' + x_7$$

$$= x_1 x_2' x_4' x_6' + x_3 x_4' x_6' + x_5 x_6' + x_7$$

$$z_2 (2^1) = x_2 x_3' x_4' x_5' x_6' x_7' + x_3 x_4' x_5' x_6' x_7' + x_6 x_7' + x_7$$

$$= x_2 x_4' x_5' + x_3 x_4' x_5' + x_6 + x_7$$

$$z_4 (2^2) = x_4 x_5' x_6' x_7' + x_5 x_6' x_7' + x_6 x_7' + x_7$$

$$= x_4 + x_5 + x_6 + x_7 \qquad (5.10)$$

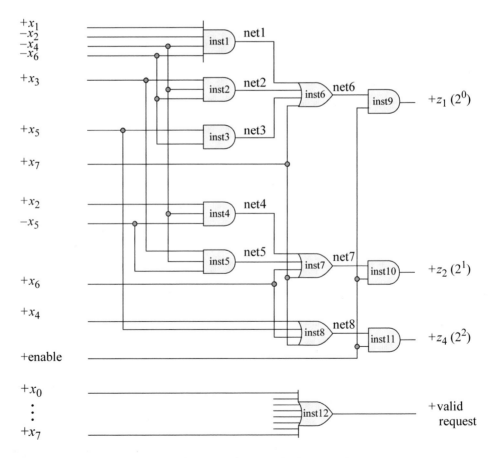

Figure 5.85   Logic diagram for the 8-input priority encoder.

```
//8-bit priority encoder
module priority_encoder (x0, x1, x2, x3, x4, x5, x6, x7,
                         enbl, z1, z2, z4, valid);

input x0, x1, x2, x3, x4, x5, x6, x7, enbl;
output z1, z2, z4, valid;


and   inst1    (net1, x1, ~x2, ~x4, ~x6);
and   inst2    (net2, x3, ~x4, ~x6);
and   inst3    (net3, x5, ~x6);
and   inst4    (net4, x2, ~x4, ~x5);
and   inst5    (net5, x3, ~x4, ~x5);


or    inst6    (net6, net1, net2, net3, x7);
or    inst7    (net7, net4, net5, x6, x7);
or    inst8    (net8, x4, x5, x6, x7);


and   inst9    (z1, net6, enbl);
and   inst10   (z2, net7, enbl);
and   inst11   (z4, net8, enbl);


or    inst12   (valid, x0, x1, x2, x3, x4, x5, x6, x7);

endmodule
```

Figure 5.86   Eight-bit priority encoder module.

```
//test bench for 8-bit priority encoder
module priority_encoder_tb;

reg x0, x1, x2, x3, x4, x5, x6, x7, enbl;
wire z1, z2, z4, valid;

//display variables
initial
$monitor ("x0x1x2x3x4x5x6x7 = %b, z4z2z1 = %b, valid = %b",
          {x0, x1, x2, x3, x4, x5, x6, x7}, {z4, z2, z1}, valid);

//continued on next page
```

Figure 5.87   Test bench for the 8-bit priority encoder module of Figure 5.86.

```
//apply input vectors

initial
begin
   #0      x0=1'b0;    x1=1'b0;    x2=1'b0;    x3=1'b0;
           x4=1'b0;    x5=1'b0;    x6=1'b0;    x7=1'b0;    enbl=1'b1;

   #10     x0=1'b0;    x1=1'b0;    x2=1'b1;    x3=1'b0;
           x4=1'b0;    x5=1'b0;    x6=1'b0;    x7=1'b0;    enbl=1'b1;

   #10     x0=1'b0;    x1=1'b0;    x2=1'b0;    x3=1'b0;
           x4=1'b0;    x5=1'b1;    x6=1'b0;    x7=1'b0;    enbl=1'b1;

   #10     x0=1'b0;    x1=1'b0;    x2=1'b1;    x3=1'b0;
           x4=1'b0;    x5=1'b0;    x6=1'b1;    x7=1'b0;    enbl=1'b1;

   #10     x0=1'b1;    x1=1'b0;    x2=1'b1;    x3=1'b1;
           x4=1'b0;    x5=1'b0;    x6=1'b0;    x7=1'b0;    enbl=1'b1;

   #10     x0=1'b1;    x1=1'b1;    x2=1'b1;    x3=1'b1;
           x4=1'b1;    x5=1'b1;    x6=1'b1;    x7=1'b1;    enbl=1'b1;

   #10     $stop;

end


//instantiate the module into the test bench
priority_encoder inst1 (
   .x0(x0),
   .x1(x1),
   .x2(x2),
   .x3(x3),
   .x4(x4),
   .x5(x5),
   .x6(x6),
   .x7(x7),
   .enbl(enbl),
   .z1(z1),
   .z2(z2),
   .z4(z4),
   .valid(valid)
   );

endmodule
```

Figure 5.87   (Continued)

```
x0x1x2x3x4x5x6x7 = 10000000, z4z2z1 = 000, valid = 1
x0x1x2x3x4x5x6x7 = 00000000, z4z2z1 = 000, valid = 0
x0x1x2x3x4x5x6x7 = 00100000, z4z2z1 = 010, valid = 1
x0x1x2x3x4x5x6x7 = 00000100, z4z2z1 = 101, valid = 1
x0x1x2x3x4x5x6x7 = 00100010, z4z2z1 = 110, valid = 1
x0x1x2x3x4x5x6x7 = 10110000, z4z2z1 = 011, valid = 1
x0x1x2x3x4x5x6x7 = 11111111, z4z2z1 = 111, valid = 1
```

Figure 5.88   Outputs for the 8-bit priority encoder module of Figure 5.86.

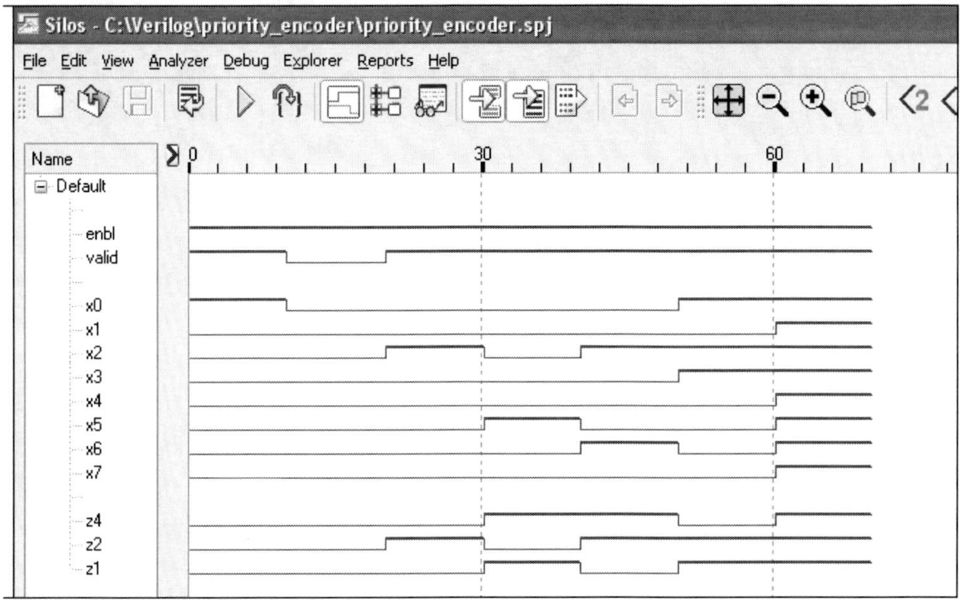

Figure 5.89   Waveforms for the 8-bit priority encoder of Figure 5.86.

5.4 Problems

5.1   Design a logic circuit using built-in primitives that will control the interior
lighting of a building. The building contains four rooms separated by remov-
able partitions. There is one switch in each room which, in conjunction with
the other switches, provides the following methods of control:

(a) All partitions are closed forming separate rooms. Each switch controls the light in its respective room only.

(b) All partitions are open forming one large room. Each switch controls all the lights in the building. That is, when the lights are on, they can be turned off by any switch. Conversely, when the lights are off, they can be turned on by any switch.

(c) Two of the partitions are closed forming three rooms. The middle partition is open, so that the middle room is larger than the other two rooms. Each switch controls the lights in its room only.

(d) The middle partition is closed forming two rooms. Each switch controls the lights in its room only.

The switch control method outlined above is referred to as *four-way switching*. This technique provides control of one set of lights by one or more switches. Design the module using built-in primitives. Obtain the test bench, outputs, and waveforms for several different input vectors.

5.2   Design the circuit for the equation shown below using built-in primitives. Obtain the module, test bench, outputs, and waveforms.

$$z_1 = [x_1 x_2 + (x_1 \oplus x_2)] \, x_3$$

5.3   Design a code converter to convert a 4-bit binary number to the corresponding 4-bit Gray code number. The inputs of the binary number are $x_1 x_2 x_3 x_4$. The outputs for the Gray code are $z_1 z_2 z_3 z_4$. Obtain the module, test bench, outputs, and waveforms.

5.4   Implement the logic function shown below using the fewest number of built-in primitives. Obtain the module, test bench, outputs, and waveforms.

$$z_1 = x_1 x_2' x_3' x_4' + x_1 x_2 x_3' x_4 + x_1' x_2 x_3 x_4 + x_1' x_2' x_3 x_4'$$

5.5   Design a logic circuit using built-in primitives that will convert a 4-bit binary code to a 4-bit excess-3 code. The excess-3 code is obtained by adding three to each binary number. Show only the low-order four bits of the excess-3 code. Obtain the module, test bench, outputs, and waveforms.

5.6   Design an 8-bit odd parity generator using built-in primitives. The output will be a logical 1 if there are an even number of 1s on the input; otherwise, the output will be a logical 0. Obtain the module, test bench, outputs, and wave-forms.

5.7   Implement the equation shown below using built-in primitives. Use x_4 and x_5 as map-entered variables.

$$z_1 = x_1'x_2'x_3'x_4x_5' + x_1'x_2 + x_1'x_2'x_3'x_4x_5 + x_1x_2'x_3'x_4x_5$$

$$+ x_1x_2'x_3 + x_1x_2'x_3'x_4 + x_1x_2'x_3'x_5'$$

5.8   Design a 4-bit binary comparator using built-in primitives. The operands are 4-bit unsigned vectors a[3:0] and b[3:0], as follows:

$$a = a[3] \ a[2] \ a[1] \ a[0]$$
$$b = b[3] \ b[2] \ b[1] \ b[0]$$

where a[0] and b[0] are the low-order bits or a and b, respectively. There are three outputs indicating the relative magnitude of the two operands: (a < b), (a = b), and (a > b). Obtain the design module and test bench. Verify the design by obtaining the outputs for all combinations of the inputs.

5.9   Design a binary-coded decimal (BCD)-to-decimal decoder using built-in primitives. All inputs and outputs are asserted high. Obtain the design module, test bench, outputs, and waveforms

5.10   Use only NAND gates and NOR gates to implement the following function:

$$z_1 = x_1'x_2 + x_3x_4 + (x_1 + x_2)' \ [x_1x_3x_4 + (x_2x_5)']$$

The inputs and output are asserted high. Design the logic without minimizing the equation. Obtain the module, test bench, and outputs. Then minimize the equation and design a new module, test bench (or use the same test bench), and outputs and compare the results.

5.11   Design the necessary logic to implement the truth table shown below. The inputs are available in both logic-high and logic-low levels. The output is asserted high. Design the module and test bench, then obtain the outputs.

x_1	x_2	x_3	x_4	z_1
0	0	0	0	0
0	0	0	1	0
0	0	1	0	1
0	0	1	1	1
0	1	0	0	1
0	1	0	1	0
0	1	1	0	0
0	1	1	1	0
1	0	0	0	1
1	0	0	1	1
1	0	1	0	1
1	0	1	1	1
1	1	0	0	0
1	1	0	1	0
1	1	1	0	0
1	1	1	1	1

5.12   Design a logic circuit using built-in primitives that will generate an output if and only if a 4-bit binary input has a value greater than 12 or less than 3. Design the module and test bench, then obtain the outputs and waveforms.

6

User-Defined Primitives

In addition to built-in primitives, Verilog provides the ability to design primitives according to user specifications. These are called *user-defined primitives* (*UDPs*) and are usually a higher-level logic function than built-in primitives. They are independent primitives and do not instantiate other primitives or modules. UDPs are instantiated into a module the same way as built-in primitives; that is, the syntax for a UDP instantiation is the same as that for a built-in primitive instantiation. A UDP is defined outside the module into which it is instantiated. There are two types of UDPs: combinational and sequential. Sequential primitives include level-sensitive and edge-sensitive circuits

6.1 Defining a User-Defined Primitive

The syntax for a UDP is similar to that for declaring a module. The definition begins with the keyword **primitive** and ends with the keyword **endprimitive**. The UDP contains a name and a list of ports, which are declared as **input** or **output**. For a sequential UDP, the output port is declared as **reg**. UDPs can have one or more scalar inputs, but only one scalar output. The output port is listed first in the terminal list followed by the input ports, in the same way that the terminal list appears in built-in primitives. UDPs do not support **inout** ports.

The UDP table is an essential part of the internal structure and defines the functionality of the circuit. It is a lookup table similar in concept to a truth table. The table begins with the keyword **table** and ends with the keyword **endtable**. The contents of the table define the value of the output with respect to the inputs. The syntax for a UDP is shown below.

```
primitive udp_name (output, input_1, input_2, . . . , input_n);
    output output;
    input input_1, input_2, . . . , input_n;
    reg sequential_output;        //for sequential UDPs

    initial                       //for sequential UDPs

    table
        state table entries
    endtable
endprimitive
```

6.2 Combinational User-Defined Primitives

To illustrate the method for defining and using combinational UDPs, several examples will be presented ranging from simple designs to designs with increasing complexity. UDPs are not compiled separately. They are saved in the same project as the module with a .v extension; for example, *udp_and.v*.

Example 6.1   A 3-input AND gate *udp_and3* will be designed using a UDP. The module is shown in Figure 6.1. The inputs in the state table must be in the same order as in the input list. The table heading is a comment for readability. The inputs and output are separated by a colon and the table entry is terminated by a semicolon. All combinations of the inputs must be entered in the table in order to obtain a correct output; otherwise, the output will be designated as **x**. To completely specify all combinations of the inputs, a value of **x** should be included in the input values where appropriate.

Thus, Figure 6.1 for a 3-input AND gate is incompletely specified. For example, if $x_1 x_2 x_3 = \mathbf{x}00$, output z_1 should equal 0. However, z_1 will equal **x** because the corresponding entry was not listed in the state table. Therefore, all combinations of the inputs — including when $x_i = \mathbf{x}$ — should be entered in the state table, together with the appropriate output value. This is true for both combinational and sequential UDPs. However, in order to avoid lengthy state tables for UDP logic gates, this chapter will provide only the logical functions when all inputs are at a known value. The value **z** is not allowed in a UDP — a **z** value is treated as an **x**.

```
//3-input AND gate as a udp
primitive udp_and3 (z1, x1, x2, x3);//output is listed first

input x1, x2, x3;
output z1;

//state table
table
//inputs are in the same order as the input list
// x1 x2 x3 :   z1;  comment is for readability
    0  0  0  :   0;
    0  0  1  :   0;
    0  1  0  :   0;
    0  1  1  :   0;
    1  0  0  :   0;
    1  0  1  :   0;
    1  1  0  :   0;
    1  1  1  :   1;
endtable
endprimitive
```

Figure 6.1   A UDP for a 3-input AND gate.

Example 6.2   In this example, a sum-of-products expression will be modeled with UDPs. The equation for the output z_1 is shown below and the logic diagram is shown in Figure 6.2.

$$z_1 = x_1 x_2 + x_3 x_4 + x_2' x_3'$$

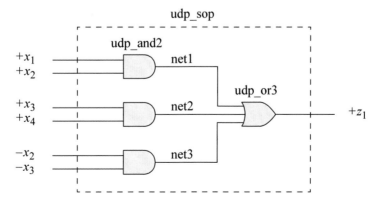

Figure 6.2   A sum-of-products circuit to be implemented with UDPs.

The Karnaugh map for the equation and logic diagram is shown in Figure 6.3, which can be used to verify the outputs obtained from the test bench. UDPs will first be designed for a 2-input AND gate and a 3-input OR gate. These UDPs will then be saved in the project folder *udp_sop* as *udp_and2.v* and *udp_or3.v*. Figure 6.4 shows the Project Properties screen that lists all the source files associated with the *udp_sop* project. The UDPs will then be instantiated into the module *udp_sop*. The Verilog code for the *udp_and2* module is shown in Figure 6.5. The Verilog code for the *udp_or3* module is shown in Figure 6.6. The design module for *udp_sop*, the test bench, the outputs, and the waveforms are shown in Figure 6.7, Figure 6.8, Figure 6.9, and Figure 6.10, respectively.

x_1x_2 \ x_3x_4	0 0	0 1	1 1	1 0
0 0	1	1	1	0
0 1	0	0	1	0
1 1	1	1	1	1
1 0	1	1	1	0

z_1

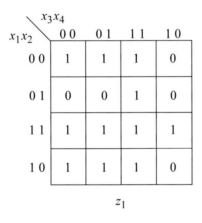

Figure 6.3 Karnaugh map for the sum-of-products implementation using UDPs.

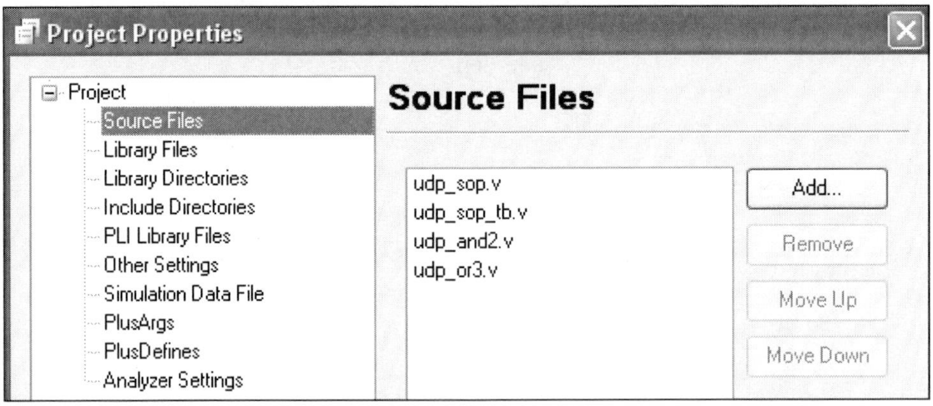

Figure 6.4 The Project Properties screen showing all the source files that are used in the project *udp_sop*.

```
//UDP for a 2-input AND gate
primitive udp_and2 (z1, x1, x2);      //output is listed first

input x1, x2;
output z1;

//define state table
table
//inputs are the same order as the input list
// x1 x2 :  z1;    comment is for readability
   0  0  :  0;
   0  1  :  0;
   1  0  :  0;
   1  1  :  1;
endtable

endprimitive
```

Figure 6.5   UDP for a 2-input AND gate.

```
//UDP for a 3-input OR gate
primitive udp_or3 (z1, x1, x2, x3);  //output is listed first

input x1, x2, x3;
output z1;

//define state table
table
//inputs are the same order as the input list
// x1 x2 x3 :  z1;    comment is for readability
   0  0  0  :  0;
   0  0  1  :  1;
   0  1  0  :  1;
   0  1  1  :  1;
   1  0  0  :  1;
   1  0  1  :  1;
   1  1  0  :  1;
   1  1  1  :  1;
endtable

endprimitive
```

Figure 6.6   UDP for a 3-input OR gate.

```
//sum of products using udps for the AND gate and OR gate
module udp_sop (x1, x2, x3, x4, z1);


input x1, x2, x3, x4;
output z1;


//define internal nets
wire net1, net2, net3;


//instantiate the udps
udp_and2 (net1, x1, x2);
udp_and2 (net2, x3, x4);
udp_and2 (net3, ~x2, ~x3);

udp_or3  (z1, net1, net2, net3);


endmodule
```

Figure 6.7   Module for the sum-of-products logic of Figure 6.2 using UDPs.

```
//test bench for sum of products using udps
module udp_sop_tb;


reg x1, x2, x3, x4;
wire z1;


//apply input vectors
initial
begin: apply_stimulus
   reg [4:0] invect;
   for (invect=0; invect<16; invect=invect+1)
      begin
         {x1, x2, x3, x4} = invect [4:0];
         #10 $display ("x1x2x3x4 = %b, z1 = %b",
                        {x1, x2, x3, x4}, z1);
      end
end

//continued on next page
```

Figure 6.8   Test bench for the sum-of-products module of Figure 6.7.

```
//instantiate the module into the test bench
udp_sop inst1 (
    .x1(x1),
    .x2(x2),
    .x3(x3),
    .x4(x4),
    .z1(z1)
    );

endmodule
```

Figure 6.8   (Continued)

```
x1x2x3x4 = 0000, z1 = 1        x1x2x3x4 = 1000, z1 = 1
x1x2x3x4 = 0001, z1 = 1        x1x2x3x4 = 1001, z1 = 1
x1x2x3x4 = 0010, z1 = 0        x1x2x3x4 = 1010, z1 = 0
x1x2x3x4 = 0011, z1 = 1        x1x2x3x4 = 1011, z1 = 1
x1x2x3x4 = 0100, z1 = 0        x1x2x3x4 = 1100, z1 = 1
x1x2x3x4 = 0101, z1 = 0        x1x2x3x4 = 1101, z1 = 1
x1x2x3x4 = 0110, z1 = 0        x1x2x3x4 = 1110, z1 = 1
x1x2x3x4 = 0111, z1 = 1        x1x2x3x4 = 1111, z1 = 1
```

Figure 6.9   Outputs for the test bench of Figure 6.8 for the sum-of-products module of Figure 6.7.

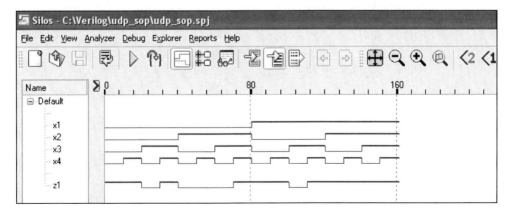

Figure 6.10   Waveforms for the test bench of Figure 6.8 for the sum-of-products module of Figure 6.7.

Example 6.3   This example will design a logic circuit to activate only segment a of the 7-segment light-emitting diode (LED) shown in Figure 6.11. The inputs to the circuit are x_1, x_2, x_3, and x_4, where x_4 is the low-order bit. The logic will be implemented in a sum-of-products form. The truth table for segment a is shown in Table 6.1 and the Karnaugh map is shown in Figure 6.12, in which the entries for minterm locations 10 through 15 must be explicitly defined even though they are "don't cares" for segment a.

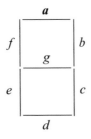

Figure 6.11   Seven-segment LED in which segment a will be implemented with UDPs.

Table 6.1 Truth Table to Activate Segment a of a 7-Segment LED

Decimal Digit	Segment a	
0	1	
1	0	
2	1	
3	1	
4	0	
5	1	
6	1	
7	1	
8	1	
9	1	
10	0	All are invalid, but must be
. . .	. . .	entered as 0s, even though they
15	0	are "don't cares."

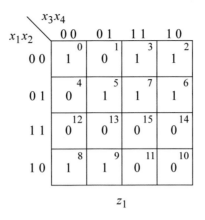

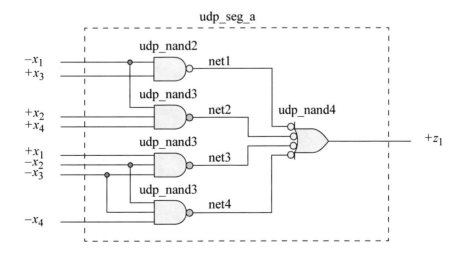

Figure 6.12   Karnaugh map representing the activation of segment *a* of a 7-segment LED.

The equation is shown in Equation 6.1 in both a sum-of-products form and a product-of-sums form. The sum-of-products form will be used in the implementation. There will be three UDPs created: one for a 2-input NAND gate *udp_nand2*, one for a 3-input NAND gate *udp_nand3*, and one for a 4-input NAND gate *udp_nand4* as shown in Figure 6.13. These will then be instantiated into the project folder *udp_seg_a*.

$$z_1 = x_1'x_3 + x_1'x_2x_4 + x_1x_2'x_3' + x_2'x_3'x_4'$$

$$z_1 = (x_1' + x_2')(x_1' + x_3')(x_2' + x_3 + x_4)(x_1 + x_2 + x_3 + x_4') \qquad (6.1)$$

Figure 6.13   Logic circuit to activate segment *a* of the 7-segment LED shown in Figure 6.11.

The UDP modules for the 2-input, 3-input, and 4-input NAND gates are shown in Figure 6.14, Figure 6.15, and Figure 6.16, respectively. The design module into which the UDPs will be instantiated is shown in Figure 6.17. The test bench, outputs, and waveforms are shown in Figure 6.18, Figure 6.19, and Figure 6.20, respectively.

```verilog
//UDP for a 2-input NAND gate
primitive udp_nand2 (z1, x1, x2);

input x1, x2;
output z1;

//define state table
table
//inputs are in the same order as the input list
// x1 x2 :  z1;   comment is for readability
   0  0  :  1;
   0  1  :  1;
   1  0  :  1;
   1  1  :  0;
endtable
endprimitive
```

Figure 6.14   UDP for a 2-input NAND gate.

```verilog
//UDP for a 3-input NAND gate
primitive udp_nand3 (z1, x1, x2, x3);

input x1, x2, x3;
output z1;

//define state table
table
//inputs are in the same order as the input list
// x1 x2 x3 :  z1;   comment is for readability
   0  0  0  :  1;
   0  0  1  :  1;
   0  1  0  :  1;
   0  1  1  :  1;
   1  0  0  :  1;
   1  0  1  :  1;
   1  1  0  :  1;
   1  1  1  :  0;
endtable
endprimitive
```

Figure 6.15   UDP for a 3-input NAND gate.

```
//UDP for a 4-input NAND gate
primitive udp_nand4 (z1, x1, x2, x3, x4);

input x1, x2, x3, x4;
output z1;

//define state table
table
//inputs are in the same order as the input list
// x1 x2 x3 x4 :   z1;   comment is for readability
   0  0  0  0  :   1;
   0  0  0  1  :   1;
   0  0  1  0  :   1;
   0  0  1  1  :   1;
   0  1  0  0  :   1;
   0  1  0  1  :   1;
   0  1  1  0  :   1;
   0  1  1  1  :   1;
   1  0  0  0  :   1;
   1  0  0  1  :   1;
   1  0  1  0  :   1;
   1  0  1  1  :   1;
   1  1  0  0  :   1;
   1  1  0  1  :   1;
   1  1  1  0  :   1;
   1  1  1  1  :   0;
endtable

endprimitive
```

Figure 6.16   UDP for a 4-input NAND gate.

```
//UDPs to design logic to activate segment a of an LED
module udp_seg_a (x1, x2, x3, x4, z1);

input x1, x2, x3, x4;
output z1;

//define internal nets
wire net1, net2, net3, net4;

//continued on next page
```

Figure 6.17   Module with UDPs to activate segment *a* of a 7-segment LED.

```
//instantiate the udps
udp_nand2 (net1, ~x1, x3);
udp_nand3 (net2, ~x1, x2, x4);
udp_nand3 (net3, x1, ~x2, ~x3);
udp_nand3 (net4, ~x2, ~x3, ~x4);

udp_nand4 (z1, net1, net2, net3, net4);

endmodule
```

Figure 6.17 (Continued)

```
//test bench for udp_seg_a
module udp_seg_a_tb;

reg x1, x2, x3, x4;
wire z1;


//apply input vectors
initial
begin: apply_stimulus
   reg [4:0] invect;
   for (invect=0; invect<16; invect=invect+1)
      begin
         {x1, x2, x3, x4} = invect [4:0];
         #10 $display ("x1x2x3x4 = %b, z1 = %b",
                       {x1, x2, x3, x4}, z1);
      end
end


//instantiate the module into the test bench
udp_seg_a inst1 (
   .x1(x1),
   .x2(x2),
   .x3(x3),
   .x4(x4),
   .z1(z1)
   );

endmodule
```

Figure 6.18   Test bench for the module of Figure 6.17.

```
x1x2x3x4 = 0000, z1 = 1        x1x2x3x4 = 1000, z1 = 1
x1x2x3x4 = 0001, z1 = 0        x1x2x3x4 = 1001, z1 = 1
x1x2x3x4 = 0010, z1 = 1        x1x2x3x4 = 1010, z1 = 0
x1x2x3x4 = 0011, z1 = 1        x1x2x3x4 = 1011, z1 = 0
x1x2x3x4 = 0100, z1 = 0        x1x2x3x4 = 1100, z1 = 0
x1x2x3x4 = 0101, z1 = 1        x1x2x3x4 = 1101, z1 = 0
x1x2x3x4 = 0110, z1 = 1        x1x2x3x4 = 1110, z1 = 0
x1x2x3x4 = 0111, z1 = 1        x1x2x3x4 = 1111, z1 = 0
```

Figure 6.19   Outputs for the test bench of Figure 6.18 for the *udp_seg_a* module of Figure 6.17. The output values match the minterm entries in the Karnaugh map of Figure 6.12.

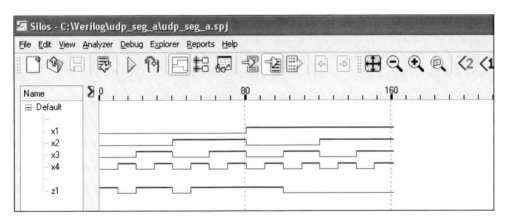

Figure 6.20   Waveforms for the test bench of Figure 6.18 for the *udp_seg_a* module of Figure 6.17.

Example 6.4   A code converter will be designed to convert a 4-bit binary number to the corresponding Gray code number using UDPs. The inputs of the binary number are $b_3 b_2 b_1 b_0$; the outputs for the Gray code are $g_3 g_2 g_1 g_0$. This repeats the code converter from Section 5.5 which used built-in primitives. A procedure for converting from the binary 8421 code to the Gray code can be formulated. Let an *n*-bit binary code word be represented as

$$b_{n-1} b_{n-2} \cdot \cdot \cdot b_1 b_0$$

and an *n*-bit Gray code word be represented as

$$g_{n-1} g_{n-2} \cdot \cdot \cdot g_1 g_0$$

where b_0 and g_0 are the low-order bits of the binary and Gray codes, respectively. The ith Gray code bit g_i can be obtained from the corresponding binary code word by the following algorithm:

$$g_{n-1} = b_{n-1}$$
$$g_i = b_i \oplus b_{i+1}$$

(6.2)

for $0 \leq i \leq n-2$, where the symbol $\oplus$ denotes modulo-2 addition defined as:

$$0 \oplus 0 = 0$$
$$0 \oplus 1 = 1$$
$$1 \oplus 0 = 1$$
$$1 \oplus 1 = 0$$

For example, using the algorithm, the 4-bit binary code word $b_3\, b_2\, b_1\, b_0 = 1010$ translates to the 4-bit Gray code word $g_3\, g_2\, g_1\, g_0 = 1111$ as follows:

$$g_3 = b_3 \qquad\qquad = 1$$
$$g_2 = b_2 \oplus b_3 = 0 \oplus 1 = 1$$
$$g_1 = b_1 \oplus b_2 = 1 \oplus 0 = 1$$
$$g_0 = b_0 \oplus b_1 = 0 \oplus 1 = 1$$

The reverse algorithm to convert from Gray code to binary 8421 code is defined as follows:

$$b_{n-1} = g_{n-1}$$
$$b_i = b_{i+1} \oplus g_i$$

Equation 6.2 indicates that the conversion process can be achieved by repetitive use of exclusive-OR gates. The binary-to-Gray code conversion table is shown in Table 6.2 and the logic diagram is shown in Figure 6.21. In order to implement the conversion module, an exclusive-OR UDP will first be designed as shown in Figure 6.22.

This will be saved as *udp_xor2.v* in the *binary_to_gray_udp* project.   Figure 6.23 shows the design module for the converter; the test bench is shown in Figure 6.24 and the outputs are shown in Figure 6.25.

Table 6.2 Binary-to-Gray Code Conversion

Binary				Gray			
b_3	b_2	b_1	b_0	g_3	g_2	g_1	g_0
0	0	0	0	0	0	0	0
0	0	0	1	0	0	0	1
0	0	1	0	0	0	1	1
0	0	1	1	0	0	1	0
0	1	0	0	0	1	1	0
0	1	0	1	0	1	1	1
0	1	1	0	0	1	0	1
0	1	1	1	0	1	0	0
1	0	0	0	1	1	0	0
1	0	0	1	1	1	0	1
1	0	1	0	1	1	1	1
1	0	1	1	1	1	1	0
1	1	0	0	1	0	1	0
1	1	0	1	1	0	1	1
1	1	1	0	1	0	0	1
1	1	1	1	1	0	0	0

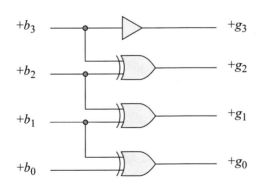

Figure 6.21   Logic diagram for a binary-to-Gray code converter.

```
//UDP for a 2-input exclusive-OR
primitive udp_xor2 (z1, x1, x2);

input x1, x2;
output z1;


//define state table
table
//inputs are in the same order as the input list
// x1 x2 :  z1;     comment is for readability
    0  0  :  0;
    0  1  :  1;
    1  0  :  1;
    1  1  :  0;
endtable

endprimitive
```

Figure 6.22   UDP for an exclusive-OR function.

```
//binary-to-Gray code converter using a UDP
module bin_to_gray_udp (b3, b2, b1, b0, g3, g2, g1, g0);

input b3, b2, b1, b0;
output g3, g2, g1, g0;


//instantiate the udps
buf        (g3, b3);
udp_xor2   (g2, b3, b2);
udp_xor2   (g1, b2, b1);
udp_xor2   (g0, b1, b0);


endmodule
```

Figure 6.23   Module for a binary-to-Gray code converter using an exclusive-OR UDP.

```
//test bench for binary-to-Gray converter
module bin_to_gray_udp_tb;

reg b3, b2, b1, b0;
wire g3, g2, g1, g0;

//apply input vectors
initial
begin: apply_stimulus
   reg [4:0] invect;
   for (invect=0; invect<16; invect=invect+1)
      begin
         {b3, b2, b1, b0} = invect [4:0];
         #10 $display ("b3b2b1b0 = %b, g3g2g1g0 = %b",
                       {b3, b2, b1, b0}, {g3, g2, g1, g0});
      end
end

//instantiate the module into the test bench
bin_to_gray_udp inst1 (
   .b3(b3),
   .b2(b2),
   .b1(b1),
   .b0(b0),
   .g3(g3),
   .g2(g2),
   .g1(g1),
   .g0(g0)
   );

endmodule
```

Figure 6.24   Test bench for the binary-to-Gray code converter of Figure 6.23.

```
b3b2b1b0=0000,  g3g2g1g0=0000     b3b2b1b0=1000,  g3g2g1g0=1100
b3b2b1b0=0001,  g3g2g1g0=0001     b3b2b1b0=1001,  g3g2g1g0=1101
b3b2b1b0=0010,  g3g2g1g0=0011     b3b2b1b0=1010,  g3g2g1g0=1111
b3b2b1b0=0011,  g3g2g1g0=0010     b3b2b1b0=1011,  g3g2g1g0=1110
b3b2b1b0=0100,  g3g2g1g0=0110     b3b2b1b0=1100,  g3g2g1g0=1010
b3b2b1b0=0101,  g3g2g1g0=0111     b3b2b1b0=1101,  g3g2g1g0=1011
b3b2b1b0=0110,  g3g2g1g0=0101     b3b2b1b0=1110,  g3g2g1g0=1001
b3b2b1b0=0111,  g3g2g1g0=0100     b3b2b1b0=1111,  g3g2g1g0=1000
```

Figure 6.25   Outputs for the binary-to-Gray code converter of Figure 6.23.

Example 6.5   This example will use a combination of built-in primitives and UDPs to design a full adder from two half adders. The truth tables for a half adder and full adder are shown in Table 6.3 and Table 6.4, respectively. A *half adder* is a combinational circuit that performs the addition of two operand bits and produces two outputs: a sum bit and a carry-out bit. The half adder does not accommodate a carry-in bit. A *full adder* is a combinational circuit that performs the addition of two operand bits plus a carry-in bit. The carry-in represents the carry-out of the previous lower-order stage. The full adder produces two outputs: a sum bit and a carry-out bit.

Table 6.3 Truth Table for a Half Adder

a	b	sum	carry-out
0	0	0	0
0	1	1	0
1	0	1	0
1	1	0	1

Table 6.4 Truth Table for a Full Adder

a	b	cin	sum	carry-out
0	0	0	0	0
0	0	1	1	0
0	1	0	1	0
0	1	1	0	1
1	0	0	1	0
1	0	1	0	1
1	1	0	0	1
1	1	1	1	1

The sum and carry-out equations for the half adder are shown in Equation 6.3. The sum and carry-out equations for the full adder are shown in Equation 6.4. The logic diagram for a full adder obtained from two half adders using Equation 6.4 is shown in Figure 6.26.

$$sum = a'b + ab'$$
$$= a \oplus b$$
$$carry\text{-}out = ab \tag{6.3}$$

$$sum = a'b'cin + a'bcin' + ab'cin' + abcin$$
$$= a \oplus b \oplus cin$$
$$carry\text{-}out = a'bcin + ab'cin + abcin' + abcin$$
$$= ab + acin + bcin \tag{6.4}$$

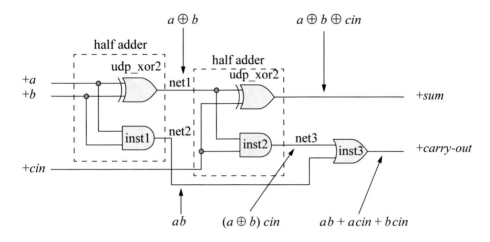

Figure 6.26   Full adder designed from two half adders.

The equation for *carry-out* can also be obtained by plotting Table 6.4 on a Karnaugh map, as shown in Figure 6.27. The equation is then easily obtained in a sum-of-products notation as: $ab + acin + bcin$.

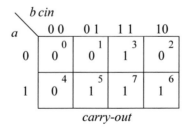

Figure 6.27   Karnaugh map for the carry-out of a full adder.

The full adder will be designed by means of a UDP for the exclusive-OR gates and built-in primitives for the AND gates and OR gate, all of which will be instantiated into the project *full_adder_udp*. The module for the *udp_xor2* is shown in Figure 6.28. The *full_adder_udp* module is shown in Figure 6.29, the test bench is shown in Figure 6.30, and the outputs are shown in Figure 6.31.

```
//UDP for a 2-input exclusive-OR
primitive udp_xor2 (z1, x1, x2);

input x1, x2;
output z1;

//define state table
table
//inputs are in the same order as the input list
// x1 x2 :   z1;      comment is for readability
   0  0  :   0;
   0  1  :   1;
   1  0  :   1;
   1  1  :   0;
endtable

endprimitive
```

Figure 6.28   Module for the *udp_xor2* to be instantiated into the full adder module *full_adder_udp*.

```
//full adder using a UDP and built-in primitives
module full_adder_udp (a, b, cin, sum, cout);

input a, b, cin;
output sum, cout;

//define internal nets
wire net1, net2, net3;

//instantiate the udps and built-in primitive
udp_xor2 (net1, a, b);
and inst1 (net2, a, b);

udp_xor2 (sum, net1, cin);
and inst2 (net3, net1, cin);

or inst3 (cout, net3, net2);

endmodule
```

Figure 6.29   Module for a full adder using a UDP and built-in primitives.

```
//test bench for full adder
module full_adder_udp_tb;

reg a, b, cin;
wire sum, cout;


//apply input vectors
initial
begin: apply_stimulus
   reg [3:0] invect;
   for (invect=0; invect<8; invect=invect+1)
      begin
         {a, b, cin} = invect [3:0];
         #10 $display ("a b cin = %b, sum cout = %b",
                         {a, b, cin}, {sum, cout});
      end
end


//instantiate the module into the test bench
full_adder_udp inst1 (
   .a(a),
   .b(b),
   .cin(cin),
   .sum(sum),
   .cout(cout)
   );

endmodule
```

Figure 6.30   Test bench for the full adder of Figure 6.29.

```
a b cin = 000, sum cout = 00
a b cin = 001, sum cout = 10
a b cin = 010, sum cout = 10
a b cin = 011, sum cout = 01
a b cin = 100, sum cout = 10
a b cin = 101, sum cout = 01
a b cin = 110, sum cout = 01
a b cin = 111, sum cout = 11
```

Figure 6.31   Outputs for the full adder of Figure 6.29.

Example 6.6 This example will design a 4:1 multiplexer as a UDP. The multiplexer will then be checked for correct functional operation by means of a test bench which will generate the outputs and waveforms. A block diagram of the multiplexer is shown in Figure 6.32 together with a table defining the output as a function of the two select inputs s_1 and s_0 and the four data inputs d_0, d_1, d_2, and d_3. The equation for the output can be written directly from the table as shown in Equation 6.5. An *Enable* input may also be associated with a multiplexer to enable the output.

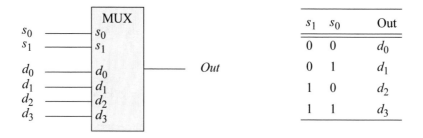

Figure 6.32 A 4:1 multiplexer to be designed as a UDP.

$$Out = s_1's_0'd_0 + s_1's_0d_1 + s_1s_0'd_2 + s_1s_0d_3 \tag{6.5}$$

The 4:1 multiplexer UDP is shown in Figure 6.33. Note the entries in the table that contain the symbol (?), which indicates a "don't care" condition. Referring to the first line in the table, if $s_1s_0 = 00$, then it does not matter what the values are for inputs $d_1d_2d_3$ because only input d_0 is selected.

The test bench for the 4:1 multiplexer is shown in Figure 6.34. The input lines are set to known values such that $d_0d_1d_2d_3 = 1010$. The input values are then displayed using the **$display** system task. The backslash (\) character is used to escape certain special characters such as \n, which is a newline character.

Beginning at 10 time units, the select lines are rotated through all four combinations of the two variables, which in turn transmit the input values to the output. For example, if $s_1s_0 = 11$, then the value of input line d_3 is transmitted to the output. When instantiating a UDP module into a test bench — when there is no design module — the ports must be instantiated by position.

The outputs are shown in Figure 6.35 and the waveforms are shown in Figure 6.36. In Figure 6.36, the lines containing a series of **x**s indicate an unknown value because the select inputs are not defined until 10 time units.

```
//4:1 multiplexer as a UDP
primitive udp_mux4 (out, s1, s0, d0, d1, d2, d3);

input s1, s0, d0, d1, d2, d3;
output out;

table     //define state table
//inputs are in the same order as the input list
// s1 s0 d0 d1 d2 d3 :  out      comment is for readability
   0  0  1  ?  ?  ?  :  1;       //? is "don't care"
   0  0  0  ?  ?  ?  :  0;

   0  1  ?  1  ?  ?  :  1;
   0  1  ?  0  ?  ?  :  0;

   1  0  ?  ?  1  ?  :  1;
   1  0  ?  ?  0  ?  :  0;

   1  1  ?  ?  ?  1  :  1;
   1  1  ?  ?  ?  0  :  0;

   ?  ?  0  0  0  0  :  0;
   ?  ?  1  1  1  1  :  1;
endtable
endprimitive
```

Figure 6.33   A UDP for a 4:1 multiplexer.

```
//test bench for the 4:1 multiplexer udp
module udp_mux4_tb;

reg s1, s0, d0, d1, d2, d3;
wire out;

initial
begin
//set the input lines to known values
   d0 = 1; d1 = 0; d2 = 1; d3 = 0;

//display the input values
   #10 $display ("d0=%b, d1=%b, d2=%b, d3=%b \n",
                 d0, d1, d2, d3);        // \n is new line
//continued on next page
```

Figure 6.34   Test bench for the UDP 4:1 multiplexer.

```
//select d0 = 1
   s1 = 0; s0 = 0;
   #10 $display ("s1=%b, s0=%b, output=%b \n",
                   s1, s0, out);

//select d1 = 0
   s1 = 0; s0 = 1;
   #10 $display ("s1=%b, s0=%b, output=%b \n",
                   s1, s0, out);

//select d2 = 1
   s1 = 1; s0 = 0;
   #10 $display ("s1=%b, s0=%b, output=%b \n",
                   s1, s0, out);

//select d3 = 0
   s1 = 1; s0 = 1;
   #10 $display ("s1=%b, s0=%b, output=%b \n",
                   s1, s0, out);

   #10 $stop;
end

//instantiate the module into the test bench.
//if instantiating only the primitive with no module,
//then instantiation must be done using positional notation

udp_mux4 inst1 (out, s1, s0, d0, d1, d2, d3);

endmodule
```

Figure 6.34   (Continued)

```
d0=1, d1=0, d2=1, d3=0

s1=0, s0=0, output=1

s1=0, s0=1, output=0

s1=1, s0=0, output=1

s1=1, s0=1, output=0
```

Figure 6.35   Outputs for the UDP 4:1 multiplexer.

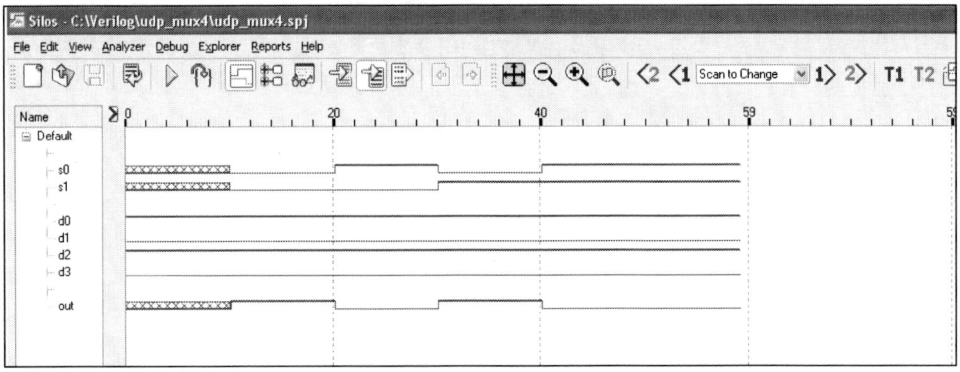

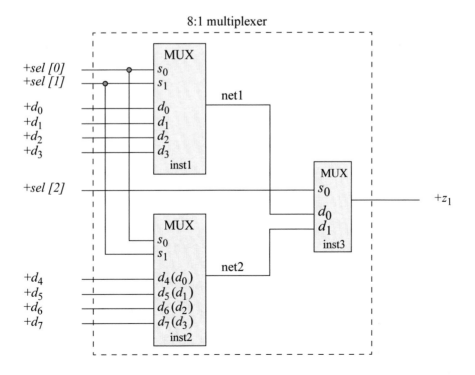

Figure 6.36   Waveforms for the UDP 4:1 multiplexer of Figure 6.33.

Example 6.7   This example will present the design of an 8:1 multiplexer using two UDPs: a 4:1 multiplexer and a 2:1 multiplexer. The logic diagram of the 8:1 multiplexer is shown in Figure 6.37.

Figure 6.37   An 8:1 multiplexer designed from two 4:1 multiplexers and one 2:1 multiplexer.

The 4:1 multiplexer UDP is the same as Figure 6.33. The 2:1 multiplexer UDP is shown in Figure 6.38. The 8:1 multiplexer module, test bench, and outputs for the 8:1 multiplexer are shown in Figure 6.39, Figure 6.40 and Figure 6.41, respectively.

```
//a 2:1 multiplexer as a udp
primitive udp_mux2 (out, s0, d0, d1);

input s0, d0, d1;
output out;

//define state table
table
//inputs are in the same order as the input list
// s0 d0 d1 :   out;  //comment is for readability
    0  0  ?  :   0;
    0  1  ?  :   1;
    1  ?  0  :   0;
    1  ?  1  :   1;
    ?  0  0  :   0;
    ?  1  1  :   1;
endtable

endprimitive
```

Figure 6.38   A UDP for a 2:1 multiplexer.

```
//8:1 multiplexer using two 4:1 multiplexer udps
//and one 2:1 multiplexer UDP

module mux8 (sel, d0, d1, d2, d3, d4, d5, d6, d7, z1);

input [2:0] sel;
input d0, d1, d2, d3, d4, d5, d6, d7;
output z1;

//instantiate the mux udps
udp_mux4 inst1 (net1, sel[1], sel[0], d0, d1, d2, d3);
udp_mux4 inst2 (net2, sel[1], sel[0], d4, d5, d6, d7);
udp_mux2 inst3 (z1, sel[2], net1, net2);

endmodule
```

Figure 6.39   Design module for an 8:1 multiplexer using UDPs.

```
//test bench for 8:1 multiplexer
module mux8_tb;

reg [2:0] sel;
reg d0, d1, d2, d3, d4, d5, d6, d7;
wire z1;

initial
begin
//set the data input lines to known values
   d0=1; d1=0; d2=1; d3=0; d4=0; d5=1; d6=0; d7=1;

//display the input values
   #0 $display ("d0=%b, d1=%b, d2=%b, d3=%b, d4=%b, d5=%b,
                 d6=%b, d7=%b",
                 d0, d1, d2, d3, d4, d5, d6, d7);

//select d0=1
   sel=3'b000;
   #10   $display ("sel=%b, z1=%b", sel, z1);

//select d1=0
   sel=3'b001;
   #10   $display ("sel=%b, z1=%b", sel, z1);

//select d2=1
   sel=3'b010;
   #10   $display ("sel=%b, z1=%b", sel, z1);

//select d3=0
   sel=3'b011;
   #10   $display ("sel=%b, z1=%b", sel, z1);

//select d4=0
   sel=3'b100;
   #10   $display ("sel=%b, z1=%b", sel, z1);

//select d5=1
   sel=3'b101;
   #10   $display ("sel=%b, z1=%b", sel, z1);

//select d6=0
   sel=3'b110;
   #10   $display ("sel=%b, z1=%b", sel, z1);

//continued on next page
```

Figure 6.40   Test bench for the 8:1 multiplexer of Figure 6.39.

```
//select d7=1
   sel=3'b111;
   #10  $display ("sel=%b, z1=%b", sel, z1);

   #10  $stop;

end


//instantiate the module into the test bench
mux8 inst1 (
   .sel(sel),
   .d0(d0),
   .d1(d1),
   .d2(d2),
   .d3(d3),
   .d4(d4),
   .d5(d5),
   .d6(d6),
   .d7(d7),
   .z1(z1)
   );

endmodule
```

Figure 6.40   (Continued)

```
d0=1, d1=0, d2=1, d3=0, d4=0, d5=1, d6=0, d7=1

sel=000, z1=1
sel=001, z1=0
sel=010, z1=1
sel=011, z1=0
sel=100, z1=0
sel=101, z1=1
sel=110, z1=0
sel=111, z1=1
```

Figure 6.41   Outputs for the test bench of Figure 6.40 for the 8:1 multiplexer of Figure 6.39.

Example 6.8 A logic circuit will be designed that is represented by the Karnaugh map shown in Figure 6.42. The equation obtained from the Karnaugh map will be in a product-of-sums notation. A product of sums is an expression in which at least one term does not contain all the variables; that is, at least one term is a proper subset of the possible variables or their complements.

The minimal product-of-sums expression can be obtained by combining the 0s to form sum terms in the same manner as the 1s were combined to form product terms. However, since 0s are being combined, each sum term must equal 0. When combining 0s to obtain sum terms, treat a variable value of 1 as false and a variable value of 0 as true. The product-of-sums expression is shown in Equation 6.6. The logic diagram is shown in Figure 6.43 and will be implemented with UDPs.

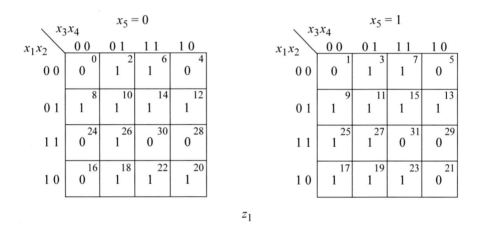

Figure 6.42 Karnaugh map for Example 6.8.

$$z_1 = (x_1' + x_2' + x_3')(x_1 + x_2 + x_4)(x_1' + x_3 + x_4 + x_5)(x_1' + x_3' + x_4 + x_5') \quad (6.6)$$

UDPs will be designed for a 3-input OR gate (*udp_or3*), a 4-input OR gate (*udp_or4*), and a 4-input AND gate (*udp_and4*). These modules are shown in Figure 6.44, Figure 6.45, and Figure 6.46 and will be instantiated into the design module (*pos_udp*), which is shown in Figure 6.47. The test bench is shown in Figure 6.48 and the outputs are shown in Figure 6.49. The values of the minterm locations specified in the outputs conform to the values of the identical minterm locations in the Karnaugh map.

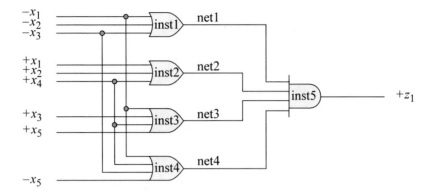

Figure 6.43   Logic diagram for Equation 6.6 of Example 6.8.

```
//a 3-input OR gate as a udp
primitive udp_or3 (z1, x1, x2, x3);


input x1, x2, x3;
output z1;


//define state table
table
//inputs are in the same order as the input list
// x1 x2 x3 :  z1;   //comment is for readability
   0  0  0  :  0;
   0  0  1  :  1;
   0  1  0  :  1;
   0  1  1  :  1;
   1  0  0  :  1;
   1  0  1  :  1;
   1  1  0  :  1;
   1  1  1  :  1;
endtable

endprimitive
```

Figure 6.44   UDP for a 3-input OR gate.

```
//a 4-input OR gate as a udp
primitive udp_or4 (z1, x1, x2, x3, x4);

input x1, x2, x3, x4;
output z1;


//define state table
table
//inputs are in the same order as the input list
// x1 x2 x3 x4 :  z1;   comment is for readability
   0  0  0  0  :  0;
   0  0  0  1  :  1;
   0  0  1  0  :  1;
   0  0  1  1  :  1;
   0  1  0  0  :  1;
   0  1  0  1  :  1;
   0  1  1  0  :  1;
   0  1  1  1  :  1;
   1  0  0  0  :  1;
   1  0  0  1  :  1;
   1  0  1  0  :  1;
   1  0  1  1  :  1;
   1  1  0  0  :  1;
   1  1  0  1  :  1;
   1  1  1  0  :  1;
   1  1  1  1  :  1;
endtable

endprimitive
```

Figure 6.45   UDP for a 4-input OR gate.

```
//a 4-input AND gate as a udp
primitive udp_and4 (z1, x1, x2, x3, x4);

input x1, x2, x3, x4;
output z1;

//continued on next page
```

Figure 6.46   UDP for a 4-input AND gate.

```
//define state table
table
//inputs are in the same order as the input list
// x1 x2 x3 x4 :   z1;    comment is for readability
   0  0  0  0  :   0;
   0  0  0  1  :   0;
   0  0  1  0  :   0;
   0  0  1  1  :   0;
   0  1  0  0  :   0;
   0  1  0  1  :   0;
   0  1  1  0  :   0;
   0  1  1  1  :   0;
   1  0  0  0  :   0;
   1  0  0  1  :   0;
   1  0  1  0  :   0;
   1  0  1  1  :   0;
   1  1  0  0  :   0;
   1  1  0  1  :   0;
   1  1  1  0  :   0;
   1  1  1  1  :   1;
endtable

endprimitive
```

Figure 6.46   (Continued)

```
//design a product-of-sums expression using udps
module pos_udp (x1, x2, x3, x4, x5, z1);

input x1, x2, x3, x4, x5;
output z1;

//define internal nets
wire net1, net2, net3, net4;

//instantiate the udps
udp_or3 inst1 (net1, ~x1, ~x2, ~x3);
udp_or3 inst2 (net2, x1, x2, x4);
udp_or4 inst3 (net3, ~x1, x3, x4, x5);
udp_or4 inst4 (net4, ~x1, ~x3, x4, ~x5);

udp_and4 inst5 (z1, net1, net2, net3, net4);

endmodule
```

Figure 6.47   Design module for a product-of-sums expression using UDPs.

```
//test bench for sum-of products expression using udps
module pos_udp_tb;

reg x1, x2, x3, x4, x5;
wire z1;

//apply input vectors
initial
begin: apply_stimulus
   reg [5:0] invect;
   for (invect=0; invect<32; invect=invect+1)
      begin
         {x1, x2, x3, x4, x5} = invect [5:0];
         #10 $display ("x1x2x3x4x5 = %b, z1 = %b",
                          {x1, x2, x3, x4, x5}, z1);
      end
end

//instantiate the module into the test bench
pos_udp inst1 (
   .x1(x1),
   .x2(x2),
   .x3(x3),
   .x4(x4),
   .x5(x5),
   .z1(z1)
   );
endmodule
```

Figure 6.48   Test bench for the product-of-sums expression of Figure 6.47.

```
x1x2x3x4x5 = 00000, z1 = 0      x1x2x3x4x5 = 01000, z1 = 1
x1x2x3x4x5 = 00001, z1 = 0      x1x2x3x4x5 = 01001, z1 = 1
x1x2x3x4x5 = 00010, z1 = 1      x1x2x3x4x5 = 01010, z1 = 1
x1x2x3x4x5 = 00011, z1 = 1      x1x2x3x4x5 = 01011, z1 = 1
x1x2x3x4x5 = 00100, z1 = 0      x1x2x3x4x5 = 01100, z1 = 1
x1x2x3x4x5 = 00101, z1 = 0      x1x2x3x4x5 = 01101, z1 = 1
x1x2x3x4x5 = 00110, z1 = 1      x1x2x3x4x5 = 01110, z1 = 1
x1x2x3x4x5 = 00111, z1 = 1      x1x2x3x4x5 = 01111, z1 = 1
                                Continued on next page
```

Figure 6.49   Outputs for the product-of-sums expression of Figure 6.47 obtained from the test bench of Figure 6.48.

```
x1x2x3x4x5 = 00000, z1 = 0        x1x2x3x4x5 = 10000, z1 = 0
x1x2x3x4x5 = 00001, z1 = 0        x1x2x3x4x5 = 10001, z1 = 1
x1x2x3x4x5 = 00010, z1 = 1        x1x2x3x4x5 = 10010, z1 = 1
x1x2x3x4x5 = 00011, z1 = 1        x1x2x3x4x5 = 10011, z1 = 1
x1x2x3x4x5 = 00100, z1 = 0        x1x2x3x4x5 = 10100, z1 = 1
x1x2x3x4x5 = 00101, z1 = 0        x1x2x3x4x5 = 10101, z1 = 0
x1x2x3x4x5 = 00110, z1 = 1        x1x2x3x4x5 = 10110, z1 = 1
x1x2x3x4x5 = 00111, z1 = 1        x1x2x3x4x5 = 10111, z1 = 1
x1x2x3x4x5 = 01000, z1 = 1        x1x2x3x4x5 = 11000, z1 = 0
x1x2x3x4x5 = 01001, z1 = 1        x1x2x3x4x5 = 11001, z1 = 1
x1x2x3x4x5 = 01010, z1 = 1        x1x2x3x4x5 = 11010, z1 = 1
x1x2x3x4x5 = 01011, z1 = 1        x1x2x3x4x5 = 11011, z1 = 1
x1x2x3x4x5 = 01100, z1 = 1        x1x2x3x4x5 = 11100, z1 = 0
x1x2x3x4x5 = 01101, z1 = 1        x1x2x3x4x5 = 11101, z1 = 0
x1x2x3x4x5 = 01110, z1 = 1        x1x2x3x4x5 = 11110, z1 = 0
x1x2x3x4x5 = 01111, z1 = 1        x1x2x3x4x5 = 11111, z1 = 0
```

Figure 6.49   (Continued)

6.2.1 Map-Entered Variables

Variables may also be entered in a Karnaugh map as *map-entered variables*, together with 1s and 0s. A map of this type is more compact than a standard Karnaugh map, but contains the same information. A map containing map-entered variables is particularly useful in analyzing and designing synchronous sequential machines. When variables are entered in a Karnaugh map, two or more squares can be combined only if the squares are adjacent and contain the same variable(s).

Example 6.9   The following Boolean equation will be minimized using a 3-variable Karnaugh map with x_4 as a map-entered variable:

$$z_1(x_1, x_2, x_3, x_4) = x_1 x_2' x_3 x_4' + x_1 x_2 + x_1' x_2' x_3' x_4' + x_1' x_2' x_3' x_4$$

The Karnaugh map of Figure 6.50 shows the above equation with x_4 as a map-entered variable. Note that instead of $2^4 = 16$ squares, the map contains only $2^3 = 8$ squares, because only three variables are used in constructing the map. To facilitate plotting the equation in the map, the variable that is to be entered is shown in parentheses as follows:

$$z_1(x_1,x_2,x_3,x_4) = x_1x_2'x_3(x_4') + x_1x_2 + x_1'x_2'x_3'(x_4') + x_1'x_2'x_3'(x_4)$$

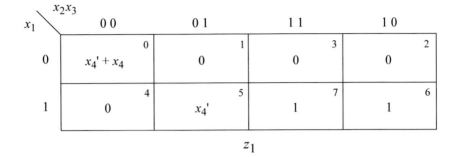

Figure 6.50   Karnaugh map using x_4 as a map-entered variable for Example 6.9.

The first term in the equation for z_1 is $x_1x_2'x_3$ (x_4') and indicates that the variable x_4' is entered in minterm location 5 $(x_1x_2'x_3)$. The second term x_1x_2 is plotted in the usual manner: 1s are entered in minterm locations 6 and 7. The third term specifies that the variable x_4' is entered in minterm location 0 $(x_1'x_2'x_3')$. The fourth term also applies to minterm 0, where x_4 is entered. The expression in minterm location 0, therefore, is $x_4' + x_4$.

To obtain the minimized equation for z_1 in a sum-of-products form, 1s are combined in the usual manner; variables are combined only if the minterm locations containing the variables are adjacent and the variables are identical. Consider the expression $x_4' + x_4$ in minterm location 0. Since $x_4' + x_4 = 1$, minterm 0 equates to x_1' $x_2'x_3'$. The entry of 1 in minterm location 7 can be restated as $1 + x_4'$ without changing the value of the entry. This allows minterm locations 5 and 7 to be combined as $x_1x_3x_4'$. Finally, minterms 6 and 7 combine to yield the term x_1x_2. The minimized equation for z_1 is shown in Equation 6.7.

$$z_1 = x_1'x_2'x_3' + x_1x_3x_4' + x_1x_2 \qquad (6.7)$$

Example 6.10   The Karnaugh map of Figure 6.51 will be implemented using a 4:1 multiplexer and any additional logic. First, the equations for the multiplexer data inputs will be obtained using E as a map-entered variable, where the multiplexer select inputs are $s_1 s_0 = x_1 x_2$. Then the circuit will be designed using UDPs for the multiplexer and associated logic gates.

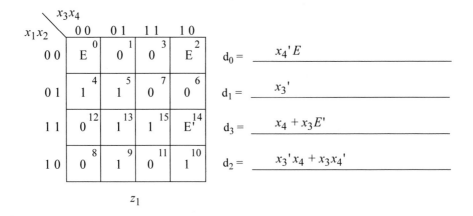

Figure 6.51   Karnaugh map for Example 6.10 using E as a map-entered variable.

To obtain the equation for data input d_0, where $s_1 s_0 = x_1 x_2 = 00$, minterm locations 0 and 2 are adjacent and contain the same variable E; therefore, the term is $x_4' E$. Data input d_1, where $s_1 s_0 = x_1 x_2 = 01$, contains 1s in minterm locations 4 and 5; therefore, $d_1 = x_3'$. To obtain the equation for d_3, where $s_1 s_0 = x_1 x_2 = 11$, minterm locations 13 and 15 combine to yield x_4. Minterm location 15 is equivalent to $1 + E'$; therefore, minterm locations 14 and 15 combine to yield the product term $x_3 E'$. The equation for d_3 is $x_4 + x_3 E'$. Data input d_2 is obtained in a similar manner.

The logic diagram is shown in Figure 6.52 using a 4:1 multiplexer (*udp_mux4*), a 2-input AND gate (*udp_and2*), a 2-input exclusive-OR function (*udp_xor2*), and a 2-input OR gate (*udp_or2*), all of which have been previously designed.

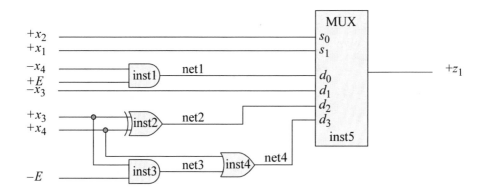

Figure 6.52   Logic diagram for the Karnaugh map of Figure 6.51.

The module for the logic diagram is shown in Figure 6.53 and the test bench is shown in Figure 6.54. The Karnaugh map of Figure 6.51 is expanded to the 5-variable map of Figure 6.55 to better visualize the minterm entries when comparing them with the outputs of Figure 6.56.

```
//logic circuit using a multiplexer udp
//together with other logic gate udps
module mux4_mev (x1, x2, x3, x4, E, z1);

input x1, x2, x3, x4, E;
output z1;

//instantiate the udps
udp_and2 inst1 (net1, ~x4, E);
udp_xor2 inst2 (net2, x3, x4);
udp_and2 inst3 (net3, x3, ~E);
udp_or2  inst4 (net4, x4, net3);

//the mux inputs are: s1, s0, d0, d1, d2, d3
udp_mux4 inst5 (z1, x1, x2, net1, ~x3, net2, net4);

endmodule
```

Figure 6.53   Module for the logic diagram of Figure 6.52.

```
//test bench for mux4_mev
module mux4_mev_tb;
reg x1, x2, x3, x4, E;
wire z1;

initial          //apply input vectors
begin: apply_stimulus
   reg [5:0] invect;
   for (invect=0; invect<32; invect=invect+1)
      begin
         {x1, x2, x3, x4, E} = invect [5:0];
         #10 $display ("x1x2x3x4E = %b, z1 = %b",
                  {x1, x2, x3, x4, E}, z1);
      end
end

//continued on next page
```

Figure 6.54   Test bench for Figure 6.53 for the logic diagram of Figure 6.52.

```
//instantiate the module into the test bench
mux4_mev inst1 (
    .x1(x1),
    .x2(x2),
    .x3(x3),
    .x4(x4),
    .E(E),
    .z1(z1)
    );

endmodule
```

Figure 6.54   (Continued)

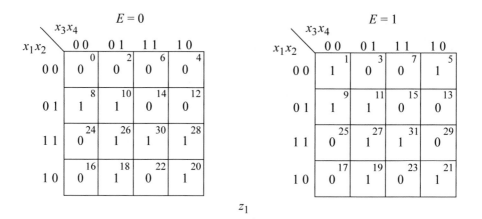

Figure 6.55   Five-variable Karnaugh map equivalent to the 4-variable map of Figure 6.51.

```
x1x2x3x4E = 00000, z1 = 0    x1x2x3x4E = 01000, z1 = 1
x1x2x3x4E = 00001, z1 = 1    x1x2x3x4E = 01001, z1 = 1
x1x2x3x4E = 00010, z1 = 0    x1x2x3x4E = 01010, z1 = 1
x1x2x3x4E = 00011, z1 = 0    x1x2x3x4E = 01011, z1 = 1
x1x2x3x4E = 00100, z1 = 0    x1x2x3x4E = 01100, z1 = 0
x1x2x3x4E = 00101, z1 = 1    x1x2x3x4E = 01101, z1 = 0
x1x2x3x4E = 00110, z1 = 0    x1x2x3x4E = 01110, z1 = 0
x1x2x3x4E = 00111, z1 = 0    x1x2x3x4E = 01111, z1 = 0
                             Continued on next page
```

Figure 6.56   Outputs obtained from the test bench of Figure 6.54 for the module of Figure 6.53.

```
x1x2x3x4E = 10000, z1 = 0      x1x2x3x4E = 11000, z1 = 0
x1x2x3x4E = 10001, z1 = 0      x1x2x3x4E = 11001, z1 = 0
x1x2x3x4E = 10010, z1 = 1      x1x2x3x4E = 11010, z1 = 1
x1x2x3x4E = 10011, z1 = 1      x1x2x3x4E = 11011, z1 = 1
x1x2x3x4E = 10100, z1 = 1      x1x2x3x4E = 11100, z1 = 1
x1x2x3x4E = 10101, z1 = 1      x1x2x3x4E = 11101, z1 = 0
x1x2x3x4E = 10110, z1 = 0      x1x2x3x4E = 11110, z1 = 1
x1x2x3x4E = 10111, z1 = 0      x1x2x3x4E = 11111, z1 = 1
```

Figure 6.56   (Continued)

6.3 Sequential User-Defined Primitives

Verilog provides a means to model sequential UDPs in much the same way as built-in primitives are modeled. Sequential UDPs can be used to model both level-sensitive and edge-sensitive sequential circuits. Level-sensitive behavior is controlled by the value of an input signal; edge-sensitive behavior is controlled by the edge of an input signal. The inputs are implied to be of type **wire**. Sequential devices have an internal state that is a 1-bit register and must be modeled as a type **reg** variable, which is the output of the device and specifies the present state. One **initial** statement can be used to initialize the output of a sequential UDP.

All combinations of the inputs must be applied to the device in order to prevent unknown values (**x**). For synchronous sequential devices, all transitions of the clock must be considered, including the following transitions: $0 \rightarrow 1$, $1 \rightarrow 0$, $0 \rightarrow x$, $1 \rightarrow x$, $x \rightarrow 0$, $x \rightarrow 1$, $0 \rightarrow ?$, $? \rightarrow 0$. If an edge is unspecified, then the output is unknown and is indicated by **x**.

There are three sections in the state table: inputs, present state, and next state, as shown below. Each of the three sections is separated by a colon (:). The first n columns represent the n inputs; the next column represents the present state of the device; the rightmost column signifies the next state.

input_1 input_2 . . . input_n : present_state : next_state;

The input variables can be specified as either levels or edge transitions. The present state is the value of the output register. The next state is a function of the present state and the present inputs and becomes the new value of the output register.

6.3.1 Level-Sensitive User-Defined Primitives

The state — and thus the output — of a level-sensitive device is a function of the input levels only, not on a low-to-high or a high-to-low transition. A latch is an example of a level-sensitive UDP.

Example 6.11 The block diagram of a gated latch is shown in Figure 6.57 and the logic diagram is shown in Figure 6.58. The UDP is shown in Figure 6.59, the test bench is shown in Figure 6.60, the outputs are shown in Figure 6.61, and the waveforms are shown in Figure 6.62.

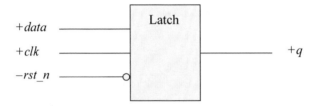

Figure 6.57   Block diagram of a gated level-sensitive latch.

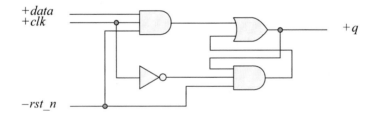

Figure 6.58   Logic diagram for a gated latch to be modeled as a sequential level-sensitive UDP.

```
//a gated latch as a level-sensitive udp
primitive udp_latch_level (q, data, clk, rst_n);
input data, clk, rst_n;
output q;
reg q;        //q is internal storage

initial
   q = 0;     //initialize output q to 0   //continued next page
```

Figure 6.59   Module for a level-sensitive gated latch UDP.

```
//define state table
table
//inputs are in the same order as the input list
// data    clk    rst_n :   q  :  q+;    q+ is next state
    ?       ?       0    :   ?  :  0;     //latch is reset
    0       0       1    :   ?  :  -;     //- means no change
    0       1       1    :   ?  :  0;     //data=0; clk=1; q+=0
    1       0       1    :   ?  :  -;
    1       1       1    :   ?  :  1;     //data=1; clk=1; q+=1
    ?       0       1    :   ?  :  -;
endtable

endprimitive
```

Figure 6.59   (Continued)

```
//test bench for level-sensitive latch
module udp_latch_level_tb;

reg data, clk, rst_n;
wire q;

//display variables
initial
$monitor ("rst_n=%b, data=%b, clk=%b, q=%b",
          rst_n, data, clk, q);

//apply input vectors
initial
begin
   #0     rst_n=1'b0;  data=1'b0;  clk=1'b0;
   #10    rst_n=1'b1;  data=1'b1;  clk=1'b1;
   #10    rst_n=1'b1;  data=1'b1;  clk=1'b0;
   #10    rst_n=1'b1;  data=1'b0;  clk=1'b1;
   #10    rst_n=1'b1;  data=1'b1;  clk=1'b1;
end

//instantiation must be done by position, not by name
udp_latch_level inst1 (q, data, clk, rst_n);

endmodule
```

Figure 6.60   Test bench for the level-sensitive gated latch of Figure 6.59.

```
rst_n=0, data=0, clk=0, q=0
rst_n=1, data=1, clk=1, q=1
rst_n=1, data=1, clk=0, q=1
rst_n=1, data=0, clk=1, q=0
rst_n=1, data=1, clk=1, q=1
```

Figure 6.61   Outputs for the level-sensitive gated latch of Figure 6.59.

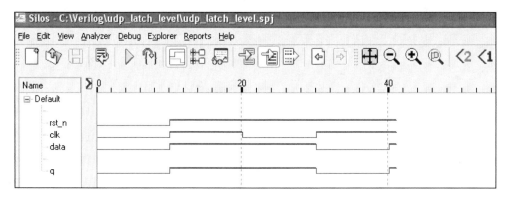

Figure 6.62   Waveforms for the level-sensitive latch of Figure 6.59.

In the primitive module of Figure 6.59, variable *q* is declared as type **reg** because it retains the state of the latch. The **initial** statement sets *q* to a value of 0. The first line of the table indicates that the reset input *rst_n* is active at a low logic level. Therefore, the state of the *data* and *clk* inputs is irrelevant and the latch is reset. In the second line, *data* and *clk* are at a logic 0 and *rst_n* is inactive; therefore, there is no change to the state of the latch. In the third line, *data* = 0, *clk* = 1, and *rst_n* = 1; therefore, the latch is set to a state of 0. In the fourth line, *data* = 1, *clk* = 0, and *rst_n* = 1. Since *clk* = 0, there is no change to the state of the latch. The latch is set to a value of 1 in line five, where *data* = 1, *clk* = 1, and *rst_n* = 1.

The test bench of Figure 6.60 verifies the operation of the latch by first resetting the latch, then setting it by assigning *data clk* = 11, then assigning *data clk* = 10, which indicates a "no change" condition, as can be seen in the outputs of Figure 6.61. The latch is then set to 0 and 1, respectively. When instantiating the primitive module into the test bench, the module ports must be instantiated by position, not by name.

Example 6.12 In this example, an 8-bit level-sensitive UDP register will be designed using the level-sensitive UDP latch of Example 6.11. The register is shown in Figure 6.63 and has the following input signals: an active-high clock signal *clk*, an active-low reset signal *rst_n*, and an 8-bit vector containing active-high data defined as *d[7:0]*. The output from the register will be an 8-bit vector containing active-high data defined as *q[7:0]*, which represents the present state of the register.

The module for the register is shown in Figure 6.64 as *reg8_udp* in which the level-sensitive latch is instantiated into the module eight times. The test bench is shown in Figure 6.65, where eight sets of vectors are applied to the inputs of the register. The outputs are shown in Figure 6.66. When a high level is applied to the *clk* input, the register is set to the value of the data inputs *d[7:0]*. When a low level is applied to the *clk* input, the register retains the value of the previous data. This can be observed in both the outputs and in the waveforms of Figure 6.67.

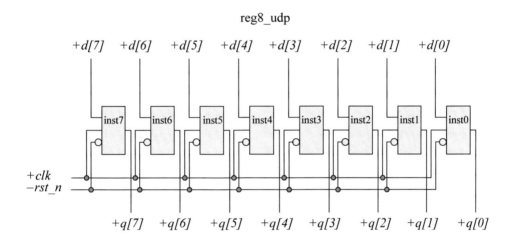

Figure 6.63   An 8-bit register using eight level-sensitive latches instantiated as *udp_latch_level*.

```
//an 8-bit register designed from a level-sensitive
//latch that is modeled as a udp
module reg8_udp (clk, rst_n, d, q);

input clk, rst_n;
input [7:0] d;
output [7:0] q;
//instantiate the udps
udp_latch_level inst7 (q[7], d[7], clk, rst_n);
udp_latch_level inst6 (q[6], d[6], clk, rst_n);
//continued on next page
```

Figure 6.64   Module for the 8-bit register of Example 6.12 using a level-sensitive latch as a UDP.

```
udp_latch_level inst5 (q[5], d[5], clk, rst_n);
udp_latch_level inst4 (q[4], d[4], clk, rst_n);
udp_latch_level inst3 (q[3], d[3], clk, rst_n);
udp_latch_level inst2 (q[2], d[2], clk, rst_n);
udp_latch_level inst1 (q[1], d[1], clk, rst_n);
udp_latch_level inst0 (q[0], d[0], clk, rst_n);

endmodule
```

Figure 6.64   (Continued)

```
//test bench for the 8-bit register using
//level-sensitive latches
module reg8_udp_tb;
reg clk, rst_n;
reg [7:0] d;
wire [7:0] q;

initial      //display variables
$monitor ("rst_n=%b, clk=%b, d=%b, q=%b", rst_n, clk, d, q);

initial      //apply input vectors
begin
    #0      rst_n=1'b0;    clk=1'b0;    d=8'b0000_0000;
    #10     rst_n=1'b1;    clk=1'b1;    d=8'b0101_0101;
    #10     rst_n=1'b1;    clk=1'b0;    d=8'b1010_1010;
    #10     rst_n=1'b1;    clk=1'b1;    d=8'b1010_1010;
    #10     rst_n=1'b1;    clk=1'b0;    d=8'b1111_0000;
    #10     rst_n=1'b1;    clk=1'b1;    d=8'b1111_0000;
    #10     rst_n=1'b1;    clk=1'b0;    d=8'b0000_1111;
    #10     rst_n=1'b1;    clk=1'b1;    d=8'b0000_1111;
    #10     $stop;
end

//instantiate the module into the test bench
reg8_udp inst1 (
    .rst_n(rst_n),
    .clk(clk),
    .d(d),
    .q(q)
    );

endmodule
```

Figure 6.65   Test bench for the 8-bit register of Figure 6.64.

```
rst_n=0, clk=0, d=00000000, q=00000000
rst_n=1, clk=1, d=01010101, q=01010101
rst_n=1, clk=0, d=10101010, q=01010101
rst_n=1, clk=1, d=10101010, q=10101010
rst_n=1, clk=0, d=11110000, q=10101010
rst_n=1, clk=1, d=11110000, q=11110000
rst_n=1, clk=0, d=00001111, q=11110000
rst_n=1, clk=1, d=00001111, q=00001111
```

Figure 6.66   Outputs for the test bench of Figure 6.65 for the 8-bit register of Figure 6.64.

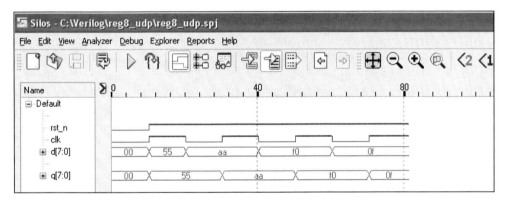

Figure 6.67   Waveforms for the 8-bit register of Figure 6.64.

6.3.2 Edge-Sensitive User-Defined Primitives

Edge-sensitive UDPs can model behavior that is triggered by either a positive edge or a negative edge. The table entries in edge-sensitive circuits are similar to those in level-sensitive circuits. The difference is that a rising or falling edge must be specified on the clock input (or any other input that triggers the circuit).

Level-sensitive and edge-sensitive behavior can be combined in the same UDP table. This allows modeling flip-flops with synchronous behavior for the clock input and asynchronous behavior for the set and reset inputs. If level-sensitive and edge-sensitive events occur simultaneously, then the level-sensitive behavior predominates, as is the case if a clock transition occurs when an asynchronous reset is asserted.

Example 6.13   A positive-edge-triggered *D* flip-flop will be modeled as an edge-sensitive UDP. There are three inputs: *data*, *clk*, and an active-low asynchronous reset *rst_n*. There is one output *q* — when the flip-flop is set, output *q* will be at a high logic level; when the flip-flop is reset, output *q* will be at a low logic level. Figure 6.68 shows the block diagram of the positive-edge-triggered *D* flip-flop. The symbol (>) on the clock input indicates a positive-edge-triggered device. A negative-edge-triggered device is indicated by the symbol (>) preceded by a low assertion symbol (○).

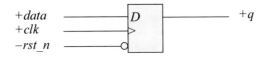

Figure 6.68   Logic diagram for a positive-edge-triggered *D* flip-flop.

The primitive module for the *D* flip-flop is shown in Figure 6.69 as *udp_dff_edge1*. The notation (01) for the clock column specifies a positive transition that will trigger the flip-flop. The notation (0x) signifies no change to the state of the device. The notation (? 0) indicates to ignore the negative edge of the clock because this is a positive-edge-triggered device. The symbol (-) indicates no change to the flip-flop output and (?) specifies that there is no clock edge

The *D* flip-flop test bench is shown in Figure 6.70 in which the clock input is toggled manually. A delay of two time units is applied to the clock to allow for the *D* input setup time. The outputs are shown in Figure 6.71. Figure 6.72 illustrates the waveforms. Refer to the outputs and waveforms for the discussion that follows. When *rst_n* is asserted at a low voltage level, the flip-flop is reset and output *q* is low. When the *d* input is at a logic 1 and *clk* is asserted at time 10, output *q* goes high and remains at a high level when *clk* goes low. At the next positive clock transition at time 30, data is low and this state is propagated to the output. Output *q* remains at a low level until the next positive clock transition at time 50.

```
//a positive-edge-sensitive D flip-flop
primitive udp_dff_edge1 (q, d, clk, rst_n);

input d, clk, rst_n;
output q;

reg q;        //q is internal storage

//continued on next page
```

Figure 6.69   Primitive module for a positive-edge-sensitive *D* flip-flop.

```
//initialize q to 0
initial
   q = 0;



//define state table
table
//inputs are in the same order as the input list
// d      clk    rst_n :  q  :  q+;     q+ is the next state
   0      (01)    1     :  ?  :  0;      //(01) is rising edge
   1      (01)    1     :  ?  :  1;      //rst_n = 1 means no rst
   1      (0x)    1     :  1  :  1;      //(0x) is no change
   0      (0x)    1     :  0  :  0;
   ?      (?0)    1     :  ?  :  -;      //ignore negative edge
//reset case when rst_n is 0 and clk has any transition
   ?      (??)    0     :  ?  :  0;      //rst_n = 0 means reset
//reset case when rst_n is 0.  d & clk can be anything, q+=0
   ?       ?      0     :  ?  :  0;
//reset case when 0 --> 1 transition on rst_n.  Hold q+ state
   ?       ?     (01)   :  ?  :  -;
//non-reset case when d has any trans, but clk has no trans
  (??)     ?      1     :  ?  :  -;      //clk = ?, means no edge

endtable

endprimitive
```

Figure 6.69   (Continued)

```
//test bench for the positive-edge-triggered D flip-flop
module udp_dff_edge1_tb;

reg d, clk, rst_n;
wire q;

//display variables
initial
$monitor ("rst_n=%b, d=%b, clk=%b, q=%b",
          rst_n, d, clk, q);

//continued on next page
```

Figure 6.70   Test bench for the *D* flip-flop primitive module of Figure 6.69.

```
//apply input vectors
initial
begin
   #0     rst_n=1'b0;  d=1'b0;      clk=1'b0;
   #10    rst_n=1'b1;  d=1'b1;  #2 clk=1'b1;
   #10    rst_n=1'b1;  d=1'b1;  #2 clk=1'b0;
   #10    rst_n=1'b1;  d=1'b0;  #2 clk=1'b1;
   #10    rst_n=1'b1;  d=1'b1;  #2 clk=1'b0;
   #10    rst_n=1'b1;  d=1'b1;  #2 clk=1'b1;
   #10    rst_n=1'b1;  d=1'b0;  #2 clk=1'b0;
   #10    $stop;
end

//instantiation must be done by position, not by name
udp_dff_edge1 inst1 (q, d, clk, rst_n);
endmodule
```

Figure 6.70   (Continued)

rst_n=0, d=0, clk=0, q=0	rst_n=1, d=1, clk=1, q=0
rst_n=1, d=1, clk=0, q=0	rst_n=1, d=1, clk=0, q=0
rst_n=1, d=1, clk=1, q=1	rst_n=1, d=1, clk=1, q=1
rst_n=1, d=1, clk=0, q=1	rst_n=1, d=0, clk=1, q=1
rst_n=1, d=0, clk=0, q=1	rst_n=1, d=0, clk=0, q=1
rst_n=1, d=0, clk=1, q=0	

Figure 6.71   Outputs for the positive-edge-sensitive D flip-flop primitive module of Figure 6.69.

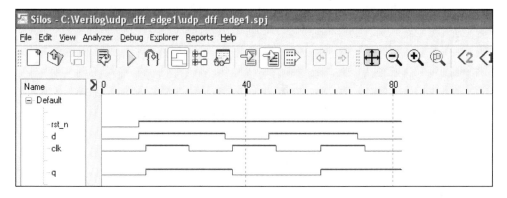

Figure 6.72   Waveforms for the positive-edge-sensitive D flip-flop primitive module of Figure 6.69.

Example 6.14 The edge-sensitive D flip-flop of Example 6.13 will now be used to design a synchronous modulo-8 counter. The counting sequence is: $y_1 y_2 y_3 = 000$, 001, 010, 011, 100, 101, 110, 111, 000, where flip-flop y_3 is the low-order bit.

The state diagram for a modulo-8 counter is shown in Figure 6.73. Unlike conventional Moore and Mealy state diagrams, which detect code words or bit sequences, the state diagram for a counter is relatively straightforward. The counter is initially reset to $y_1 y_2 y_3 = 000$, then increments by one at each positive clock transition until state h ($y_1 y_2 y_3 = 111$) is reached. At the next positive clock transition, the counter sequences to state a ($y_1 y_2 y_3 = 000$).

Table 6.5 lists all possible present states and next states. The next state $y_{k(t+1)}$ for a D flip-flop is the same as the present value of the D input before the flip-flop is clocked. Thus, $y_{i(t+1)} = Dy_{i(t)}$. No binary input variables are required and the counter outputs are taken directly from the state flip-flop outputs.

When designing a modulo-m counter, if the modulus is an integral power of 2, such as $2, 4, 8, 16, \ldots$, then the number of flip-flops required to implement the counter is

$$n = \log_2 m \qquad (6.8)$$

where n is the number of flip-flops and m is the modulus. The logarithm of a positive number with a positive base greater than 1 is the exponent of the power to which the base must be raised to equal the number. From this definition of a logarithm, the following equation is obtained:

$$m = 2^n \qquad (6.9)$$

Equation 6.8 and Equation 6.9 state the same relation in different ways. Thus, for a modulo-8 counter, three flip-flops ($y_1 y_2 y_3$) are required since $8 = 2^3$.

The input maps, or excitation maps, represent the δ next-state logic for the flip-flop inputs. The input maps can be obtained from either the state diagram or the next-state table. Since the counter contains three flip-flops, there will be three input maps, one for each flip-flop. The input maps are shown in Figure 6.74 and specify the input logic for flip-flops y_1, y_2, and y_3.

Referring to the state diagram of Figure 6.73 and beginning in state a ($y_1 y_2 y_3 = 000$), the next state for y_1 is 0. Therefore, a 0 is entered in minterm location 000 for flip-flop y_1. Zeros are also entered in locations 001 and 010 for y_1. In state d ($y_1 y_2 y_3 = 011$), the next state for y_1 is 1. Thus, a 1 is entered in minterm location 011 for y_1. In the same manner, next-state values are entered in the remaining minterm locations for y_1. Similarly, entries for the input maps representing flip-flops y_2 and y_3 are derived. Because the modulus for a modulo-8 counter is an integral power of 2, there are no unused ("don't care") states in the input maps. The input equations for flip-flops y_1, y_2, and y_3 are presented in Equation 6.10.

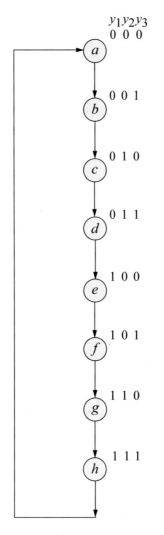

$y_1 y_2 y_3$
0 0 0

0 0 1

0 1 0

0 1 1

1 0 0

1 0 1

1 1 0

1 1 1

Figure 6.73   State diagram for a modulo-8 counter.

Table 6.5 Next-State Table for a Modulo-8 Counter

Present State $y_1 y_2 y_3$	Next State $y_1 y_2 y_3$	Flip-Flop Inputs Dy_1 Dy_2 Dy_3		
0 0 0	0 0 1	0	0	1
0 0 1	0 1 0	0	1	0
0 1 0	0 1 1	0	1	1
0 1 1	1 0 0	1	0	0

Continued on next page

Table 6.5 Next-State Table for a Modulo-8 Counter

Present State $y_1y_2y_3$	Next State $y_1y_2y_3$	Flip-Flop Inputs		
		Dy_1	Dy_2	Dy_3
1 0 0	1 0 1	1	0	1
1 0 1	1 1 0	1	1	0
1 1 0	1 1 1	1	1	1
1 1 1	0 0 0	0	0	0

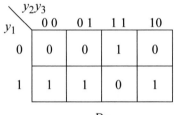

$Dy_1 = y_1y_2' + y_1y_3' + y_1'y_2y_3$

Dy_1

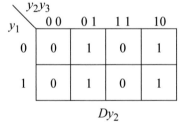

$Dy_2 = y_2'y_3 + y_2y_3' = y_2 \oplus y_3$

Dy_2

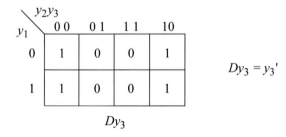

$Dy_3 = y_3'$

Dy_3

Figure 6.74   Input maps for the modulo-8 counter of Example 6.14.

$$Dy_1 = y_1y_2' + y_1y_3' + y_1'y_2y_3$$

$$Dy_2 = y_2'y_3 + y_2y_3'$$

$$= y_2 \oplus y_3$$

$$Dy_3 = y_3' \tag{6.10}$$

The logic diagram is obtained from the input equations of Equation 6.10 and is shown in Figure 6.75. The flip-flop outputs are fed back to the input logic in both a true and complemented form. Note that flip-flop y_3 is connected in toggle mode and that flip-flop y_2 is set to 1 only when $y_2 \neq y_3$. The clock pulse is applied to the clock input of all flip-flops simultaneously.

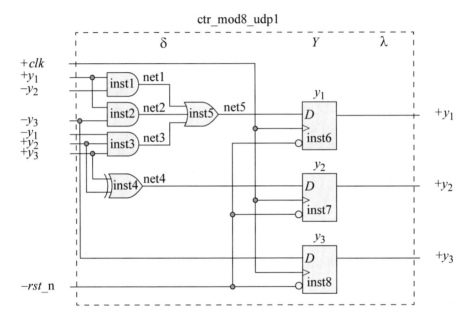

Figure 6.75   Logic diagram for the modulo-8 counter using D flip-flops, where y_3 is the low-order stage.

Figure 6.76 shows the module for the modulo-8 counter using the edge-sensitive UDP D flip-flop of Example 6.13, together with combinational UDPs for the input logic gates. The Verilog code corresponds to the logic diagram of Figure 6.75. Figure 6.77 contains the test bench which asserts and deasserts the clock every 10 time units for a clock period of 20 time units. The length of simulation is determined by the **repeat** keyword, which repeats the clock cycles for 10 clock periods using the positive edge of the clock (**posedge**) as a reference. Figure 6.78 shows the outputs obtained from the test bench, and Figure 6.79 illustrates the waveforms for the modulo-8 counter.

```
//a modulo-8 counter using an edge-sensitive udp
//for a D flip-flop
module ctr_mod8_udp1 (clk, rst_n, y1, y2, y3);

input clk, rst_n;
output y1, y2, y3;

//instantiate the udp logic gates
//for the input logic of the counter
udp_and2 inst1 (net1, y1, ~y2);
udp_and2 inst2 (net2, y1, ~y3);
udp_and3 inst3 (net3, ~y1, y2, y3);
udp_xor2 inst4 (net4, y2, y3);
udp_or3  inst5 (net5, net1, net2, net3);

//instantiate the udp D flip-flops
//the D flip-flop inputs are: d, clk, rst_n
udp_dff_edge1 inst6 (y1, net5, clk, rst_n);
udp_dff_edge1 inst7 (y2, net4, clk, rst_n);
udp_dff_edge1 inst8 (y3, ~y3, clk, rst_n);
endmodule
```

Figure 6.76   Module for the modulo-8 counter using combinational UDPs and edge-sensitive UDPs.

```
//test bench for the modulo-8 counter using a udp
//for a D flip-flop
module ctr_mod8_udp1_tb;

reg clk, rst_n;
wire y1, y2, y3;

//display variables
initial
$monitor ("{y1 y2 y3} = %b", {y1, y2, y3});

//generate reset
initial
begin
   #0 rst_n = 1'b1;
   #2 rst_n = 1'b0;
   #5 rst_n = 1'b1;
end                         //continued on next page
```

Figure 6.77   Test bench for the modulo-8 counter of Figure 6.76.

```
initial                    //generate clock
begin
   clk = 1'b0;
   forever
      #10 clk = ~clk;
end

initial                    //determine length of simulation
begin
   repeat (10) @ (posedge clk);
   $stop;
end

ctr_mod8_udp1 inst1 (    //instantiate the module
   .clk(clk),
   .rst_n(rst_n),
   .y1(y1),
   .y2(y2),
   .y3(y3)
   );
endmodule
```

Figure 6.77   (Continued)

{y1 y2 y3} = 000	{y1 y2 y3} = 100	{y1 y2 y3} = 000
{y1 y2 y3} = 001	{y1 y2 y3} = 101	{y1 y2 y3} = 001
{y1 y2 y3} = 010	{y1 y2 y3} = 110	
{y1 y2 y3} = 011	{y1 y2 y3} = 111	

Figure 6.78   Outputs for the modulo-8 counter of Figure 6.76 for the logic diagram of Figure 6.75.

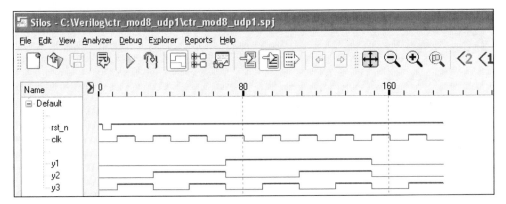

Figure 6.79   Waveforms for the modulo-8 counter of Figure 6.76.

Example 6.15   A 4-bit parallel-in, parallel-out shift register will be designed, which can also rotate right the contents of the register. The storage elements will be the positive-edge-sensitive *D* flip-flop from Example 6.13, labeled *y[3:0]*, where *y[0]* is the low-order flip-flop. There will be four data inputs: *d[3:0]*, where *d[0]* is the low-order bit.

If the asynchronous *load* input is at a high logic level, then the next positive clock transition will load the data inputs *d[3:0]* into flip-flops *y[3:0]*. When the *load* input is at a low logic level, then the clock pulse shifts the contents right one bit position and rotates the contents of the low-order flip-flop into the high-order flip-flop. The register will shift-rotate according to the following sequence:

$$y[3] \rightarrow y[2] \rightarrow y[1] \rightarrow y[0] \rightarrow y[3]$$

There is also an asynchronous reset input *rst_n*. The logic diagram is illustrated in Figure 6.80, showing the instantiation names and net names. The design module, test bench, outputs, and waveforms are shown in Figure 6.81, Figure 6.82, Figure 6.83, and Figure 6.84, respectively. The test bench applies one input vector for *d[3:0]* = 1100, which is loaded into the register, then shifted right one bit position for each clock pulse. This is clearly seen in the outputs and the waveforms.

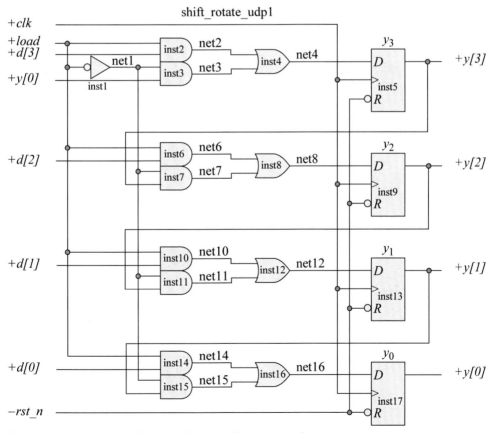

Figure 6.80   Logic diagram for a shift-rotate register.

```
//a parallel-in, parallel-out register that shifts right and
//rotates right

module shift_rotate_udp1 (rst_n, clk, load, d, y);


input rst_n, clk, load;
input [3:0] d;
output [3:0] y;


//flip-flop inputs are: q, d, clk, rst_n

//instantiate flip-flop y[3] *****************************
udp_not        inst1 (net1, load);
udp_and2       inst2 (net2, load, d[3]);
udp_and2       inst3 (net3, net1, y[0]);
udp_or2        inst4 (net4, net2, net3);
udp_dff_edge1  inst5 (y[3], net4, clk, rst_n);


//instantiate flip-flop y[2] *****************************
udp_and2       inst6 (net6, load, d[2]);
udp_and2       inst7 (net7, net1, y[3]);
udp_or2        inst8 (net8, net6, net7);
udp_dff_edge1  inst9 (y[2], net8, clk, rst_n);


//instantiate flip-flop y[1] *****************************
udp_and2       inst10 (net10, load, d[1]);
udp_and2       inst11 (net11, net1, y[2]);
udp_or2        inst12 (net12, net10, net11);
udp_dff_edge1  inst13 (y[1], net12, clk, rst_n);


//instantiate flip-flop y[0] *****************************
udp_and2       inst14 (net14, load, d[0]);
udp_and2       inst15 (net15, net1, y[1]);
udp_or2        inst16 (net16, net14, net15);
udp_dff_edge1  inst17 (y[0], net16, clk, rst_n);


endmodule
```

Figure 6.81   Module for the shift-rotate register of Figure 6.80.

```verilog
//test bench for the shift-rotate register
module shift_rotate_udp1_tb;

reg rst_n, clk, load;
reg [3:0] d;
wire [3:0] y;

//display variables
initial
$monitor ("y = %b", y);

//define reset
initial
begin
   #0 rst_n = 1'b0;
      load = 1'b0;

   #2 rst_n = 1'b1;
      load = 1'b1;
      d = 4'b1100;

   #4 load = 1'b0;
end

//generate clock
initial
begin
   clk = 1'b0;
   forever
      #5 clk = ~clk;
end

initial         //determine length of simulation
begin
   repeat (8) @ (posedge clk);
   $stop;
end

//instantiate the module into the test bench
shift_rotate_udp1 inst1 (
   .rst_n(rst_n),
   .clk(clk),
   .load(load),
   .d(d),
   .y(y)
   );
endmodule
```

Figure 6.82   Test bench for the shift-rotate register module of Figure 6.81.

```
y = 0000
y = 1100
y = 0110
y = 0011
y = 1001
y = 1100
y = 0110
y = 0011
y = 1001
```

Figure 6.83   Outputs for the shift-rotate test bench of Figure 6.82 for the module of Figure 6.81.

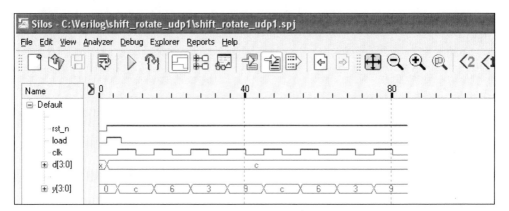

Figure 6.84   Waveforms for the shift-rotate test bench of Figure 6.82 for the module of Figure 6.81.

Figure 6.85 more clearly illustrates the generation of *rst_n*, *load*, *clk* and *d* for the first 12 time units of simulation. Refer to Figure 6.84 and the test bench of Figure 6.82 for the discussion that follows. In the test bench of Figure 6.82, at time 0 *rst_n* is asserted (*rst_n* = 1'b0;) and *load* is deasserted (*load* = 1'b0;). At two time units, *rst_n* is deasserted (*rst_n* = 1'b1;), *load* is asserted (*load* = 1'b1;), and *d[3:0]* = 1100. At six time units (#2 + #4), *load* is deasserted (*load* = 1'b0;).

In the clock block at time 0, *clk* is assigned a value of 0 (*clk* = 1'b0;). Then at five time units, *clk* is set to 1 (#5 *clk* = ~*clk*) and is toggled for eight cycles [**repeat** (8) @ (**posedge** *clk*);].

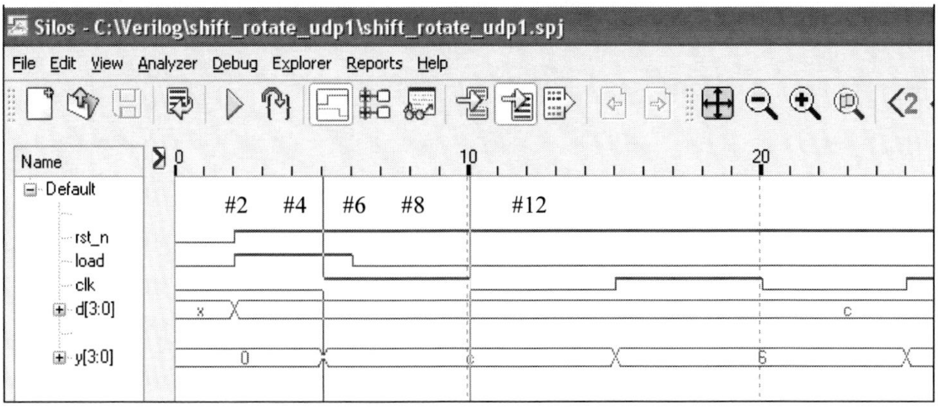

Figure 6.85   Waveforms showing more detail of Figure 6.84 to initialize variables.

Example 6.16   As a final example in this chapter, a synchronous Moore finite-state machine will be designed using the positive-edge-triggered D flip-flop UDP from Example 6.13 together with UDPs for the combinational logic gates.

The state diagram for the machine is shown in Figure 6.86, which depicts the operation of the Moore machine as it sequences through different states depending on the value of the input variables x_1 and x_2. The state flip-flop variables are $y_1 y_2 y_3$ and are initialized to a state code of $y_1 y_2 y_3 = 000$. Outputs z_1, z_2, and z_3 are asserted for one clock period in their respective states.

The Karnaugh maps that represent the input equations for the flip-flops are shown in Figure 6.87 using x_1 and x_2 as map-entered variables. Refer to the state diagram and the Karnaugh map for Dy_1 for the discussion that follows. In state ⓐ ($y_1 y_2 y_3 = 000$), the machine remains in state ⓐ if $x_1 = 0$ and sequences to state ⓑ ($y_1 y_2 y_3 = 011$) if $x_1 = 1$. In both cases, flip-flop y_1 has a next state of $y_1 = 0$. Thus, minterm location 0 contains a value of 0.

In state ⓑ ($y_1 y_2 y_3 = 011$), the machine sequences to state ⓓ ($y_1 y_2 y_3 = 110$) or state ⓒ ($y_1 y_2 y_3 = 101$) depending on the value of x_2. In both cases, flip-flop y_1 has a next value of $y_1 = 1$. Therefore, a value of 1 is entered in minterm location 3 of the map for Dy_1.

Consider the map for Dy_2. In state ⓐ ($y_1 y_2 y_3 = 000$), flip-flop y_2 has a next state of $y_2 = 1$ only if $x_1 = 1$. Therefore, a value of x_1 is entered in minterm location 0 of the map for Dy_2. In state ⓑ ($y_1 y_2 y_3 = 011$), flip-flop y_2 has a next value of 1 only if $x_2 = 0$. Therefore, x_2' is entered in minterm location 3 of the map for Dy_2. The remaining entries for the three maps are determined in a similar manner, with "don't cares" placed in the unused states.

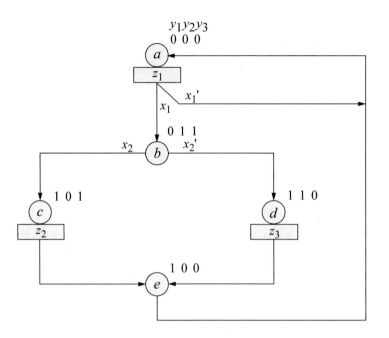

Figure 6.86   State diagram for the Moore synchronous sequential machine of Example 6.16.

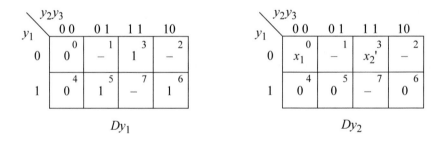

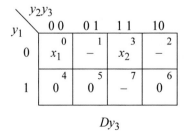

Figure 6.87   Karnaugh maps for the synchronous Moore machine of Example 6.16.

   The equations for the D inputs and the outputs are shown in Equation 6.11 and the logic diagram is shown in Figure 6.88. The design module, test bench, outputs, and waveforms are shown in Figure 6.89, Figure 6.90, Figure 6.91, and Figure 6.92, respectively.

$$Dy_1 = y_2 + y_3$$

$$Dy_2 = y_1'y_2'x_1 + y_1'y_2\,x_2'$$

$$Dy_3 = y_1'y_2'x_1 + y_1'y_2x_2$$

$$z_1 = y_1'y_2'y_3'$$

$$z_2 = y_1y_2'\,y_3$$

$$z_3 = y_1y_2\,y_3' \tag{6.11}$$

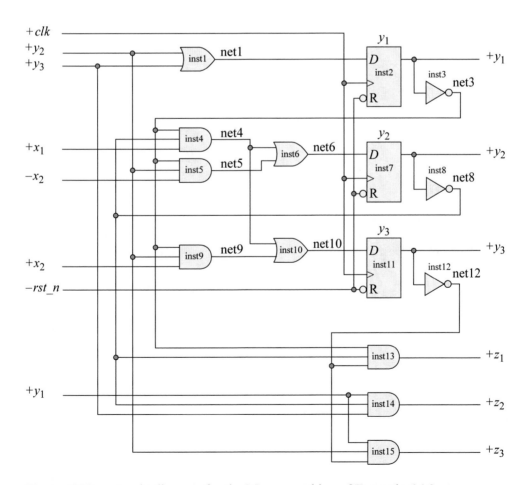

Figure 6.88   Logic diagram for the Moore machine of Example 6.16.

```
//synchronous Moore state machine
module moore1_udp (clk, rst_n, x1, x2, y1, y2, y3, z1, z2, z3);

input clk, rst_n, x1, x2;
output y1, y2, y3, z1, z2, z3;

//D flip-flop ports are: q, d, clk, rst_n

//instantiate the udps for flip-flop y1
udp_or2        inst1     (net1, y2, y3);
udp_dff_edge1  inst2     (y1, net1, clk, rst_n);
udp_not        inst3     (net3, y1);

//instantiate the udps for flip-flop y2
udp_and3       inst4     (net4, net3, net8, x1);
udp_and3       inst5     (net5, net3, y2, ~x2);
udp_or2        inst6     (net6, net4, net5);
udp_dff_edge1  inst7     (y2, net6, clk, rst_n);
udp_not        inst8     (net8, y2);

//instantiate the udps for flip-flop y3
udp_and3       inst9     (net9, net3, y2, x2);
udp_or2        inst10    (net10, net4, net9);
udp_dff_edge1  inst11    (y3, net10, clk, rst_n);
udp_not        inst12    (net12, y3);

//instantiate the output logic
udp_and3       inst13    (z1, net3, net8, net12);
udp_and3       inst14    (z2, y1, net8, y3);
udp_and3       inst15    (z3, y1, y2, net12);

endmodule
```

Figure 6.89   Design module for the Moore machine of Example 6.16.

```
//test bench for the Moore state machine
module moore1_udp_tb;

reg x1, x2, clk, rst_n;
wire y1, y2, y3, z1, z2, z3;

//continued on next page
```

Figure 6.90   Test bench for the synchronous Moore machine of Example 6.16.

```
//display variables
initial
$monitor ("x1=%b, x2=%b, y1y2y3=%b, z1z2z3=%b",
          x1, x2, {y1, y2, y3}, {z1, z2, z3});

//define reset
initial
begin
   #0    rst_n = 1'b0;   //reset
   #5    rst_n = 1'b1;   //no reset
end

//define clock
initial
begin
   clk = 1'b0;
   forever
      #10   clk = ~clk;
end

//define input sequence
initial
begin
   x1 = 1'b0;   x2 = 1'b0;
   @ (posedge clk)      //go to state_a (000)
                        //assert z1

   x1 = 1'b1;   x2 = 1'b0;
   @ (posedge clk)      //go to state_b (011)

   x2 = 1'b1;   x1 = 1'b0;
   @ (posedge clk)      //go to state_c (101)
                        //assert z2

   x2 = 1'b1;   x1 = 1'b0;//sequence is independent of x1, x2
   @ (posedge clk)      //go to state_e (100)

   x1 = 1'b1;   x2 = 1'b0;//sequence is independent of x1, x2
   @ (posedge clk)      //go to state_a (000)
                        //assert z1

   x1 = 1'b1;   x2 = 1'b0;
   @ (posedge clk)      //go to state_b (011)

//continued on next page
```

Figure 6.90   (Continued)

```
    x2 = 1'b0;   x1 = 1'b0;
    @ (posedge clk)        //go to state_d (110)
                           //assert z3

    x1 = 1'b0;   x2 = 1'b0;//sequence is independent of x1, x2
    @ (posedge clk)        //go to state_e (100)

    x2 = 1'b0;   x1 = 1'b1;//sequence is independent of x1, x2
    @ (posedge clk)        //go to state_a (000)
                           //assert z1
    #10    $stop;
end

//instantiate the module into the test bench
moore1_udp inst1 (
    .clk(clk),
    .rst_n(rst_n),
    .x1(x1),
    .x2(x2),
    .y1(y1),
    .y2(y2),
    .y3(y3),
    .z1(z1),
    .z2(z2),
    .z3(z3)
    );

endmodule
```

Figure 6.90   (Continued)

```
x1=0,  x2=0,  y1y2y3=000,  z1z2z3=100
x1=1,  x2=0,  y1y2y3=000,  z1z2z3=100
x1=0,  x2=1,  y1y2y3=011,  z1z2z3=000
x1=0,  x2=1,  y1y2y3=101,  z1z2z3=010
x1=1,  x2=0,  y1y2y3=100,  z1z2z3=000
x1=1,  x2=0,  y1y2y3=000,  z1z2z3=100
x1=0,  x2=0,  y1y2y3=011,  z1z2z3=000
x1=0,  x2=0,  y1y2y3=110,  z1z2z3=001
x1=1,  x2=0,  y1y2y3=100,  z1z2z3=000
x1=1,  x2=0,  y1y2y3=000,  z1z2z3=100
```

Figure 6.91   Outputs for the synchronous Moore machine of Example 6.16.

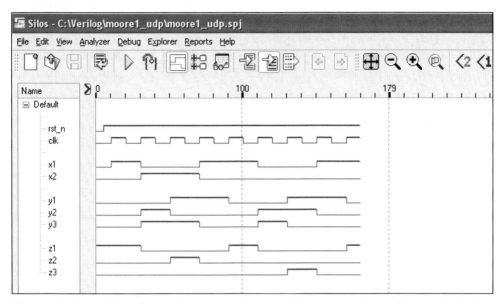

Figure 6.92   Waveforms for the synchronous Moore machine of Example 6.16.

In the test bench of Figure 6.90, inputs x_1 and x_2 are assigned appropriate values to sequence the machine through the various states. The machine is reset to state $\textcircled{a}$ $(y_1y_2y_3 = 000)$. Initially, the inputs are set to values of $x_1 = 0, x_2 = 0$. Then, at the next positive edge of the clock, the machine remains in state $\textcircled{a}$ because $x_1 = 0$.

Then $x_1 = 1, x_2 = 0$ and the machine sequences to state $\textcircled{b}$ at the next positive edge of the clock. The test bench continues to assign values to the inputs until all paths in the state diagram have been tested. The machine sequencing can be seen clearly in the outputs of Figure 6.91.

6.4 Problems

6.1   Design an 8-bit odd parity generator using UDPs. The output will be a logical 1 if there are an even number of 1s on the input; otherwise, the output will be a logical 0. Design the module and test bench, then obtain the outputs and waveforms. Check only 6 of the 256 combinations.

6.2   Design a built-in primitive for a 5-input majority circuit. Create a test bench and obtain the outputs for all combinations of the inputs.

6.3   Design a code converter to convert a BCD number (*abcd*) to the corresponding excess-3 number (*wxyz*). Use NAND logic as user-defined primitives. Obtain the design module, test bench, and outputs. Binary values above 10 are invalid for BCD and should produce outputs of *wxyz* = 0000 for excess-3. Four Karnaugh maps are required, one each for *w*, *x*, *y*, and *z* as a function of *a*, *b*, *c*, and *d*.

6.4   Design the logic circuit shown below using UDPs. Obtain the design module, test bench, and outputs. Verify the outputs by obtaining the equation for z_1 directly from the logic diagram, then use Boolean algebra to obtain the minimal sum-of-products expression.

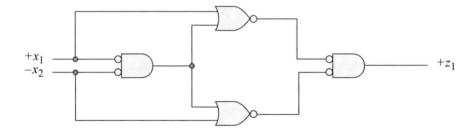

6.5   Design a 3:1 multiplexer using two 2:1 multiplexers that will operate according to the table shown below. Obtain the module, test bench, and outputs.

a	b	Input Selected
0	0	*In_0*
0	1	*In_1*
1	0	*In_2*
1	1	*In_2*

6.6   Obtain the minimal Boolean expression for a logic circuit that generates an output z_1 whenever a 4-bit unsigned binary number *N* meets the following requirements:

N is an even number or N is evenly divisible by three.

The format for *N* is $N = n_3 \, n_2 \, n_1 \, n_0$, where n_0 is the low-order bit. Obtain the design module, test bench, and outputs.

6.7 The logic block shown below generates a logic 1 output if the four inputs contain an odd number of 1s. Use only blocks of this type to design a 9-bit (eight data bits plus parity) odd parity checker. The output of the resulting circuit will be a logic 1 (indicating a parity error) if the parity of the nine bits is even. All inputs must be used. Use the least amount of logic.

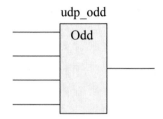

6.8 The logic block shown below generates an output of a logic 1 when the inputs contain an odd number of 1s. Use only this type of logic block to generate a parity bit for an 8-bit byte of data. The parity bit will be a logic 1 when there are an even number of 1s in the byte of data. All inputs must be used. Using UDPs, obtain the design module, test bench, and outputs for eight combinations of the input variables.

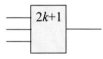

6.9 Obtain the equation for a logic circuit that will generate a logic 1 whenever a 4-bit unsigned binary number N satisfies the following criteria:

$$N = x_1 x_2 x_3 x_4 \text{ (low order)}$$

$$2 < N \le 6$$

$$11 \le N < 14$$

Using UDPs of NOR logic, obtain the design module, test bench, and outputs. Output z_1 will be active high if the above conditions are met.

6.10 Given the Karnaugh map shown below for the function z_1, obtain the minimum expression for z_1 in a sum-of-products form. Use NAND logic as UDPs. Obtain the design module, test bench, and waveforms.

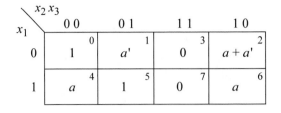

6.11   Design a negative-edge-triggered toggle flip-flop as a UDP. There are two inputs: *clk* and an active-high reset *rst_n*; there is one active-high output *q*. Then use the flip-flop to design a modulo-16 asynchronous counter. Obtain the *T* flip-flop primitive, the design module, test bench, outputs, and waveforms.

6.12   Design a synchronous modulo-10 counter using the *D* flip-flop UDP of Section 6.3.2. Obtain the design module, test bench, outputs, and waveforms.

6.13   Design a synchronous modulo-16 counter using the positive-edge-triggered *D* flip-flop of Section 6.3.2. There will be two scalar inputs: *clk* and *rst_n*; and one vector output: *y[3:0]*. Obtain the design module, test bench, outputs, and waveforms.

6.14   Design a synchronous 4-bit Johnson counter using the positive-edge-triggered *D* flip-flop of Section 6.3.2. A Johnson counter counts in the following sequence: $y_1y_2y_3y_4$ = 0000, 1000, 1100, 1110, 1111, 0111, 0011, 0001, 0000, where y_4 is the low-order bit. Obtain the design module, test bench, outputs, and waveforms.

6.15   Use the positive-edge-triggered UDP *D* flip-flop of Section 6.3.2 to design a Moore machine that operates according to the following equations:

$$Dy_1 = y_2'y_3'x_1 + y_1y_3x_1$$

$$Dy_2 = y_3$$

$$Dy_3 = y_2'$$

Then obtain the test bench, outputs, and waveforms. The variable x_1 is an input to the machine. Test for all variations of input x_1 as shown in the state diagram that follows.

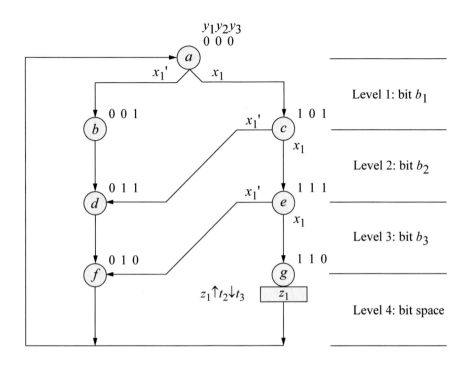

7

Dataflow Modeling

Gate-level modeling is an intuitive approach to digital design because it corresponds one-to-one with conventional digital logic design at the gate level. Dataflow modeling, however, is at a higher level of abstraction than gate-level modeling. Design automation tools are used to create gate-level logic from dataflow modeling by a process called *logic synthesis*. Register transfer level (RTL) is a combination of dataflow modeling and behavioral modeling and characterizes the flow of data through logic circuits.

7.1 Continuous Assignment

The *continuous assignment* statement models dataflow behavior and is used to design combinational logic without using gates and interconnecting nets. Continuous assignment statements provide a Boolean correspondence between the right-hand side expression and the left-hand side target. The continuous assignment statement uses the keyword **assign** and has the following syntax with optional drive strength and delay:

assign [drive_strength] [delay] left-hand side target = right-hand side expression

The continuous assignment statement assigns a value to a net (**wire**) that has been previously declared — it cannot be used to assign a value to a register. Therefore, the left-hand target must be a scalar or vector net or a concatenation of scalar and vector nets. The operands on the right-hand side can be registers, nets, or function calls. The registers and nets can be declared as either scalars or vectors.

The following are examples of continuous assignment statements for scalar nets:

$$\textbf{assign } z_1 = x_1 \text{ \& } x_2 \text{ \& } x_3;$$
$$\textbf{assign } z_1 = x_1 \text{ \textasciicircum } x_2;$$
$$\textbf{assign } z_1 = (x_1 \text{ \& } x_2) \,|\, x_3;$$

The following are examples of continuous assignment statements for vector and scalar nets, where *sum* is a 9-bit vector to accommodate the *sum* and carry-out, *a* and *b* are 8-bit vectors, and *cin* is a scalar:

$$\textbf{assign } \text{sum} = \text{a} + \text{b} + \text{cin}$$
$$\textbf{assign } \text{sum} = \text{a} \wedge \text{b} \wedge \text{cin}$$

The following is an example of a continuous assignment statement for vector nets and a concatenation of a scalar net and a vector net, where *a* and *b* are 4-bit vectors, and *cin* and *cout* are scalars:

$$\textbf{assign } \{\text{cout, sum}\} = \text{a} + \text{b} + \text{cin};$$

The **assign** statement continuously monitors the right-hand side expression. If a variable changes value, then the expression is evaluated and the result is assigned to the target after any specified delay. If no delay is specified, then the default delay is zero. The drive strength defaults to **strong0** and **strong1**. The continuous assignment statement can be considered to be a form of behavioral modeling, because the behavior of the circuit is specified, not the implementation.

7.1.1 Three-Input AND Gate

The AND function of two variables x_1 and x_2 is also called the *conjunction* of x_1 and x_2 and is stated as x_1 *and* x_2. The AND operator, which corresponds to the Boolean product, is indicated by the symbol "•" ($x_1 \bullet x_2$), "∧" ($x_1 \wedge x_2$), or by no symbol $x_1 x_2$ if the operation is unambiguous. Thus, $x_1 x_2$, $x_1 \bullet x_2$, and $x_1 \wedge x_2$ are all read as "x_1 AND x_2". The truth table for a 2-input AND gate is shown in Table 7.1. In Verilog, however, the caret symbol "∧" indicates the exclusive-OR operation.

**Table 7.1 Truth Table
for a 2-Input AND Gate**

x_1	x_2	$x_1 x_2$
0	0	0
0	1	0
1	0	0
1	1	1

Example 7.1 This example uses continuous assignment to design a 3-input AND gate. The inputs are x_1, x_2, and x_3; the output is z_1. The module, test bench, outputs, and waveforms are shown in Figure 7.1, Figure 7.2, Figure 7.3, and Figure 7.4, respectively.

```
//and3 dataflow
module and3_df (x1, x2, x3, z1);

input x1, x2, x3;
output z1;

wire x1, x2, x3;   //define signals as wire for dataflow
wire z1;

//continuous assignment for dataflow
assign z1 = x1 & x2 & x3;

endmodule
```

Figure 7.1 Module for a 3-input AND gate using continuous assignment.

```
//test bench for the 3-input AND gate
module and3_df_tb;

reg x1, x2, x3;         //inputs are reg for test bench
wire z1;                //outputs are wire for test bench

//apply input vectors and display variables
initial
begin: apply_stimulus
   reg [3:0] invect;
      for (invect = 0; invect < 8; invect = invect + 1)
      begin
         {x1, x2, x3} = invect [3:0];
         #10 $display ("x1 x2 x3 = %b, z1 = %b",
                        {x1, x2, x3}, z1);
      end
end

//continued on next page
```

Figure 7.2 Test bench for the 3-input AND gate using continuous assignment.

```
//instantiate the module into the test bench
and3_df inst1 (
    .x1(x1),
    .x2(x2),
    .x3(x3),
    .z1(z1)
    );

endmodule
```

Figure 7.2   (Continued)

```
x1 x2 x3 = 000, z1 = 0
x1 x2 x3 = 001, z1 = 0
x1 x2 x3 = 010, z1 = 0
x1 x2 x3 = 011, z1 = 0
x1 x2 x3 = 100, z1 = 0
x1 x2 x3 = 101, z1 = 0
x1 x2 x3 = 110, z1 = 0
x1 x2 x3 = 111, z1 = 1
```

Figure 7.3   Outputs for the 3-input AND gate using continuous assignment.

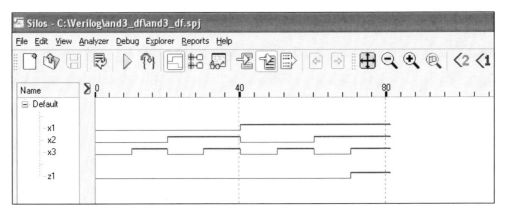

Figure 7.4   Waveforms for the 3-input AND gate using continuous assignment.

7.1.2 Sum of Products

A sum of products is an expression in which at least one term does not contain all the variables; that is, at least one term is a proper subset of the possible variables or their complements. For example,

$$z_1(x_1, x_2, x_3) = x_1'x_2x_3 + x_2'x_3' + x_1\,x_2x_3$$

is a sum of products for the function z_1 because the second term does not contain the variable x_1.

Example 7.2 Given the Karnaugh map shown in Figure 7.5, the logic diagram will be obtained and then implemented using continuous assignment statements. Output z_1 is shown in Equation 7.1. The logic diagram is shown in Figure 7.6. The design module is shown in two implementations in Figure 7.7 and Figure 7.8. The test bench, outputs, and waveforms are shown in Figure 7.9, Figure 7.10, and Figure 7.11, respectively.

x_1x_2 \ x_3x_4	0 0	0 1	1 1	1 0
0 0	(0) 0	(1) 1	(3) 0	(2) 0
0 1	(4) 0	(5) 1	(7) 1	(6) 1
1 1	(12) 0	(13) 1	(15) 1	(14) 1
1 0	(8) 1	(9) 1	(11) 1	(10) 1

z_1

Figure 7.5 Karnaugh map for a logic circuit to be implemented using continuous assignment statements.

$$z_1 = x_1x_2' + x_3'x_4 + x_2x_3 \tag{7.1}$$

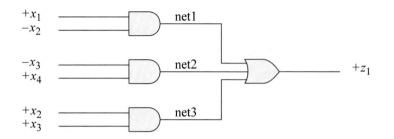

Figure 7.6   Karnaugh map for a circuit to be implemented using continuous assignment statements.

```
//a sum-of-products expression using
//continuous assignment statements
module log_eqn_sop11 (x1, x2, x3, x4, z1);

input x1, x2, x3, x4;
output z1;

wire x1, x2, x3, x4;
wire z1;

assign z1 = (x1 & ~x2) | (~x3 & x4) | (x2 & x3);
endmodule
```

Figure 7.7   Module to implement the logic diagram of Figure 7.6.

```
//a sum-of-products expression using
//continuous assignment statements
module log_eqn_sop11a (x1, x2, x3, x4, z1);
input x1, x2, x3, x4;
output z1;

wire x1, x2, x3, x4;
wire net1, net2, net3;     //define internal nets
wire z1;

assign net1 = x1 & ~x2;
assign net2 = ~x3 & x4;
assign net3 = x2 & x3;
assign z1 = net1 | net2 | net3;
endmodule
```

Figure 7.8   Alternative method to implement the logic diagram of Figure 7.6.

```
//test bench for the sum-of-products expression
module log_eqn_sop11a_tb;

reg x1, x2, x3, x4;
wire z1;


//apply input vectors and display variables
initial
begin: apply_stimulus
   reg [4:0] invect;
   for (invect = 0; invect < 16; invect = invect + 1)
      begin
         {x1, x2, x3, x4} = invect [4:0];
         #10 $display ("x1 x2 x3 x4 = %b, z1 = %b",
                        {x1, x2, x3, x4}, z1);
      end
end


//instantiate the module into the test bench
log_eqn_sop11a inst1 (
   .x1(x1),
   .x2(x2),
   .x3(x3),
   .x4(x4),
   .z1(z1)
   );

endmodule
```

Figure 7.9   Test bench for the modules of Figure 7.7 and Figure 7.8.

```
x1 x2 x3 x4 = 0000, z1 = 0      x1 x2 x3 x4 = 1000, z1 = 1
x1 x2 x3 x4 = 0001, z1 = 1      x1 x2 x3 x4 = 1001, z1 = 1
x1 x2 x3 x4 = 0010, z1 = 0      x1 x2 x3 x4 = 1010, z1 = 1
x1 x2 x3 x4 = 0011, z1 = 0      x1 x2 x3 x4 = 1011, z1 = 1
x1 x2 x3 x4 = 0100, z1 = 0      x1 x2 x3 x4 = 1100, z1 = 0
x1 x2 x3 x4 = 0101, z1 = 1      x1 x2 x3 x4 = 1101, z1 = 1
x1 x2 x3 x4 = 0110, z1 = 1      x1 x2 x3 x4 = 1110, z1 = 1
x1 x2 x3 x4 = 0111, z1 = 1      x1 x2 x3 x4 = 1111, z1 = 1
```

Figure 7.10   Outputs for the modules of Figure 7.7 and Figure 7.8.

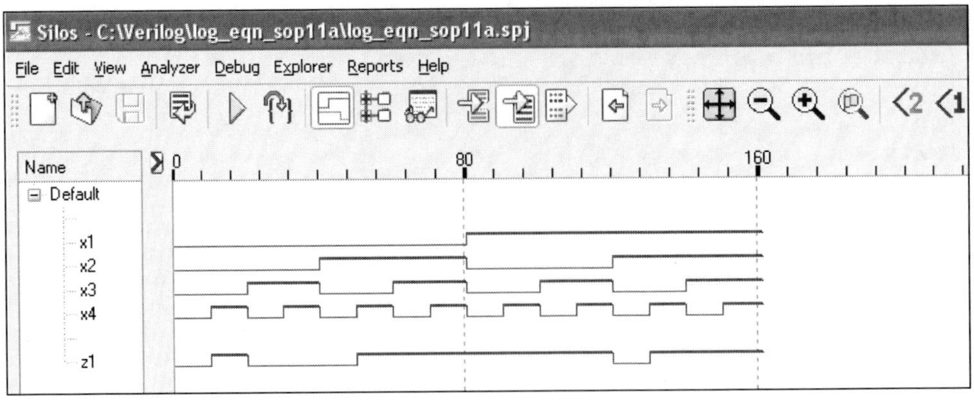

Figure 7.11   Waveforms for the modules of Figure 7.7 and Figure 7.8.

7.1.3 Reduction Operators

The reduction operators are: AND (&), NAND (~&), OR (|), NOR (~|), exclusive-OR (^), and exclusive-NOR (^~ or ~^). Reduction operators are unary operators; that is, they operate on a single vector and produce a single-bit result. Reduction operators perform their respective operations on a bit-by-bit basis from right to left.

Example 7.3   This example illustrates the continuous assignment statement to demonstrate the functionality of the reduction operators. Figure 7.12 contains the design module for the reduction operators. If no delays are specified for the continuous assignment statement, then only one **assign** keyword is required. Only the final statement is terminated by a semicolon; all other statements are terminated by a colon. The test bench, outputs, and waveforms are shown in Figure 7.13, Figure 7.14, and Figure 7.15, respectively.

```
//module to illustrate the use of reduction operators
module reduction1 (a, red_and, red_nand, red_or, red_nor,
                   red_xor, red_xnor);
input [7:0] a;
output red_and, red_nand, red_or, red_nor, red_xor, red_xnor;

wire [7:0] a;
wire red_and, red_nand, red_or, red_nor, red_xor, red_xnor;

//continued on next page
```

Figure 7.12   Module using continuous assignment to demonstrate the reduction operators.

```
assign  red_and  = &a,      //reduction AND
        red_nand = ~&a,     //reduction NAND
        red_or   = |a,      //reduction OR
        red_nor  = ~|a,     //reduction NOR
        red_xor  = ^a,      //reduction exclusive-OR
        red_xnor = ^~a;     //reduction exclusive-NOR

endmodule
```

Figure 7.12   (Continued)

```
//test bench for reduction module
module reduction1_tb;

reg [7:0] a;
wire red_and, red_nand, red_or, red_nor, red_xor, red_xnor;

initial
$monitor ("a=%b, red_and=%b, red_nand=%b, red_or=%b,
          red_nor=%b, red_xor=%b, red_xnor=%b",
          a, red_and, red_nand, red_or, red_nor, red_xor,
          red_xnor);

//apply input vectors
initial
begin
   #0   a = 8'b1100_0011;

   #10  a = 8'b1001_0111;

   #10  a = 8'b0000_0000;

   #10  a = 8'b0100_1111;

   #10  a = 8'b1111_1111;

   #10  $stop;

end

//continued on next page
```

Figure 7.13   Test bench for the continuous assignment module of Figure 7.12.

```
//instantiate the module into the test bench

reduction1 inst1 (
    .a(a),
    .red_and(red_and),
    .red_nand(red_nand),
    .red_or(red_or),
    .red_nor(red_nor),
    .red_xor(red_xor),
    .red_xnor(red_xnor)
    );

endmodule
```

Figure 7.13   (Continued)

```
a=11000011,
red_and=0, red_nand=1, red_or=1,
red_nor=0, red_xor=0, red_xnor=1

a=10010111,
red_and=0, red_nand=1, red_or=1,
red_nor=0, red_xor=1, red_xnor=0

a=00000000,
red_and=0, red_nand=1, red_or=0,
red_nor=1, red_xor=0, red_xnor=1

a=01001111,
red_and=0, red_nand=1, red_or=1,
red_nor=0, red_xor=1, red_xnor=0

a=11111111,
red_and=1, red_nand=0, red_or=1,
red_nor=0, red_xor=0, red_xnor=1
```

Figure 7.14   Outputs for the test bench of Figure 7.13.

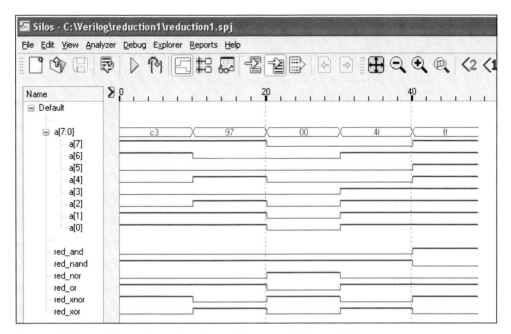

Figure 7.15 Waveforms for the test bench of Figure 7.13.

7.1.4 Octal-to-Binary Encoder

An encoder is a macro logic circuit with n mutually exclusive inputs and m binary outputs, where $n \leq 2^m$. The function of an encoder can be considered to be the inverse of a decoder; that is, the mutually exclusive inputs are encoded into a corresponding binary number.

A general block diagram for an $n{:}m$ encoder is shown in Figure 7.16. An encoder is also referred to as a code converter. In the label of Figure 7.16, X corresponds to the input code and Y corresponds to the output code. The general qualifying label X/Y is replaced by the input and output codes, respectively, such as OCT/BIN for an octal-to-binary code converter. Only one input x_i is asserted at a time. The decimal value of x_i is encoded as a binary number which is specified by the m outputs. An 8:3 octal-to-binary encoder is shown in Figure 7.17. If input 7 is asserted, then the output values are $b[2{:}0] = 111$, where the binary weight of the outputs is $b[2]\ b[1]\ b[0] = 2^2 2^1 2^0$.

The truth table for an octal-to-binary encoder is shown in Table 7.2, where the binary outputs represent the octal number of the active input. The equations for the encoder are obtained directly from the truth table and are shown in Equation 7.2. The logic diagram is shown in Figure 7.18 using three OR gates.

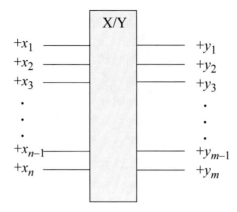

Figure 7.16   General block diagram of an *n:m* encoder or code converter.

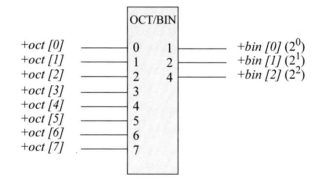

Figure 7.17   Octal-to-binary encoder.

Table 7.2 Truth Table for an Octal-to-Binary Encoder

Inputs								Outputs		
d[0]	d[1]	d[2]	d[3]	d[4]	d[5]	d[6]	d[7]	b[2]	b[1]	b[0]
1	0	0	0	0	0	0	0	0	0	0
0	1	0	0	0	0	0	0	0	0	1
0	0	1	0	0	0	0	0	0	1	0
0	0	0	1	0	0	0	0	0	1	1
0	0	0	0	1	0	0	0	1	0	0
0	0	0	0	0	1	0	0	1	0	1
0	0	0	0	0	0	1	0	1	1	0
0	0	0	0	0	0	0	1	1	1	1

$$b[0] = d[1] + d[3] + d[5] + d[7]$$
$$b[1] = d[2] + d[3] + d[6] + d[7]$$
$$b[2] = d[4] + d[5] + d[6] + d[7] \qquad (7.2)$$

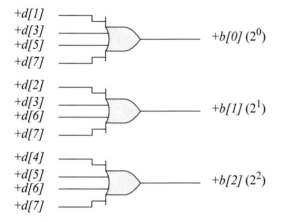

Figure 7.18   Logic diagram for an 8:3 encoder.

The design module is shown in Figure 7.19 using three continuous assignment statements, each one terminated by a semicolon. The test bench, outputs, and waveforms are shown in Figure 7.20, Figure 7.21, and Figure 7.22, respectively.

```
//an octal-to-binary encoder
module encoder_8_to_3 (oct, bin);

input [0:7] oct;
output [2:0] bin;

wire [0:7] oct;
wire [2:0] bin;

assign bin[0] = oct[1] | oct[3] | oct[5] | oct[7];
assign bin[1] = oct[2] | oct[3] | oct[6] | oct[7];
assign bin[2] = oct[4] | oct[5] | oct[6] | oct[7];

endmodule
```

Figure 7.19   Module for the octal-to-binary encoder using continuous assignment statements.

```
//test bench for the 8-to-3 encoder
module encoder_8_to_3_tb;

reg [0:7] oct;
wire [2:0] bin;

//display variables
initial
$monitor ("octal = %b, binary = %b", oct [0:7], bin [2:0]);

//apply input vectors
initial
begin
    #0      oct [0:7] = 8'b1000_0000;
    #10     oct [0:7] = 8'b0100_0000;
    #10     oct [0:7] = 8'b0010_0000;
    #10     oct [0:7] = 8'b0001_0000;
    #10     oct [0:7] = 8'b0000_1000;
    #10     oct [0:7] = 8'b0000_0100;
    #10     oct [0:7] = 8'b0000_0010;
    #10     oct [0:7] = 8'b0000_0001;
    #10     $stop;
end

//instantiate the module into the test bench
encoder_8_to_3 inst1 (
    .oct(oct),
    .bin(bin)
    );

endmodule
```

Figure 7.20   Test bench for the octal-to-binary encoder.

```
octal = 10000000, binary = 000
octal = 01000000, binary = 001
octal = 00100000, binary = 010
octal = 00010000, binary = 011
octal = 00001000, binary = 100
octal = 00000100, binary = 101
octal = 00000010, binary = 110
octal = 00000001, binary = 111
```

Figure 7.21   Outputs for the test bench of Figure 7.20 for the octal-to-binary encoder.

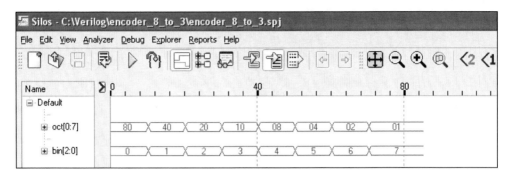

Figure 7.22 Waveforms for the octal-to-binary encoder.

7.1.5 Four-to-One Multiplexer

A multiplexer is a logic macro device that allows digital information from two or more data inputs to be directed to a single output. Data input selection is controlled by a set of select inputs that determine which data input is gated to the output. The select inputs are labeled $s_0, s_1, s_2, \ldots, s_i, \ldots, s_{n-1}$, where s_0 is the low-order select input with a binary weight of 2^0 and s_{n-1} is the high-order select input with a binary weight of 2^{n-1}. The data inputs are labeled $d_0, d_1, d_2, \ldots, d_j, \ldots, d_{2^n-1}$. Thus, if a multiplexer has n select inputs, then the number of data inputs will be 2^n and will be labeled d_0 through d_{2^n-1}. For example, if $n = 2$, then the multiplexer has two select inputs s_0 and s_1 and four data inputs $d_0, d_1, d_2,$ and d_3.

Figure 7.23 shows a 4:1 multiplexer drawn in the ANSI/IEEE Std. 91-1984 format. The truth table is shown in Table 7.3. If $s_1 s_0 = 00$, then data input d_0 is selected and its value is propagated to the multiplexer output z_1. Similarly, if $s_1 s_0 = 01$, then data input d_1 is selected and its value is directed to the multiplexer output.

The equation that represents output z_1 in the 4:1 multiplexer of Figure 7.23 is shown in Equation 7.3 with an implied *enable* input. Output z_1 assumes the value of d_0 if $s_1 s_0 = 00$, as indicated by the term $s_1's_0'd_0$. Likewise, z_1 assumes the value of d_1 when $s_1 s_0 = 01$, as indicated by the term $s_1's_0 d_1$.

Example 7.4 The design module for a 4:1 multiplexer is shown in Figure 7.24 using a continuous assignment statement for Equation 7.3. The test bench, outputs, and waveforms are shown in Figure 7.25, Figure 7.26, and Figure 7.27, respectively.

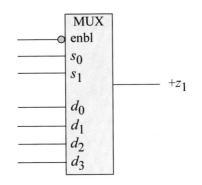

Figure 7.23   Block diagram of a 4:1 multiplexer.

Table 7.3 Truth Table for a 4:1 Multiplexer

Select Inputs $s_1 s_0$	Data Input Selected
0 0	d_0
0 1	d_1
1 0	d_2
1 1	d_3

$$z_1 = s_1' s_0' d_0 + s_1' s_0 d_1 + s_1 s_0' d_2 + s_1 s_0 d_3 \qquad (7.3)$$

```
module mux4_df (s, d, enbl, z1);//dataflow 4:1 multiplexer
input [1:0] s;
input [3:0] d;
input enbl;
output z1;

wire [1:0] s;
wire [3:0] d;
wire enbl;
wire z1;

assign z1 =(~s[1] & ~s[0] & d[0] & enbl) |
           (~s[1] &  s[0] & d[1] & enbl) |
           ( s[1] & ~s[0] & d[2] & enbl) |
           ( s[1] &  s[0] & d[3] & enbl);
endmodule
```

Figure 7.24   Design module for a 4:1 multiplexer using a continuous assignment statement.

```verilog
//dataflow mux4_df test bench
module mux4_df_tb;


reg [1:0] s;
reg [3:0] d;
reg enbl;
wire z1;


//display variables
initial
$monitor ("select: s=%b, data: d=%b, out: z1=%b", s, d, z1);


//apply input vectors

initial
begin

#0      s = 2'b00;
        d = 4'b1111;
        enbl = 1'b0;


#10     s = 2'b00;
        d = 4'b0000;
        enbl = 1'b1;


#10     s = 2'b00;
        d = 4'b0001;
        enbl = 1'b1;


#10     s = 2'b01;
        d = 4'b0001;
        enbl = 1'b1;


#10     s = 2'b01;
        d = 4'b0011;
        enbl = 1'b1;


#10     s = 2'b10;
        d = 4'b1001;
        enbl = 1'b1;

//continued on next page
```

Figure 7.25   Test bench for the 4:1 multiplexer of Figure 7.24.

```
#10     s = 2'b10;
        d = 4'b0101;
        enbl = 1'b1;


#10     s = 2'b11;
        d = 4'b0111;
        enbl = 1'b1;


#10     s = 2'b11;
        d = 4'b1001;
        enbl = 1'b1;


#10     $stop;

end

//instantiate the module into the test bench
mux4_df inst1 (
    .s(s),
    .d(d),
    .enbl(enbl),
    .z1(z1)
    );

endmodule
```

Figure 7.25   (Continued)

```
select: s = 00, data: d = 1111, out: z1 = 0
select: s = 00, data: d = 0000, out: z1 = 0
select: s = 00, data: d = 0001, out: z1 = 1
select: s = 01, data: d = 0001, out: z1 = 0
select: s = 01, data: d = 0011, out: z1 = 1
select: s = 10, data: d = 1001, out: z1 = 0
select: s = 10, data: d = 0101, out: z1 = 1
select: s = 11, data: d = 0111, out: z1 = 0
select: s = 11, data: d = 1001, out: z1 = 1
```

Figure 7.26   Outputs of the 4:1 multiplexer of Figure 7.24.

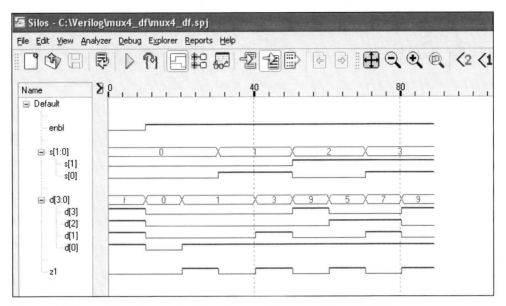

Figure 7.27   Waveforms for the 4:1 multiplexer of Figure 7.24.

7.1.6 Four-to-One Multiplexer Using the Conditional Operator

The conditional operator (**? :**) has three operands, as shown in the syntax below. The *conditional_expression* is evaluated. If the result is true (1), then the *true_expression* is evaluated; if the result is false (0), then the *false_expression* is evaluated. The conditional operator can be used when one of two expressions is to be selected.

conditional_expression **?** true_expression **:** false_expression;

Example 7.5   A conditional operator can be used in a continuous assignment statement to design a multiplexer. A 4:1 multiplexer will be designed using a conditional operator. This design will declare the multiplexer inputs as scalars instead of vectors like those used in Example 7.4. The select inputs are: s_0 and s_1; the data inputs are: in_0, in_1, in_2, and in_3. The design module is shown in Figure 7.28. The test bench and outputs are shown in Figure 7.29 and Figure 7.30, respectively.

```
//dataflow 4:1 mux using conditional operator
//compare to module of Example 7.5
module mux4to1_cond (out, in0, in1, in2, in3, s0, s1);

input s0, s1;
input in0, in1, in2, in3;
output out;

//use nested conditional operator
assign out = s1 ? (s0 ? in3 : in2) : (s0 ? in1 : in0);

endmodule

//s1          s0            out
//1(true)     1(true)       in3
//0(false)    1(true)       in1
//1(true)     0(false)      in2
//0(false)    0(false)      in0
```

Figure 7.28   Module for a 4:1 multiplexer using a conditional operator and a continuous assignment statement.

```
//mux4to1_cond test bench
module mux4to1_cond_tb;

reg in0, in1, in2, in3, s0, s1;   //inputs are reg
wire out;                         //outputs are wire

//display signals
initial
$monitor ("s1s0 = %b, in0in1in2in3 = %b, out = %b",
          {s1, s0}, {in0, in1, in2, in3}, out);
//apply stimulus
initial
begin
    #0    s1  = 1'b0;
          s0  = 1'b0;
          in0 = 1'b0;    //out = 0
          in1 = 1'b1;
          in2 = 1'b1;
          in3 = 1'b1;    //continued on next page
```

Figure 7.29   Test bench for the conditional operator multiplexer of Figure 7.28.

```
    #10    s1  = 1'b0;
           s0  = 1'b1;
           in0 = 1'b0;
           in1 = 1'b1;       //out = 1
           in2 = 1'b1;
           in3 = 1'b1;

    #10    s1  = 1'b1;
           s0  = 1'b0;
           in0 = 1'b0;
           in1 = 1'b1;
           in2 = 1'b0;       //out = 0
           in3 = 1'b1;

    #10    s1  = 1'b1;
           s0  = 1'b1;
           in0 = 1'b0;
           in1 = 1'b1;
           in2 = 1'b0;
           in3 = 1'b1;       //out = 0

    #10    $stop;
end

//instantiate the module into the test bench
mux4to1_cond inst1 (
    .s0(s0),
    .s1(s1),
    .in0(in0),
    .in1(in1),
    .in2(in2),
    .in3(in3),
    .out(out)
    );

endmodule
```

Figure 7.29   (Continued)

```
s1s0 = 00, in0in1in2in3 = 0111, out = 0
s1s0 = 01, in0in1in2in3 = 0111, out = 1
s1s0 = 10, in0in1in2in3 = 0101, out = 0
s1s0 = 11, in0in1in2in3 = 0101, out = 1
```

Figure 7.30   Outputs for the 4:1 conditional operator multiplexer of Figure 7.28.

7.1.7 Four-Bit Adder

A parallel adder that adds two n-bit operands requires n full adders. A *full adder* for stage$_i$ is a combinational circuit that has three inputs: an augend a_i, an addend b_i, and a carry-in c_{i-1}. There are two outputs: a sum s_i and a carry-out c_i. The truth table for the sum and carry-out functions is shown in Table 7.4 for adding two equally weighted bits a_i and b_i in vectors

$$A = a_{n-1}\, a_{n-2} \ldots a_1\, a_0$$
$$B = b_{n-1}\, b_{n-2} \ldots b_1\, b_0$$

Table 7.4 Truth Table for Binary Addition

a_i	b_i	c_{i-1}	s_i	c_i
0	0	0	0	0
0	0	1	1	0
0	1	0	1	0
0	1	1	0	1
1	0	0	1	0
1	0	1	0	1
1	1	0	0	1
1	1	1	1	1

Each stage of the addition algorithm must be able to accommodate carry-in bit c_{i-1} from the immediately preceding lower-order stage. The carry-out of the ith stage is c_i. The sum and carry equations for the full adder are shown in Equation 7.4. The resulting equation for c_i can also be written as $c_i = a_i b_i + (a_i \oplus b_i) c_{i-1}$, although this requires more gate delays.

$$s_i = a_i' b_i' c_{i-1} + a_i' b_i c_{i-1}' + a_i b_i' c_{i-1}' + a_i b_i c_{i-1}$$

$$= c_{i-1}' (a_i \oplus b_i) + c_{i-1} (a_i \oplus b_i)'$$

$$= a_i \oplus b_i \oplus c_{i-1}$$

$$c_i = a_i' b_i c_{i-1} + a_i b_i' c_{i-1} + a_i b_i c_{i-1}' + a_i b_i c_{i-1}$$

$$= a_i' b_i c_{i-1} + a_i b_i' c_{i-1} + a_i b_i$$

$$= a_i b_i + a_i c_{i-1} + b_i c_{i-1} \tag{7.4}$$

Example 7.6   A 4-bit adder will be designed in a dataflow module using a single continuous assignment statement. The module does specify the implementation of the adder — only the functional operation of the adder. The are two vector operands: the augend *a[3:0]* and the addend *b[3:0]* with a scalar input *cin*, which is the carry-in to the low-order stage of the adder. There are two outputs: a 4-bit vector *sum[3:0]* and a scalar *cout*, which is the carry-out of the adder.

The design module is shown in Figure 7.31 in which *cout* and *sum* are concatenated. The test bench, outputs, and waveforms are shown in Figure 7.32, Figure 7.33, and Figure 7.34, respectively. The waveforms display the values of the augend, addend, and sum in hexadecimal notation as well as individual bits.

```
//dataflow for a 4-bit adder
module adder4_df (a, b, cin, sum, cout);

//list inputs and outputs
input [3:0] a, b;
input cin;
output [3:0] sum;
output cout;

//define signals as wire for dataflow (or default to wire)
wire [3:0] a, b;
wire cin, cout;
wire [3:0] sum;

//continuous assignment for dataflow
//implement the 4-bit adder as a logic equation
//concatenating cout and sum
assign {cout, sum} = a + b + cin;
endmodule
```

Figure 7.31   Dataflow module for a 4-bit adder using continuous assignment.

```
//dataflow 4-bit adder test bench
module adder4_df_tb;

reg [3:0] a, b;
reg cin;
wire [3:0] sum;
wire cout;          //continued on next page
```

Figure 7.32   Test bench for Figure 7.31 for a 4-bit adder using continuous assignment.

```
//display signals
initial
$monitor ("a = %b, b = %b, cin = %b, cout = %b, sum = %b",
          a, b, cin, cout, sum);

//apply stimulus
initial
begin
   #0 a = 4'b0000;
      b = 4'b0000;
      cin = 1'b0;     //sum = 0000

   #10a = 4'b0001;
      b = 4'b0010;
      cin = 1'b0;     //sum = 0011

   #10a = 4'b0010;
      b = 4'b0110;
      cin = 1'b0;     //sum = 1000

   #10a = 4'b0111;
      b = 4'b0111;
      cin = 1'b0;     //sum = 1110

   #10a = 4'b1001;
      b = 4'b0110;
      cin = 1'b0;     //sum = 1111

   #10a = 4'b1100;
      b = 4'b0011;
      cin = 1'b0;     //sum = 1111

   #10a = 4'b1111;
      b = 4'b0000;
      cin = 1'b0;     //sum = 1111

   #10a = 4'b1100;
      b = 4'b0001;
      cin = 1'b1;     //sum = 1110

   #10a = 4'b0011;
      b = 4'b1000;
      cin = 1'b1;     //sum = 1100

//continued on next page
```

Figure 7.32   (Continued)

```
   #10a = 4'b1011;
      b = 4'b1011;
      cin = 1'b1;    //sum = 1_0111


   #10a = 4'b1111;
      b = 4'b1111;
      cin = 1'b1;    //sum = 1_1111


   #10    $stop;

end

//instantiate the module into the test bench
adder4_df inst1 (
   .a(a),
   .b(b),
   .cin(cin),
   .sum(sum),
   .cout(cout)
   );

endmodule
```

Figure 7.32   (Continued)

```
a = 0000, b = 0000, cin = 0, cout = 0, sum = 0000
a = 0001, b = 0010, cin = 0, cout = 0, sum = 0011
a = 0010, b = 0110, cin = 0, cout = 0, sum = 1000
a = 0111, b = 0111, cin = 0, cout = 0, sum = 1110
a = 1001, b = 0110, cin = 0, cout = 0, sum = 1111
a = 1100, b = 0011, cin = 0, cout = 0, sum = 1111
a = 1111, b = 0000, cin = 0, cout = 0, sum = 1111
a = 1100, b = 0001, cin = 1, cout = 0, sum = 1110
a = 0011, b = 1000, cin = 1, cout = 0, sum = 1100
a = 1011, b = 1011, cin = 1, cout = 1, sum = 0111
a = 1111, b = 1111, cin = 1, cout = 1, sum = 1111
```

Figure 7.33   Outputs for the 4-bit adder of Figure 7.31.

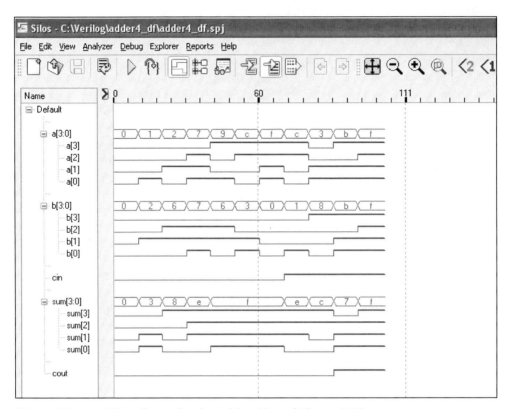

Figure 7.34   Waveforms for the 4-bit adder of Figure 7.31.

7.1.8 Carry Lookahead Adder

The speed limitation in the ripple adder arises from specifying c_i as a function of the carry-out from the previous lower-order stage c_{i-1}. A considerable increase in speed can be realized by expressing the carry-out c_i of any stage $_i$ as a function of the two operand bits a_i and b_i and the carry-in to the low-order stage $_0$ of the adder c_{-1}, where the adder is an n-bit adder $n_{-1} n_{-2} \ldots n_1 n_0$. The carry-out c_i from Equation 7.4 can be restated as shown in Equation 7.5.

$$c_i = a_i' b_i c_{i-1} + a_i b_i' c_{i-1} + a_i b_i$$

$$= a_i b_i + (a_i \oplus b_i) c_{i-1} \qquad (7.5)$$

Equation 7.5 states that a carry will be generated whenever $a_i = b_i = 1$, or when either $a_i = 1$ or $b_i = 1$ with $c_{i-1} = 1$. A technique will now be presented that increases the speed of the carry propagation in a parallel adder. The carries entering all the bit positions of the adder can be generated simultaneously by a *carry lookahead* generator. This results in a constant addition time that is independent of the length of the adder. Two auxiliary functions will be defined as follows:

$$\text{Generate} \quad G_i = a_i b_i$$
$$\text{Propagate} \quad P_i = a_i \oplus b_i$$

The carry *generate* function G_i reflects the condition where a carry is generated at the ith stage. The carry *propagate* function P_i is true when the ith stage will pass (or propagate) the incoming carry c_{i-1} to the next higher stage $_{i+1}$. Equation 7.5 can now be restated as Equation 7.6.

$$c_i = a_i b_i + (a_i \oplus b_i)\, c_{i-1}$$
$$= G_i + P_i\, c_{i-1} \qquad (7.6)$$

Equation 7.6 indicates that the generate G_i and propagate P_i functions for any carry c_i can be obtained independently and in parallel when the operand inputs are applied to the n-bit adder. The equation can be applied recursively to obtain the following set of carry equations in terms of the variables G_i, P_i, and c_{-1} for a 4-bit adder, where c_{-1} is the carry-in to the low-stage of the adder.

$$c_0 = G_0 + P_0\, c_{-1}$$
$$c_1 = G_1 + P_1\, c_0$$
$$= G_1 + P_1\, (G_0 + P_0\, c_{-1})$$
$$= G_1 + P_1 G_0 + P_1 P_0\, c_{-1}$$
$$c_2 = G_2 + P_2\, c_1$$
$$= G_2 + P_2\, (G_1 + P_1 G_0 + P_1 P_0\, c_{-1})$$
$$= G_2 + P_2 G_1 + P_2 P_1 G_0 + P_2 P_1 P_0\, c_{-1}$$
$$c_3 = G_3 + P_3\, c_2$$
$$= G_3 + P_3 (G_2 + P_2 G_1 + P_2 P_1 G_0 + P_2 P_1 P_0\, c_{-1})$$
$$= G_3 + P_3\, G_2 + P_3 P_2 G_1 + P_3 P_2 P_1 G_0 + P_3 P_2 P_1 P_0\, c_{-1} \qquad (7.7)$$

Example 7.7   A 4-bit carry lookahead adder will be designed using Equation 7.7 and continuous assignment statements. The block diagram of the adder is shown in Figure 7.35, where the augend is *a[3:0]*, the addend is *b[3:0]*, and the sum is *sum[3:0]*. There are also four internal carries: *c3, c2, c1,* and *c0*.

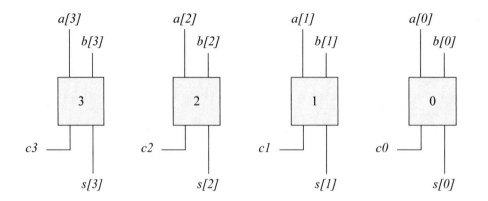

Figure 7.35   Block diagram of a 4-bit adder to be implemented as a carry look-ahead adder.

The design module is shown in Figure 7.36. Internal wires are defined for the *generate*, *propagate*, and *carry* functions. Multiple statements can be declared using one continuous assignment. The test bench is shown in Figure 7.37 utilizing only a few of the $2^9 = 512$ input vectors. The outputs and waveforms are shown in Figure 7.38 and Figure 7.39, respectively.

```
//dataflow 4-bit adder using carry lookahead.
//This is another way to design an adder--
//by designing individual stages, rather than
//by designing the adder as one unit of 4 stages.
//Using this method, the propagate functions must
//be defined using the exclusive-or function,
//not the or function.

module adder4_cla (a, b, cin, sum, cout);

//continued on next page
```

Figure 7.36   Design module for a 4-bit carry lookahead adder.

```verilog
//i/o port declaration
input [3:0] a, b;
input cin;
output [3:0] sum;
output cout;


//define internal wires
wire g3, g2, g1, g0;
wire p3, p2, p1, p0;
wire c3, c2, c1, c0;


//define generate functions
assign g0 = a[0] & b[0],//multiple statements using 1 assign
       g1 = a[1] & b[1],
       g2 = a[2] & b[2],
       g3 = a[3] & b[3];


//define propagate functions
assign p0 = a[0] ^ b[0],//multiple statements using 1 assign
       p1 = a[1] ^ b[1],
       p2 = a[2] ^ b[2],
       p3 = a[3] ^ b[3];


//obtain the carry equations
assign c0 = g0| (p0 & cin),
       c1 = g1| (p1 & g0) | (p1 & p0 & cin),
       c2 = g2| (p2 & g1) | (p2 & p1 & g0) | (p2 & p1 & p0 & cin),
       c3 = g3| (p3 & g2) | (p3 & p2 & g1) | (p3 & p2 & p1 & g0) |
                (p3 & p2 & p1 & p0 & cin);


//obtain the sum equations
assign sum[0] = p0 ^ cin,
       sum[1] = p1 ^ c0,
       sum[2] = p2 ^ c1,
       sum[3] = p3 ^ c2;


//obtain cout
assign cout = c3;

endmodule
```

Figure 7.36   (Continued)

```verilog
//test bench for the dataflow 4-bit carry lookahead adder
module adder4_cla_tb;


reg [3:0] a, b;
reg cin;
wire [3:0] sum;
wire cout;


//display signals
initial
$monitor ("a = %b, b = %b, cin = %b, cout = %b, sum = %b",
            a, b, cin, cout, sum);


//apply input vectors
initial
begin
   #0     a = 4'b0000;
          b = 4'b0000;
          cin = 1'b0;        //cout = 0, sum = 0000

   #10    a = 4'b0001;
          b = 4'b0010;
          cin = 1'b0;        //cout = 0, sum = 0011

   #10    a = 4'b0010;
          b = 4'b0110;
          cin = 1'b0;        //cout = 0, sum = 1000

   #10    a = 4'b0111;
          b = 4'b0111;
          cin = 1'b0;        //cout = 0, sum = 1110

   #10    a = 4'b1001;
          b = 4'b0110;
          cin = 1'b0;        //cout = 0, sum = 1111

   #10    a = 4'b1100;
          b = 4'b1100;
          cin = 1'b0;        //cout = 1, sum = 1000

//continued on next page
```

Figure 7.37   Test bench for the 4-bit carry lookahead adder of Figure 7.36.

```
    #10    a = 4'b1111;
           b = 4'b1110;
           cin = 1'b0;          //cout = 1, sum = 1101


    #10    a = 4'b1110;
           b = 4'b1110;
           cin = 1'b1;          //cout = 1, sum = 1101


    #10    a = 4'b1111;
           b = 4'b1111;
           cin = 1'b1;          //cout = 1, sum = 1111


    #10    $stop;

end

//instantiate the module into the test bench
adder4_cla inst1 (
   .a(a),
   .b(b),
   .cin(cin),
   .sum(sum),
   .cout(cout)
   );

endmodule
```

Figure 7.37   (Continued)

```
a = 0000, b = 0000, cin = 0, cout = 0, sum = 0000
a = 0001, b = 0010, cin = 0, cout = 0, sum = 0011
a = 0010, b = 0110, cin = 0, cout = 0, sum = 1000
a = 0111, b = 0111, cin = 0, cout = 0, sum = 1110
a = 1001, b = 0110, cin = 0, cout = 0, sum = 1111
a = 1100, b = 1100, cin = 0, cout = 1, sum = 1000
a = 1111, b = 1110, cin = 0, cout = 1, sum = 1101
a = 1110, b = 1110, cin = 1, cout = 1, sum = 1101
a = 1111, b = 1111, cin = 1, cout = 1, sum = 1111
```

Figure 7.38   Outputs for the 4-bit carry lookahead adder of Figure 7.36.

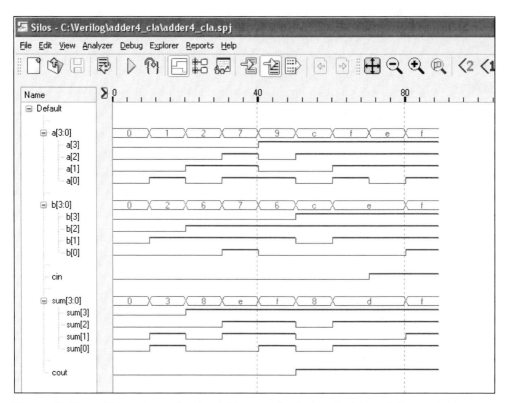

Figure 7.39   Waveforms for the 4-bit carry lookahead adder of Figure 7.36.

7.1.9 Asynchronous Sequential Machine

The design of asynchronous sequential machines is one of the most interesting and certainly the most challenging concepts of sequential machine design. In many situations, a synchronous clock is not available. The design procedure for asynchronous sequential machines is similar in many respects to that for synchronous sequential machines. This section develops a systematic method for the design of fundamental-mode asynchronous sequential machines using *SR* latches as the storage elements. The machine operation is specified as a timing diagram and/or verbal statements. The design procedure is summarized below.

1. **State diagram** The machine specifications are converted into a state diagram. A timing diagram and/or a verbal statement of the machine specifications is converted into a precise delineation which specifies the machine's operation for all applicable input sequences. This step is not a necessary requirement and is usually omitted; however, the state diagram characterizes the

machine's operation in a graphical representation and adds completeness to the design procedure.

2. **Primitive flow table**   The machine specifications are converted to a state transition table called a primitive flow table. This is the least methodical step in the synthesis procedure and the most important. The primitive flow table depicts the state transition sequences and output assertions for all valid input vectors. The flow table must correctly represent the machine's operation for all applicable input sequences, even those that are not initially apparent from the machine specifications.

3. **Equivalent states**   The primitive flow table may have an inordinate number of rows. The number of rows can be reduced by finding equivalent states and then eliminating redundant states. If the machine's operation is indistinguishable whether commencing in state Y_i or state Y_j, then one of the states is redundant and can be eliminated. The flow table thus obtained is a *reduced primitive flow table*. In order for two stable states to be equivalent, all three of the following conditions must be satisfied:

    1. The same input vector.
    2. The same output value.
    3. The same, or equivalent, next state for all valid input sequences.

4. **Merger diagram**   The merger diagram graphically portrays the result of the merging process in which an attempt is made to combine two or more rows of the reduced primitive flow table into a single row. The result of the merging technique is analogous to that of finding equivalent states; that is, the merging process can also reduce the number of rows in the table and, hence, reduce the number of feedback variables that are required. Fewer feedback variables will result in a machine with less logic and, therefore, less cost. Two rows can merge into a single row if the entries in the same column of each row satisfy one of the following three merging rules:

    1. Identical state entries, either stable or unstable.
    2. A state entry and a "don't care."
    3. Two "don't care" entries.

5. **Merged flow table**   The merged flow table is constructed from the merger diagram. The table represents the culmination of the merging process in which two or more rows of a primitive flow table are replaced by a single equivalent row which contains one stable state for each merged row.

6. **Excitation maps and equations**   An excitation map is generated for each excitation variable. Then the transient states are encoded, where applicable, to

avoid critical race conditions. Appropriate assignment of the excitation variables for the transient states can minimize the δ next-state logic for the excitation variables. The operational speed of the machine can also be established
at this step by reducing the number of transient states through which the machine must sequence during a cycle. Then the excitation equations are derived
from the excitation maps. All static-1 and static-0 hazards are eliminated
from the network for a sum-of-products or product-of-sums implementation,
respectively.

7. **Output maps and equations** An output map is generated for each machine
output. Output values are assigned for all nonstable states so that no transient
signals will appear on the outputs. In this step, the speed of circuit operation
can also be established. Then the output equations are derived from the output
maps, assuring that all outputs will be free of static-1 and static-0 hazards.

8. **Logic diagram** The logic diagram is implemented from the excitation and
output equations using an appropriate logic family.

The fundamental-mode model is shown in Figure 7.40. The input alphabet X consists of n binary variables $x_1, x_2, \ldots, x_n$ that change value in an asynchronous manner.
Since the machine is characterized by a fundamental-mode operation, only one input
variable x_i is allowed to change state at a time and no other input variable x_j will
change state until the machine has sequenced to a stable state. This operating characteristic precludes the possibility of an inherent race condition. A single delay element
that represents the total delay of the entire machine is placed in series with the feedback path. This method simplifies the analysis and design by specifying zero delay for
all logic gates and interconnecting wires.

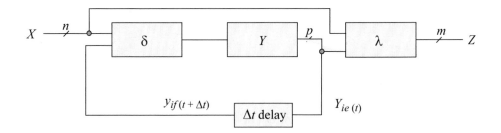

Figure 7.40   General block diagram of a fundamental-mode asynchronous
sequential machine.

The state alphabet Y contains p storage elements in the form of SR latches. The outputs of the latches represent the excitation variables $Y_{1e}, Y_{2e}, \ldots, Y_{pe}$. After a delay of Δt, these same outputs represent the feedback variables $y_{1f}, y_{2f}, \ldots, y_{pf}$ and connect to the AND gates of the SR latches. Thus, when the feedback variables become equal to the excitation variables, the machine enters a stable state where $y_{if} = Y_{ie}$. The output alphabet Z is represented by $z_1, z_2, \ldots, z_m$.

Example 7.8 An asynchronous sequential machine will be designed which has two inputs x_1 and x_2 and one output z_1. Output z_1 will be asserted coincident with the assertion of x_2, but only if x_1 is already asserted. The deassertion of x_2 causes the deassertion of z_1. Input x_1 will not become deasserted while x_2 is asserted. The timing diagram of Figure 7.41 further illustrates the operation of the machine and shows the various states through which the machine sequences.

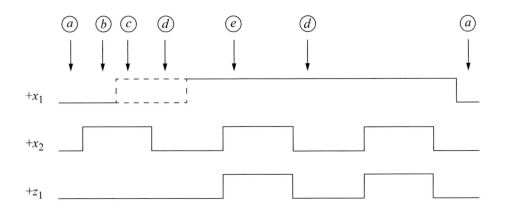

Figure 7.41 Timing diagram for the asynchronous sequential machine of Example 7.8.

The primitive flow table is shown in Figure 7.42 and is a tabular representation of the sequential operation of the machine. In state $\textcircled{a}$, the machine sequences through transient state b to stable state $\textcircled{b}$ when $x_1 x_2 = 01$. Since simultaneous input changes are not allowed in asynchronous sequential machines, the inputs cannot change from $x_1 x_2 = 00$ to $x_1 x_2 = 11$ or vice versa, or from $x_1 x_2 = 01$ to $x_1 x_2 = 10$ or vice versa.

In stable state $\textcircled{b}$, the next state is $\textcircled{c}$ if x_1 is asserted. Output z_1 remains at $z_1 = 0$, since x_1 was not asserted when x_2 was asserted. From state $\textcircled{c}$ the machine proceeds to stable state $\textcircled{d}$ when x_2 is deasserted; output z_1 remains inactive.

When x_2 again becomes asserted with x_1 already asserted, the machine proceeds to state $\textcircled{e}$ where output z_1 is asserted. The machine cannot sequence to column $x_1 x_2 = 01$ because the machine specifications state that x_1 will not become deasserted while x_2 is asserted.

x_1x_2	00	01	11	10	z_1
	ⓐ	b	–	d	0
	a	ⓑ	c	–	0
	–	–	ⓒ	d	0
	a	–	e	ⓓ	0
	–	–	ⓔ	d	1

Figure 7.42   Primitive flow table for the asynchronous sequential machine of Figure 7.41.

Figure 7.42 is also a reduced primitive flow table because there are no equivalent states. In order for two stable states to be equivalent, they must have the same input vector, the same output value, and the same, or equivalent, next state for all valid input sequences.

The merging process is facilitated by means of a merger diagram. The merger diagram depicts all rows of the reduced primitive flow table in a graphical representation. Each row of the table is portrayed as a vertex in the merger diagram. The rows in which merging is possible are connected by lines.

A set of rows in a reduced primitive flow table can merge into a single row if and only if the rows are strongly connected. That is, every row in the set must merge with all other rows in the set. For example, Figure 7.43 illustrates strongly connected sets of three, four, and five rows each. Each set in Figure 7.43 is a maximal compatible set, in which compatible pairs {ⓐ, ⓑ} and {ⓑ, ⓒ} imply the compatibility of {ⓐ, ⓒ}. Thus, the transitive property applies to maximal compatible sets. Adding another state to a maximal compatible set negates the transitive property on the new set, unless the set remains strongly connected.

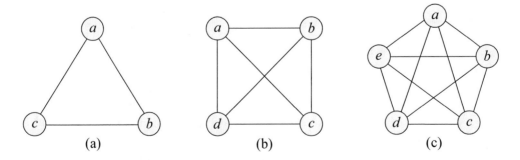

(a)   (b)   (c)

Figure 7.43   Examples of merger diagrams.

Figure 7.44 illustrates the merger diagram for the asynchronous sequential machine of Figure 7.41.

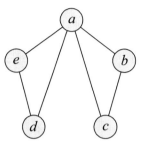

Figure 7.44   Merger diagram for the asynchronous sequential machine of Figure 7.41.

The next step in the design procedure is the generation of a merged flow table. The merged flow table specifies the operational characteristics of the machine in a manner analogous to that of the primitive flow table and the reduced primitive flow table, but in a more compact form. Each row in a merged flow table represents a set of maximal compatible rows.

Most unspecified entries in the reduced primitive flow table are replaced with either a stable or an unstable state entry in the merged flow table. Since more than one stable state is usually present in each row of a merged flow table, many state transition sequences do not cause a change to the feedback variables. Thus, faster operational speed is realized.

The merged flow table is derived from the merger diagram in conjunction with the reduced primitive flow table. The partition of sets of maximal compatible rows obtained from the merger diagram dictates the minimal number of rows in the merged flow table. The merged flow table for this example is shown in Figure 7.45.

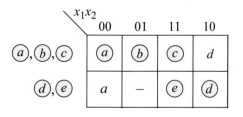

Figure 7.45   Merged flow table for the asynchronous sequential machine of Figure 7.41.

Rows a, b, and c are transcribed from the reduced primitive flow table to the merged flow table. Each of the three rows is transferred, column by column, to the same row in the merged flow table. Row a is transferred first, then row b is superimposed on row a, and then row c is superimposed on previously transferred rows a and b. Rows d and e are then transferred to the merged flow table using the same process.

The excitation map is shown in Figure 7.46(a). The machine is stable when the feedback variable y_{1f} is equal to the excitation variable Y_{1e}. Therefore, because states a, b, and c are stable, the excitation variables are 0 for all three states, due to the feedback variable being equal to 0, as shown in the first row of the excitation map. Stable states d and e are similarly defined as an excitation value of 1. The output map is shown in Figure 7.46(b) and indicates an output value of $z_1 = 1$ in state e.

The logic diagram is shown in Figure 7.47 and is obtained from the excitation equations and the output equation of Equation 7.8. The design module is shown in Figure 7.48 using two continuous assignment statements. The test bench is shown in Figure 7.49 in which the inputs are applied in the correct sequence to assert output z_1. The waveforms are shown in Figure 7.50 and correctly replicate the machine specifications.

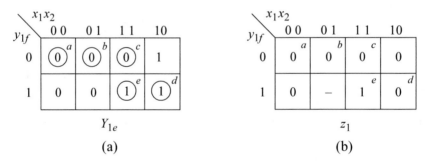

Figure 7.46   Karnaugh maps for Example 7.8: (a) excitation map and (b) output map.

$$Y_{1e} = x_1 x_2' + y_{1f} x_1$$

$$z_1 = y_{1f} x_2 \tag{7.8}$$

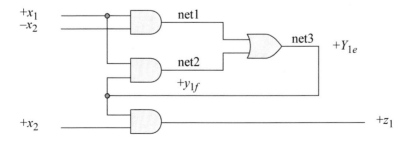

Figure 7.47   Logic diagram for the asynchronous sequential machine of Example 7.8.

```verilog
//asynchronous sequential machine
module asm (x1, x2, z1);

input x1, x2;
output z1;
wire net1, net2, net3;     //define internal nets

assign net1 = x1 & ~x2,
       net2 = x1 & net3,
       net3 = net1 | net2;

assign z1 = net3 & x2;
endmodule
```

Figure 7.48   Design module for the asynchronous sequential machine of Example 7.8.

```verilog
// test bench for the asynchronous sequential machine
module asm_tb;
reg x1, x2;
wire z1;

initial          //display variables
$monitor ("x1 = %b, x2 = %b, z1 = %b", x1, x2, z1);

initial          //apply input vectors
begin
  #0    x1 = 1'b0;   x2 = 1'b0;
  #10   x1 = 1'b0;   x2 = 1'b1;
  #10   x1 = 1'b1;   x2 = 1'b0;
  #10   x1 = 1'b1;   x2 = 1'b1;
  #10   x1 = 1'b1;   x2 = 1'b0;
  #10   x1 = 1'b1;   x2 = 1'b1;
  #10   x1 = 1'b1;   x2 = 1'b0;
  #10   x1 = 1'b0;   x2 = 1'b0;
  #10   $stop;
end

asm inst1 (    //instantiate the module into the test bench
  .x1(x1),
  .x2(x2),
  .z1(z1)
  );
endmodule
```

Figure 7.49   Test bench for the asynchronous sequential machine module of Figure 7.48.

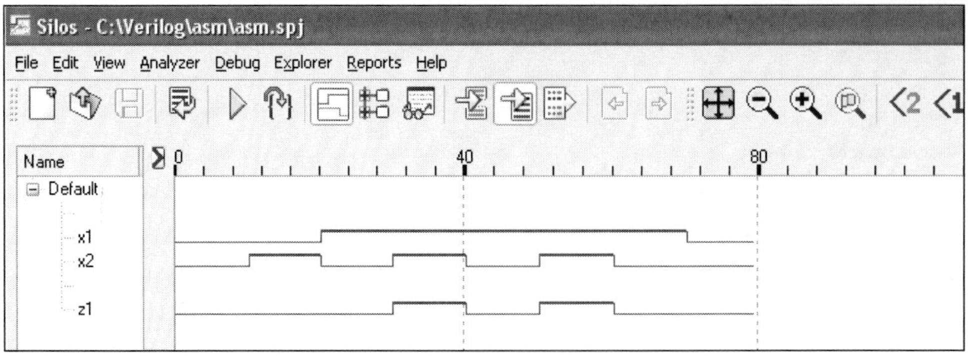

Figure 7.50   Waveforms for the asynchronous sequential machine module of Figure 7.48.

Example 7.9   An asynchronous sequential machine has two inputs x_1 and x_2 and one output z_1. The machine operates according to the following specifications:

If $x_1 x_2 = 00$, then the state of z_1 is unchanged.
If $x_1 x_2 = 01$, then z_1 is deasserted.
If $x_1 x_2 = 10$, then z_1 is asserted.
If $x_1 x_2 = 11$, then z_1 changes state.

A representative timing diagram is shown in Figure 7.51, which sequences the machine through several states. Timing diagrams may not show all possible state transition sequences due to space limitations, but all possible state transition sequences must be accounted for in the primitive flow table. The primitive flow table is shown in Figure 7.52 and is derived directly from the word description and the timing diagram.

There are no equivalent states because not all the rules for equivalence apply. In order to be equivalent, stable states must have the same input vector, the same outputs, and the same, or equivalent, next states. Therefore, stable states $\textcircled{a}$ and $\textcircled{c}$ may be equivalent because they both have inputs of $x_1 x_2 = 00$; however, they have different outputs. The same is true for stable states $\textcircled{e}$ and $\textcircled{f}$ — they have the same input vector, but have different output values.

Unless the number of rows in the reduced primitive flow table can be decreased, there will be three feedback variables for the six rows — including two rows of "don't cares." The merger diagram may reduce the number of rows, and thus, the number of feedback variables. To obtain the merger diagram, the six stable states are arranged in a circle as shown in Figure 7.53.

Rows $\textcircled{a}$ and $\textcircled{b}$ cannot merge because there is a conflict in state names for the entries in column $x_1 x_2 = 00$ of the reduced primitive flow table. Rows $\textcircled{a}$ and $\textcircled{c}$ cannot merge for the same reason. Rows $\textcircled{a}$ and $\textcircled{d}$, however, can merge into one row because there is no conflict in state names for the entries in all columns. The remaining rows of the reduced primitive flow table are processed in a similar manner.

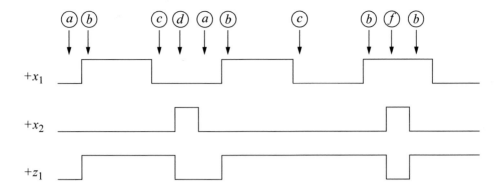

Figure 7.51   Timing diagram for the asynchronous sequential machine of Example 7.9.

x_1x_2	00	01	11	10	z_1
ⓐ	d	$-$	b	0
c	$-$	f	ⓑ	1
ⓒ	d	$-$	b	1
a	ⓓ	e	$-$	0
$-$	d	ⓔ	b	1
$-$	d	ⓕ	b	0

Figure 7.52   Primitive flow table for the asynchronous sequential machine of Example 7.9.

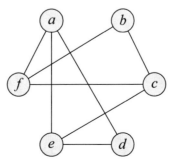

Figure 7.53   Merger diagram for Example 7.9.

The merged flow table is obtained from the merger diagram and the reduced primitive flow table. From the merger diagram it is apparent that rows ⓐ, ⓓ, and ⓔ are strongly connected; therefore, they can merge into a single row as shown in the merged flow table of Figure 7.54. Also, rows ⓑ, ⓒ, and ⓕ can merge into a single row. Thus, the merged flow table contains two rows yielding one feedback variable y_{1f} as shown in the excitation map of Figure 7.55.

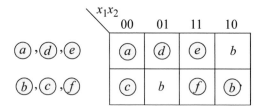

Figure 7.54   Merged flow table for Example 7.9.

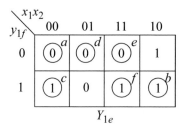

Figure 7.55   Excitation map for Example 7.9.

The equation for Y_{1e} is shown in Equation 7.9 and the output map for z_1 is shown in Figure 7.56. The equation for z_1 is shown in Equation 7.10 and the logic diagram for the asynchronous sequential machine of Example 7.9 is shown in Figure 7.57.

$$Y_{1e} = x_1 x_2' + y_{1f} x_1 + y_{1f} x_2' \tag{7.9}$$

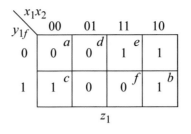

Figure 7.56 Output map for Example 7.9.

$$z_1 = y_{1f}'x_1 + y_{1f}x_2' + x_1x_2' \tag{7.10}$$

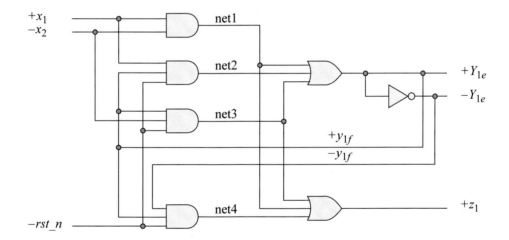

Figure 7.57 Logic diagram for the asynchronous sequential machine of Example 7.9.

The module is shown in Figure 7.58 and is generated from the logic diagram using continuous assignment statements. The test bench is shown in Figure 7.59, which sequences the machine through the states depicted in the timing diagram and the reduced primitive flow table. The results of the test bench are shown in the outputs of Figure 7.60 and the waveforms of Figure 7.61.

```
//asynchronous sequential machine
module asm6 (x1, x2, rst_n, y1e, z1);

input x1, x2, rst_n;
output y1e, z1;

//define internal nets
wire net1, net2, net3, net4;

//design for Y1e
assign    net1 = (x1 & ~x2),
          net2 = (x1 & y1e & rst_n),
          net3 = (y1e & ~x2 & rst_n),
          y1e  = (net1 | net2 | net3);

//design for z1
assign    net4 = (~y1e & x1 & rst_n),
          z1   = (net1 | net3 | net4);

endmodule
```

Figure 7.58   Design module for the asynchronous machine of Example 7.9.

```
//test bench for the asynchronous sequential machine
module asm6_tb;

reg x1, x2, rst_n;
wire y1e, z1;

//display variables
initial
$monitor ("x1=%b, state=%b, z1=%b",
           x1, y1e, z1);

//apply stimulus
initial
begin
   #0    rst_n=1'b0;
         x1=1'b0;
         x2=1'b0;
   #5    rst_n=1'b1;

//continued on next page
```

Figure 7.59   Test bench for the asynchronous machine of Example 7.9.

```
    #10     x1=1'b1;  x2=1'b0;
    #10     x1=1'b0;  x2=1'b0;
    #10     x1=1'b0;  x2=1'b1;
    #10     x1=1'b0;  x2=1'b0;
    #10     x1=1'b1;  x2=1'b0;
    #10     x1=1'b0;  x2=1'b0;
    #10     x1=1'b1;  x2=1'b0;
    #10     x1=1'b1;  x2=1'b1;
    #10     x1=1'b1;  x2=1'b0;
    #10     x1=1'b0;  x2=1'b0;
    #10     $stop;
end

//instantiate the module into the test bench
asm6 inst1 (
    .x1(x1),
    .x2(x2),
    .rst_n(rst_n),
    .y1e(y1e),
    .z1(z1)
    );

endmodule
```

Figure 7.59   (Continued)

In Figure 7.58, net2, net3, and net4 indicate that *rst_n* must be at a high logic level; that is, *rst_n* must be inactive. The test bench of Figure 7.59 assigns *rst_n* to be active (*rst_n = 1'b0;*) at #0, then at #5, the test bench assigns *rst_n* to be inactive (*rst_n = 1'b1*).

```
x1=0,  x2=0,  state=0,  z1=0
x1=1,  x2=0,  state=1,  z1=1
x1=0,  x2=0,  state=1,  z1=1
x1=0,  x2=1,  state=0,  z1=0
x1=0,  x2=0,  state=0,  z1=0
x1=1,  x2=0,  state=1,  z1=1
x1=0,  x2=0,  state=1,  z1=1
x1=1,  x2=0,  state=1,  z1=1
x1=1,  x2=1,  state=1,  z1=0
x1=1,  x2=0,  state=1,  z1=1
x1=0,  x2=0,  state=1,  z1=1
```

Figure 7.60   Outputs for the asynchronous machine of Example 7.9.

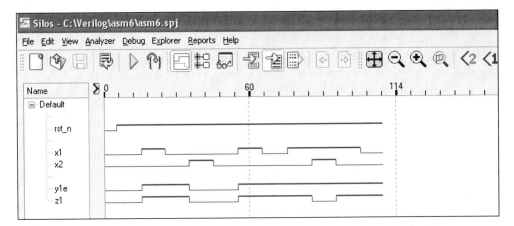

Figure 7.61   Waveforms for the asynchronous machine of Example 7.9.

7.1.10 Pulse-Mode Asynchronous Sequential Machine

Many situations are encountered in digital engineering where the input signals occur as pulses and in which there is no periodic clock signal to synchronize the operation of the sequential machine. Typical examples that use the principles of *pulse-mode* techniques are vending machines, demand-access road intersections, and automatic toll booths.

In the design of synchronous sequential machines, the data input signals are asserted as voltage levels. A periodic clock input is also required such that state changes occurred on the active clock transition. In the design of asynchronous sequential machines, the input variables were also considered as voltage levels; however, there was no machine clock to synchronize state changes. The operation of pulse-mode machines is similar, in some respects, to both synchronous and asynchronous sequential machines. State changes occur on the application of input pulses that trigger the storage elements, rather than on a clock signal. The input pulses, however, occur randomly in an asynchronous manner and more than one input pulse can generate an output.

A typical example of a pulse-mode circuit is a vending machine in which coins of various denominations produce pulses that determine — together with a switch — the selection criteria for the product.

Figure 7.62 illustrates a general block diagram for a pulse-mode sequential machine. The machine is similar in structure to a Moore machine if $\lambda(Y)$ or to a Mealy machine if $\lambda(X,Y)$. The input variables are conditioned by the δ next-state combinational logic to provide inputs to the storage elements. The inputs may also be connected to the λ output logic, providing characteristics of a Mealy machine.

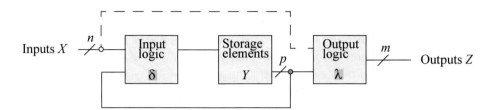

Figure 7.62   Block diagram of a pulse-mode sequential machine.

Since the storage element inputs are in the form of pulses, every term in the input equations will contain an input variable x_i. For example, the input equation to set a storage element labeled y_1 may be defined as shown in Equation 7.11, where x_1 and x_2 are pulsed inputs and y_1, y_2, and y_3 are storage elements. The outputs of a pulse-mode machine may be either voltage levels or pulses, representing Moore or Mealy outputs, respectively. A Moore output z_1 is represented as a Boolean product of two or more storage elements, as shown in Equation 7.12. A Mealy output z_2 is also represented as a Boolean product of storage elements with the inclusion of an input variable, as shown in Equation 7.13.

$$Sy_1 = x_2 + y_1'y_2y_3'x_1 \qquad (7.11)$$

$$z_1 = y_1y_2'y_3 \qquad (7.12)$$

$$z_2 = y_2'y_3x_1 \qquad (7.13)$$

The storage elements in pulse-mode machines are usually level-sensitive rather than edge-triggered devices. Thus, *SR* latches using NAND or NOR logic are typically used in the implementation of pulse-mode machines. In order for the operation of the machine to be deterministic, some restrictions apply to the input pulses:

1. Input pulses must be of sufficient duration to trigger the storage elements.

2. The time duration of the pulses must be shorter than the minimal propagation delay through the combinational input logic and the storage elements, so that the pulses are deasserted before the storage elements can again change state.

3. The time duration between successive input pulses must be sufficient to allow the machine to stabilize before application of the next pulse.

4. Only one input pulse can be active at a time.

If the input pulse is of insufficient duration, then the storage elements may not be triggered and the machine will not sequence to the next state. If the pulse duration is too long, then the pulse will still be active when the machine changes from the present state $Y_{j(t)}$ to the next state $Y_{k(t+1)}$. The storage elements may then be triggered again and sequence the machine to an incorrect next state. If the time between consecutive pulses is too short, then the machine will be triggered while in an unstable condition, resulting in unpredictable behavior.

Since pulse inputs cannot occur simultaneously, a pulse-mode machine with n input signals can have only $n+1$ combinations of the input alphabet, instead of 2^n combinations as in synchronous sequential machines and asynchronous sequential machines that are not inherently characterized by pulse-mode operation. For example, for a pulse-mode machine with two inputs x_1 and x_2, three possible valid combinations can occur: $x_1 x_2 = 00, 10,$ or 01. However, since no changes are initiated by an input vector of $x_1 x_2 = 00$, it is necessary to consider only the vectors $x_1 x_2 = 10$ and 01 when analyzing or designing a pulse-mode asynchronous sequential machine.

Similarly, for a machine with three inputs x_1, x_2, and x_3, only the following input vectors need be considered: $x_1 x_2 x_3 = 100, 010,$ and 001. The absence of a clock signal implies that state transitions occur only when an input is asserted.

The pulse width restrictions that are dominant in pulse-mode sequential machines can be eliminated by including D flip-flops in the feedback path from the SR latches to the δ next-state logic. Providing edge-triggered D flip-flops as a constituent part of the implementation negates the requirement of precisely controlled input pulse durations. This is by far the most reliable means of synthesizing pulse-mode machines. The SR latches, in conjunction with the D flip-flops, form a master-slave configuration as shown in Figure 7.63.

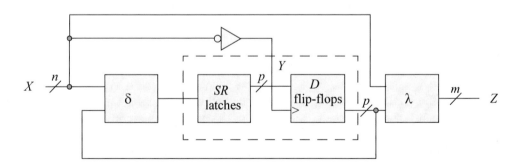

Figure 7.63   General block diagram of a pulse-mode sequential machine using SR latches and D flip-flops in a master-slave configuration.

The output of each latch connects to the D input of its associated flip-flop which in turn connects to the δ next-state logic. The flip-flops are clocked on the complemented trailing edge of the active input variable. For example, if the input pulses are active high, then the D flip-flops are triggered on the inverted negative transition of the active input pulse; that is, the flip-flops are triggered on a positive transition, as shown in Figure 7.63. JK flip-flops may also be used as the slave storage elements. The output alphabet Z of pulse-mode machines can be generated as either levels for Moore machines or as pulses for Mealy machines.

The set and reset equations for each latch will be of the form shown in Equation 7.14. The set equation SLy_j for latch Ly_j is a function f of the present input variables $X_{i(t)}$ and the present state $Y_{j(t)}$. The reset equation RLy_j is a function g of $X_{i(t)}$ and $Y_{j(t)}$. Each term of the input equations must contain a pulsed input variable because pulses trigger the latches. Since the latches are level-sensitive devices and connect directly to the D input of their corresponding slave flip-flops, the set and reset equations also serve as the input equations for the flip-flops.

$$SLy_j = f(X_{i(t)}, Y_{j(t)})$$

$$RLy_j = g(X_{i(t)}, Y_{j(t)})$$

$$Z_r = h(Y_{j(t)}) \qquad \text{Moore}$$

$$Z_r = h'(X_{i(t)}, Y_{j(t)})\text{Mealy} \qquad\qquad (7.14)$$

Equation 7.14 also lists the λ output equations for the output vector Z_r. The output equations will be a function h of the present state for Moore-type outputs, as indicated by $Z_r = h(Y_{j(t)})$. Since Moore-type outputs are not a function of the input vector $X_{i(t)}$, the outputs will be represented as levels. The outputs for a Mealy machine, however, will be a function h' of the input vector and the present state, as indicated by the equation $Z_r = h'(X_{i(t)}, Y_{j(t)})$. Mealy-type outputs, therefore, will be generated as pulses.

An example will be presented in which D flip-flops are utilized to introduce delay characteristics in the output of the SR latches.

Example 7.10 A Moore pulse-mode asynchronous sequential machine will be designed which has two inputs x_1 and x_2 and one output z_1. The deassertion of every second consecutive x_1 pulse will assert output z_1 as a level. The output will remain set for all following contiguous x_1 pulses. The output will be deasserted at the trailing edge of the second of two consecutive x_2 pulses. NAND logic will be used for the SR latches and their associated input logic. A D flip-flop will be designed, then instantiated into the design module. A state diagram is shown in Figure 7.64 that represents the complete sequencing for this Moore machine.

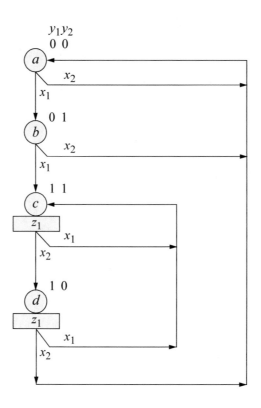

Figure 7.64   State diagram for the Moore pulse-mode asynchronous sequential machine of Example 7.10.

Each latch requires two input maps, one each for x_1 and x_2, as shown in Figure 7.65. The maps are arranged such that the maps corresponding to each latch are in the same row, and each column of maps corresponds to a unique input. The map entries are defined as follows:

> S indicates that the latch will be set.
> s indicates that the latch will remain set.
> R indicates that the latch will be reset.
> r indicates that the latch will remain reset.

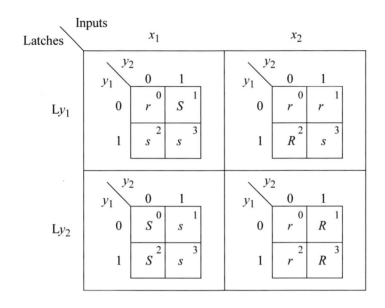

Figure 7.65   Input maps for the Moore pulse-mode asynchronous sequential machine of Example 7.10.

The map entries are obtained as in previous examples. Refer to the state diagram and minterm location 0 of the map in row Ly_1, column $x1$. In state a $(y_1y_2 = 00)$, if x_1 is pulsed, then the machine sequences to state b $(y_1y_2 = 01)$ where flip-flop y_1 remains reset. Thus, the letter r is inserted in minterm location 0. In the same map, minterm location 1 contains the entry S. That is, in state b $(y_1y_2 = 01)$, flip-flop y_1 is set if x_1 is pulsed.

The map in row Ly_2, column x_1 contains the letter S in minterm location 0. In state a $(y_1y_2 = 00)$, if x_1 is pulsed, then the machine sequences to state b $(y_1y_2 = 01)$ and flip-flop y_2 is set. In minterm location 1 of the same map, the letter s is entered. In state b $(y_1y_2 = 01)$, if x_1 is pulsed, then the machine proceeds to state c $(y_1y_2 = 11)$ where flip-flop y_2 remains set. In a similar manner, the remaining input maps are derived.

When obtaining the equations for the latches from the input maps, only the uppercase letters must be considered. The lowercase letters and any unused states are used only if they contribute to a minimized equation. The set and reset input equations are listed in Equation 7.15, where SLy_1, RLy_1 and SLy_2, RLy_2 are the set and reset equations for latches Ly_1 and Ly_2, respectively. Note that all equations contain an input variable x_i because the machine is triggered by input pulses. The output map for z_1 is shown in Figure 7.66 and the equation for z_1 is shown in Equation 7.16.

$$SLy_1 = y_2 x_1$$

$$RLy_1 = y_2' x_2$$

$$SLy_2 = x_1$$

$$RLy_2 = x_2 \qquad (7.15)$$

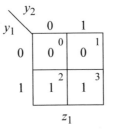

Figure 7.66   Output map for z_1 for the Moore pulse-mode asynchronous sequential machine of Example 7.10.

$$z_1 = y_1 \qquad (7.16)$$

The logic diagram is shown in Figure 7.67 using NAND logic for the SR latches and associated input logic. The latches and D flip-flops form a master-slave configuration. An active-low set ($-set\_n$) and reset ($-rst\_n$) signals are available for the D flip-flops and latches.

The module for a D flip-flop is shown in Figure 7.68 using behavioral modeling. This module will be instantiated two times into the design module for the Moore machine, which is shown in Figure 7.69. The test bench is shown in Figure 7.70, which sequences the machine through several paths of the state diagram. The outputs and waveforms are shown in Figure 7.71 and Figure 7.72, respectively.

Refer to the logic diagram and the waveforms for the discussion that follows. The assertion of the first x_1 pulse sets the Ly_2 latch (net7), which is the D input to flip-flop y_2. When x_1 is deasserted, flip-flop y_2 is set. On the assertion of the second x_1 pulse, latch Ly_1 is set (net5), which is the D input to flip-flop y_1. On the deassertion of x_1, flip-flop y_1 is set, which also asserts output z_1. Thus, the master-slave relationship of the latches in conjunction with the D flip-flops guarantees that the pulse width of the input pulses is not a factor for deterministic machine operation.

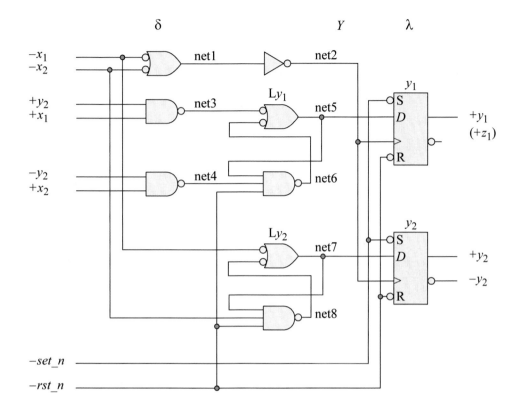

Figure 7.67   Logic diagram for the Moore pulse-mode asynchronous sequential machine of Example 7.10.

```
//behavioral d flip-flop
module d_ff (d, clk, q, q_n, set_n, rst_n);


input d, clk, set_n, rst_n;
output q, q_n;


wire d, clk, set_n, rst_n;
reg q, q_n;


//continued on next page
```

Figure 7.68   Module for a D flip-flop to be instantiated into the module for the Moore machine of Example 7.10.

```
always @ (posedge clk or negedge rst_n or negedge set_n)
begin
   if (rst_n == 0)
      begin
         q = 1'b0;
         q_n = 1'b1;
      end

   else if (set_n == 0)
      begin
         q = 1'b1;
         q_n = 1'b0;
      end

   else
      begin
         q = d;
         q_n = ~q;
      end
end

endmodule
```

Figure 7.68   (Continued)

```
//pulse-mode asynchronous sequential machine
module pm_asm3 (x1, x2, set_n, rst_n, y, y_n, z1);

input x1, x2, set_n, rst_n;
output [1:2] y, y_n;
output z1;

//define internal nets
wire net1, net2, net3, net4, net5, net6, net7, net8;

//design for D flip-flop clock
assign   net1 = ~(~x1 & ~x2),
         net2 = ~(net1);

//continued on next page
```

Figure 7.69   Design module for the Moore machine of Example 7.10.

```
//design for latch Ly1
assign    net3 = ~(y[2] & x1),
          net4 = ~(~y[2] & x2),
          net5 = (~net3 | ~net6),
          net6 = ~(net5 & net4 & rst_n);

//design for latch Ly2
assign    net7 = (x1 | ~net8),
          net8 = ~(net7 & ~x2 & rst_n);

//instantiate D flip-flop for y1
d_ff inst1 (
    .d(net5),
    .clk(net2),
    .q(y[1]),
    .q_n(y_n[1]),
    .set_n(set_n),
    .rst_n(rst_n)
    );

//instantiate D flip-flop for y2
d_ff inst2 (
    .d(net7),
    .clk(net2),
    .q(y[2]),
    .q_n(y_n[2]),
    .set_n(set_n),
    .rst_n(rst_n)
    );

//design for z1
assign z1 = y[1];
endmodule
```

Figure 7.69   (Continued)

```
//test bench for pulse-mode asynchronous sequential machine
module pm_asm3_tb;

reg x1, x2;
reg set_n, rst_n;
wire [1:2] y, y_n;
wire z1;

//continued on next page
```

Figure 7.70   Test bench for the Moore machine of Example 7.10.

```
initial                //display inputs and outputs
$monitor ("rst_n = %b, x1x2 = %b, state = %b, z1 = %b",
          rst_n, {x1, x2}, y, z1);

initial                //define input sequence
begin
   #0 set_n = 1'b1;
      rst_n = 1'b0;//reset to state_a(00); no output
      x1    = 1'b0;
      x2    = 1'b0;
   #5 rst_n = 1'b1;

   #10    x1 = 1'b1; //go to state_b(01) on posedge of 1st x1
   #10    x1 = 1'b0; //no output

   #10    x1 = 1'b1; //go to state_c(11) on posedge of 2nd x1
   #10    x1 = 1'b0; //assert output z1 on negedge of 2nd x1

   #10    x1 = 1'b1; //remain in state_c(11)
   #10    x1 = 1'b0; //output z1 remains asserted

   #20    x2 = 1'b1; //go to state_d(10) on posedge of 1st x2
   #10    x2 = 1'b0; //output z1 remains asserted

   #10    x1 = 1'b1; //go to state_c(11)
   #10    x1 = 1'b0; //output z1 remains asserted

   #10    x2 = 1'b1; //go to state_d(10) on posedge of 1st x2
   #10    x2 = 1'b0; //output z1 remains asserted

   #10    x2 = 1'b1; //go to state_a(00) on posedge of 2nd x2
   #10    x2 = 1'b0; //deassert output z1 on negedge of 2nd x2

   #30    $stop;
end

pm_asm3 inst1 (     //instantiate the module into test bench
   .x1(x1),
   .x2(x2),
   .set_n(set_n),
   .rst_n(rst_n),
   .y(y),
   .y_n(y_n),
   .z1(z1)
   );

endmodule
```

Figure 7.70   (Continued)

```
rst_n = 0, x1x2 = 00, state = 00, z1 = 0
rst_n = 1, x1x2 = 00, state = 00, z1 = 0
rst_n = 1, x1x2 = 10, state = 00, z1 = 0
rst_n = 1, x1x2 = 00, state = 01, z1 = 0
rst_n = 1, x1x2 = 10, state = 01, z1 = 0
rst_n = 1, x1x2 = 00, state = 11, z1 = 1
rst_n = 1, x1x2 = 10, state = 11, z1 = 1
rst_n = 1, x1x2 = 00, state = 11, z1 = 1
rst_n = 1, x1x2 = 01, state = 11, z1 = 1
rst_n = 1, x1x2 = 00, state = 10, z1 = 1
rst_n = 1, x1x2 = 10, state = 10, z1 = 1
rst_n = 1, x1x2 = 00, state = 11, z1 = 1
rst_n = 1, x1x2 = 01, state = 11, z1 = 1
rst_n = 1, x1x2 = 00, state = 10, z1 = 1
rst_n = 1, x1x2 = 01, state = 10, z1 = 1
rst_n = 1, x1x2 = 00, state = 00, z1 = 0
```

Figure 7.71   Outputs for the Moore machine of Example 7.10.

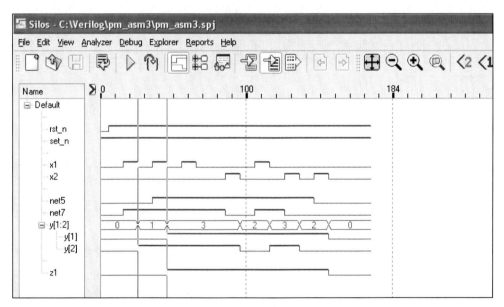

Figure 7.72   Waveforms for the Moore machine of Example 7.10.

7.2 Implicit Continuous Assignment

Continuous assignment statements specify values that are driven onto nets. Verilog provides a method of declaring a net and a continuous assignment in one statement.

This is called a net declaration assignment or *implicit continuous assignment*. Since a net is declared only once, there can be only one implicit declaration assignment per net. The example below shows a regular continuous assignment and an implicit continuous assignment. Both achieve the same result. The implicit continuous assignment is a combined net and continuous assignment.

> **wire** z_1;
> **assign** $z_1 = x_1$ & x_2;   //regular continuous assignment
>
> **wire** $z_1 = x_1$ & x_2;   //implicit continuous assignment

7.3 Delays

Delays determine the time that an assignment is made to the left-hand side when a change occurs on the right-hand side of an expression. When a delay is assigned to a net declaration, the delay is added to the inertial delay of the driver. If no delay is specified in a continuous assignment, then the right-hand side expression is assigned to the left-hand side target after zero delay.

A delay can be assigned explicitly or implicitly to a continuous assignment. An explicit assignment is shown below. The delay value is specified after the keyword **assign**.

> **assign** #5 z1 = x1 & x2;   //explicit assignment

If x_1 or x_2 changes values, then the right-hand side expression will be evaluated and the result assigned to the left-hand side target after five time units. A pulse on either x_1 or x_2 that is shorter than the delay of the assignment statement will not propagate to the output because the *inertial delay* of the gate is longer than the input pulse width. An implicit delay assignment is shown below.

> **wire** #5 z1 = x1 ^ x2;   //implicit assignment

A delay can be assigned to a net when it is declared without using a continuous assignment on the net. An example is shown below.

> **wire** #5 z1;
> **assign** z1 = x1 | x2;

Example 7.11   This example demonstrates the use of an implicit continuous assignment in an exclusive-OR circuit. There are two operands: *a[7:0]* and *b[7:0]*, where *a[0]* and *b[0]* are the low-order bits of *a* and *b*, respectively. There is one 8-bit result *rslt[7:0]*. The design module, test bench, outputs, and waveforms are shown in Figure 7.73, Figure 7.74, Figure 7.75, and Figure 7.76, respectively. There is a delay of four time units before the result is displayed.

```
//demonstrating an implicit continuous assignment
module impl_cont_assign (a, b, rslt);

input [7:0] a, b;
output [7:0] rslt;

wire [7:0] #4 rslt = a ^ b;    //implicit continuous assignment

endmodule
```

Figure 7.73   Module demonstrating an implicit continuous assignment statement.

```
//test bench for implicit continuous assignment
module impl_cont_assign_tb;

reg [7:0] a, b;
wire [7:0] rslt;

initial //apply stimulus and display variables
begin
   #0     a = 8'b1010_0011;b = 8'b1101_1111;
          #10 $display ("a=%b, b=%b, rslt=%b", a, b, rslt);

   #10    a = 8'b1111_0000;b = 8'b0000_1111;
          #10 $display ("a=%b, b=%b, rslt=%b", a, b, rslt);

   #10    a = 8'b0101_0101;b = 8'b0101_0101;
          #10 $display ("a=%b, b=%b, rslt=%b", a, b, rslt);

   #10    a = 8'b0000_0000;b = 8'b0100_0000;
          #10 $display ("a=%b, b=%b, rslt=%b", a, b, rslt);

//continued on next page
```

Figure 7.74   Test bench for the implicit continuous assignment module of Figure 7.73.

```
   #10    a = 8'b0000_0000;b = 8'b0010_0000;
          #10 $display ("a=%b, b=%b, rslt=%b", a, b, rslt);

   #10    a = 8'b0000_0000;b = 8'b0001_0000;
          #10 $display ("a=%b, b=%b, rslt=%b", a, b, rslt);

   #10    a = 8'b0000_0000;b = 8'b0000_1000;
          #10 $display ("a=%b, b=%b, rslt=%b", a, b, rslt);

   #10    $stop;
end

//instantiate the module into the test bench
impl_cont_assign inst1 (
   .a(a),
   .b(b),
   .rslt(rslt)
   );
endmodule
```

Figure 7.74   (Continued)

```
a = 10100011, b = 11011111, rslt = 01111100
a = 11110000, b = 00001111, rslt = 11111111
a = 01010101, b = 01010101, rslt = 00000000
a = 00000000, b = 01000000, rslt = 01000000
a = 00000000, b = 00100000, rslt = 00100000
a = 00000000, b = 00010000, rslt = 00010000
a = 00000000, b = 00001000, rslt = 00001000
```

Figure 7.75   Outputs for the test bench of Figure 7.74.

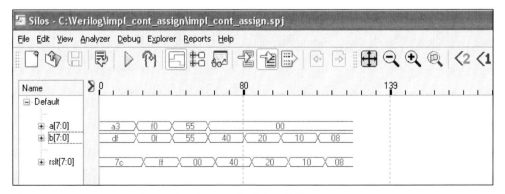

Figure 7.76   Waveforms for the test bench of Figure 7.74.

Example 7.12   A 2-bit decoder will be designed using NAND gates and an active-low *enable* input. The operands are *a* and *b*; the outputs are a 4-bit active-low vector *z[3:0]*. A delay of three time units will be specified for the continuous assignment statements that generate the outputs. When delays are used, individual continuous assignment statements must be used. Although this is a simple decoder, the same approach can be used to design any size decoder. The logic diagram is shown in Figure 7.77.

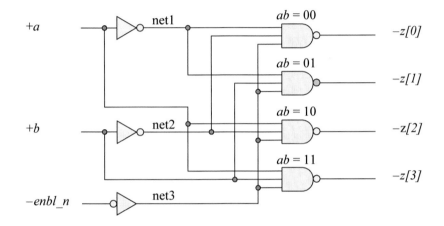

Figure 7.77   Logic diagram for a 2:4 decoder.

The design module is shown in Figure 7.78 using continuous assignment statements. The test bench and waveforms are shown in Figure 7.79 and Figure 7.80, respectively.

```
//2-to-4 decoder using NAND gates
module decoder_2to4 (a, b, enbl_n, z);

input a, b, enbl_n;
output [3:0] z;

//define internal nets
wire net1, net2, net3;

assign net1 = ~a;
assign net2 = ~b;
assign net3 = ~enbl_n;
//continued on next page
```

Figure 7.78   Module for the 2:4 decoder of Example 7.12.

```verilog
//design for decoder with delay
assign #3 z[0] = ~(net1 & net2 & net3);
assign #3 z[1] = ~(net1 & b & net3);
assign #3 z[2] = ~(a & net2 & net3);
assign #3 z[3] = ~(a & b & net3);
endmodule
```

Figure 7.78   (Continued)

```verilog
//test bench for 2-to-4 decoder
module decoder_2to4_tb;

reg a, b, enbl_n;
wire [3:0] z;

//display variables
initial
$monitor ("ab = %b, z = %b", {a, b}, z);

//apply input vectors
initial
begin
   #0    enbl_n = 1'b0;                //enable is asserted
         a = 1'b0;    b = 1'b0;
   #10   a = 1'b0;    b = 1'b0;
   #10   a = 1'b0;    b = 1'b1;
   #10   a = 1'b1;    b = 1'b0;
   #10   a = 1'b1;    b = 1'b1;
   #10   a = 1'b0;    b = 1'b0;
   #10   a = 1'b0;    b = 1'b0;
   #10   enbl_n = 1'b1;                //enable is deasserted
   #10   a = 1'b0;    b = 1'b1;
   #10   a = 1'b1;    b = 1'b0;
   #10   $stop;
end

//instantiate the module into the test bench
decoder_2to4 inst1 (
   .a(a),
   .b(b),
   .enbl_n(enbl_n),
   .z(z)
   );
endmodule
```

Figure 7.79   Test bench for the 2:4 decoder of Example 7.12.

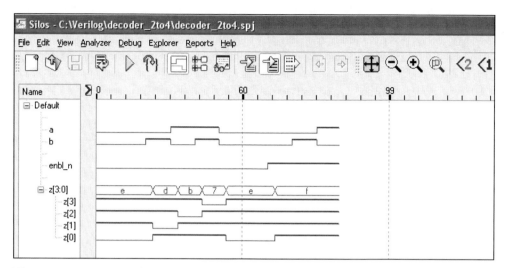

Figure 7.80   Waveforms for the 2:4 decoder. The output vector z is active low.

7.4 Problems

7.1   Design the logic for a D flip-flop with an active-high reset using three SR latches. Then implement the design using dataflow modeling. Then use the D flip-flop to design a T flip-flop. Then use the T flip-flop to design modulo-10 4-bit ripple counter. Obtain the module, test bench, outputs, and waveforms.

7.2   Given the logic diagram shown below, use dataflow modeling to obtain the design module, test bench, and waveforms for several combinations of the clock and data.

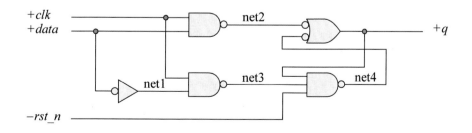

7.3   Use dataflow modeling to design a 3-bit comparator. The operands are unsigned binary numbers: *a[2:0]* and *b[2:0]*, where *a[0]* and *b[0]* are the low-order bits of *a* and *b*, respectively. Obtain the design module, test bench module, outputs, and waveforms for eight combinations of the input vectors.

7.4   Design a full adder from two half adders using dataflow modeling. There are three scalar inputs: *a*, *b*, and *cin*; there are two scalar outputs: *sum* and *cout*. Obtain the design module, test bench, outputs, and waveforms for all combinations of the inputs.

7.5   Given the Karnaugh map shown below, write a dataflow module that specifies function z_1 in a minimal sum-of-products form. Obtain the test bench and outputs for all combinations of the inputs.

x_3x_4

x_1x_2	0 0	0 1	1 1	1 0
0 0	1	1	0	1
0 1	0	0	0	0
1 1	0	1	1	1
1 0	0	0	0	0

z_1

7.6   Given the Karnaugh map shown below, write a dataflow module that specifies function z_1 in a minimal product-of-sums form. Obtain the test bench and outputs for all combinations of the inputs.

x_3x_4

x_1x_2	0 0	0 1	1 1	1 0
0 0	0	0	0	0
0 1	1	1	0	1
1 1	1	1	1	1
1 0	1	0	0	1

z_1

7.7   Given the logic diagram shown below, use dataflow modeling to obtain the design module, the test bench module, and the outputs to determine the equation for output z_1.

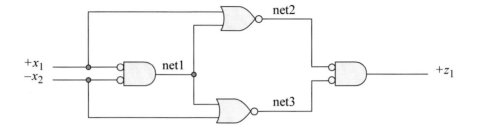

7.8   Design a dataflow module that will generate a logic 1 when two 4-bit unsigned operands are unequal. The operands are: $a[3:0]$ and $b[3:0]$; the output is z_1. Obtain the test bench and outputs for eight of the $2^8 = 256$ combinations of the inputs.

7.9   Using dataflow modeling, design an 8-bit odd parity generator. Obtain the design module, test bench module, and outputs for eight of the $2^8 = 256$ combinations of the inputs. There will be one output labeled z_1.

7.10   Use D flip-flops to design a counter that counts in the following sequence: $y_1 y_2 = 00, 10, 01, 11, 00, \ldots$, where y_2 is the low-order stage of the counter. Use the D flip-flop from Problem 7.1. The negative edge of the clock triggers the counter. The reset is active high. The Karnaugh maps for flip-flops y_1 and y_2 are shown below. Obtain the design module, test bench, outputs, and waveforms for the complete counting sequence.

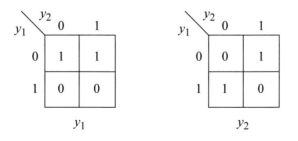

7.11   Use D flip-flops to design a counter that counts in the following sequence: $y_1y_2y_3y_4 = 0000, 1000, 0100, 0010, 0001, 1000, \ldots$, where y_4 is the low-order stage of the counter. Use the D flip-flop from Problem 7.1. The negative edge of the clock triggers the counter. The reset is active high. The Karnaugh maps are shown below. Obtain the design module, test bench, outputs, and waveforms for the complete counting sequence. This counter is useful for obtaining four discrete time intervals. Alternatively, the counting sequence can be $y_1y_2y_3y_4 = 1000, 0100, 0010, 0001, 1000, \ldots$ if the D flip-flop has a *set* input. The counter would be initialized to $y_1y_2y_3y_4 = 1000$.

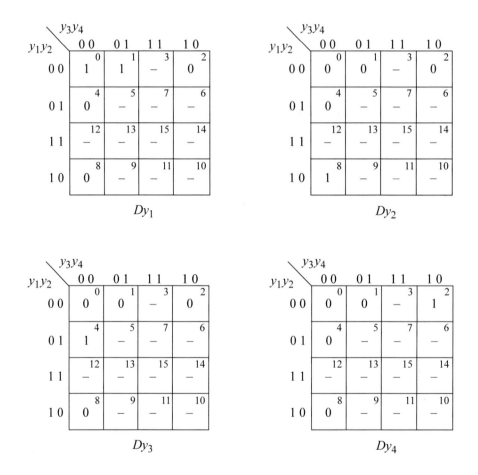

7.12   The waveforms for an asynchronous sequential machine are shown on the next page. There is one input and two outputs. Output z_1 toggles on the rising edge of input x_1. Output z_2 toggles on the falling edge of x_1. This represents a toggle flip-flop with two outputs. Use dataflow modeling to obtain the module, test bench, outputs, and waveforms. A reset signal must be connected to each AND gate of the latches; otherwise, the outputs will be designated as **x**.

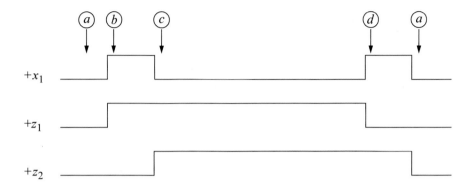

7.13 Design an asynchronous sequential machine that has one input x_1 and one output z_1 that operates according to the timing diagram shown below. Obtain the design module, test bench module, outputs, and waveforms.

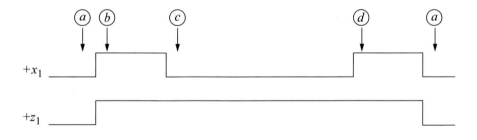

7.14 Design a Mealy pulse-mode asynchronous sequential machine which has two inputs x_1 and x_2 and one output z_1. For a Mealy machine, the outputs are a function of the present states and the present inputs. Output z_1 is asserted coincident with the x_2 pulse if the x_2 pulse is immediately preceded by a pair of x_1 pulses. Use *SR* latches and *D* flip-flops in a master-slave configuration. Use NAND logic for all gates and latches except for output z_1, which will be an AND gate.

Derive the state diagram, input maps, output map, and logic diagram. Then use continuous assignment to design the module. Obtain the test bench, outputs, and waveforms.

8

Behavioral Modeling

In previous chapters, built-in primitives, user-defined primitives (UDPs), and data-flow modeling were used to design hardware primarily at the gate level. This chapter describes the *behavior* of a digital system and is not concerned with the direct implementation of logic gates but more on the architecture of the system. This is an algorithmic approach to hardware implementation and represents a higher level of abstraction than previous modeling methods. A Verilog module may contain a mixture of built-in primitives, UDPs, dataflow constructs, and behavioral constructs. The constructs in behavioral modeling closely resemble those used in the C programming language.

8.1   Procedural Constructs

A *procedure* is series of operations taken to design a module. A Verilog module that is designed using behavioral modeling contains no internal structural details, it simply defines the behavior of the hardware in an abstract, algorithmic description. Verilog contains two structured procedure statements or behaviors: **initial** and **always**. A behavior may consist of a single statement or a block of statements delimited by the keywords **begin . . . end**. A module may contain multiple **initial** and **always** statements. These statements are the basic statements used in behavioral modeling and execute concurrently starting at time zero in which the order of execution is not important. All other behavioral statements are contained inside these structured procedure statements.

8.1.1 Initial Statement

All statements within an **initial** statement comprise an **initial** block. An **initial** statement executes only once beginning at time zero, then suspends execution. An **initial** statement provides a method to initialize and monitor variables before the variables are used in a module; it is also used to generate waveforms. For a given time unit, all statements within the **initial** block execute sequentially. Execution or assignment is controlled by the # symbol. The syntax for an **initial** statement is shown below.

> **initial** [optional timing control] procedural statement or
> block of procedural statements

Examples of procedural statements are shown below.

    Procedural assignment
        Blocking
        Nonblocking
    Conditional statement
    Case statement
    Loop statement
        For
        While
        Repeat
        Forever
    Block statement
        Sequential
        Parallel
    Procedural continuous assignment
        Assign
        Deassign
        Force
        Release

Each **initial** block executes concurrently at time zero and each block ends execution independently. If there is only one procedural statement, then the statement does not require the keywords **begin** . . . **end**. However, if there are two or more procedural statements, they are delimited by the keywords **begin** . . . **end**.

Example 8.1   A module showing the use of the **initial** statement is shown in Figure 8.1, where the variables x_1, x_2, x_3, x_4, and x_5 are initialized to specific values. Multiple **initial** statements are used for both a single procedural statement and a block of procedural statements. The outputs and waveforms are shown in Figure 8.2 and Figure 8.3, respectively.

```verilog
//module showing use of the initial keyword
module initial_ex (x1, x2, x3, x4, x5);

output x1, x2, x3, x4, x5;

reg x1, x2, x3, x4, x5;

//display variables
initial
$monitor ($time, " x1x2x3x4x5 = %b", {x1, x2, x3, x4, x5});

//initialize variables to 0
//multiple statements require begin . . . end
initial
begin
   #0    x1 = 1'b0
         x2 = 1'b0;
         x3 = 1'b0;
         x4 = 1'b0;
         x5 = 1'b0;
end

//set x1
//single statement requires no begin . . . end
initial
   #10   x1 = 1'b1;

//set x2 and x3
initial
begin
   #10   x2 = 1'b1;
   #10   x3 = 1'b1;
end

//set x4 and x5
initial
begin
   #10   x4 = 1'b1;
   #10   x5 = 1'b1;
end

//continued on next page
```

Figure 8.1   Module to illustrate the use of the **initial** statement.

```
//reset variables
initial
begin
    #20    x1 = 1'b0;
    #10    x2 = 1'b0;
    #10    x3 = 1'b0;
    #10    x4 = 1'b0;
    #10    x5 = 1'b0;
end

//determine length of simulation
initial
    #70    $finish;

endmodule
```

Figure 8.1   (Continued)

```
0   x1x2x3x4x5 = 00000
10  x1x2x3x4x5 = 11010
20  x1x2x3x4x5 = 01111
30  x1x2x3x4x5 = 00111
40  x1x2x3x4x5 = 00011
50  x1x2x3x4x5 = 00001
60  x1x2x3x4x5 = 00000
```

Figure 8.2   Outputs for the module of Figure 8.1.

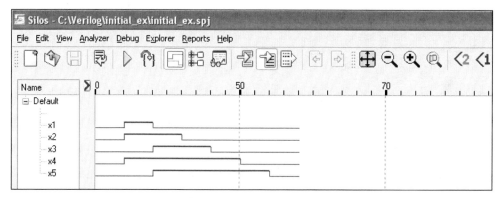

Figure 8.3   Waveforms for the module of Figure 8.1.

Refer to the module of Figure 8.1 and the waveforms of Figure 8.3 for the discussion that follows. Figure 8.1 contains seven **initial** statements. The first **initial** statement invokes the system task **$monitor**, which causes the specified string (enclosed in quotation marks) to be printed whenever a variable changes in the argument list (enclosed in braces). The **$time** system function returns the simulation time as a 64-bit number.

The second **initial** statement initializes all variables to zero. The third **initial** statement sets x_1 at 10 time units. Since all **initial** statements begin execution at time zero, the fourth **initial** statement sets x_2 at 10 time units also, and sets x_3 at time 20 time units (#10 plus #10). This can be seen in the waveforms of Figure 8.3. Variable x_4 is set at 10 time units by the fifth **initial** statement, which also sets x_5 at 20 time units. The sixth **initial** statement resets all variables.

Example 8.2 The **initial** statement can also be used to generate waveforms for vectors as shown in the module of Figure 8.4, the outputs of Figure 8.5, and the waveforms of Figure 8.6.

```
//module showing the use of initial keyword
module initial_ex2 (z1);

output [7:0] z1;

reg [7:0] z1;


//display variables
initial
$monitor ("z1 = %b", z1);


//apply stimulus
initial
begin
   #0      z1 = 8'h00;
   #10     z1 = 8'hc0;
   #10     z1 = 8'he0;
   #10     z1 = 8'hfc;
   #10     z1 = 8'hff;
   #10     z1 = 8'h00;
end

endmodule
```

Figure 8.4   Vector waveforms generated by an **initial** statement.

```
z1 = 00000000
z1 = 11000000
z1 = 11100000
z1 = 11111100
z1 = 11111111
z1 = 00000000
```

Figure 8.5   Outputs for the module of Figure 8.4.

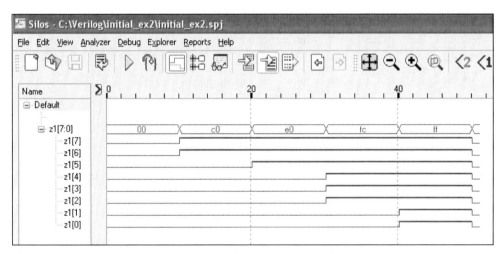

Figure 8.6   Waveforms for the module of Figure 8.4.

8.1.2 Always Statement

The **always** statement executes the behavioral statements within the **always** block repeatedly in a looping manner and begins execution at time zero. Execution of the statements continues indefinitely until the simulation is terminated. The keywords **initial** and **always** specify a behavior and the statements within a behavior are classified as *behavioral* or *procedural*. The syntax for the **always** statement is shown below.

> **always** [optional timing control] procedural statement or
> block of procedural statements

A typical application of the **always** statement is shown in Figure 8.7 to generate a series of clock pulses as used in test bench constructs. There are two **initial** statements and one **always** statement. The clock is first initialized to zero, then the **always** statement cycles the clock every 10 time units for a clock period of 20 time units. The clock stops cycling after 100 time units. The system task **$finish** causes the simulator to exit the module and return control to the operating system. The clock waveform is shown in Figure 8.8.

```
//clock generation using initial and always statements
module clk_gen2 (clk);

output clk;
reg clk;

//initialize clock to 0
initial
    clk = 1'b0;

//toggle clock every 10 time units
always
    #10    clk = ~clk;

//determine length of simulation
initial
    #100    $finish;

endmodule
```

Figure 8.7   Clock waveform generation.

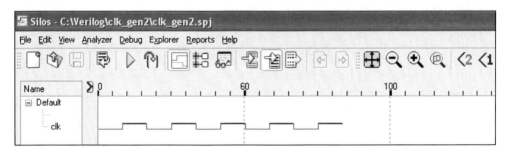

Figure 8.8   Waveform for the clock generation module of Figure 8.7.

An alternative method to generate clock pulses is shown in Figure 8.9 using the **initial** statement and the **forever** loop statement. The waveform is shown in Figure 8.10.

```
//clock generation using initial and forever
module clk_gen3 (clk);

output clk;
reg clk;

//define clock
initial
begin
   clk = 1'b0;
   forever
      #10  clk = ~clk;
end

//define length of simulation
initial
   #100  $finish;

endmodule
```

Figure 8.9   Alternative method to generate clock pulses.

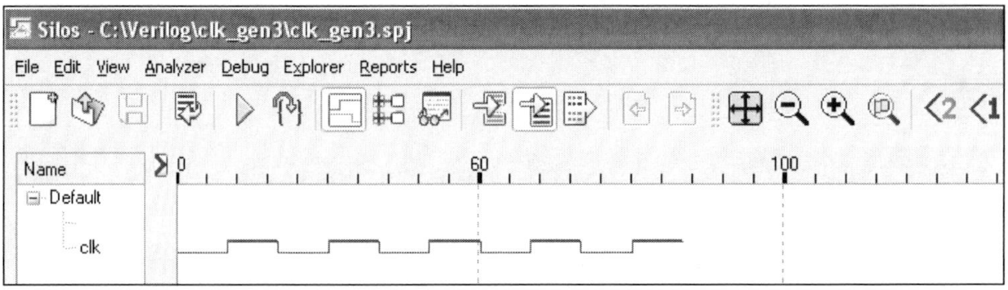

Figure 8.10   Waveforms for the clock generation module of Figure 8.9.

Example 8.3   An **always** statement is often used with an *event control list* — or *sensitivity list* — to execute a sequential block. When a change occurs to a variable in the sensitivity list, the statement or block of statements in the **always** block is executed. The keyword **or** is used to indicate multiple events as shown in Figure 8.11 for a 3-input AND gate. When one or more inputs change state, the statement in the **always** block is executed. The **begin** . . . **end** keywords are not necessary in this example because there is only one behavioral statement. Target variables used in an **always**

statement are declared as type **reg**. Output z_1 is delayed by five time units to allow for the propagation delay — inertial delay — of the AND gate.

   The test bench is shown in Figure 8.12 in which an input vector *invect*, containing one more bit than the number of input variables, is applied. This allows for complete testing of the AND gate by applying all eight combinations of the inputs. The outputs are shown in Figure 8.13 and the waveforms are shown in Figure 8.14. Output z_1 is in an unknown state for the first five time units until the inputs propagate through the AND gate. Note that at time 42 all inputs are asserted, but output z_1 is not asserted until time 47, a delay of five time units.

```
//behavioral 3-input AND gate
module and3_bh (x1, x2, x3, z1);

input x1, x2, x3;
output z1;

wire x1, x2, x3;      //alternatively do not declare wires
reg z1;               //because inputs are wire by default

always @ (x1 or x2 or x3)
     z1 = #5 (x1 & x2 & x3);

endmodule
```

Figure 8.11   Module to illustrate the use of an **always** statement with an event control list.

```
//test bench for behavioral 3-input AND gate
module and3_bh_tb;

reg x1, x2, x3;
wire z1;

//generate stimulus and display variables
initial
begin: apply_stimulus
   reg [3:0] invect;
   for (invect=0; invect<8; invect=invect+1)
      begin
         {x1, x2, x3} = invect [2:0];
         #6 $display ($time, "x1x2x3=%b%b%b, z1=%b",
                       x1, x2, x3, z1);
      end
end           //continued on next page
end
```

Figure 8.12   Test bench for the 3-input AND gate of Figure 8.11.

```
//instantiate the module into the test bench
and3_bh inst1 (
    .x1(x1),
    .x2(x2),
    .x3(x3),
    .z1(z1)
    );

endmodule
```

Figure 8.12   (Continued)

```
 6 x1x2x3 = 000, z1 = 0
12x1x2x3 = 001, z1 = 0
18x1x2x3 = 010, z1 = 0
24x1x2x3 = 011, z1 = 0
30x1x2x3 = 100, z1 = 0
36x1x2x3 = 101, z1 = 0
42x1x2x3 = 110, z1 = 0
48x1x2x3 = 111, z1 = 1
```

Figure 8.13   Outputs for the behavioral model of a 3-input AND gate of Figure 8.11.

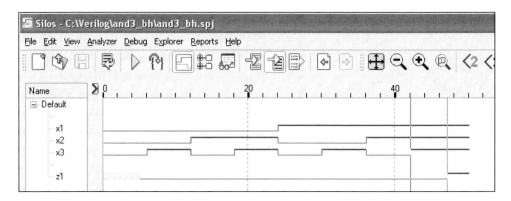

Figure 8.14   Waveforms for the behavioral model of a 3-input AND gate of Figure 8.11.

Example 8.4   This example models the design of a 4-bit carry lookahead adder using an expanded version of the *generate* and *propagate* functions. Recall that the generate and propagate functions were defined as

$$\text{Generate:} \quad G_i = a_i b_i$$

$$\text{Propagate:} \quad P_i = a_i + b_i$$

for any stage $_i$ of the adder. For a 4-bit adder, the augend and addend are *a[3:0]* and *b[3:0]*, where *a[0]* and *b[0]* are the low-order bits of *a* and *b*, respectively. The sum is *sum[3:0]* with a carry-out *cout*.

The behavioral module is shown in Figure 8.15 using the **always** statement with a sensitivity list of *a*, *b*, and carry-in *cin*. Whenever any of the three variables in the sensitivity list changes values, the statements in the **begin** . . . **end** block execute. Notice that *blocking* procedural assignments are used for *sum* and *cout* because the adder is designed with combinational logic. Blocking assignments complete execution before the following statement executes. They are executed in the order in which they are specified in a sequential block.

The variables *sum* and *cout* are used as left-hand side targets in the **always** block; therefore, they are declared as type **reg**. The first term in the expression for *cout* is (*a[3]* & *b[3]*), which equates to P_3; the second term is ((*a[3]* | *b[3]*) & (*a[2]* & *b[2]*)), which is equivalent to $P_3 G_2$.

The test bench is shown in Figure 8.16 for 10 values of *a*, *b*, and *cin*. The outputs are shown in Figure 8.17 and the waveforms are shown in Figure 8.18 in hexadecimal notation.

```
//behavioral model for a 4-bit adder
module adder4 (a, b, cin, sum, cout);

input [3:0] a, b;
input cin;
output [3:0] sum;
output cout;

wire [3:0] a, b;
wire cin;

reg [3:0] sum;
reg cout;

//continued on next page
```

Figure 8.15   Module for the 4-bit adder of Example 8.4.

```
always @ (a or b or cin)
begin
  sum  = a + b + cin;
  cout = (a[3] & b[3]) |
         ((a[3] | b[3]) & (a[2] & b[2])) |
         ((a[3] | b[3]) & (a[2] | b[2]) & (a[1] & b[1])) |
         ((a[3] | b[3]) & (a[2] | b[2]) & (a[1] | b[1])
              & (a[0] & b[0])) |
         ((a[3] | b[3]) & (a[2] | b[2]) & (a[1] | b[1])
              & (a[0] | b[0]) & cin);
end

endmodule
```

Figure 8.15   (Continued)

```
//test bench for the 4-bit adder
module adder4_tb;

reg [3:0] a, b;
reg cin;
wire [3:0] sum;
wire cout;

initial          //display variables
$monitor ("a=%b, b=%b, cin=%b, cout=%b, sum=%b",
           a, b, cin, cout, sum);

initial          //apply input vectors
begin
   #0    a=4'b0000;   b=4'b0000;   cin=1'b0;
   #10   a=4'b0001;   b=4'b0001;   cin=1'b0;
   #10   a=4'b0001;   b=4'b0011;   cin=1'b0;
   #10   a=4'b0101;   b=4'b0001;   cin=1'b0;
   #10   a=4'b0111;   b=4'b0001;   cin=1'b0;
   #10   a=4'b0101;   b=4'b0101;   cin=1'b0;
   #10   a=4'b1001;   b=4'b0101;   cin=1'b1;
   #10   a=4'b1000;   b=4'b1000;   cin=1'b1;
   #10   a=4'b1011;   b=4'b1110;   cin=1'b1;
   #10   a=4'b1111;   b=4'b1111;   cin=1'b1;
   #10   $stop;
end

//continued on next page
```

Figure 8.16   Test bench for the 4-bit adder of Figure 8.15.

```
//instantiate the module into the test bench
adder4 inst1 (
    .a(a),
    .b(b),
    .cin(cin),
    .sum(sum),
    .cout(cout)
    );

endmodule
```

Figure 8.16   (Continued)

```
a=0000, b=0000, cin=0, cout=0, sum=0000
a=0001, b=0001, cin=0, cout=0, sum=0010
a=0001, b=0011, cin=0, cout=0, sum=0100
a=0101, b=0001, cin=0, cout=0, sum=0110
a=0111, b=0001, cin=0, cout=0, sum=1000
a=0101, b=0101, cin=0, cout=0, sum=1010
a=1001, b=0101, cin=1, cout=0, sum=1111
a=1000, b=1000, cin=1, cout=1, sum=0001
a=1011, b=1110, cin=1, cout=1, sum=1010
a=1111, b=1111, cin=1, cout=1, sum=1111
```

Figure 8.17   Outputs for the 4-bit adder of Figure 8.15.

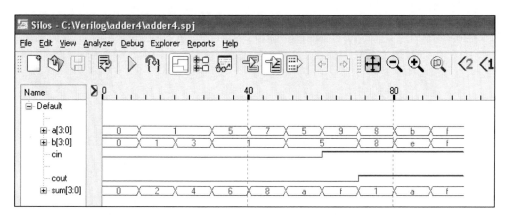

Figure 8.18   Waveforms for the 4-bit adder of Figure 8.15.

Example 8.5   This example presents the design of a simple add-shift unit that illustrates behavioral modeling for an 8-bit adder together with shift left and shift right capabilities. The sensitivity list in the **always** statement contains the augend *a* and the addend *b*. When either *a* or *b* changes value, the sum is obtained and then shifted left four bit positions for an equivalent multiply-by-16 operation; the sum is also shifted right four bit positions for an equivalent divide-by-16 operation.

The add operation produces a 9-bit result to accommodate the sum plus carry-out. The *left_shft_rslt* and the *right_shft_rslt* are 16-bit registers to allow for larger shift amounts. The module provides no information on the detailed design or architecture of the unit — it simply states the behavior.

The design module is shown in Figure 8.19 and the test bench is shown in Figure 8.20, providing five different input vectors. The outputs are shown in Figure 8.21 and the waveforms are shown in Figure 8.22.

```
//add shift operations
module add_shift (a, b, sum, left_shft_rslt, right_shft_rslt);

input [7:0] a, b;
output [8:0] sum;
output [15:0] left_shft_rslt, right_shft_rslt;

wire [7:0] a, b;
reg [8:0] sum;
reg [15:0] left_shft_rslt, right_shft_rslt;

always @ (a or b)
begin
   sum = a + b;
   left_shft_rslt  = sum << 4;
   right_shft_rslt = sum >> 4;
end
endmodule
```

Figure 8.19   Behavioral design module for an 8-bit add-shift unit.

```
//add shift test bench
module add_shift_tb;

reg [7:0] a, b;
wire [8:0] sum;
wire [15:0] left_shft_rslt, right_shft_rslt;
//continued on next page
```

Figure 8.20   Test bench for the add-shift unit of Figure 8.19.

```
initial                 //display variables
$monitor ("a=%b, b=%b, sum=%b, left_shft_rslt=%b,
                        right_shft_rslt=%b",
         a, b, sum, left_shft_rslt, right_shft_rslt);

initial                 //apply input vectors
begin
   #0    a = 8'b0101_0101;
         b = 8'b0101_0101;

   #10   a = 8'b0000_1100;
         b = 8'b0000_0100;

   #10   a = 8'b1111_0000;
         b = 8'b0000_1111;

   #10   a = 8'b1010_0000;
         b = 8'b0000_1111;

   #10   a = 8'b1111_1111;
         b = 8'b1111_1111;
   #10   $stop;
end

add_shift inst1 (    //instantiate the module
   .a(a),
   .b(b),
   .sum(sum),
   .left_shft_rslt(left_shft_rslt),
   .right_shft_rslt(right_shft_rslt)
   );

endmodule
```

Figure 8.20   (Continued)

```
a=01010101, b=01010101, sum=010101010,
left_shft_rslt=0000101010100000,
right_shft_rslt=0000000000001010

a=00001100, b=00000100, sum=000010000,
left_shft_rslt=0000000100000000,
right_shft_rslt=0000000000000001

//continued on next page
```

Figure 8.21   Outputs for the add-shift unit of Figure 8.19.

```
a=11110000, b=00001111, sum=011111111,
left_shft_rslt=0000111111110000,
right_shft_rslt=0000000000001111

a=10100000, b=00001111, sum=010101111,
left_shft_rslt=0000101011110000,
right_shft_rslt=0000000000001010

a=11111111, b=11111111, sum=111111110,
left_shft_rslt=0001111111100000,
right_shft_rslt=0000000000011111
```

Figure 8.21   (Continued)

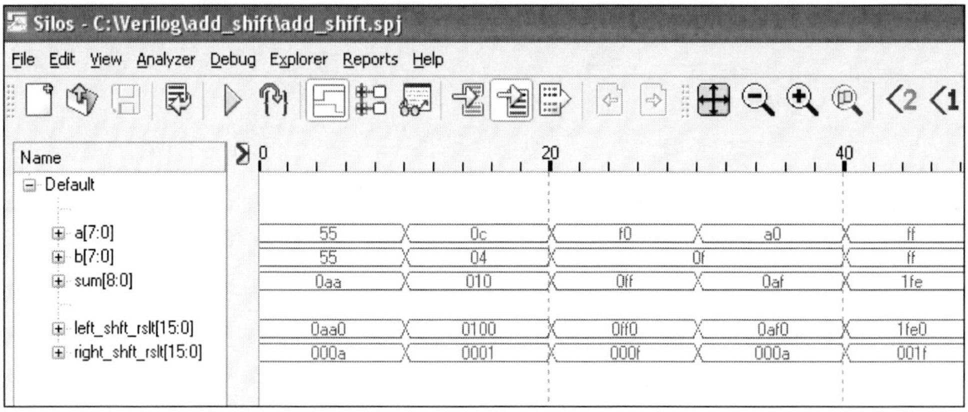

Figure 8.22   Waveforms for the add-shift unit of Figure 8.19.

Example 8.6   Although multiplexers are typically designed using a **case** statement, this example shows another method to design a 4:1 multiplexer using behavioral modeling together with continuous assignment statements. There are two select inputs $s[1:0]$, four data inputs $d[3:0]$, and one enable input *enbl*.

The logic diagram is shown in Figure 8.23 with four internal nets which will be ORed together in the **always** block. The select inputs are available in both high and low assertion. The data inputs and the enable input are active high. Because output z_1 is a target in the **always** block, it must be declared as type **reg**. Any change to the data inputs will affect the internal nets in the sensitivity list and thus cause the procedural statement for z_1 to be executed.

The design module, test bench module, outputs, and waveforms are shown in Figure 8.24, Figure 8.25, Figure 8.26, and Figure 8.27, respectively.

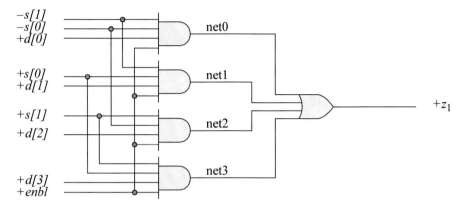

Figure 8.23   Logic diagram for the 4:1 multiplexer of Example 8.6.

```
//behavioral 4:1 multiplexer
module mux_4to1_behav (d, s, enbl, z1);


input [3:0] d;
input [1:0] s;
input enbl;
output z1;


//define internal nets
wire net0, net1, net2, net3;

reg z1;        //z1 is used in the always statement
               //and must be declared as type reg


//define AND gates
assign    net0 = (d[0] & ~s[1] & ~s[0] & enbl),
          net1 = (d[1] & ~s[1] &  s[0] & enbl),
          net2 = (d[2] &  s[1] & ~s[0] & enbl),
          net3 = (d[3] &  s[1] &  s[0] & enbl);

always @ (net0 or net1 or net2 or net3)
   z1 = (net0 || net1 || net2 || net3);

endmodule
```

Figure 8.24   Behavioral module for the 4:1 multiplexer of Example 8.6.

```
//test bench for behavioral 4:1 multiplexer
module mux_4to1_behav_tb;

reg [3:0] d;
reg [1:0] s;
reg enbl;
wire z1;

initial          //display variables
$monitor ("s=%b, d=%b, z1=%b", s, d, z1);

initial          //apply input vectors
begin
    #0     s=2'b00; d=4'b0000; enbl=1'b1;    //d[0]=0; z1=0
    #10    s=2'b00; d=4'b1001; enbl=1'b1;    //d[0]=1; z1=1
    #10    s=2'b01; d=4'b1010; enbl=1'b1;    //d[1]=1; z1=1
    #10    s=2'b10; d=4'b0011; enbl=1'b1;    //d[2]=0; z1=0
    #10    s=2'b01; d=4'b1011; enbl=1'b1;    //d[1]=1; z1=1
    #10    s=2'b11; d=4'b1000; enbl=1'b1;    //d[3]=1; z1=1
    #10    s=2'b11; d=4'b0110; enbl=1'b1;    //d[3]=0; z1=0
    #10    $stop;
end

//instantiate the module into the test bench
mux_4to1_behav inst1 (
    .d(d),
    .s(s),
    .enbl(enbl),
    .z1(z1)
    );

endmodule
```

Figure 8.25   Test bench for the 4:1 multiplexer of Figure 8.24.

```
s=00, d=0000, z1=0
s=00, d=1001, z1=1
s=01, d=1010, z1=1
s=10, d=0011, z1=0
s=01, d=1011, z1=1
s=11, d=1000, z1=1
s=11, d=0110, z1=0
```

Figure 8.26   Outputs for the 4:1 multiplexer of Figure 8.24.

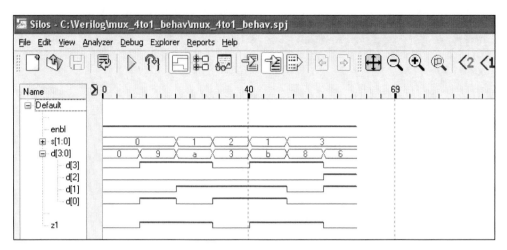

Figure 8.27   Waveforms for the 4:1 multiplexer of Figure 8.24.

Example 8.7   An 8-bit odd parity generator can also be designed using behavioral modeling. The inputs consist of an 8-bit vector $x[7:0]$ and a scalar output z_1. Output z_1 is declared as type **reg** because it is used in the **always** block. Any change to the input vector causes the **always** statement to be executed.

The design module is shown in Figure 8.28 and the test bench is shown in Figure 8.29, providing several different input vectors. The outputs are shown in Figure 8.30 and the waveforms in Figure 8.31 show the input vector in hexadecimal notation together with the individual bits.

```
//behavioral 8-bit odd parity generator
module par_gen8_behav (x, z1);

input [7:0] x;
output z1;

//inputs are wire by default

reg z1;

always @ (x)
   z1 = ~(x[0] ^ x[1] ^ x[2] ^ x[3] ^ x[4] ^ x[5] ^ x[6] ^ x[7]);

endmodule
```

Figure 8.28   Behavioral module for an 8-bit odd parity generator.

```verilog
//test bench for behavioral 4:1 multiplexer
module par_gen8_behav_tb;

reg [7:0] x;
wire z1;

//display variables
initial
$monitor ("x=%b, z1=%b", x, z1);

//apply input vectors
initial
begin
    #0      x = 8'b0000_0000;
    #10     x = 8'b0001_1010;
    #10     x = 8'b1001_1010;
    #10     x = 8'b1001_1111;
    #10     x = 8'b1101_0011;
    #10     x = 8'b1001_1010;
    #10     x = 8'b1101_1010;
    #10     x = 8'b1011_1111;
    #10     x = 8'b1111_1111;
    #10     $stop;
end

//instantiate the module into the test bench
par_gen8_behav inst1 (
    .x(x),
    .z1(z1)
    );

endmodule
```

Figure 8.29   Test bench for the 8-bit odd parity generator of Figure 8.28.

```
x=00000000, z1=1
x=00011010, z1=0
x=10011010, z1=1
x=10011111, z1=1
x=11010011, z1=0
x=10011010, z1=1
x=11011010, z1=0
x=10111111, z1=0
x=11111111, z1=1
```

Figure 8.30   Outputs for the 8-bit odd parity generator of Figure 8.28.

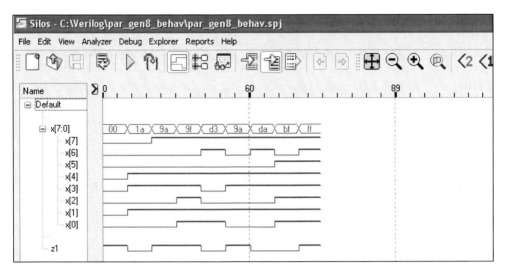

Figure 8.31   Waveforms for the 8-bit odd parity generator of Figure 8.28.

8.2 Procedural Assignments

A statement that assigns values to register variables is called a procedural assignment. Procedural assignments are performed sequentially and assign values only to register variables of type **reg** (including memory elements), **integer**, **real**, **realtime**, and **time**. The left-hand target of a procedural assignment can also be a bit-select, a part-select, or a concatenation of the two. A procedural assignment is an assignment that takes place within an **initial** or an **always** block. A sequential block may contain procedural assignments with timing controls as shown below.

```
reg [15:0] bus_a;
    . . .
initial
begin
    #0  bus_a = 16'h00ff;
    #5  bus_a = 16'hff00;
    #5  bus_a = 16'habdc;
    #5  bus_a = 16'h65ab;
end
```

Procedural assignments (used in behavioral modeling) and continuous assignments (used in dataflow modeling) differ in the following areas:

Procedural Assignment	Continuous Assignment
Occurs within an **initial** or an **always** statement	Not used in an **initial** or an **always** statement
Operates with respect to other statements in the **initial** or **always** block	Executes concurrently with other statements
Drives registers	Drives nets
Uses blocking (=) or nonblocking (<=) symbols	Uses the assignment (=) symbol

When the right-hand expression of a continuous assignment changes, the expression is evaluated and scheduled to be assigned to the left-hand target at a time specified by an optional timing control. Register variables, however, are assigned a value only when a procedural statement executes. If control is not passed to the procedural statement, then the statement is not executed. This can occur when using the **if** . . . **else** conditional statement.

A net is used to indicate a connection between elements within a module or between modules. A net should not be assigned a value within a behavior, task, or function except in the case of **force** . . . **release** procedural continuous assignments. A net variable within a module must be driven by a built-in primitive, a UDP, a continuous assignment, a **force** . . . **release** procedural continuous assignment, or a module port. A register can be assigned a value only within a procedural statement, task, or function.

8.2.1 Intrastatement Delay

A procedural assignment may have an optional delay. A delay appearing to the right of an assignment operator is called an intrastatement delay. It is the delay by which the right-hand result is delayed before assigning it to the left-hand target. In the example below, the expression $(x_1 \ \& \ x_2)$ is evaluated, a delay of five time units is taken, then the result is assigned to z_1.

$$z_1 = \#5 \ (x_1 \ \& \ x_2);$$

One purpose for an intrastatement delay is to simulate the delay through a logic gate. In the above example, the propagation delay through the AND gate is five time units. Although an **initial** statement is used to model test benches, it is used here to demonstrate intrastatement delays for waveform generation as shown in the modules of Figure 8.32 and Figure 8.33.

```
//behavioral intrastatement delay example
module intra_stmt_dly (x1, x2);

//assign timescale
`timescale 10ns / 1ns


output x1, x2;
reg x1, x2;

//apply input signals
initial
begin
   x1 = #0    1'b0;
   x2 = #0    1'b0;

   x1 = #1    1'b1;
   x2 = #0.5  1'b1;

   x1 = #1    1'b0;
   x2 = #2    1'b0;

   x1 = #1    1'b1;
   x2 = #2    1'b1;

   x1 = #2    1'b0;
   x2 = #1    1'b0;
end

endmodule
```

Figure 8.32   Module to generate waveforms using intrastatement delays.

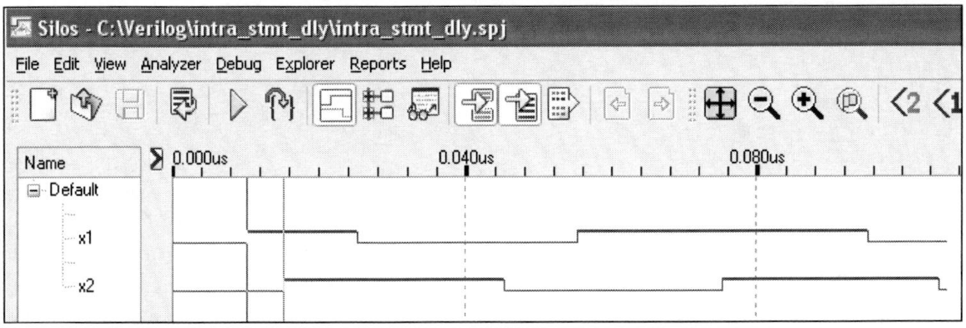

Figure 8.33   Waveforms for the intrastatement delay module of Figure 8.32.

Example 8.8   This example will illustrate intrastatement delay for three operations: a statement consisting of two AND gates and an exclusive-OR gate; a statement using the conditional operator; and a statement for a 4-bit odd parity generator. There are four inputs: x_1, x_2, x_3, and x_4. There are three outputs: z_1, z_2, and z_3, which are defined as follows:

$$z_1 = \#2 \ (x_1 \ \& \sim x_2) \ \wedge (\sim x_3 \ \& \ x_4);$$
$$z_2 = \#3 \ (x_1 >= x_2) \ ? \ x_3 : x_4;$$
$$z_3 = \#4 \sim (x_1 \wedge x_2 \wedge x_3 \wedge x_4);$$

The equation for output z_1 can be expanded into a sum-of-products expression as follows:

$$
\begin{aligned}
z_1 &= x_1 x_2' \oplus x_3' x_4 \\
&= (x_1 x_2')(x_3' x_4)' + (x_1 x_2')'(x_3' x_4) \\
&= (x_1 x_2')(x_3 + x_4') + (x_1' + x_2)(x_3' x_4) \\
&= x_1 x_2' x_3 + x_1 x_2' x_4' + x_1' x_3' x_4 + x_2 x_3' x_4
\end{aligned}
$$

In order to compare the values for z_1 with those obtained from the test bench module, the Karnaugh map shown in Figure 8.34 is provided.

$x_1 x_2$ \\ $x_3 x_4$	0 0	0 1	1 1	1 0
0 0	0	1	0	0
0 1	0	1	0	0
1 1	0	1	0	0
1 0	1	0	1	1

z_1

Figure 8.34   Karnaugh map for output z_1 for the intrastatement delay of Example 8.8.

The behavioral module is shown in Figure 8.35 in which intrastatement delays are assigned to the statements that generate z_1, z_2, and z_3. The test bench module is shown in Figure 8.36 for all combinations of the inputs. Figure 8.37 shows the outputs for z_1, z_2, and z_3 based upon the definitions stated above. The waveforms are shown in Figure 8.38, which show the delays for each output. The values for the outputs are unknown until their respective delays have taken place. Since blocking assignments are used, the delays are cumulative; that is, z_1 receives its value two time units after the inputs change, z_2 receives its value at five time units, and z_3 receives its value at nine time units after the inputs change.

```verilog
//behavioral model to demonstrate intrastatement delay
module intra_stmt_dly2 (x1, x2, x3, x4, z1, z2, z3);

input x1, x2, x3, x4;
output z1, z2, z3;

reg z1, z2, z3;

always @ (x1 or x2 or x3 or x4)
begin
   z1 = #2 (x1 & ~x2) ^ (~x3 & x4);
   z2 = #3 (x1 >= x2) ? x3 : x4;
   z3 = #4 ~(x1 ^ x2 ^ x3 ^ x4);
end

endmodule
```

Figure 8.35   Module to illustrate intrastatement delay.

```verilog
//test bench for intrastatement delay
module intra_stmt_dly2_tb;

reg x1, x2, x3, x4;
wire z1, z2, z3;

//apply input vectors and display variables
initial
begin: apply_stimulus
   reg [4:0] invect;
   for (invect=0; invect<16; invect=invect+1)
      begin
         {x1, x2, x3, x4} = invect [4:0];
         #10 $display ("x1 x2 x3 x4 = %b, z1=%b, z2=%b, z3=%b",
                       {x1, x2, x3, x4}, z1, z2, z3);
      end
end

//continued on next page
```

Figure 8.36   Test bench for the intrastatement delay module of Figure 8.35.

```
//instantiate the module into the test bench
intra_stmt_dly2 inst1 (
    .x1(x1),
    .x2(x2),
    .x3(x3),
    .x4(x4),
    .z1(z1),
    .z2(z2),
    .z3(z3)
    );

endmodule
```

Figure 8.36   (Continued)

$$z_1 = \#2 \ (x_1 x_2' x_3 + x_1 x_2' x_4' + x_1' x_3' x_4 + x_2 x_3' x_4)$$

$$z_2 = \#3 \ (x_1 >= x_2) \ ? \ x_3 \ : \ x_4)$$

$$z_3 = \#4 \ (x_1 \oplus x_2 \oplus x_3 \oplus x_4)'$$

```
x1 x2 x3 x4 = 0000, z1 = 0, z2 = 0, z3 = 1
x1 x2 x3 x4 = 0001, z1 = 1, z2 = 0, z3 = 0
x1 x2 x3 x4 = 0010, z1 = 0, z2 = 1, z3 = 0
x1 x2 x3 x4 = 0011, z1 = 0, z2 = 1, z3 = 1
x1 x2 x3 x4 = 0100, z1 = 0, z2 = 0, z3 = 0
x1 x2 x3 x4 = 0101, z1 = 1, z2 = 1, z3 = 1
x1 x2 x3 x4 = 0110, z1 = 0, z2 = 0, z3 = 1
x1 x2 x3 x4 = 0111, z1 = 0, z2 = 1, z3 = 0
x1 x2 x3 x4 = 1000, z1 = 1, z2 = 0, z3 = 0
x1 x2 x3 x4 = 1001, z1 = 0, z2 = 0, z3 = 1
x1 x2 x3 x4 = 1010, z1 = 1, z2 = 1, z3 = 1
x1 x2 x3 x4 = 1011, z1 = 1, z2 = 1, z3 = 0
x1 x2 x3 x4 = 1100, z1 = 0, z2 = 0, z3 = 1
x1 x2 x3 x4 = 1101, z1 = 1, z2 = 0, z3 = 0
x1 x2 x3 x4 = 1110, z1 = 0, z2 = 1, z3 = 0
x1 x2 x3 x4 = 1111, z1 = 0, z2 = 1, z3 = 1
```

Figure 8.37   Outputs for the intrastatement delay module of Figure 8.35.

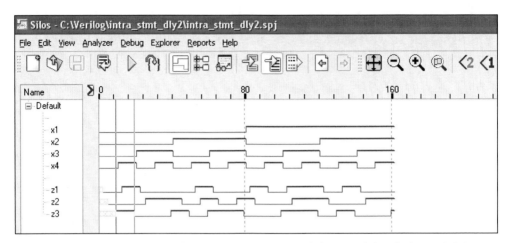

Figure 8.38   Waveforms for the intrastatement delay module of Figure 8.35.

8.2.2 Interstatement Delay

Interstatement delay is the delay taken before a statement is executed. In the code segment shown below, the delay given in the second statement specifies that when the first statement has finished executing, wait five time units before executing the second statement.

$$z_1 = (x_1 + x_2)\, x_3;$$
$$\#5\ z_2 = x_4 \wedge x_5;$$

If no delays are specified in a procedural assignment, then there is zero delay in the assignment. Example 8.9 illustrates the use of an interstatement delay.

Example 8.9   To illustrate an interstatement delay, a module will be designed that has three inputs: x_1, x_2, and x_3. There are two outputs, z_1 and z_2, which are defined as follows:

$$z_1 = (x_1\ \&\ x_2\ \&\ {\sim}x_3);$$
$$\#5\ z_2 = {\sim}(x_1 \wedge x_2 \wedge x_3);$$

The design module is shown in Figure 8.39 using behavioral modeling. The test bench is shown in Figure 8.40 and the outputs are shown in Figure 8.41. The outputs show the function of z_1 and z_2. Output z_1 is a logic 1 when $x_1 x_2 x_3 = 110$; output z_2 is a logic 1 to maintain odd parity over $x_1 x_2 x_3 z_2$. The waveforms are shown in Figure 8.42 in which the interstatement delay is clearly seen for output z_2. When the statement for z_1 has executed, a delay of five time units is taken and then the statement for z_2 executes. The right-hand side is evaluated and then z_2 is assigned the result with zero delay.

```verilog
//behavioral module to illustrate interstatement delay
module inter_stmt_dly (x1, x2, x3, z1, z2);

input x1, x2, x3;
output z1, z2;

reg z1, z2;

always @ (x1 or x2 or x3)
begin
      z1 = (x1 & x2 & ~x3);
   #5 z2 = ~(x1 ^ x2 ^ x3);
end

endmodule
```

Figure 8.39   Module to demonstrate interstatement delay.

```verilog
//test bench for interstatement delay
module inter_stmt_dly_tb;

reg x1, x2, x3;
wire z1, z2;

//apply input vectors and display variables
initial
begin: apply_stimulus
   reg [3:0] invect;
   for (invect=0; invect<8; invect=invect+1)
      begin
         {x1, x2, x3} = invect [3:0];
         #10  $display ("x1 x2 x3 = %b, z1 = %b, z2 = %b",
                        {x1, x2, x3}, z1, z2);
      end
end

//continued on next page
```

Figure 8.40   Test bench for the interstatement delay module of Figure 8.39.

```
//instantiate the module into the test bench
inter_stmt_dly inst1 (
    .x1(x1),
    .x2(x2),
    .x3(x3),
    .z1(z1),
    .z2(z2)
    );

endmodule
```

Figure 8.40   (Continued)

```
        z1 = (x1 & x2 & ~x3);
    #5 z2 = ~(x1 ^ x2 ^ x3);

x1 x2 x3 = 000, z1 = 0, z2 = 1
x1 x2 x3 = 001, z1 = 0, z2 = 0
x1 x2 x3 = 010, z1 = 0, z2 = 0
x1 x2 x3 = 011, z1 = 0, z2 = 1
x1 x2 x3 = 100, z1 = 0, z2 = 0
x1 x2 x3 = 101, z1 = 0, z2 = 1
x1 x2 x3 = 110, z1 = 1, z2 = 1
x1 x2 x3 = 111, z1 = 0, z2 = 0
```

Figure 8.41   Outputs for the interstatement delay module of Figure 8.39.

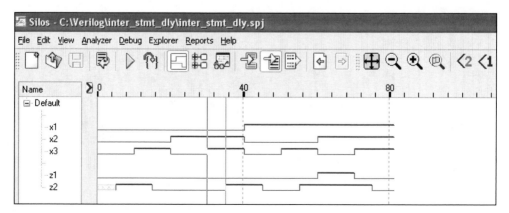

Figure 8.42   Waveforms for the interstatement delay module of Figure 8.39.

8.2.3 Blocking Assignments

A blocking procedural assignment completes execution before the next statement executes. The assignment operator $(=)$ is used for blocking assignments. The right-hand expression is evaluated, then the assignment is placed in an internal temporary register called the event queue and scheduled for assignment. If no time units are specified, the scheduling takes place immediately. The event queue is covered in Appendix A.

In the code segment below, an interstatement delay of one time unit is specified. Evaluation of the assignment is delayed by the timing control. The right-hand expression for z_1 is evaluated and the assignment is scheduled to take place at time units $t +$ 1. The execution of any following statements is blocked until the assignment occurs.

```
initial
   begin
      #1  z1 = x1 & x2;
          z2 = x2 | x3;
   end
```

Example 8.10 The module of Figure 8.43 shows delayed blocking assignments for a block of three statements, each with an interstatement delay of one time unit. The blocking statement for z_1 is assigned to be executed one time unit later than the current simulation time t at $t + 1$. The right-hand side expression is evaluated at time $t + 1$ and assigned to z_1 at time $t + 1$. The statement for z_2 is evaluated at time $t + 2$, then assigned to z_2. The statement for z_3 is evaluated at time $t + 3$, then assigned to z_3.

The test bench is shown in Figure 8.44 and the waveforms are shown in Figure 8.45, which show the delay for each blocking statement. At 10 time units, x_2 changes from 0 to 1. Two time units later, output z_2 as asserted; one time unit later, z_3 is asserted.

```
//example of blocking procedural assignment
module blocking1 (x1, x2, z1, z2, z3);

input x1, x2;
output z1, z2, z3;
reg z1, z2, z3;

always @ (x1 or x2)
begin
   #1 z1 = x1 & x2;
   #1 z2 = x1 | x2;
   #1 z3 = x1 ^ x2;
end
endmodule
```

Figure 8.43   Module showing delayed blocking assignments.

```
//test bench for blocking assignment
module blocking1_tb;

reg x1, x2;
wire z1, z2, z3;

//display variables
initial
$monitor ("x1 x2 = %b, z1 = %b, z2 = %b, z3 = %b",
          {x1, x2}, z1, z2, z3);

//apply input vectors
initial
    #0     x1 = 1'b0;   x2 = 1'b0;
    #10    x1 = 1'b0;   x2 = 1'b1;
    #10    x1 = 1'b1;   x2 = 1'b0;
    #10    x1 = 1'b1;   x2 = 1'b1;
    #10    $stop;
end

//instantiate the module into the test bench
blocking1 inst1 (
    .x1(x1),
    .x2(x2),
    .z1(z1),
    .z2(z2),
    .z3(z3)
    );

endmodule
```

Figure 8.44   Test bench for the blocking assignment module of Figure 8.43.

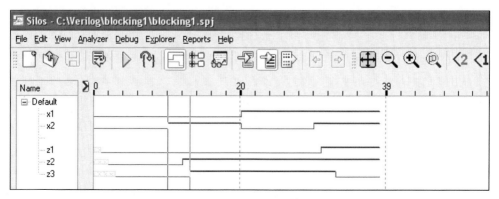

Figure 8.45   Waveforms for the blocking assignment module of Figure 8.43.

Example 8.11   Figure 8.46 shows an example of blocking assignments using intra-statement delays. Evaluations of $((x_1 \mid x_2) \& x_3)$ and $\sim(x_1 \char`^ x_2 \char`^ x_3)$ are blocked until $(x_1 \& x_2) \mid x_3$ has been assigned to z_1, which occurs at $t+1$ time units. When the statement for z_2 is reached, it will be scheduled in the event queue at time $t+1$, but the assignment to z_2 will not occur until time $t+2$ time units. The evaluation of

$$z_3 = \#1 \sim (x_1 \char`^ x_2 \; x_3)$$

is blocked until the assignment is made to z_2.

The test bench is shown in Figure 8.47 and the waveforms are shown in Figure 8.48. When the inputs change, the waveforms show that the assignments to z_1, z_2, and z_3 are made at time $t+1$, $t+2$, and $t+3$, respectively.

```
//blocking intrastatement delay
module blocking2 (x1, x2, x3, z1, z2, z3);

input x1, x2, x3;
output z1, z2, z3;

reg z1, z2, z3;

always @ (x1 or x2 or x3)
begin
   z1 = #1 (x1 & x2) | x3;
   z2 = #1 (x1 | x2) & x3;
   z3 = #1 ~(x1 ^ x2 ^ x3);
end
endmodule
```

Figure 8.46   Module showing blocking assignments using intrastatement delays.

```
//test bench for blocking intrastatement delay
module blocking2_tb;

reg x1, x2, x3;
wire z1, z2, z3;

//display variables
initial
$monitor ("x1 x2 x3 = %b, z1 = %b, z2 = %b, z3 = %b",
          {x1, x2, x3}, z1, z2, z3);

//continued on next page
```

Figure 8.47   Test bench for the module using blocking assignments for intrastatement delays.

```
//apply input vectors and display variables

initial
begin: apply_stimulus
   reg [3:0] invect;
   for (invect=0; invect<8; invect=invect+1)
      begin
         {x1, x2, x3} = invect [3:0];
         #10  $display ("x1 x2 x3 = %b, z1=%b, z2=%b, z3=%b",
                        {x1, x2, x3}, z1, z2, z3);
      end
end


//instantiate the module into the test bench
blocking2 inst1 (
   .x1(x1),
   .x2(x2),
   .x3(x3),
   .z1(z1),
   .z2(z2),
   .z3(z3)
   );

endmodule
```

Figure 8.47   (Continued)

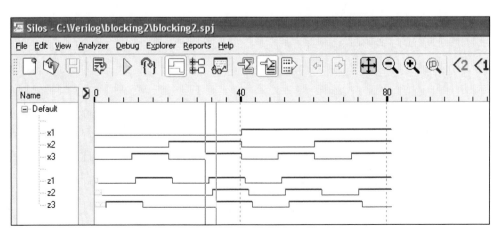

Figure 8.48   Waveforms for Figure 8.46.

Example 8.12 Blocking procedural assignments with intrastatement delays can also be used to generate waveforms as shown in the module of Figure 8.49. The variable *clk* is unknown until time unit four. The first statement executes at time zero, but *clk* is not assigned the value 0 until four time units. Then the second statement executes, which assigns a value of 1 to *clk* after eight time units — eight time units from time zero. The waveform is shown in Figure 8.50.

```
//blocking assignment to generate waveform
module blocking3 (clk);

output clk;

reg clk;

initial
begin
   clk = #4  1'b0;
   clk = #4  1'b1;
   clk = #8  1'b0;
   clk = #10 1'b1;
   clk = #10 1'b0;
   clk = #20 1'b1;
   clk = #30 1'b0;
end

endmodule
```

Figure 8.49   Clock generation using blocking procedural assignment with intrastatement delays.

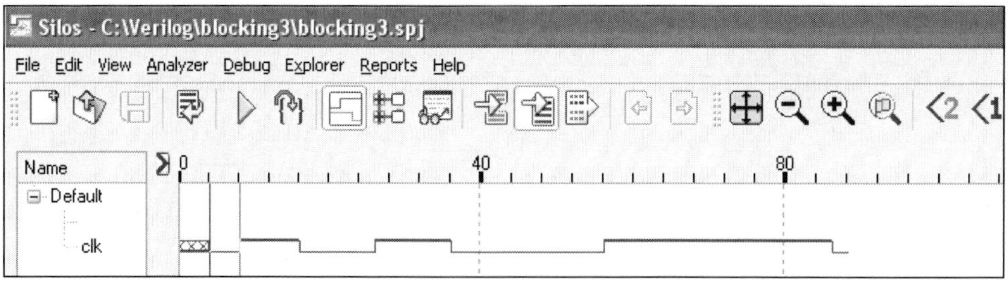

Figure 8.50   Waveform for Figure 8.49.

8.2.4 Nonblocking Assignments

The assignment symbol ($<=$) is used to represent a nonblocking procedural assignment. Nonblocking assignments allow the scheduling of assignments without blocking execution of the following statements in a sequential procedural block. A nonblocking assignment is used to synchronize assignment statements so that they appear to execute at the same time. In the code segment shown below using blocking assignments, the result is indeterminate because both **always** blocks execute concurrently resulting in a race condition. Depending on the simulator implementation, either $x_1 = x_2$ would be executed before $x_2 = x_3$ or vice versa.

> **always** @ (posedge clk)
> $x_1 = x_2$;
>
> **always** @ (posedge clk)
> $x_2 = x_3$;

The race condition is solved by using nonblocking assignments as shown below.

> **always** @ (posedge clk)
> $x_1 <= x_2$;
>
> **always** @ (posedge clk)
> $x_2 <= x_3$;

The Verilog simulator schedules a nonblocking assignment statement to execute, then proceeds to the next statement in the block without waiting for the previous nonblocking statement to complete execution. That is, the right-hand expression is evaluated and the value is stored in the event queue and is *scheduled* to be assigned to the left-hand target. The assignment is made at the end of the current time step if there are no intrastatement delays specified.

Nonblocking assignments are typically used to model several concurrent assignments that are caused by a common event such as @ **posedge** clk. The order of the assignments is irrelevant because the right-hand side evaluations are stored in the event queue before any assignments are made.

Example 8.13 This example uses nonblocking assignments and intrastatement delays to create a waveform. Because nonblocking assignments are used, the delays are not cumulative. All the statements in Figure 8.51 begin execution at time zero. The execution of the first statement results in a value of 0 scheduled to be assigned to *clk* at time unit 5. The second statement schedules a value of 1 to be assigned to *clk* at time unit 10. The third statement schedules *clk* to receive a value of 0 at time unit 15. The assignments continue relative to time zero. The waveforms are shown in Figure 8.52.

```
//example of nonblocking statements
module nonblock (clk);

output clk;
reg clk;

initial
begin
  clk <= #5   1'b0;
  clk <= #10  1'b1;
  clk <= #15  1'b0;
  clk <= #20  1'b1;
  clk <= #30  1'b0;
end
endmodule
```

Figure 8.51   Module to illustrate the use of nonblocking assignments with intra-statement delays.

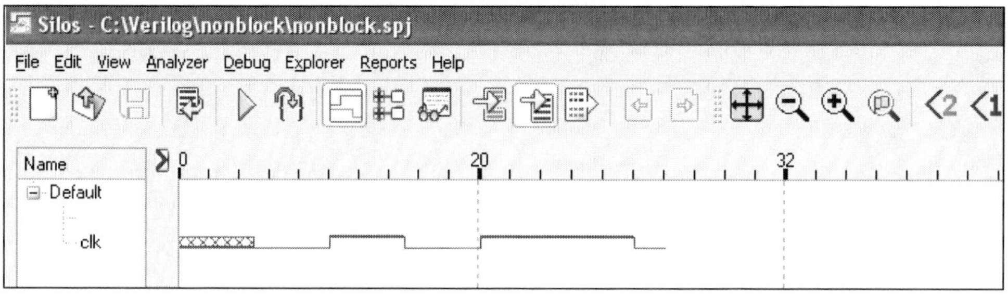

Figure 8.52   Waveform for Figure 8.51.

Example 8.14 This example will model register assignments using blocking and nonblocking constructs with intrastatement delays. The first three statements in the **initial** block of the module shown in Figure 8.53 use blocking assignments and execute sequentially at time 0. Because of the nonblocking behavioral construct in the **initial** block, the next three statements are processed at the same simulation time, but are scheduled to execute at different times due to the intrastatement delays.

The statement *data_reg_a[1:0] <= #10 2'b11;* is scheduled to execute at time unit 10. The statement *data_reg_b[7:0] <= #5 {data_reg_a[3:0], 4'b0011};* is scheduled to execute at time unit 5. The last statement *index <= index + 1;* executes at time unit 0 because there is no intrastatement delay. The waveforms of Figure 8.54 show the assignments to the registers based on their scheduling in the event queue.

```
//example of nonblocking assignment
module nonblock3 (data_reg_a, data_reg_b, index);

output [7:0] data_reg_a, data_reg_b;
output [3:0] index;

reg [7:0] data_reg_a, data_reg_b;
reg [3:0] index;

initial
begin
   data_reg_a = 8'h84;
   data_reg_b = 8'h0f;
   index = 4'b0;

   data_reg_a [1:0] <= #10 2'b11;
   data_reg_b [7:0] <= #5 {data_reg_a [3:0], 4'b0011};
   index <= index +1;
end

endmodule
```

Figure 8.53   Module to model blocking and nonblocking assignments.

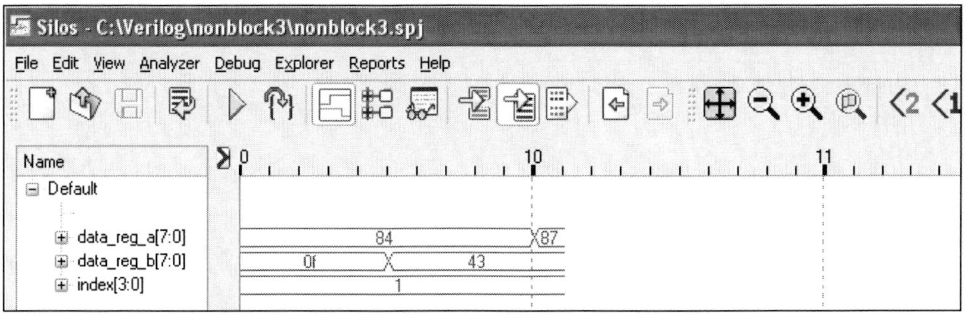

Figure 8.54   Waveforms for Figure 8.53.

Example 8.15   This example will emulate a 4-bit parallel-in, serial-out shift register. The load/shift operation is controlled by the state of a *load* input. When the *load* input is at a high logic level, the register loads data from an input bus *x[1:4]*; when the *load* input is at a low logic level, the contents are shifted right one bit position and a zero is shifted into the vacated leftmost position. There is a 4-bit internal register *y[1:4]*. The serial output z_1 is the output of *y[4]*. The design module, test bench module, outputs, and waveforms are shown in Figure 8.55, Figure 8.56, Figure 8.57, and Figure 8.58, respectively.

```
//parallel-in serial-out shift register
module shift_reg_piso (clk, load, x, y, z1);


input clk, load;
input [1:4] x;
output [1:4] y;
output z1;

reg [1:4] y;            //must be reg if used in always
assign z1 = y[4];


always @ (posedge clk)
begin
   y[1] <= ((load && x[1]) || (~load && 1'b0));
   y[2] <= ((load && x[2]) || (~load && y[1]));
   y[3] <= ((load && x[3]) || (~load && y[2]));
   y[4] <= ((load && x[4]) || (~load && y[3]));
end

endmodule
```

Figure 8.55   Behavioral module for a 4-bit parallel-in, serial-out shift register using nonblocking assignments.

```
//test bench for piso shift register
module shift_reg_piso_tb;
reg clk, load;
reg [1:4] x;
wire [1:4] y;
wire z1;
initial          //define clock
begin
   clk = 1'b0;
   forever
      #10  clk = ~clk;
end

initial
$monitor ("load=%b, x=%b, y=%b, z1=%b", load, x, y, z1);

initial                      //apply input vectors
begin
//continued on next page
```

Figure 8.56   Test bench for the 4-bit shift register using behavioral modeling.

```
    #0      load = 1'b0;    x = 4'b0000;
    #5      load = 1'b1;    x = 4'b0101;
    #10     load = 1'b0;
    #30     load = 1'b1;    x = 4'b1100;
    #20     load = 1'b0;
    #10     load = 1'b1;    x = 4'b1111;
    #10     load = 1'b0;
    #20     load = 1'b1;    x = 4'b1111;
    #10     load = 1'b0;
    #30     $stop;
end
shift_reg_piso inst1 (      //instantiate the module
    .clk(clk),
    .load(load),
    .x(x),
    .y(y),
    .z1(z1)
    );
endmodule
```

Figure 8.56   (Continued)

```
load=0, x=0000, y=xxxx, z1=x    load=0, x=1100, y=0110, z1=0
load=1, x=0101, y=xxxx, z1=x    load=1, x=1111, y=0110, z1=0
load=1, x=0101, y=0101, z1=1    load=0, x=1111, y=0110, z1=0
load=0, x=0101, y=0101, z1=1    load=0, x=1111, y=0011, z1=1
load=0, x=0101, y=0010, z1=0    load=1, x=1111, y=0011, z1=1
load=1, x=1100, y=0010, z1=0    load=1, x=1111, y=1111, z1=1
load=1, x=1100, y=1100, z1=0    load=0, x=1111, y=1111, z1=1
load=0, x=1100, y=1100, z1=0    load=0, x=1111, y=0111, z1=1
```

Figure 8.57   Outputs for the shift register of Figure 8.55.

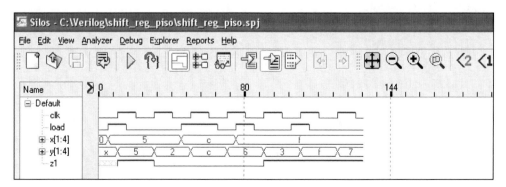

Figure 8.58   Waveforms for the shift register of Figure 8.55.

8.3 Conditional Statements

Conditional statements alter the flow within a behavior based upon certain conditions. The choice among alternative statements depends on the Boolean value of an expression. The alternative statements can be a single statement or a block of statements delimited by the keywords **begin** . . . **end**. The keywords **if** and **else** are used in conditional statements. There are three categories of the conditional statement as shown below. A true value is 1 or any nonzero value; a false value is 0, **x**, or **z**. If the evaluation is false, then the next expression in the activity flow is evaluated.

//no **else** statement
if (expression) statement1;   //if expression is true, then statement1 is executed.

//one **else** statement   //choice of two statements. Only one is executed.
if (expression) statement1;   //if expression is true, then statement1 is executed.
else statement2;   //if expression is false, then statement2 is executed.

//nested **if-else if**   //choice of multiple statements. Only one is executed.
if (expression1) statement1;   //if expression1 is true, then statement1 is executed.
else if (expression2) statement2; //if expression2 is true, then statement2 is executed.
else if (expression3) statement3; //if expression3 is true, then statement3 is executed.
else default statement;

Examples of the three categories are shown below.

//no **else** statement
if $(x_1 \ \& \ x_2) \, z_1 = 1;$

//one **else** statement
if (rst_n == 0)
    ctr = 3'b000;
else ctr = next_count;

//nested **if-else if**
if (opcode == 00)
    $z_1 = x_1 + x_2;$
else if (opcode == 01)
    $z_1 = x_1 - x_2;$
else if (opcode == 10)
    $z_1 = x_1 * x_2;$
else
    $z_1 = x_1 / x_2;$

Several examples will now be presented to illustrate the use of conditional statements. The examples will range from relatively simple to moderately complex. They will include logic circuits, a D flip-flop, a JK flip-flop, and counters of different moduli.

Example 8.16   This example illustrates the implementation of a sum-of-products expression for two AND gates and one OR gate. There are four inputs: x_1, x_2, x_3, and x_4 and one output z_1. Output z_1 is asserted when $x_1 x_2 = 11$ or when $x_3 x_4 = 11$. Figure 8.59 contains the behavioral module using conditional statements with intrastatement delays. The test bench, outputs, and waveforms are shown in Figure 8.60, Figure 8.61, and Figure 8.62, respectively.

```
//sum-of-products equation using if - else
module sop_eqn_if_else (x1, x2, x3, x4, z1);

input x1, x2, x3, x4;
output z1;

reg z1;

always @ (x1 or x2 or x3 or x4)
begin
   if ((x1 && x2) || (x3 && x4))
      z1 = #2 1;
   else
      z1 = #2 0;
end

endmodule
```

Figure 8.59   Module to illustrate the use of **if . . . else** conditional statements.

```
//test bench for sop_eqn_if_else
module sop_eqn_if_else_tb;

reg x1, x2, x3, x4;
wire z1;

//continued on next page
```

Figure 8.60   Test bench for the behavioral module of Figure 8.59.

```
//apply input vectors and display variables
initial
begin: apply_stimulus and display variables
    reg [4:0] invect;
    for (invect = 0; invect < 16; invect = invect + 1)
        begin
            {x1, x2, x3, x4} = invect [4:0];
            #10 $display ("{x1x2x3x4} = %b, z1 = %b",
                          {x1, x2, x3, x4}, z1);
        end
end


//instantiate the module into the test bench
sop_eqn_if_else inst1 (
    .x1(x1),
    .x2(x2),
    .x3(x3),
    .x4(x4),
    .z1(z1)
    );

endmodule
```

Figure 8.60   (Continued)

```
{x1x2x3x4} = 0000, z1 = 0
{x1x2x3x4} = 0001, z1 = 0
{x1x2x3x4} = 0010, z1 = 0
{x1x2x3x4} = 0011, z1 = 1
{x1x2x3x4} = 0100, z1 = 0
{x1x2x3x4} = 0101, z1 = 0
{x1x2x3x4} = 0110, z1 = 0
{x1x2x3x4} = 0111, z1 = 1
{x1x2x3x4} = 1000, z1 = 0
{x1x2x3x4} = 1001, z1 = 0
{x1x2x3x4} = 1010, z1 = 0
{x1x2x3x4} = 1011, z1 = 1
{x1x2x3x4} = 1100, z1 = 1
{x1x2x3x4} = 1101, z1 = 1
{x1x2x3x4} = 1110, z1 = 1
{x1x2x3x4} = 1111, z1 = 1
```

Figure 8.61   Outputs for the sum-of-products module of Figure 8.59.

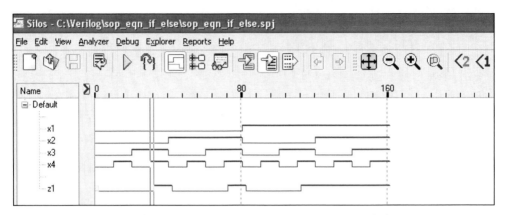

Figure 8.62   Waveforms for the sum-of-products module of Figure 8.59.

Example 8.17   A modulo-16 counter will be designed in this example. The internal organization of the counter is not specified nor are the types of storage elements. The only declaration is the counting sequence defined by the modulus operator. The counting sequence is 0000, 0001, 0010, 0011, . . ., 1111, 0000. There are two inputs, *clk* and *rst_n*, and one output, *count*. The behavioral module is shown in Figure 8.63. The test bench, outputs, and waveforms are shown in Figure 8.64, Figure 8.65, and Figure 8.66, respectively.


```
//behavioral modulo-16 counter
module ctr_mod_16 (clk, rst_n, count);

input clk, rst_n;
output [3:0] count;

wire clk, rst_n;
reg [3:0] count;

//define counting sequence
always @ (posedge clk or negedge rst_n)
begin
  if (rst_n == 0)
     count <= 4'b0000;
  else
     count <= (count + 1) % 16;
end

endmodule
```

Figure 8.63   Behavioral module for a modulo-16 counter using conditional statements.

```verilog
//modulo-16 counter test bench
module ctr_mod_16_tb;


reg clk, rst_n;
wire [3:0] count;


//display count
initial
$monitor ("count = %b", count);



//define reset
initial
begin
   #0 rst_n = 1'b0;       //assert reset
   #5 rst_n = 1'b1;       //deassert reset
end



//define clock
initial
begin
   #0  clk = 1'b0;

   forever
      #10 clk = ~clk;
end



//define length of simulation
initial
begin
   #330 $stop;
end


//instantiate the module into the test bench
ctr_mod_16 inst1 (
   .clk(clk),
   .rst_n(rst_n),
   .count(count)
   );

endmodule
```

Figure 8.64   Test bench for the modulo-16 counter of Figure 8.63.

```
count = 0000
count = 0001
count = 0010
count = 0011
count = 0100
count = 0101
count = 0110
count = 0111
count = 1000
count = 1001
count = 1010
count = 1011
count = 1100
count = 1101
count = 1110
count = 1111
count = 0000
```

Figure 8.65 Outputs for the modulo-16 counter of Figure 8.63.

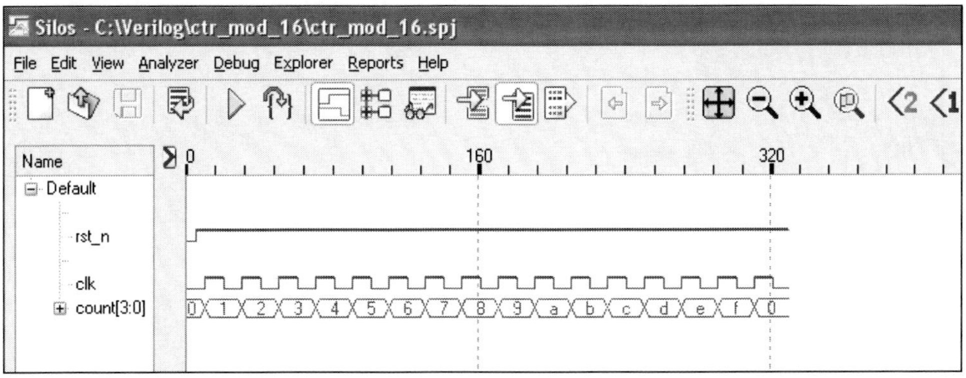

Figure 8.66 Waveforms for the modulo-16 counter of Figure 8.63.

Example 8.18 A positive-edge-triggered *D* flip-flop will be designed using behavioral modeling and conditional statements. There will be six ports: a data input *d*, a clock input *clk*, an active-low asynchronous set input *set_n*, an active-low asynchronous reset input *rst_n*, an active-high output *q*, and an active-low output *q_n* that is the complement of q. The logic diagram is shown below. The sensitivity list triggers execution of the **always** block when a change occurs on *clk*, *rst_n*, or *set_n*.

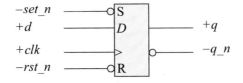

One way to model a *D* flip-flop is shown in the behavioral model of Figure 8.67 in which the equality operator (==) is used to determine the value of *rst_n* and *set_n*. If the reset input is active (*rst_n* == 0), then the flip-flop is reset to a logic 0. The conditional statements **if**, **else if**, and **else** are also used in the module to control the flow through the **always** block. The test bench, outputs, and waveforms are shown in Figure 8.68, Figure 8.69, and Figure 8.70, respectively.

```
//behavioral D flip-flop
module d_ff (d, clk, q, q_n,
             set_n, rst_n);

input d, clk, set_n, rst_n;
output q, q_n;

wire d, clk, set_n, rst_n;
reg q, q_n;

always @ (posedge clk or
          negedge rst_n or
          negedge set_n)
begin
   if (rst_n == 0)
      begin
         q <= 1'b0;
         q_n <= 1'b1;
      end
   else if (set_n == 0)
      begin
         q <= 1'b1;
         q_n <= 1'b0;
      end
   else
      begin
         q <= d;
         q_n <= ~q;
      end
end

endmodule
```

Figure 8.67   Behavioral model of a *D* flip-flop.

```
//behavioral d_ff test bench
module d_ff_tb;


reg d, clk, set_n, rst_n;
wire q, q_n;


initial
$monitor ($time, " set_n=%b, rst_n=%b, d=%b, clk=%b, q = %b",
                    set_n, rst_n, d, clk, q);


//define clock
initial
begin
   #5 clk = 1'b0;
      forever
         #5 clk = ~clk;
end


//define set and reset
initial
begin
   #0 rst_n = 1'b0;
      d     = 1'b0;
      clk   = 1'b0;
      set_n = 1'b1;
   #5 rst_n = 1'b1;

   #5 set_n = 1'b0;
   #5 set_n = 1'b1;

   #5 rst_n = 1'b0;
   #5 rst_n = 1'b1;
end


//define data input
initial
begin
   #32   d = 1'b0;
   #10   d = 1'b1;
   #10   d = 1'b0;


//continued on next page
```

Figure 8.68   Test bench for the *D* flip-flop module of Figure 8.67.

```
   #10    d = 1'b1;
   #10    d = 1'b1;
   #10    d = 1'b0;
   #10    $stop;
end


//instantiate the module into the test bench
d_ff inst1 (
   .d(d),
   .clk(clk),
   .set_n(set_n),
   .rst_n(rst_n),
   .q(q),
   .q_n(q_n)
   );

endmodule
```

Figure 8.68   (Continued)

```
0  set_n=1, rst_n=0, d=0, clk=0, q = 0
5  set_n=1, rst_n=1, d=0, clk=0, q = 0
10 set_n=0, rst_n=1, d=0, clk=0, q = 1
15 set_n=1, rst_n=1, d=0, clk=0, q = 1
20 set_n=1, rst_n=0, d=0, clk=0, q = 0
25 set_n=1, rst_n=1, d=0, clk=0, q = 0
35 set_n=1, rst_n=1, d=0, clk=1, q = 0
40 set_n=1, rst_n=1, d=0, clk=0, q = 0
42 set_n=1, rst_n=1, d=1, clk=0, q = 0
45 set_n=1, rst_n=1, d=1, clk=1, q = 1
50 set_n=1, rst_n=1, d=1, clk=0, q = 1
52 set_n=1, rst_n=1, d=0, clk=0, q = 1
55 set_n=1, rst_n=1, d=0, clk=1, q = 0
60 set_n=1, rst_n=1, d=0, clk=0, q = 0
62 set_n=1, rst_n=1, d=1, clk=0, q = 0
65 set_n=1, rst_n=1, d=1, clk=1, q = 1
70 set_n=1, rst_n=1, d=1, clk=0, q = 1
75 set_n=1, rst_n=1, d=1, clk=1, q = 1
80 set_n=1, rst_n=1, d=1, clk=0, q = 1
82 set_n=1, rst_n=1, d=0, clk=0, q = 1
85 set_n=1, rst_n=1, d=0, clk=1, q = 0
90 set_n=1, rst_n=1, d=0, clk=0, q = 0
```

Figure 8.69   Outputs for the *D* flip-flop module of Figure 8.67.

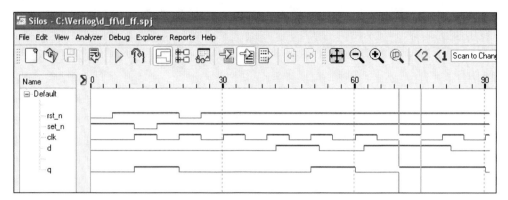

Figure 8.70   Waveforms for the *D* flip-flop module of Figure 8.67.

Refer to the test bench, outputs, and waveforms for the discussion that follows. In each **initial** block in the test bench, the times are cumulative. The flip-flop is reset initially at time 0, then set at time 10 at which time *q* is asserted. The flip-flop is then reset at time 20. The clock begins toggling at time 35 at which time $d = 0, q = 0$. At time 42, $d = 1$ providing the setup time for *clk* at time 45 which asserts *q*. At time 50, *clk* = 0, but *q* remains asserted. At time 62, $d = 1$. At time 65, *clk* = 1 and $d = 1$, which again asserts output *q*.

Example 8.19   This example shows the versatility of Verilog by offering a simplified alternative method to design a *D* flip-flop. This is a behavioral module, but in this version there is no set input and the reset input is not considered as an edge in the sensitivity list of the **always** statement. Also, a continuous assignment construct is used. The behavioral module is shown in Figure 8.71. The test bench, outputs, and waveforms are shown in Figure 8.72, Figure 8.73, and Figure 8.74, respectively.

```
//d flip-flop behavioral
module d_ff_bh (rst_n, clk, d, q, q_n);

input rst_n, clk, d;
output q, q_n;

wire rst_n, clk, d;
reg q;

assign q_n = ~q;
//continued on next page
```

Figure 8.71   Alternative method to design a *D* flip-flop.

```
always @ (rst_n or posedge clk)
begin
   if (rst_n == 0)
        q <= 1'b0;
   else q <= d;
end

endmodule
```

Figure 8.71 (Continued)

```
//d_ff_bh test bench
module d_ff_bh_tb;

reg rst_n, clk, d;
wire q;

initial
$monitor ("rst_n=%b, clk=%b, d=%b, q=%b", rst_n, clk, d, q);

initial
begin
   clk = 1'b0;
   forever
       #10 clk = ~clk;
end

initial
begin
   #0 rst_n = 1'b0;
   #0 d      = 1'b0;
   #5 rst_n = 1'b1;

   #10 d = 1'b1;
   #10 d = 1'b1;
   #10 d = 1'b0;
   #10 d = 1'b0;
   #10 d = 1'b0;

   #10 $stop;
end

//continued on next page
```

Figure 8.72 Test bench for the module of Figure 8.71.

```
//instantiate the module into the test bench
d_ff_bh inst1 (
    .rst_n(rst_n),
    .clk(clk),
    .d(d),
    .q(q)
    );

endmodule
```

Figure 8.72   (Continued)

```
rst_n = 0, clk = 0, d = 0, q = 0
rst_n = 1, clk = 0, d = 0, q = 0
rst_n = 1, clk = 1, d = 0, q = 0
rst_n = 1, clk = 1, d = 1, q = 0
rst_n = 1, clk = 0, d = 1, q = 0
rst_n = 1, clk = 1, d = 1, q = 1
rst_n = 1, clk = 1, d = 0, q = 1
rst_n = 1, clk = 0, d = 0, q = 1
rst_n = 1, clk = 1, d = 0, q = 0
rst_n = 1, clk = 0, d = 0, q = 0
```

Figure 8.73   Outputs for the *D* flip-flop module of Figure 8.71.

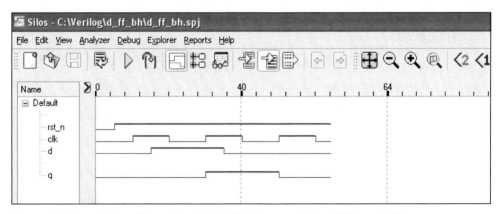

Figure 8.74   Waveforms for the *D* flip-flop module of Figure 8.71.

Example 8.20   This example presents the design of a positive-edge-triggered *JK* flip-flop using behavioral modeling and conditional statements. There are five inputs: *clk*, *j*, *k*, an active-low *set_n*, and an active-low *rst_n*. There are two complementary outputs: *q* and *q_n*. The logic symbol for a *JK* flip-flop is shown below.

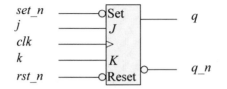

The functional characteristics of the *JK* data inputs are defined in Table 8.1. Table 8.2 lists the next state $Y_{k(t+1)}$ for each combination of *J*, *K*, and the present state $Y_{j(t)}$ based on the functional characteristics of *J* and *K*. Table 8.3 shows an excitation table in which a particular state transition predicates a set of values for *J* and *K*.

For example, if the state transition is $0 \rightarrow 0$, the *J* must be 0, but *K* can be 0 or 1. If $K = 0$, then this indicates no change; if $K = 1$, then this indicates a reset condition. The same rationale is used for the remaining state transitions by assigning values of 0 and 1 to the "don't care" entries. This table is especially useful in the design of synchronous sequential machines.

Table 8.1 *JK* Flip-Flop Functional Characteristics

J K	Function
0 0	No change
0 1	Reset
1 0	Set
1 1	Toggle

Table 8.3 Excitation Table for a *JK* Flip-Flop

Present State $Y_{j(t)}$	Next State $Y_{k(t+1)}$	Data Inputs J K
0	0	0 –
0	1	1 –
1	0	– 1
1	1	– 0

Table 8.2 *JK* Flip-Flop Characteristic Table

Data Inputs J K	Present State $Y_{j(t)}$	Next State $Y_{k(t+1)}$
0 0	0	0
0 0	1	1
0 1	0	0
0 1	1	0
1 0	0	1
1 0	1	1
1 1	0	1
1 1	1	0

The excitation equation for a *JK* flip-flop is derived from Table 8.3 and is shown in Equation 8.1. The behavioral module is shown in Figure 8.75. The test bench is shown in Figure 8.76 using the **$time** system function which returns the current simulation time. The outputs and waveforms are shown in Figure 8.77 and Figure 8.78, respectively.

$$Y_{k(t+1)} = Y_{j(t)}' J + Y_{j(t)} K' \tag{8.1}$$

```
//behavioral jkff
module jkff (clk, j, k, set_n, rst_n, q, q_n);

input clk, j, k, set_n, rst_n;
output q, q_n;
wire clk, j, k, set_n, rst_n;
reg q, q_n;

initial
   q = 1'b0;
always @ (posedge clk or negedge set_n or negedge rst_n)
begin
   if(~rst_n)
      begin
         q   <= 1'b0;
         q_n <= 1'b1;
      end

   else if (~set_n)
      begin
         q   <= 1'b1;
         q_n <= 1'b0;
      end

   else if (j==1'b0 && k==1'b0)
      begin
         q <= q;
         q_n <= q_n;
      end

   else if (j==1'b0 && k==1'b1)
      begin
         q   <= 1'b0;
         q_n <= 1'b1;
      end
//continued on next page
```

Figure 8.75 Behavioral module for a *JK* flip-flop.

```
    else if (j==1'b1 && k==1'b0)
        begin
            q   <= 1'b1;
            q_n <= 1'b0;
        end

    else if (j==1'b1 && k==1'b1)
        begin
            q   <= q_n;
            q_n <= q;
        end
end
endmodule
```

Figure 8.75   (Continued)

```
//jk flip-flop test bench
module jkff_tb;

reg clk, j, k, rst_n, set_n;
wire q, q_n;

//display outputs at simulation time
initial
$monitor ($time, " q = %b", q);

initial
begin
    set_n = 1'b1;
    rst_n = 1'b0;
    j     = 1'b0;
    k     = 1'b0;
    clk = 1'b0;
    #3 rst_n = 1'b1;
    forever
        #5 clk = ~clk;
end

initial
begin
    #10 j = 1'b1;
        k = 1'b0;       //set

//continued on next page
```

Figure 8.76   Test bench for the *JK* flip-flop behavioral module of Figure 8.75.

```
   #10 j = 1'b1;
       k = 1'b1;        //toggle reset
   #10 j = 1'b1;
       k = 1'b1;        //toggle set
   #10 j = 1'b0;
       k = 1'b0;        //no change (set)
   #10 j = 1'b0;
       k = 1'b1;        //reset
   #10 $stop;
end

jkff inst1 (
   .clk(clk),
   .j(j),
   .k(k),
   .set_n(set_n),
   .rst_n(st_n,
   .q(q),
   .q_n(q_n)
   );
endmodule
```

Figure 8.76 (Continued)

0 q = 0	28 q = 0	58 q = 0
18 q = 1	38 q = 1	

Figure 8.77 Outputs for the *JK* flip-flop behavioral module of Figure 8.75.

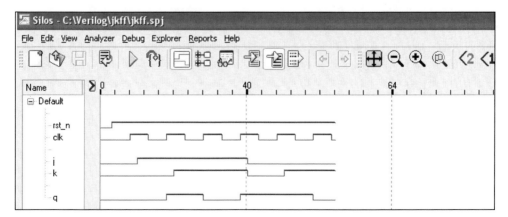

Figure 8.78 Waveforms for the *JK* flip-flop behavioral module of Figure 8.75.

Example 8.21   A 4-bit counter that counts even numbers first, then odd numbers will be designed using D flip-flops. The counting sequence is: $y_3 y_2 y_1 y_0 = 0000, 0010,$ $0100, 0110, 1000, 1010, 1100, 1110, 0001, 0011, 0101, 0111, 1001, 1011, 1101, 1111,$ $0000, \ldots$ The counting sequence will be controlled by conditional statements only. The same counter will be designed in Section 8.4 using the **case** statement to illustrate an alternative design method. The Karnaugh maps are shown in Figure 8.79 for the D flip-flop input equations. The equations for the D inputs are shown in Equation 8.2.

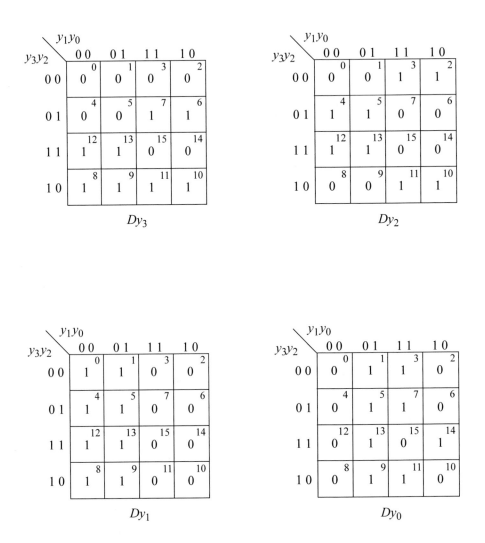

Figure 8.79   Karnaugh maps for the even-odd counter of Example 8.21.

$$Dy_3 = y_3 y_1' + y_3 y_2' + y_3' y_2 y_1$$

$$Dy_2 = y_2 y_1' + y_2' y_1$$

$$Dy_1 = y_1'$$

$$Dy_0 = y_1' y_0 + y_3' y_0 + y_2' y_0 + y_3 y_2 y_1 y_0' \qquad (8.2)$$

The design module and test bench are shown in Figure 8.80 and Figure 8.81, respectively. In Figure 8.80, there are five **always** constructs. The first **always** either resets the counter or assigns *d[3:0]* to *y[3:0]* using nonblocking statements. The next four **always** statements implement the *D* flip-flop input equations of Equation 8.2. The *D* inputs are set to a logic 1 if any term in their respective equations is true.

In Figure 8.81, reset is asserted for one time unit beginning at time 3. The clock cycles are repeated 17 times, then a delay of two time units is taken before simulation ends. The outputs and waveforms are shown in Figure 8.82 and Figure 8.83, respectively.

```
//behavioral even-odd counter
module ctr_evn_odd3 (rst_n, clk, y);

input rst_n, clk;
output [3:0] y;
wire rst_n, clk; //inputs are wire

reg [3:0] y;      //outputs are reg
reg [3:0] d;      //an internal next state

//reset counter and set y
always @ (posedge clk or negedge rst_n)
begin
  if (rst_n == 0)
     y <= 4'b0000;
  else
     y <= d;
end

//continued on next page
```

Figure 8.80   Behavioral module for the even-odd counter of Example 8.21 using conditional statements only.

```verilog
//Dy[3] = y[3] y[1]' + y[3] y[2]' + y[3]' y[2] y[1]
always @ (y)
begin
   if ((y[3]==1 && y[1]==0) ||
       (y[3]==1 && y[2]==0) ||
       (y[3]==0 && y[2]==1 && y[1]==1))
      d[3] = 1'b1;
   else
      d[3] = 1'b0;
end


//Dy[2] = y[2] y[1]' + y[2]' y[1]
always @ (y)
begin
   if ((y[2]==1 && y[1]==0) ||
       (y[2]==0 && y[1]==1))
      d[2] = 1'b1;
   else
      d[2] = 1'b0;
end


//Dy[1] = y[1]'
always @ (y)
begin
   if (y[1]==0)
      d[1] = 1'b1;
   else
      d[1] = 1'b0;
end


//Dy[0] = y[1]' y[0] + y[3]' y[0] + y[2]' y[0] +
//y[3] y[2] y[1] y[0]'
always @ (y)
begin
   if ((y[1]==0 && y[0]==1) ||
       (y[3]==0 && y[0]==1) ||
       (y[2]==0 && y[0]==1) ||
       (y[3]==1 && y[2]==1 && y[1]==1 && y[0]==0))
      d[0] = 1'b1;
   else
      d[0] = 1'b0;
end

endmodule
```

Figure 8.80   (Continued)

```
//test bench for even-odd counter
module ctr_evn_odd3_tb;


reg rst_n, clk;
wire [3:0] y;


//display count
initial
$monitor ("count = %b", y);


//define clock
initial
begin
   clk = 1'b0;

   forever
      #5 clk = ~clk;
end


//define length of simulation
initial
begin
   #3 rst_n = 1'b0;
   #1 rst_n = 1'b1;

   repeat (17) @ (posedge clk);
   #2;

   $stop;
end


//instantiate the module into the test bench
ctr_evn_odd3 inst1 (
   .rst_n(rst_n),
   .clk(clk),
   .y(y)
   );

endmodule
```

Figure 8.81   Test bench for the even-odd counter of Figure 8.80.

count = xxxx	count = 0001
count = 0000	count = 0011
count = 0010	count = 0101
count = 0100	count = 0111
count = 0110	count = 1001
count = 1000	count = 1011
count = 1010	count = 1101
count = 1100	count = 1111
count = 1110	count = 0000

Figure 8.82   Outputs for the even-odd counter of Figure 8.80.

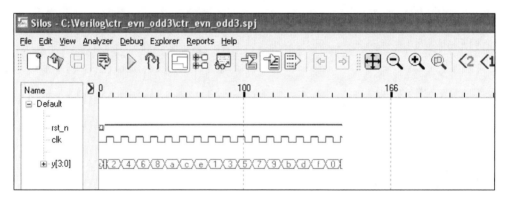

Figure 8.83   Waveforms for the even-odd counter of Figure 8.80.

8.4 Case Statement

The **case** statement is an alternative to the **if** . . . **else if** construct and may simplify the readability of the Verilog code. The **case** statement is a multiple-way conditional branch. It executes one of several different procedural statements depending on the comparison of an expression with a case item. The expression and the case item are compared bit-by-bit and must match exactly. The statement that is associated with a case item may be a single procedural statement or a block of statements delimited by the keywords **begin** . . . **end.** The **case** statement has the following syntax:

```
case (expression)
    case_item1 : procedural_statement1;
    case_item2 : procedural_statement2;
    case_item3 : procedural_statement3;
                .
                .
                .
    case_itemn : procedural_statementn;
    default : default_statement;
endcase
```

The case expression may be an expression or a constant. The case items are evaluated in the order in which they are listed. If a match occurs between the case expression and the case item, then the corresponding procedural statement, or block of statements, is executed. If no match occurs, then the optional default statement is executed.

There are two other variations of the **case** statement that handle "don't cares": **casex** and **casez**. In the **casex** construct, the values of **x** and **z** that appear in either the case expression or in the case item are treated as "don't cares." In the **casez** construct, a value of **z** that appears in either the case expression or the case item is treated as a "don't care." The **casex** and **casez** statements are useful for comparing only certain bits of the case expression and case item. Wherever an **x** or **z** appears, that bit is ignored. Bit positions that correspond to **z** can be replaced by a **?** in those bit positions. An example of the use of "don't cares" is shown below in a **casex** construct. The **casez** construct can be used in a similar situation.

```
casex (opcode)
    4'b1xxx : a + b;
    4'bx1xx : a − b;
    4'bxx1x : a * b;
    4'bxxx1 : a / b;
endcase
```

Several examples will now be presented that demonstrate the use of **case** statements using behavioral modeling. As in previous sections, the examples will increase in complexity. Different types of counters will be presented, plus a simple arithmetic and logic unit, and various types of Moore and Mealy sequential machines — both synchronous and asynchronous.

Example 8.22   A 4-bit Gray code counter will be designed that counts in the following sequence: 0000, 0001, 0011, 0010, 0110, 0111, 0101, 0100, 1100, 1101, 1111, 1110, 1010, 1011, 1001, 1000, 0000, . . . The Gray code has the unique characteristic that only one bit changes between successive code words. The **case** statement will be used to determine the next count from any current count. For example, if the current count is 0110, then the expression *count* is compared with the *case item* 0110 yielding a next count of 0111. The flow then exits the **case** statement and continues with the next statement in the module.

The design module is shown in Figure 8.84 using behavioral modeling. The expression *count* in the **always** statement represents the event control or sensitivity list. Whenever a change occurs to *count*, the code in the **begin** . . . **end** block executes. Each count is then compared to the value of the expression *count*. The test bench is shown in Figure 8.85. The outputs and waveforms are shown in Figure 8.86 and Figure 8.87, respectively.

```
//behavioral 4-bit Gray code counter
module ctr_gray4_case (clk, rst_n, count);

input clk, rst_n;
output [3:0] count;

wire clk, rst_n;//inputs are wire
reg [3:0] count;//outputs are reg
reg [3:0] next_count;//define internal reg

//set next count
always @ (posedge clk or negedge rst_n)
begin
   if (~rst_n)
      count <= 4'b0000;
   else
      count <= next_count;
end

//determine next count
always @ (count)
begin
   case (count)
      4'b0000 : next_count = 4'b0001;
      4'b0001 : next_count = 4'b0011;
      4'b0011 : next_count = 4'b0010;
      4'b0010 : next_count = 4'b0110;
      4'b0110 : next_count = 4'b0111;
      4'b0111 : next_count = 4'b0101;
      4'b0101 : next_count = 4'b0100;
      4'b0100 : next_count = 4'b1100;

//continued next page
```

Figure 8.84   Behavioral module for a Gray code counter using the **case** statement.

```
      4'b1100 : next_count = 4'b1101;
      4'b1101 : next_count = 4'b1111;
      4'b1111 : next_count = 4'b1110;
      4'b1110 : next_count = 4'b1010;
      4'b1010 : next_count = 4'b1011;
      4'b1011 : next_count = 4'b1001;
      4'b1001 : next_count = 4'b1000;
      4'b1000 : next_count = 4'b0000;

      default : next_count = 4'b0000;

  endcase
  end

  endmodule
```

Figure 8.84   (Continued)

```
//test bench for 4-bit Gray code counter using case
module ctr_gray4_case_tb;

reg clk, rst_n;
wire [3:0] count;

initial
$monitor ("count = %b", count);

//define reset
initial
begin
   #0 rst_n = 1'b0;
   #5 rst_n = 1'b1;
end

initial        //define clock
begin
   #0 clk = 1'b0;
   forever
      #10clk = ~clk;
end

//continued on next page
```

Figure 8.85   Test bench for the Gray code counter of Figure 8.84.

```
initial           //define length of simulation
begin
   #330 $stop;
end

//instantiate the module into the test bench
ctr_gray4_case inst1 (
   .clk(clk),
   .rst_n(rst_n),
   .count(count)
   );

endmodule
```

Figure 8.85   (Continued)

count = 0000	count = 0101	count = 1010
count = 0001	count = 0100	count = 1011
count = 0011	count = 1100	count = 1001
count = 0010	count = 1101	count = 1000
count = 0110	count = 1111	count = 0000
count = 0111	count = 1110	

Figure 8.86   Outputs for the Gray code counter of Figure 8.84.

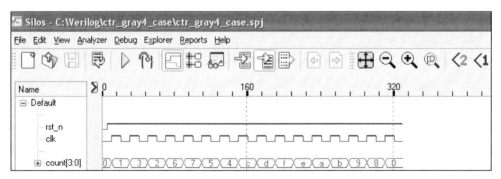

Figure 8.87   Waveforms for the Gray code counter of Figure 8.84.

Example 8.23 A 4-bit counter will be designed that counts even numbers and then odd numbers. The counting sequence in decimal is: 0, 2, 4, 6, 8, 10, 12, 14, 1, 3, 5, 7, 9, 11, 13, 15, 0, . . . The counting sequence will be controlled by a **case** statement as shown in the behavioral module of Figure 8.88. When using the **case** statement, the flip-flop input equations are not required, simply indicate the counting sequence. The design details are left to the synthesis tools. The test bench, outputs, and waveforms are shown in Figure 8.89, Figure 8.90, and Figure 8.91, respectively.

```verilog
//behavioral even-odd counter
module ctr_evn_odd (clk, rst_n, count);

input clk, rst_n;
output [3:0] count;

wire clk, rst_n;                          //inputs are wire
reg [3:0] count;                          //outputs are reg
reg [3:0] next_count;                     //define internal reg

always @ (posedge clk or negedge rst_n)   //set next count
begin
   if (~rst_n)                            //rst_n is active-low
      count = 4'b0000;
   else
      count = next_count;
end

always @ (count)                          //determine next count
begin
   case (count)
      4'b0000 : next_count = 4'b0010;
      4'b0010 : next_count = 4'b0100;
      4'b0100 : next_count = 4'b0110;
      4'b0110 : next_count = 4'b1000;
      4'b1000 : next_count = 4'b1010;
      4'b1010 : next_count = 4'b1100;
      4'b1100 : next_count = 4'b1110;
      4'b1110 : next_count = 4'b0001;
      4'b0001 : next_count = 4'b0011;
      4'b0011 : next_count = 4'b0101;
      4'b0101 : next_count = 4'b0111;
      4'b0111 : next_count = 4'b1001;
      4'b1001 : next_count = 4'b1011;
      4'b1011 : next_count = 4'b1101;
      4'b1101 : next_count = 4'b1111;
      4'b1111 : next_count = 4'b0000;
      default : next_count = 4'b0000;
   endcase
end

endmodule
```

Figure 8.88   Behavioral module for an even-odd counter.

```
//ctr_evn_odd test bench
module ctr_evn_odd_tb;

reg clk, rst_n;        //inputs are reg for tb
wire [3:0] count;      //outputs are wire for tb

initial
$monitor ("count = %b", count);

initial                //define clk input
begin
   clk = 1'b0;
   forever
      #10 clk = ~clk;
end

initial                //define reset input
begin
   #0 rst_n = 1'b0;
   #5 rst_n = 1'b1;
   #320 $stop;
end

//instantiate the module into the test bench
ctr_evn_odd inst1 (
   .clk(clk),
   .rst_n(rst_n),
   .count(count)
   );

endmodule
```

Figure 8.89   Test bench for the even-odd counter of Figure 8.88.

count = 0000	count = 0011
count = 0010	count = 0101
count = 0100	count = 0111
count = 0110	count = 1001
count = 1000	count = 1011
count = 1010	count = 1101
count = 1100	count = 1111
count = 1110	count = 0000
count = 0001	

Figure 8.90   Outputs for the even-odd counter of Figure 8.88.

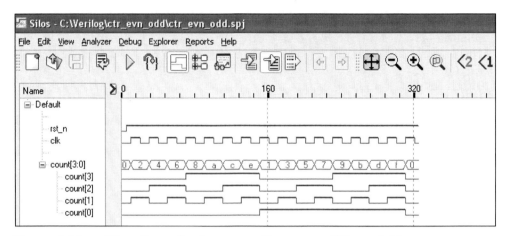

Figure 8.91   Waveforms for the even-odd counter of Figure 8.88.

Example 8.24   A 4-function arithmetic and logic unit (ALU) will be designed in this example using the **case** construct. There are two 4-bit inputs: operands *a[3:0]* and *b[3:0]* and one 2-bit input *opcode[1:0]*. There is one 8-bit output *z[7:0]* which contains the result of the operations and is declared to be eight bits to accommodate the 2*n*-bit product. The **parameter** keyword will declare and assign values to the operation codes.

A block diagram is shown in Figure 8.92. The behavioral module is shown in Figure 8.93. The test bench is shown in Figure 8.94 in which input vectors are applied for each operation code. Notice that the results are correctly represented in 2s complement representation; thus, 4 − 8 = −4 (11111100) and 6 − 15 = −9 (11110111). The outputs and waveforms are shown in Figure 8.95 and Figure 8.96, respectively.

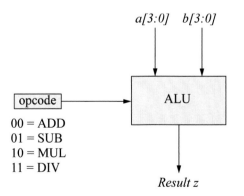

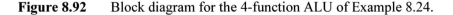

Figure 8.92   Block diagram for the 4-function ALU of Example 8.24.

```
//behavioral 4-bit ALU
module alu4 (a, b, opcode, z);

input [3:0] a, b;
input [1:0] opcode;
output [7:0] z;

wire [3:0] a, b;        //inputs are wire
wire [1:0] opcode;
reg [7:0] z;            //outputs are reg

//define operation codes
parameter    addop = 2'b00,
             subop = 2'b01,
             mulop = 2'b10,
             divop = 2'b11;

//perform operations
always @(a or b or opcode)
begin
case (opcode)
     addop: z = a + b;
     subop: z = a - b;
     mulop: z = a * b;
     divop: z = a / b;
endcase
end

endmodule
```

Figure 8.93   Behavioral module for the 4-function ALU of Example 8.24.

```
//test bench for 4-bit ALU
module alu4_tb;

reg [3:0] a, b;
reg [1:0] opcode;
wire [7:0] z;

//display variables
initial
$monitor ("a=%b, b=%b, opcode=%b, result=%b",
          a, b, opcode, z);
//continued on next page
```

Figure 8.94   Test bench for the 4-function ALU of Figure 8.93.

```verilog
//apply input vectors
initial
begin
//add operation
  #0   a=4'b0001; b=4'b0001; opcode=2'b00;   //sum=2

  #10  a=4'b0010; b=4'b1101; opcode=2'b00;   //sum= 15(f)

  #10  a=4'b1111; b=4'b1111; opcode=2'b00;   //sum=30(1e)


//subtract operation
  #10  a=4'b1000; b=4'b0100; opcode=2'b01;   //difference=4

  #10  a=4'b1111; b=4'b0101; opcode=2'b01;   //difference=10(a)

  #10  a=4'b1110; b=4'b0011; opcode=2'b01;   //difference=11(b)

  #10  a=4'b0100; b=4'b1000; opcode=2'b01;   //difference=-4(fc)

  #10  a=4'b0110; b=4'b1111; opcode=2'b01;   //difference=-9(f7)


//multiply operation
  #10  a=4'b0100; b=4'b0111; opcode=2'b10;   //product=28(1c)

  #10  a=4'b0101; b=4'b0011; opcode=2'b10;   //product=15(f)

  #10  a=4'b1111; b=4'b1111; opcode= 'b10;   //product=225(e1)


//divide operation
  #10  a=4'b1111; b=4'b0101; opcode=2'b11;   //quotient=3

  #10  a=4'b1100; b=4'b0011; opcode=2'b11;   //quotient=4

  #10  a=4'b1110; b=4'b0010; opcode=2'b11;   //quotient=7

  #10  a=4'b0011; b=4'b1100; opcode=2'b11;   //quotient=0

  #10  $stop;

end

//continued on next page
```

Figure 8.94 (Continued)

```
//instantiate the module into the test bench
alu4 inst1 (
    .a(a),
    .b(b),
    .opcode(opcode),
    .z(z)
    );

endmodule
```

Figure 8.94   (Continued)

```
a=0001, b=0001, opcode=00, result=00000010
a=0010, b=1101, opcode=00, result=00001111
a=1111, b=1111, opcode=00, result=00011110
a=1000, b=0100, opcode=01, result=00000100
a=1111, b=0101, opcode=01, result=00001010
a=1110, b=0011, opcode=01, result=00001011
a=0100, b=1000, opcode=01, result=11111100
a=0110, b=1111, opcode=01, result=11110111
a=0100, b=0111, opcode=10, result=00011100
a=0101, b=0011, opcode=10, result=00001111
a=1111, b=1111, opcode=10, result=11100001
a=1111, b=0101, opcode=11, result=00000011
a=1100, b=0011, opcode=11, result=00000100
a=1110, b=0010, opcode=11, result=00000111
a=0011, b=1100, opcode=11, result=00000000
```

Figure 8.95   Outputs for the 4-function ALU of Figure 8.93.

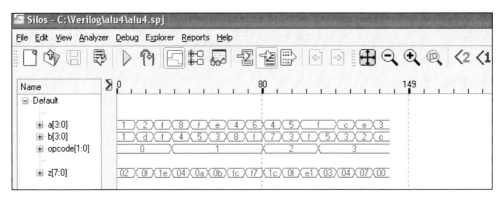

Figure 8.96   Waveforms for the 4-function ALU of Figure 8.93.

Example 8.25   A Moore synchronous sequential machine will be designed according to the state diagram shown in Figure 8.97. Since it is a Moore machine, the outputs are a function of the present state only. The outputs will be asserted for the entire clock cycle. The state codes are selected to provide glitch-free operation. This is only one of several viable state code assignments that will produce glitch-free outputs for z_1 and z_2. There are three unused states: $y_1 y_2 y_3 = 010$, 100, and 110.

It is important that no state transition sequence passes through states d or e as a transient state. This can result when two or more storage elements change state at different times, resulting in an output glitch on z_1 or z_2.

With the state code assignment given, no state transition will pass through states d or e; therefore, there will be no output glitches. If only one variable changes during a state transition sequence, the machine cannot pass through a transient state. The transition from state a to state b has only one change of variable. Likewise, the transition from state a to state c has only one change of variable.

However, the transition from state b to state d has two changes. The state codes change from $y_1 y_2 y_3 = 101$ to $y_1 y_2 y_3 = 000$. The machine may pass through state $y_1 y_2 y_3 = 001$ (state a), which has no output, if y_1 changes before y_3 changes. Or the machine may pass through state $y_1 y_2 y_3 = 100$, which is an unused state. The path from state d to state a has only one change of variable.

The path from state c to state d has two changes. The machine may pass through state $y_1 y_2 y_3 = 001$, which has no output, or state $y_1 y_2 y_3 = 010$, which is an unused state. The only remaining state transition that has two changes is from state e to state a. The machine may sequence through transient states b or c, both of which have no output.

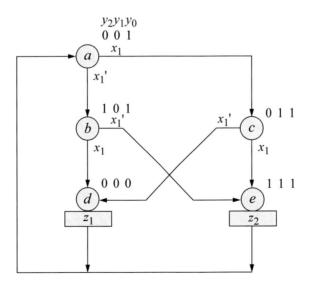

Figure 8.97   State diagram for the Moore machine of Example 8.25.

The behavioral module is shown in Figure 8.98 using the **parameter** keyword, conditional statements, and the **case** statement. The **parameter** keyword assigns state codes. The **if** . . . **else** constructs set the next state, determine the outputs, and determine the next state in conjunction with the **case** statement. Figure 8.99 shows the test bench, which sequences the machine through the various states at the positive edge of the clock, depending upon input x_1.

The waveforms of Figure 8.100 clearly show the state transition sequences and the outputs as the machine proceeds through the different states depicted in the state diagram. Output z_1 is asserted for the entire clock period in state d ($y_2 y_1 y_0 = 000$). Output z_2 is asserted for the entire clock period in state e ($y_2 y_1 y_0 = 111$).

```
//behavioral moore ssm
module moore_ssm (clk, rst_n, x1, y, z1, z2);


//specify inputs and outputs
input clk, rst_n, x1;
output [2:0] y;                  //y is an array of 3 bits
output z1, z2;


reg [2:0] y, next_state;         //outputs are reg in behav
reg z1, z2;


//assign state codes
parameter   state_a = 3'b001,    //param defines a constant
            state_b = 3'b101,    //state names must have
            state_c = 3'b011,    //at least 2 characters
            state_d = 3'b000,
            state_e = 3'b111;


//set next state
always @ (posedge clk)            //rst synched to posedge clk
begin
   if (~rst_n)                    //if (~rst_n) is true,
      y <= state_a;               //y <= state_a
   else
      y <= next_state;
end

//continued on next page
```

Figure 8.98   Behavioral module for the Moore synchronous sequential machine of Figure 8.97.

```
//determine outputs
always @ (y)
begin
   if (y == state_d)              //== specifies logical
      z1 = 1'b1;                  //equality or compare
   else
      z1 = 1'b0;

   if (y == state_e)
      z2 = 1'b1;
   else
      z2 = 1'b0;
end


//determine next state
always @ (y or x1)
begin
   case (y)                       //case is a multiple-way
      state_a:                    //conditional branch
         if (x1)                  //if y = state_a, then
            next_state = state_c; //do if . . . else
         else
            next_state = state_b;

      state_b:
         if (x1)
            next_state = state_d;
         else
            next_state = state_e;

      state_c:
         if (x1)
            next_state = state_e;
         else
            next_state = state_d;

      state_d:next_state = state_a;

      state_e:next_state = state_a;

      default: next_state = state_a;
   endcase
end

endmodule
```

Figure 8.98   (Continued)

```verilog
//test bench for moore ssm
module moore_ssm_tb;


reg clk, rst_n, x1;
wire [2:0] y;
wire z1, z2;


initial
$monitor ("x1 = %b, state = %b, z1 = %b, z2 = %b",
          x1, y, z1, z2);


//define clock
initial
begin
   clk = 1'b0;
   forever
      #10 clk = ~clk;
end


//define input sequence
initial
begin
   #0    rst_n = 1'b0;   //rst to state_a (001)
   #15   rst_n = 1'b1;

   x1 = 1'b0;
   @ (posedge clk)        //go to state_b (101)

   x1 = 1'b1;
   @ (posedge clk)        //go to state_d (000); assert z1

   x1 = 1'b0;
   @ (posedge clk)        //go to state_a (001)

   x1 = 1'b1;
   @ (posedge clk)        //go to state_c (011)

   x1 = 1'b1;
   @ (posedge clk)        //go to state_e (111); assert z2

   x1 = 1'b0;
   @ (posedge clk)        //go to state_a (001)

//continued on next page
```

Figure 8.99   Test bench for the Moore synchronous sequential machine of Figure 8.98.

```
   x1 = 1'b0;
   @ (posedge clk)      //go to state_b (101)

   x1 = 1'b0;
   @ (posedge clk)      //go to state_e (111); assert z2

   x1 = 1'b0;
   @ (posedge clk)      //go to state_a (001)

   x1 = 1'b1;
   @ (posedge clk)      //go to state_c (011)

   x1 = 1'b0;
   @ (posedge clk)      //go to state_d (000); assert z1

   x1 = 1'b0;
   @ (posedge clk)      //go to state_a (001)
   #10 $stop;
end

moore_ssm inst1 (       //instantiate the module
   .clk(clk),
   .rst_n(rst_n),
   .x1(x1),
   .y(y),
   .z1(z1),
   .z2(z2)
   );
endmodule
```

Figure 8.99   (Continued)

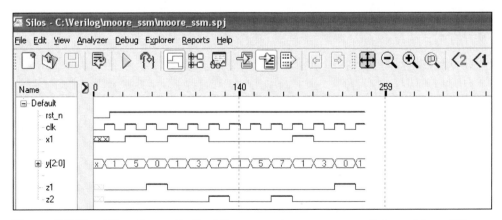

Figure 8.100   Waveforms for the Moore synchronous sequential machine of Figure 8.98.

Example 8.26   A Mealy synchronous sequential machine will be designed from the state diagram of Figure 8.101. There is only one storage element y_1 and two inputs: x_1 and x_2. Output z_1 is asserted only in state a ($y_1 = 0$) if $x_1 = 1$, thus meeting the Mealy requirement in which the outputs are a function of the present state and the present inputs. The behavioral module is shown in Figure 8.102 and the test bench is shown in Figure 8.103. The outputs and waveforms are shown in Figure 8.104 and Figure 8.105, respectively, which show output z_1 asserted only in state a if input x_1 is asserted.

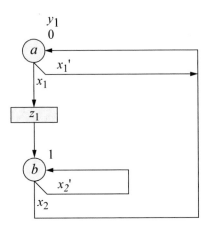

Figure 8.101   State diagram for the Mealy synchronous sequential machine of Example 8.26.

```
//behavioral mealy ssm
module mealy_ssm (clk, rst_n, x1, x2, y1, z1);

input clk, rst_n, x1, x2;

output y1;
output z1;

reg y1, next_state;
reg z1;

//continued on next page
```

Figure 8.102   Behavioral module for the Mealy synchronous sequential machine of Figure 8.101.

```
//assign state codes
parameter    state_a = 1'b0,
             state_b = 1'b1;


//set next state
always @ (posedge clk)
begin
   if (~rst_n)
      y1 <= state_a;
   else
      y1 <= next_state;
end


//determine outputs
always @(x1)
begin
   if ((y1 == state_a) && (x1 == 1'b1))
      z1 = 1'b1;
   else
      z1 = 1'b0;
end


//determine next state
always @ (y1 or x1 or x2)
begin
   case (y1)
      state_a:
         if (x1)
            next_state = state_b;
         else
            next_state = state_a;

      state_b:
         if (x2)
            next_state = state_a;
         else
            next_state = state_b;

      default: next_state = state_a;
   endcase
end

endmodule
```

Figure 8.102   (Continued)

```verilog
//test bench for mealy ssm
module mealy_ssm_tb;

reg clk, rst_n, x1, x2;
wire y1, z1;

//display variables
initial
$monitor ("x1=%b, x2=%b, state=%b, z1=%b",
            x1, x2, y1, z1);

//define clock
initial
begin
   clk = 1'b0;
   forever
      #10 clk = ~clk;
end

initial                    //define input sequence
begin
   #0    rst_n = 1'b0;   //rst to state_a
   #15   rst_n = 1'b1;

   x1 = 1'b0;   x2 = 1'b0;
   @ (posedge clk)        //go to state_a

   x1 = 1'b1;   x2 = 1'b0;
   @ (posedge clk)        //assert z1; go to state_b

   x1 = 1'b0;   x2 = 1'b0;
   @ (posedge clk)        //go to state_b

   x1 = 1'b0;   x2 = 1'b1;
   @ (posedge clk)        //go to state_a

   x1 = 1'b0;   x2 = 1'b0;
   @ (posedge clk)        //go to state_a

   x1 = 1'b1;   x2 = 1'b0;
   @ (posedge clk)        //assert z1; go to state_b

   #10    $stop;
end
//continued on next page
```

Figure 8.103   Test bench for the Mealy synchronous sequential machine of Figure 8.101.

```
//instantiate the module into the test bench
mealy_ssm inst1 (
   .clk(clk),
   .rst_n(rst_n),
   .x1(x1),
   .x2(x2),
   .y1(y1),
   .z1(z1)
   );
endmodule
```

Figure 8.103   (Continued)

```
x1=x,  x2=x,  state=x,  z1=x
x1=x,  x2=x,  state=0,  z1=x
x1=0,  x2=0,  state=0,  z1=0
x1=1,  x2=0,  state=0,  z1=1
x1=0,  x2=0,  state=1,  z1=0
x1=0,  x2=1,  state=1,  z1=0
x1=0,  x2=0,  state=0,  z1=0
x1=1,  x2=0,  state=0,  z1=1
x1=0,  x2=0,  state=1,  z1=0
```

Figure 8.104   Outputs for the Mealy synchronous sequential machine of Figure 8.101.

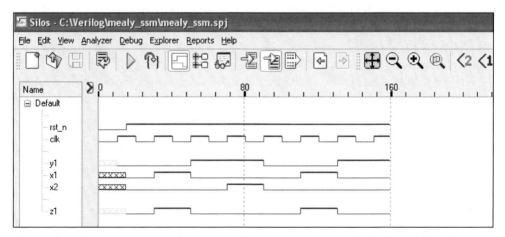

Figure 8.105   Waveforms for the Mealy synchronous sequential machine of Figure 8.101.

Example 8.27 A Moore synchronous sequential machine receives three 1-bit words on a parallel input bus as shown below, where x_3 is the low-order bit. The input words have values of $x_1x_2x_3 = 001, 010, 011, 100$, or 101. There are five outputs: z_1, z_2, z_3, z_4, and z_5, where the subscripts indicate the decimal value of the corresponding unsigned binary input word.

Thus, the following input and output sequences are valid:

If $x_1x_2x_3 = 001$, then z_1 is asserted.

If $x_1x_2x_3 = 010$, then z_2 is asserted.

If $x_1x_2x_3 = 011$, then z_3 is asserted.

If $x_1x_2x_3 = 100$, then z_4 is asserted.

If $x_1x_2x_3 = 101$, then z_5 is asserted.

There must be no transient glitches on the outputs as the machine passes through transient states. The simplest and most inexpensive method of eliminating output glitches is to include the complement of the machine clock in the implementation of the λ output logic. The output logic will consist of an AND gate which decodes the p-tuple state codes. One input of the AND gate is connected to the complement of the machine clock; that is, the negation of the clock signal which drives the state flip-flops. This will generate an output signal that is only one-half the duration of the clock cycle, but guarantees that the output is free from any erroneous assertions. The output is asserted during the last half of the machine clock.

Figure 8.106(a) shows a general block diagram for a Moore machine using the complement of the machine clock as an output gating function. Glitches are possible when two or more storage elements change state at different times upon application of the machine clock. The time duration from the active edge of the clock until the storage elements are stable is indicated in Figure 8.106(b) as Δt. The machine may pass through transient states that contain Moore-type outputs during this time.

A glitch that is caused by a state transition in which two or more flip-flops change state has no effect on the output because the glitch has returned to a logic 0 before the active level of the complemented clock occurs. The timing diagram of Figure 8.106(b) shows output vector Z_r asserted at time t_2 as a result of the high level of the machine clock complement. This method also applies to Mealy machines; the inputs, however, must remain active during the last half of the clock cycle.

The state diagram is shown in Figure 8.107 with state codes assigned for states a through f. Glitches are possible for the following state transition sequences:

State d ($y_2y_1y_0 = 011$) to state a ($y_2y_1y_0 = 000$) may pass through state b ($y_2y_1y_0 = 001$) if y_1 changes before y_0 changes.

State d ($y_2y_1y_0 = 011$) to state a ($y_2y_1y_0 = 000$) may pass through state c ($y_2y_1y_0 = 010$) if y_0 changes before y_1 changes.

State f ($y_2y_1y_0 = 101$) to state a ($y_2y_1y_0 = 000$) may pass through state b ($y_2y_1y_0 = 001$) if y_2 changes before y_0 changes.

State f ($y_2y_1y_0 = 101$) to state a ($y_2y_1y_0 = 000$) may pass through state e ($y_2y_1y_0 = 100$) if y_0 changes before y_2 changes.

There are two unused states: $y_2y_1y_0 = 110$ and $y_2y_1y_0 = 111$. The state diagram does not specify the type of storage element to be used in the design — only the state transitions are specified. The behavioral module is shown in Figure 8.108 which uses continuous assignment to generate the outputs for the appropriate states with the complement of the machine clock.

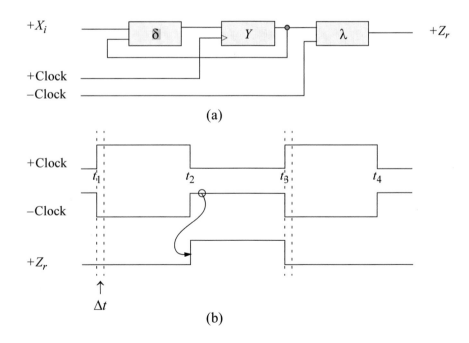

Figure 8.106   A generalized Moore model which uses the complement of the machine clock as a gating function to eliminate output glitches. The output vector Z_r is asserted at time t_2 and deasserted at time t_3: (a) block diagram and (b) timing diagram.

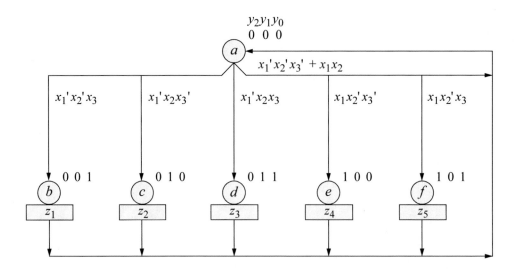

Figure 8.107   State diagram for the Moore synchronous sequential machine of Example 8.27.

```
//behavioral moore ssm2
module moore_ssm2 (clk, rst_n, x1, x2, x3, y,
                   z1, z2, z3, z4, z5);

//specify inputs and outputs
input clk, rst_n, x1, x2, x3;
output [2:0] y;                  //y is an array of 3 bits
output z1, z2, z3, z4, z5;

reg [2:0] y, next_state;         //must be reg for always
wire z1, z2, z3, z4, z5;

//assign state codes
parameter   state_a = 3'b000,    //param defines a constant
            state_b = 3'b001,    //state names must have
            state_c = 3'b010,    //at least 2 characters
            state_d = 3'b011,
            state_e = 3'b100,
            state_f = 3'b101;

//continued on next page
```

Figure 8.108   Behavioral module for the Moore sequential machine of Figure 8.107.

```
//set next state
always @ (posedge clk)        //reset synched to posedge clk
begin
   if (~rst_n)                //if (~rst_n) is true,
      y <= state_a;           //y <= state_a
   else
      y <= next_state;
end



//determine outputs
assign z1 = (~y[2] & ~y[1] &  y[0] & ~clk);
assign z2 = (~y[2] &  y[1] & ~y[0] & ~clk);
assign z3 = (~y[2] &  y[1] &  y[0] & ~clk);
assign z4 = ( y[2] & ~y[1] & ~y[0] & ~clk);
assign z5 = ( y[2] & ~y[1] &  y[0] & ~clk);



//determine next state
always @ (y or x1 or x2 or x3)
begin
   case (y)                            //case is a multiple-way
      state_a:                         //conditional branch
         if (x1==0 & x2==0 & x3==1) //if y = state_a, then
            next_state = state_b;   //do if-else if-else
         else if (x1==0 & x2==1 & x3==0)
            next_state = state_c;
         else if (x1==0 & x2==1 & x3==1)
            next_state = state_d;
         else if (x1==1 & x2==0 & x3==0)
            next_state = state_e;
         else if (x1==1 & x2==0 & x3==1)
            next_state = state_f;
         else next_state = state_a;
      state_b: next_state = state_a;
      state_c: next_state = state_a;
      state_d: next_state = state_a;
      state_e: next_state = state_a;
      state_f: next_state = state_a;
      default next_state = state_a;
   endcase
end

endmodule
```

Figure 8.108   (Continued)

The test bench is shown in Figure 8.109 which sequences the machine through all states in the state diagram. The waveforms are shown in Figure 8.110 and show the outputs asserted in their respective states during the last half of the clock cycle.

```
//test bench for moore ssm
module moore_ssm_tb;

reg clk, rst_n, x1;
wire [2:0] y;
wire z1, z2;

initial
$monitor ("x1 = %b, state = %b, z1 = %b, z2 = %b",
     x1, y, z1, z2);

//define clock
initial
begin
   clk = 1'b0;
   forever
      #10 clk = ~clk;
end

//define input sequence
initial
begin
   #0    rst_n = 1'b0;   //rst to state_a (001)
   #15   rst_n = 1'b1;

   x1=1'b0;    x2=1'b0;    x3=1'b1;
   @ (posedge clk)             //go to state_b (001); assert z1
   @ (posedge clk)             //go to state_a (000)
   x1=1'b0;    x2=1'b1;    x3=1'b0;
   @ (posedge clk)             //go to state_c (010); assert z2
   @ (posedge clk)             //go to state_a (000)
   x1=1'b0;    x2=1'b1;    x3=1'b1;
   @ (posedge clk)             //go to state_d (011); assert z3
   @ (posedge clk)             //go to state_a (000)
   x1=1'b1;    x2=1'b0;    x3=1'b0;
   @ (posedge clk)             //go to state_e (100); assert z4
   @ (posedge clk)             //go to state_a (000)
   x1=1'b1; x2=1'b0; x3=1'b1;
   @ (posedge clk)             //go to state_f (101); assert z5
//continued on next page
```

Figure 8.109   Test bench for the Moore synchronous sequential machine of Figure 8.108.

```
     @ (posedge clk)              //go to state_a (000)
     #10 $stop;
end


//instantiate the behav module into the test bench
moore_ssm2 inst1 (
   .clk(clk),
   .rst_n(rst_n),
   .x1(x1),
   .x2(x2),
   .x3(x3),
   .y(y),
   .z1(z1),
   .z2(z2),
   .z3(z3),
   .z4(z4),
   .z5(z5)
   );

endmodule
```

Figure 8.109   (Continued)

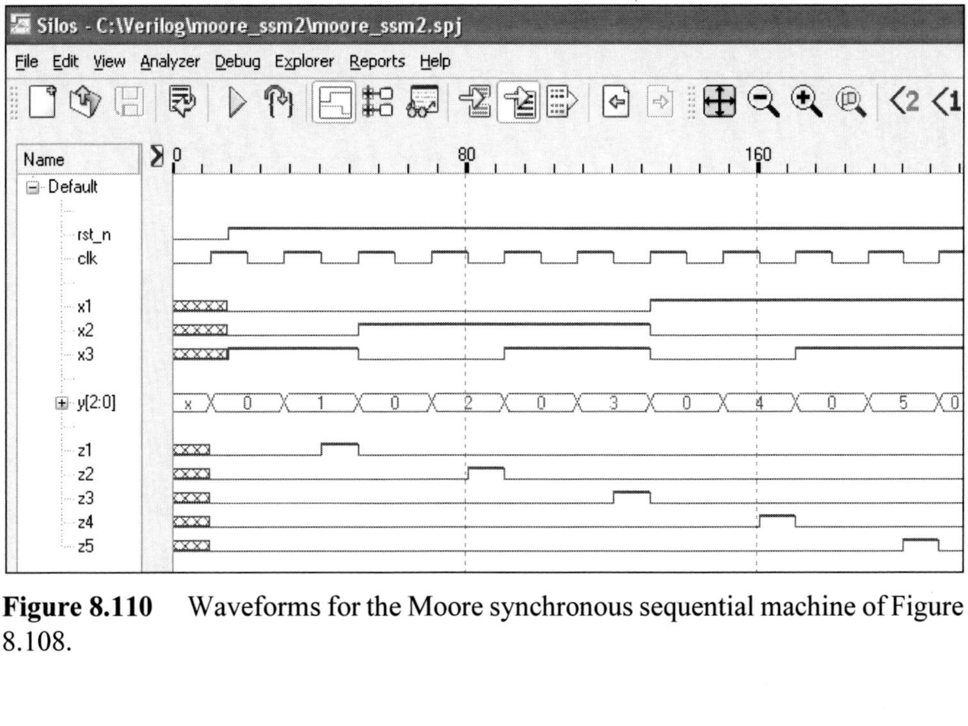

Figure 8.110   Waveforms for the Moore synchronous sequential machine of Figure 8.108.

Example 8.28 In this example, a Mealy synchronous sequential machine will be designed that will generate an output z_1 whenever the sequence 1001 is detected on a serial data input line x_1. Overlapping sequences are valid. For example, the bit sequence . . . 0110$\underline{1001}$000$\underline{11001}$0010 . . . will assert z_1 three times. The outputs will be generated during the first half of the clock period. This type of sequence detector can be used to detect the *start frame* and *end frame* flags in the high-level data link control (HDLC) format. The start frame and end frame flags are each 1 byte containing a bit sequence of 01111110. The state diagram is shown in Figure 8.111. The behavioral module, test bench, and waveforms are shown in Figure 8.112, Figure 8.113, and Figure 8.114, respectively.

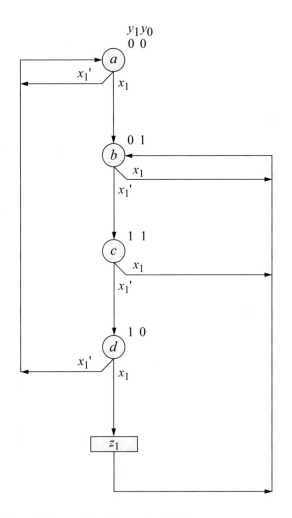

Figure 8.111   State diagram for Example 8.28.

```
//behavioral module for a mealy ssm
module mealy_ssm2 (clk, rst_n, x1, y, z1);

input clk, rst_n, x1;
output [1:0] y;
output z1;

reg [1:0] y, next_state;
wire z1;

parameter   state_a = 2'b00,     //assign state codes
            state_b = 2'b01,
            state_c = 2'b11,
            state_d = 2'b10;

always @ (posedge clk)            //set next state
begin
   if (~rst_n)
      y <= state_a;
   else
      y <= next_state;
end

assign z1 = ((y[1]) && (~y[0]) && (x1)  && (clk));

always @ (y or x1)               //determine next state
begin
   case (y)
      state_a:
         if (x1)
            next_state = state_b;
         else
            next_state = state_a;

      state_b:
         if (x1)
            next_state = state_b;
         else
            next_state = state_c;

      state_c:
         if (x1)
            next_state = state_b;
         else
            next_state = state_d;
//continued on next page
```

Figure 8.112   Behavioral module for the Mealy machine of Example 8.28.

```
        state_d:
            if (x1)
                next_state = state_b;
            else
                next_state = state_a;
        default: next_state = state_a;
    endcase
end

endmodule
```

Figure 8.112   (Continued)

```
//test bench for mealy ssm2
module mealy_ssm2_tb;


reg clk, rst_n, x1;
wire [1:0] y;
wire z1;


//display variables
initial
$monitor ("x1=%b, state=%b, z1=%b", x1, y, z1);


//define clock
initial
begin
    clk = 1'b0;
    forever
        #10 clk = ~clk;
end


//define input sequence
initial
begin
    #0      rst_n = 1'b0;                //rst to state_a
    #15     rst_n = 1'b1;

    x1 = 1'b0;   @ (posedge clk)         //go to state_a
    x1 = 1'b1;   @ (posedge clk)         //go to state_b
    x1 = 1'b1;   @ (posedge clk)         //go to state_b
//continued on next page
```

Figure 8.113   Test bench for the Mealy machine of Figure 8.112.

```
   x1 = 1'b0;   @ (posedge clk)        //go to state_c
   x1 = 1'b1;   @ (posedge clk)        //go to state_b
   x1 = 1'b0;   @ (posedge clk)        //go to state_c
   x1 = 1'b0;   @ (posedge clk)        //go to state_d
   x1 = 1'b1;   @ (posedge clk)        //go to state_b
   x1 = 1'b0;   @ (posedge clk)        //go to state_c
   x1 = 1'b0;   @ (posedge clk)        //go to state_d
   x1 = 1'b0;   @ (posedge clk)        //go to state_a
   x1 = 1'b1;   @ (posedge clk)        //go to state_b
   x1 = 1'b1;   @ (posedge clk)        //go to state_b
   x1 = 1'b0;   @ (posedge clk)        //go to state_c
   x1 = 1'b0;   @ (posedge clk)        //go to state_d
   x1 = 1'b1;   @ (posedge clk)        //go to state_b
   x1 = 1'b0;   @ (posedge clk)        //go to state_c
   x1 = 1'b0;   @ (posedge clk)        //go to state_d
   x1 = 1'b1;   @ (posedge clk)        //go to state_b
   x1 = 1'b0;   @ (posedge clk)        //go to state_c
   #10     $stop;
end

mealy_ssm2 inst1 (    //instantiate the module
   .clk(clk),
   .rst_n(rst_n),
   .x1(x1),
   .y(y),
   .z1(z1)
   );
endmodule
```

Figure 8.113   (Continued)

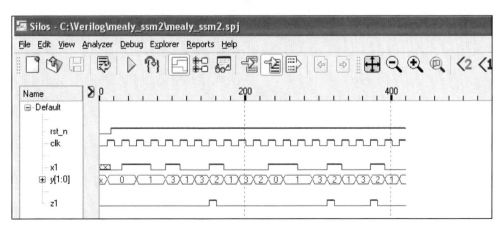

Figure 8.114   Waveforms for the Mealy synchronous sequential machine of Figure 8.112.

Example 8.29 This example will design a behavioral module of an asynchronous sequential machine using the **case** construct. There is one input x_1 and two outputs: z_1 and z_2. The machine functions as a 2-output bistable multivibrator, whose operation is characterized by the timing diagram of Figure 8.115. Output z_1 toggles on the positive transition of x_1 and output z_2 toggles on the negative transition of x_1. Figure 8.116 shows the primitive flow table. There are no equivalent states and no rows can merge. The design module, test bench, and waveforms are shown in Figure 8.117, Figure 8.118, and Figure 8.119, respectively. There are two excitation variables $y_e[1:0]$ as shown in the behavioral module.

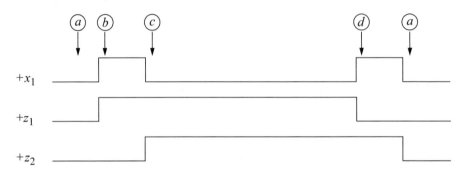

Figure 8.115 Timing diagram for the asynchronous sequential machine of Example 8.29.

x_1 = 0	x_1 = 1	z_1	z_2
(a)	b	0	0
c	(b)	1	0
(c)	d	1	1
a	(d)	0	1

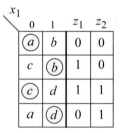

Figure 8.116 Primitive flow table for the asynchronous sequential machine of Example 8.29.

```
//behavioral asynchronous sequential machine
module asm8 (x1, rst_n, ye, z1, z2);

input x1, rst_n;
output [1:0] ye;
output z1, z2;             //continued on next page
```

Figure 8.117 Behavioral module for the asynchronous sequential machine of Example 8.29.

```verilog
wire x1, rst_n;
reg [1:0] ye, next_state;      //must be reg for always
reg z1, z2;

//assign state codes
parameter    state_a = 2'b00,  //parameter defines a constant
             state_b = 2'b01,
             state_c = 2'b11,
             state_d = 2'b10;

always @ (x1 or rst_n)         //latch next state
begin
   if (~rst_n)
      ye <= state_a;
   else
      ye <= next_state;
end

always @ (ye)                  //define outputs for each state
begin                          //sensitivity list must contain
   if (ye == state_a)          //the state variables
      begin
         z1 = 1'b0;
         z2 = 1'b0;
      end

   if (ye == state_b)          //== is logical equality
      begin
         z1 = 1'b1;
         z2 = 1'b0;
      end

   if (ye == state_c)
      begin
         z1 = 1'b1;
         z2 = 1'b1;
      end

   if (ye == state_d)
      begin
         z1 = 1'b0;
         z2 = 1'b1;
      end
end

//continued on next page
```

Figure 8.117   (Continued)

```
//determine next state
always @ (x1)
begin
   case (ye)
      state_a:
         if (x1)
            next_state = state_b;
         else
            next_state = state_a;

      state_b:
         if (~x1)
            next_state = state_c;
         else
            next_state = state_b;

      state_c:
         if (x1)
            next_state = state_d;
         else
            next_state = state_c;

      state_d:
         if (~x1)
            next_state = state_a;
         else
            next_state = state_d;

      default:
         next_state = state_a;
   endcase
end
endmodule
```

Figure 8.117   (Continued)

```
//test bench for asynchronous sequential machine
module asm8_tb;

reg x1, rst_n;          //inputs are reg in test bench
wire [1:0] ye;          //outputs are wire in test bench
wire z1, z2;

//continued on next page
```

Figure 8.118   Test bench for the asynchronous sequential machine of Figure 8.117.

```
initial
begin
   #0    rst_n = 1'b0;
         x1 = 1'b0;
   #10   rst_n = 1'b1;

   #10   x1 = 1'b1; //state_a to state_b; assert z1
   #10   x1 = 1'b0; //state_b to state_c; assert z1, z2
   #10   x1 = 1'b1; //state_c to state_d; deassert z1; assert z2
   #10   x1 = 1'b0; //state_d to state_a; no outputs
   #10   x1 = 1'b1; //state_a to state_b; assert z1
   #10   x1 = 1'b0; //state_b to state_c; assert z1, z2
   #10   x1 = 1'b1; //state_c to state_d; deassert z1; assert z2
   #10   x1 = 1'b0; //state_d to state_a; no outputs
   #40      $stop;
end

//instantiate the module into the test bench
asm8 inst1 (
   .x1(x1),
   .rst_n(rst_n),
   .ye(ye),
   .z1(z1),
   .z2(z2)
   );

endmodule
```

Figure 8.118   (Continued)

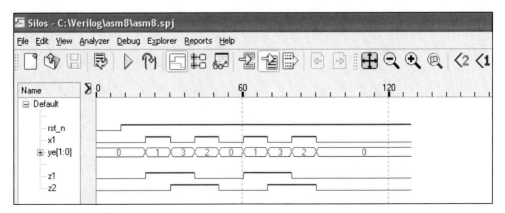

Figure 8.119   Waveforms for the asynchronous sequential machine of Figure 8.117.

Example 8.30   A 16-function ALU will be designed using behavioral modeling with the **case** statement to execute the following 1-byte [7:0] operations:

ADD	Addition	Sum = $a + b$
SUB	Subtraction	Difference = $a - b$
MUL	Multiply	Product = $a * b$
DIV	Divide	Quotient = a / b. No remainder.
AND	Logical AND	Result = $a \& b$
OR	Logical OR	Result = $a \mid b$
XOR	Logical exclusive-OR	Result = $a \wedge b$
NOT	Logical invert	Result = $\sim a$
SRA	Shift right arithmetic (For signed operands)	The sign bit [7] is propagated right one bit position. All other bits shift right one bit position.
SRL	Shift right logical (For unsigned operands)	All bits shift right one bit position. A zero is shifted into the vacated bit [7] position.
SLA	Shift left arithmetic (For signed operands)	The sign bit [7] does not shift. One bit is shifted out of bit position [6]. All other bits shift left one bit position. A zero is shifted into bit [0].
SLL	Shift left logical (For unsigned operands)	All bits shift left one bit position. A zero is shifted into the vacated bit [0] position.
ROR	Rotate right	Bit [0] rotates to the bit [7] position. All other bits shift right one bit position.
ROL	Rotate left	Bit [7] rotates to the bit [0] position. All other bits shift left one bit position.
INC	Increment	Add one to operand a.
DEC	Decrement	Subtract one from operand a.

Operands $a[7:0]$ and $b[7:0]$ are eight bit operands; the operation code *op-code[3:0]* is four bits; the multiply result *result_mul[15:0]* is 16 bits to accommodate the $2n$-bit product; all other results *result[7:0]* are eight bits. The block diagram for the 16-function ALU is shown in Figure 8.120. The behavioral module, test bench, outputs, and waveforms are shown in Figure 8.121, Figure 8.122, Figure 8.123, and Figure 8.124, respectively.

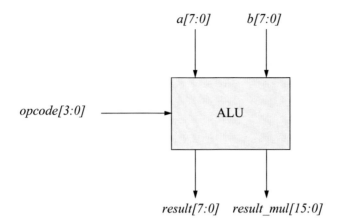

Figure 8.120   Block diagram for the 16-function ALU of Example 8.30.

```
module alu_16fctn (a, b, opcode, result, result_mul);

//list which are input or output ------------------
input  [7:0] a, b;
input  [3:0] opcode;
output [7:0] result;
output [15:0] result_mul;

//specify wire for input and reg for output -------
wire [7:0] a, b;//inputs are wire
wire [3:0] opcode;
reg  [7:0] result;//outputs are reg
reg  [15:0] result_mul;

//define the opcodes ------------------------------
parameter   add_op = 4'b0000,
            sub_op = 4'b0001,
            mul_op = 4'b0010,
            div_op = 4'b0011,
//-------------------------------------------------
            and_op = 4'b0100,
            or_op  = 4'b0101,
            xor_op = 4'b0110,
            not_op = 4'b0111,
//-------------------------------------------------
//continued on next page
```

Figure 8.121   Behavioral module for the 16-function ALU of Example 8.30.

```
//-------------------------------------------------
            sra_op = 4'b1000,
            srl_op = 4'b1001,
            sla_op = 4'b1010,
            sll_op = 4'b1011,
//-------------------------------------------------
            ror_op = 4'b1100,
            rol_op = 4'b1101,
            inc_op = 4'b1110,
            dec_op = 4'b1111;
//-------------------------------------------------

//execute operations
always @ (a or b or opcode)
begin
   case (opcode)
      add_op : result = a + b;
      sub_op : result = a - b;
      mul_op : result_mul = a * b;
      div_op : result = a / b;
//-------------------------------------------------
      and_op : result = a & b;
      or_op  : result = a | b;
      xor_op : result = a ^ b;
      not_op : result = ~a;
//-------------------------------------------------
      sra_op : result = {a[7], a[7], a[6], a[5],
                         a[4], a[3], a[2], a[1]};
      srl_op : result = a >> 1;
      sla_op : result = {a[7], a[5], a[4], a[3],
                         a[2], a[1], a[0], 1'b0};
      sll_op : result = a << 1;
//-------------------------------------------------
      ror_op : result = {a[0], a[7], a[6], a[5],
                         a[4], a[3], a[2], a[1]};
      rol_op : result = {a[6], a[5], a[4], a[3],
                         a[2], a[1], a[0], a[7]};
      inc_op : result = a + 1;
      dec_op : result = a - 1;
//-------------------------------------------------
      default : result = 0;
   endcase

end

endmodule
```

Figure 8.121   (Continued)

```
//alu_16fctn test bench
module alu_16fctn_tb;

reg [7:0] a, b;              //inputs are reg for test bench
reg [3:0] opcode;
wire [7:0] result;           //outputs are wire for test bench
wire [15:0] result_mul;

initial
$monitor ("a=%h, b=%h, opcode=%h, rslt=%b, rslt=%h,
           rslt_mul=%h",
           a, b, opcode, result, result_mul);
initial
begin
//add op -----------------------------------------
   #0    a = 8'b00000000;
         b = 8'b00000000;
         opcode = 4'b0000;

   #10   a = 8'b01100010;            //a = 98d = 62h
         b = 8'b00011100;            //b = 28d = 1Ch
         opcode = 4'b0000;           //result = 126d = 007eh

   #10   a = 8'b11111111;            //a = 255d = ffh
         b = 8'b11111111;            //b = 255d = ffh
         opcode = 4'b0000;           //result = 510d = 01feh

//sub op -----------------------------------------
   #10   a = 8'b10000000;            //a = 128d = 80h
         b = 8'b01100011;            //b = 99d = 63h
         opcode = 4'b0001;           //result = 29d = 001dh

   #10   a = 8'b11111111;            //a = 255d = ffh
         b = 8'b00001111;            //b = 15d = fh
         opcode = 4'b0001;           //result = 240d = 00f0h

//mul op -----------------------------------------
   #10   a = 8'b00011001;            //a = 25d = 19h
         b = 8'b00011001;            //b = 25d = 19h
         opcode = 4'b0010;           //result_mul = 625d = 0271h

   #10   a = 8'b10000100;            //a = 132d = 84h
         b = 8'b11111010;            //b = 250d = fah
         opcode = 4'b0010;           //result_mul = 33000d = 80e8h


//continued on next page
```

Figure 8.122   Test bench for the 16-function ALU of Figure 8.121.

```
   #10   a = 8'b11110000;       //a = 240d = f0h
         b = 8'b00010100;       //b = 20d = 14h
         opcode = 4'b0010;      //result_mul = 4800d = 12c0h

//div op ----------------------------------------
   #10   a = 8'b11110000;       //a = 240d = f0h
         b = 8'b00001111;       //b = 15d = 0fh
         opcode = 4'b0011;      //result = 16d = 10h

   #10   a = 8'b00010000;       //a = 16d = 10h
         b = 8'b00010010;       //b = 18d = 12h
         opcode = 4'b011;       //result = 0d = 0h

//and op ----------------------------------------
   #10   a = 8'b11111111;
         b = 8'b11110000;
         opcode = 4'b0100;      //result = 11110000

//or op -----------------------------------------
   #10   a = 8'b10101010;
         b = 8'b11110101;
         opcode = 4'b0101;      //result = 11111111

//xor op ----------------------------------------
   #10   a = 8'b00001111;
         b = 8'b10101010;
         opcode = 4'b0110;      //result = 10100101

//not op ----------------------------------------
   #10   a = 8'b00001111;
         opcode = 4'b0111;      //result = 11110000

//sra op ----------------------------------------
   #10   a = 8'b10001110;
         opcode = 4'b1000;      //result = 11000111

//srl op ----------------------------------------
   #10   a = 8'b11110011;
         opcode = 4'b1001;      //result = 01111001

//sla op ----------------------------------------
   #10   a = 8'b10001111;
         opcode = 4'b1010;      //result = 10011110


//continued on next page
```

Figure 8.122   (Continued)

```
//sll op -----------------------------------------
   #10   a = 8'b01110111;
         opcode = 4'b1011;        //result = 11101110

//ror op -----------------------------------------
   #10   a = 8'b01010101;
         opcode = 4'b1100;        //result = 10101010

//rol op -----------------------------------------
   #10   a = 8'b01010101;
         opcode = 4'b1101;        //result = 10101010

//inc op -----------------------------------------
   #10   a = 8'b11111101;
         opcode = 4'b1110;        //result = 11111110

//dec op -----------------------------------------
   #10   a = 8'b11111110;
         opcode = 4'b1111;        //result = 11111101
//-------------------------------------------------
   #10      $stop;
end

//instantiate the module into the test bench
alu_16fctn inst1 (
   .a(a),
   .b(b),
   .opcode(opcode),
   .result(result),
   .result_mul(result_mul)
   );
endmodule
```

Figure 8.122   (Continued)

```
a=00, b=00, opcode=0, rslt=00000000, rslt=00, rslt_mul=xxxx
a=62, b=1c, opcode=0, rslt=01111110, rslt=7e, rslt_mul=xxxx
a=ff, b=ff, opcode=0, rslt=11111110, rslt=fe, rslt_mul=xxxx
a=80, b=63, opcode=1, rslt=00011101, rslt=1d, rslt_mul=xxxx
a=ff, b=0f, opcode=1, rslt=11110000, rslt=f0, rslt_mul=xxxx
a=19, b=19, opcode=2, rslt=11110000, rslt=f0, rslt_mul=0271
a=84, b=fa, opcode=2, rslt=11110000, rslt=f0, rslt_mul=80e8
a=f0, b=14, opcode=2, rslt=11110000, rslt=f0, rslt_mul=12c0
//continued on next page
```

Figure 8.123   Outputs for the 16-function ALU of Figure 8.121.

```
a=f0, b=0f, opcode=3, rslt=00010000, rslt=10, rslt_mul=12c0
a=10, b=12, opcode=3, rslt=00000000, rslt=00, rslt_mul=12c0
a=ff, b=f0, opcode=4, rslt=11110000, rslt=f0, rslt_mul=12c0
a=aa, b=f5, opcode=5, rslt=11111111, rslt=ff, rslt_mul=12c0
a=0f, b=aa, opcode=6, rslt=10100101, rslt=a5, rslt_mul=12c0
a=0f, b=aa, opcode=7, rslt=11110000, rslt=f0, rslt_mul=12c0
a=8e, b=aa, opcode=8, rslt=11000111, rslt=c7, rslt_mul=12c0
a=f3, b=aa, opcode=9, rslt=01111001, rslt=79, rslt_mul=12c0
a=8f, b=aa, opcode=a, rslt=10011110, rslt=9e, rslt_mul=12c0
a=77, b=aa, opcode=b, rslt=11101110, rslt=ee, rslt_mul=12c0
a=55, b=aa, opcode=c, rslt=10101010, rslt=aa, rslt_mul=12c0
a=55, b=aa, opcode=d, rslt=10101010, rslt=aa, rslt_mul=12c0
a=fd, b=aa, opcode=e, rslt=11111110, rslt=fe, rslt_mul=12c0
a=fe, b=aa, opcode=f, rslt=11111101, rslt=fd, rslt_mul=12c0
```

Figure 8.123   (Continued)

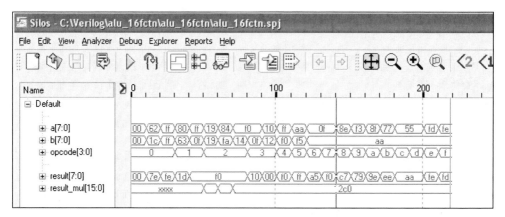

Figure 8.124   Waveforms for the 16-function ALU of Figure 8.121.

Example 8.31 As a final example in this section, a Moore machine that accepts serial data in the form of 3-bit words on an input line x_1 will be designed. There is one bit space between contiguous words, as shown below, where $b_i = 0$ or 1.

$$x_1 = \dots \ \left| b_1 b_2 b_3 \right| \ \ \left| b_1 b_2 b_3 \right| \ \ \left| b_1 b_2 b_3 \right| \dots$$

Whenever a word contains the bit pattern $b_1 b_2 b_3 = 111$, the machine will assert output z_1 during the bit time between words according to the following assertion/deassertion statement: $z_1 \uparrow t_2 \downarrow t_3$, where the time $t_2 - t_3$ is the last half of the clock cycle. An example of a valid word in a series of words is shown below. Notice that the output signal is displaced in time with respect to the input sequence and occurs one state time later.

$$x_1 = \dots \ \left| 0\ 0\ 1 \right| \ \ \left| 1\ 0\ 1 \right| \ \ \left| 0\ 1\ 1 \right| \ \ \left| 1\ 1\ 1 \right| \ \ \left| 0\ 1\ 0 \right| \dots$$

$$\text{Output } z_1 \uparrow t_2 \downarrow t_3 \ \underline{\qquad}|$$

The first step is to generate a state diagram. This is an extremely important step because all remaining steps depend upon a state diagram that correctly represents the machine specifications. Generating an accurate state diagram is thus a pivotal step in the design of synchronous sequential machines. The state diagram for this example is illustrated in Figure 8.125, which graphically describes the behavior of the machine. Seven states are required, providing four state levels — one level for each bit in the 3-bit words and one level for the bit space between words.

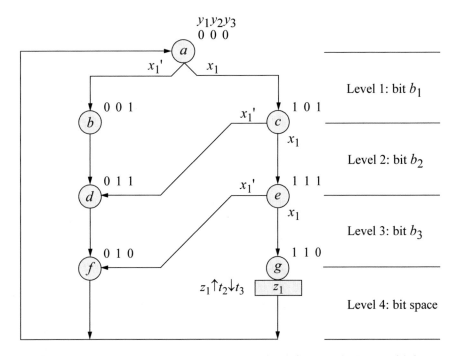

Figure 8.125 State diagram for the Moore machine of Example 8.31, which generates an output z_1 whenever a 3-bit word $x_1 = 111$. There is one unused state: 100.

The machine is reset to an initial state which is labeled a in Figure 8.125. Since x_1 is a serial input line, the state transition can proceed in only one of two directions from state a, depending upon the value of x_1: to state b if $x_1 = 0$ or to state c if $x_1 = 1$. Both paths represent a test of the first bit (b_1) of a word and both state transitions occur on the rising edge of the clock signal. Since the state transition from a to b occurs only if $x_1 = 0$, any bit sequence consisting of $b_1 b_2 b_3 = 000$ through 011 will proceed from state a to state b. Since this path will never generate an output (the first bit is invalid), there is no need to test the value of x_1 in states b or d. States b and d, together with state f, are required only to maintain four state levels in the machine, where the first three levels represent one bit each of the 3-bit word as follows: states a, b, and d correspond to bits b_1, b_2, and b_3, respectively. State f, which is the fourth level, corresponds to the bit space between words. This assures that the clocking will remain synchronized for the following word. From state d, the machine proceeds to state f and then returns to state a where the first bit of the next word is checked.

The second path from state a takes the machine to state c if $x_1 = 1$. Because this is the first bit of a possible valid 3-bit sequence, state c must also test the value of x_1, in this case bit b_2. If x_1 (b_2) = 0 in state c, then this represents an invalid word and output z_1 will not be asserted. Thus, any bit sequence consisting of $b_1 b_2 b_3 = 100$ or 101 will not assert z_1, but must still maintain four state levels. For either of the previous sequences, the state transition will be a, c, d, f, a.

If x_1 (b_2) = 1 in state c, then the machine proceeds to state e. This corresponds to a bit sequence of $x_1 = b_1 b_2 b_3 = 11-$, where b_2 is the second bit of a possible valid 3-bit sequence. Finally, in state e, x_1 (b_3) = 0 or 1. If $x_1 = 0$, then this represents an invalid sequence. Output z_1 is not asserted and the transition is to state f and then to state a where the first bit of the next word is checked. If, however, $x_1 = 1$ in state e, then this completes a valid sequence for a 3-bit word and the machine proceeds to state g where z_1 is asserted. The machine then returns to state a. In order to preserve synchronization between state a and the first bit of each word, it is important that any path taken in the state diagram always maintains four state levels.

The path a, c, e, g depicts a valid word which culminates in the assertion of output z_1 in state g. All other sequences of three bits will not generate an output, but each path must maintain four state levels. This guarantees that state a will always test the first bit of each 3-bit word. Output z_1 is a function of the state alphabet only, and thus, the state diagram represents a Moore machine.

By carefully considering each state transition and by efficiently utilizing existing states while maintaining four state levels, the state diagram of Figure 8.125 is obtained in which there are no equivalent states. States b and c are not equivalent because the next state for c is different than the next state for b if $x_1 = 1$. Thus, the second requirement for equivalent states is not met. The same is true for states d and e. Also, states f and g cannot be equivalent because the outputs are different (the first requirement for equivalency). The absence of equivalent states can be verified by using either the row-matching technique or an implication table. In this example, however, it is intuitively obvious that there are no redundant states.

Whenever possible, state codes should be assigned such that there are a maximal number of adjacent 1s in the flip-flop input maps. This allows more minterm locations to be combined, resulting in minimized input equations in a sum-of-products form.

State codes are adjacent if they differ in only one variable. For example, state codes $y_1y_2y_3 = 101$ and 100 are adjacent because only y_3 changes. Thus, minterm locations 101 and 100 can be combined into one term. However, state codes $y_1y_2y_3 = 101$ and 110 are not adjacent because two variables change: flip-flops y_2 and y_3.

The behavioral module is shown in Figure 8.126 using the **parameter** keyword, conditional statements, and the **case** statement. The test bench and waveforms are shown in Figure 8.127 and Figure 8.128, respectively. The system task **$random** is used in the test bench to randomly select a value for x_1 from the values 0 and 1 because the transitions $b \rightarrow d$, $d \rightarrow f$, and $f \rightarrow a$ are independent of the value for x_1.

```
//moore synchronous sequential machine to detect
//a sequence of 111 on a serial data line
module moore_ssm3 (clk, rst_n, x1, y, z1);


input clk, rst_n, x1;
output [2:0] y;                 //y is an array of 3 bits
output z1;


wire clk, rst_n, x1;
reg [2:0] y, next_state;        //outputs are reg in behavioral
reg z1;


//assign state codes
parameter    state_a = 3'b000,  //parameter defines a constant
             state_b = 3'b001,  //state names must have at
             state_c = 3'b101,  //least two characters
             state_d = 3'b011,
             state_e = 3'b111,
             state_f = 3'b010,
             state_g = 3'b110;


//set next state
always @ (posedge clk or negedge rst_n)
begin
   if (~rst_n)                  //if (~rst_n) is true (1),
      y <= #3 state_a;          //then y <= state_a #3 later
   else
      y <= #3 next_state;
end

//continued on next page
```

Figure 8.126 Behavioral module for the Moore machine of Example 8.31.

```
//determine output
always @ (y or clk)
begin
   if (y == state_g)
      begin
         if (~clk)
            z1 = 1'b1;
         else
            z1 = 1'b0;
      end
   else
      z1 = 1'b0;
end


//determine next state
always @ (y or x1)
begin
   case (y)                          //case is a multiple-way
      state_a:                       //conditional branch.
         if (x1)                     //if y = state_a, then
            next_state = state_c;    //execute if . . . else
         else
            next_state = state_b;
      state_b: next_state = state_d;
      state_c:
         if (x1)
            next_state = state_e;
         else
            next_state = state_d;
      state_d: next_state = state_f;
      state_e:
         if (x1)
            next_state = state_g;
         else
            next_state = state_f;
      state_f: next_state = state_a;
      state_g: next_state = state_a;
      default: next_state = state_a;
   endcase

end

endmodule
```

Figure 8.126   (Continued)

```
//test bench for moore_ssm3
module moore_ssm3_tb;


reg clk, x1, rst_n;
wire [2:0] y;
wire z1;


//define clock
initial
begin
  clk = 1'b0;
  forever              //forever continually executes
    #10 clk = ~clk;    //the procedural statement
end


//define input vectors
initial
begin
  #0 x1 = 1'b0;
     rst_n = 1'b0;
  #5 rst_n = 1'b1;

  @ (posedge clk)
  //if x1=0 in state_a, go to state_b (001)

  x1 = $random;   @ (posedge clk)
     //if x1=0/1 in state_b, go to state_d (011)

  x1 = $random;   @ (posedge clk)
     //if x1=0/1 in state_d, go to state_f (010)

  x1 = $random;   @ (posedge clk)
     //if x1=0/1 in state_f, go to state_a (000)

  x1 = 1'b1;      @ (posedge clk)
     //if x1=1 in state_a, go to state_c (101)

  x1 = 1'b1;      @ (posedge clk)
     //if x1=1 in state_c, go to state_e (111)

  x1 = 1'b1;      @ (posedge clk)
     //if x1=1 in state_e, go to state_g (110); z1=1

//continued on next page
```

Figure 8.127   Test bench for the Moore synchronous sequential machine of Figure 8.126.

```verilog
   x1 = $random;       @(posedge clk)
      //if x1=0/1 in state_g, go to state_a (000)

   x1 = 1'b1;          @(posedge clk)
      //if x1=1 in state_a, go to state_c (101)

   x1 = 1'b0;          @(posedge clk)
      //if x1=0 in state_c, go to state_d (011)

   x1 = $random;       @(posedge clk)
      //if x1=0/1 in state_d, go to state_f (010)

   x1 = $random;       @(posedge clk)
      //if x1=0/1 in state_f, go to state_a (000)

   x1 = 1'b1;          @(posedge clk)
      //if x1=1 in state_a, go to state_c (101)

   x1 = 1'b1;          @(posedge clk)
      //if x1=1 in state_c, go to state_e (111)

   x1 = 1'b0;          @(posedge clk)
      //if x1=0 in state_e, go to state_f (010)

   x1 = $random;       @(posedge clk)
      //if x1=0/1 in state_f, go to state_a (000)

   #150   $stop;

end

//instantiate the module into the test bench
moore_ssm3 inst1 (
 .clk(clk),
 .rst_n(rst_n),
 .x1(x1),
 .y(y),
 .z1(z1)
 );

endmodule
```

Figure 8.127   (Continued)

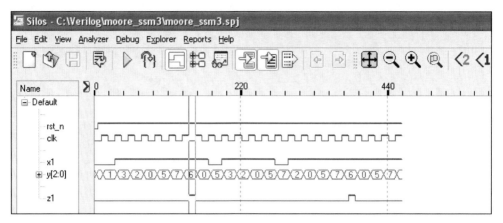

Figure 8.128   Waveforms for the Moore synchronous sequential machine of Figure 8.126.

8.5 Loop Statements

There are four types of loop statements in Verilog: **for**, **while**, **repeat**, and **forever**. Loop statements must be placed within an **initial** or an **always** block and may contain delay controls. The loop constructs allow for repeated execution of procedural statements within an **initial** or an **always** block.

8.5.1 For Loop

The **for** loop contains three parts:

1. An *initial* condition to assign a value to a register control variable. This is executed once at the beginning of the loop to initialize a register variable that controls the loop.

2. A *test* condition to determine when the loop terminates. This is an expression that is executed before the procedural statements of the loop to determine if the loop should execute. The loop is repeated as long as the expression is true. If the expression is false, the loop terminates and the activity flow proceeds to the next statement in the module.

3. An *assignment* to modify the control variable, usually an increment or a decrement. This assignment is executed after each execution of the loop and before the next test to terminate the loop.

The syntax of a **for** loop is shown below. The body of the loop can be a single procedural statement or a block of procedural statements.

for (initial control variable assignment; test expression; control variable assignment)
    procedural statement or block of procedural statements

The **for** loop is generally used when there is a known beginning and an end to a loop. The **for** loop is similar in function to the **for** loop in the C programming language and has been used in the test bench of several previous examples. A code segment is shown in Figure 8.129 using a **for** loop which provides all 16 combinations of the four bits x_1, x_2, x_3, and x_4.

```
//apply input vectors and display variables
initial
begin: apply_stimulus
   reg [4:0] invect;
   for (invect = 0; invect < 16; invect = invect + 1)
      begin
         {x1, x2, x3, x4} = invect [4:0];
         #10 $display ("{x1x2x3x4} = %b, z1 = %b",
                       {x1, x2, x3, x4}, z1);
      end
end
```

Figure 8.129   Code segment using a **for** loop.

8.5.2 While Loop

The **while** loop executes a procedural statement or a block of procedural statements as long as a Boolean expression returns a value of true. When the procedural statements are executed, the Boolean expression is reevaluated. The loop is executed until the expression returns a value of false. If the evaluation of the expression is false, then the **while** loop is terminated and control is passed to the next statement in the module. If the expression is false before the loop is initially entered, then the **while** loop is never executed.

The Boolean expression may contain any of the following types: arithmetic, logical, relational, equality, bitwise, reduction, shift, concatenation, replication, or conditional. If the **while** loop contains multiple procedural statements, then they are contained within the **begin** . . . **end** keywords. The syntax for a **while** statement is as follows:

> **while** (expression)
>   procedural statement or block of procedural statements

Example 8.32   This example demonstrates the use of the **while** construct to count the number of 1s in a 16-bit register *reg_a*. The module is shown in Figure 8.130. The variable *count* is declared as type **integer** and is used to obtain the cumulative count of the number of 1s. The first **begin** keyword must have a name associated with the keyword because this declaration is allowed only with named blocks.

The register is initialized to contain ten 1s (*16'haabb*). Alternatively, the register can be loaded from any other register. If *reg_a* contains a 1 bit in any bit position, then the **while** loop is executed. If *reg_a* contains all zeroes, then the **while** loop is terminated.

The low-order bit position (*reg_a[0]*) is tested for a 1 bit. If a value of 1 (true) is returned, *count* is incremented by one and the register is shifted right one bit position. There is only one procedural statement following the **if** statement; therefore, if a value of 0 (false) is returned, then *count* is not incremented and the register is shifted right one bit position.

The **$display** system task then displays the number of 1s that were contained in the register *reg_a* as shown in Figure 8.131. Notice that the count changes value only when there is a 1 bit in the low-order bit position of *reg_a*. If *reg_a[0]* = 0, then the count is not incremented, but the total count is still displayed.

```verilog
//example of a while loop.
//count the number of 1s in a 16-bit register
module while_loop;

integer count;

initial
begin: number_of_1s
   reg [16:0] reg_a;

   count = 0;

   reg_a = 16'haabb;              //set reg_a to a known value

   while (reg_a)                  //do while reg_a contains 1s
      begin
         if (reg_a[0])            //check low-order bit
            count = count + 1;    //if true, add one to count
         reg_a = reg_a >> 1;      //shift right 1 bit position
         $display ("count = %d", count);
      end
end

endmodule
```

Figure 8.130   Module to illustrate the use of the **while** construct.

```
count = 1                         count = 6
count = 2                         count = 7
count = 2                         count = 7
count = 3                         count = 8
count = 4                         count = 8
count = 5                         count = 9
count = 5                         count = 9
count = 6                         count = 10
```

Figure 8.131   Outputs for the module of Figure 8.129 illustrating the use of the **while** loop.

8.5.3 Repeat Loop

The **repeat** loop executes a procedural statement or a block of procedural statements a specified number of times. The **repeat** construct can contain a constant, an expression, a variable, or a signed value. The syntax for the **repeat** loop is as follows:

> **repeat** (loop count expression)
>   procedural statement or block of procedural statements

If the loop count is **x** or **z**, then the loop count is treated as zero. The value of the loop count expression is evaluated once at the beginning of the loop.

Example 8.33   An example of the **repeat** loop, which increments a variable *count* 16 times, is shown in Figure 8.132. The integer *count* is initialized to zero; therefore, the count will stop when *count* is equal to 15. The outputs are shown in Figure 8.133.

```verilog
//example of the repeat keyword
module repeat_example;

integer count;

initial
begin
   count = 0;
   repeat (16)
   begin
      $display ("count = %d", count);
      count = count + 1;
   end
end
endmodule
```

Figure 8.132   Module to illustrate the use of the **repeat** construct.

count = 0	count = 8
count = 1	count = 9
count = 2	count = 10
count = 3	count = 11
count = 4	count = 12
count = 5	count = 13
count = 6	count = 14
count = 7	count = 15

Figure 8.133   Outputs for the module of Figure 8.132.

8.5.4 Forever Loop

The **forever** loop executes the procedural statement continuously until the system tasks **$finish** or **$stop** are encountered. It can also be terminated by the **disable** statement. The **disable** statement is a procedural statement; therefore, it must be used within an **initial** or an **always** block. It is used to prematurely terminate a block of procedural statements or a system task. When a **disable** statement is executed, control is transferred to the statement immediately following the procedural block or task. The **forever** loop is similar to a **while** loop in which the expression always evaluates to true (1). A timing control must be used with the **forever** loop; otherwise, the simulator would execute the procedural statement continuously without advancing the simulation time. The syntax of the **forever** loop is as follows:

> **forever**
> procedural statement

The **forever** statement is typically used for clock generation as shown in Figure 8.134 together with the system task **$finish**. The variable *clk* will toggle every 10 time units for a period of 20 time units. The length of simulation is 100 time units.

```verilog
//define clock
initial
begin
   clk = 1'b0;
   forever
      #10  clk = ~clk;
end

//define length of simulation
initial
   #100  $finish;
```

Figure 8.134   Clock generation using a **forever** statement.

8.6 Block Statements

Block statements provide a means to group multiple statements together so that they function as one statement. There are two types of blocks used in Verilog: sequential and parallel. A block can have an optional name associated with it (preceded by a colon) and registers can be declared locally within the block.

8.6.1 Sequential Blocks

Sequential blocks are delimited by the keywords **begin** . . . **end**. The statements in a sequential block execute in the order in which they are listed in the block, except for nonblocking assignments; that is, they execute sequentially. A delay value in a statement is relative to the execution of the immediately preceding statement. The syntax for a sequential block is as follows:

> **begin** [: optional name]
>   procedural statements
> **end**

Example 8.34   Sequential blocks have been used several times in previous sections; however, this example presents a sequential block using intrastatement delays and the concatenation of inputs. In Figure 8.135, the design module is shown containing three inputs: x_1, x_2, and x_3 together with six outputs: z_1, z_2, z_3, z_4, z_5, and z_6. The test bench, outputs, and waveforms are shown in Figure 8.136, Figure 8.137, and Figure 8.138, respectively.

```
//sequential block
module seq_block (x1, x2, x3, z1, z2, z3, z4, z5, z6);

input x1, x2, x3;
output z1, z2, z3;
output [2:0] z4, z5, z6;
reg z1, z2, z3;
reg [2:0] z4, z5, z6;

always @ (x1 or x2 or x3)
begin
   z1 = #2 (x1 & x2) | x3;
   z2 = #3 (x1 | x2) & x3;
   z3 = #4 ~(x1 ^ x2 ^ x3);
   z4 = {x1, x2, x3};
   z5 = {x3, x1, x2};
   z6 = {x2, x3, x1};
end
endmodule
```

Figure 8.135   Module to illustrate a sequential block.

```
//test bench for sequential block
module seq_block_tb;



reg x1, x2, x3;
wire z1, z2, z3;
wire [2:0] z4, z5, z6;



//display variables
initial
$monitor ("x1 x2 x3= %b, z1=%b,z2=%b,z3=%b,z4=%b,z5=%b,z6=%b",
          {x1, x2, x3}, z1, z2, z3, z4, z5, z6);



//apply input vectors
initial
begin
   #0    x1 = 1'b0;   x2 = 1'b0;   x3 = 1'b1;
   #10   x1 = 1'b0;   x2 = 1'b1;   x3 = 1'b0;
   #10   x1 = 1'b1;   x2 = 1'b0;   x3 = 1'b1;
   #10   x1 = 1'b1;   x2 = 1'b1;   x3 = 1'b1;

   #20   $stop;
end



//instantiate the module into the test bench
seq_block inst1 (
   .x1(x1),
   .x2(x2),
   .x3(x3),
   .z1(z1),
   .z2(z2),
   .z3(z3),
   .z4(z4),
   .z5(z5),
   .z6(z6)
   );

endmodule
```

Figure 8.136 Test bench for the module of Figure 8.135.

```
x1 x2 x3 = 001, z1=x, z2=x, z3= x, z4=xxx, z5=xxx, z6=xxx
x1 x2 x3 = 001, z1=1, z2=x, z3= x, z4=xxx, z5=xxx, z6=xxx
x1 x2 x3 = 001, z1=1, z2=0, z3= x, z4=xxx, z5=xxx, z6=xxx
x1 x2 x3 = 001, z1=1, z2=0, z3= 0, z4=001, z5=100, z6=010
x1 x2 x3 = 010, z1=1, z2=0, z3= 0, z4=001, z5=100, z6=010
x1 x2 x3 = 010, z1=0, z2=0, z3= 0, z4=001, z5=100, z6=010
x1 x2 x3 = 010, z1=0, z2=0, z3= 0, z4=010, z5=001, z6=100
x1 x2 x3 = 101, z1=0, z2=0, z3= 0, z4=010, z5=001, z6=100
x1 x2 x3 = 101, z1=1, z2=0, z3= 0, z4=010, z5=001, z6=100
x1 x2 x3 = 101, z1=1, z2=1, z3= 0, z4=010, z5=001, z6=100
x1 x2 x3 = 101, z1=1, z2=1, z3= 1, z4=101, z5=110, z6=011
x1 x2 x3 = 111, z1=1, z2=1, z3= 1, z4=101, z5=110, z6=011
x1 x2 x3 = 111, z1=1, z2=1, z3= 0, z4=111, z5=111, z6=111
```

Figure 8.137   Outputs for the module of Figure 8.135.

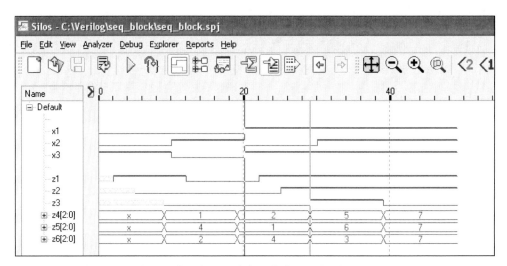

Figure 8.138   Waveforms for the module of Figure 8.135.

In the outputs of Figure 8.137, there are several occurrences in which the outputs do not change. This is due to the intrastatement delays. In the waveforms of Figure 8.138, the intrastatement delays are shown graphically and are readily apparent. For example, output z_1 is not asserted until two time units after a change occurs on the inputs. Output z_2 will not be asserted until five time units (#2 + #3) and output z_3 will not be asserted until nine time units (#2 + #3 + #4).

8.6.2 Parallel Blocks

Parallel blocks are delimited by the keywords **fork** . . . **join** and cause the statements in the corresponding block to execute concurrently. If a delay or event control is specified, it is relative to the time that the block began execution, not to the execution of the immediately preceding statement. Thus, the sequence in which the statements are written is not important. When execution of the block ends, the statement immediately following the parallel block is executed. The keyword **fork** can be considered as separating a single flow into multiple flows. The keyword **join** then combines the multiple flows into a single flow. Sequential and parallel blocks can be nested and mixed; that is, a parallel block can be nested within a sequential block.

The **fork** . . . **join** construct is used in test benches and is not supported by synthesis tools. It is suitable for waveform generation in test benches. Figure 8.139 shows a module to generate waveforms using the **fork** . . . **join** construct in which the sequential order of the statements has been altered. The waveforms are shown in Figure 8.140. Notice that simulation ends at time unit 60 which is relative to the time that the **fork** . . . **join** block began execution at time unit 0, not the accumulated time of 210 time units as in sequential blocks.

```
//parallel block
module par_block (z1, z2, z3);

output z1, z2, z3;
reg z1, z2, z3;

//initialize variables
initial
begin
   #0      z1 = 1'b0;
           z2 = 1'b0;
           z3 = 1'b0;
end

initial
   fork
       #10     z1 = 1'b1;
       #30     z2 = 1'b1;
       #20     z1 = 1'b0;
       #40     z2 = 1'b0;
       #60     z3 = 1'b0;
       #50     z3 = 1'b1;
   join
endmodule
```

Figure 8.139   Module to illustrate the use of the keywords **fork** . . . **join** to form a parallel block.

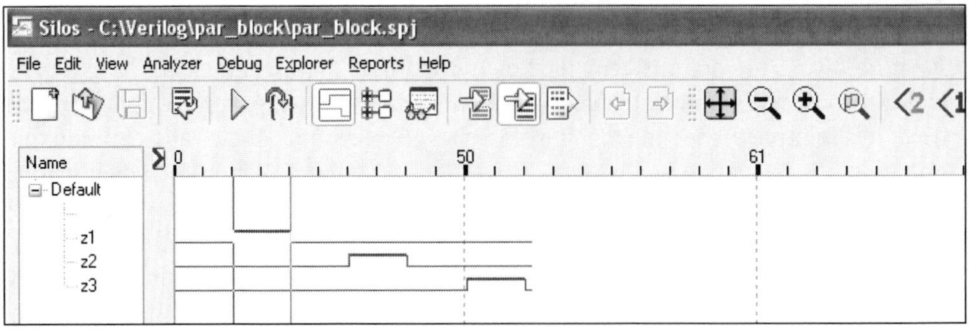

Figure 8.140   Waveforms for the module of Figure 8.139.

8.7 Procedural Continuous Assignment

A *procedural continuous assignment* is a procedural assignment that is contained within an **initial** statement or an **always** statement. Procedural assignments assign values to registers where the values remain until changed by another procedural assignment. Procedural continuous assignments, however, continuously assign values to nets and registers for a specified time period and override all other assignments. Continuous assignments were presented in other chapters and assigned values only to nets, not to registers. Verilog provides two types of procedural continuous assignments: **assign . . . deassign** and **force . . . release**.

8.7.1 Assign . . . Deassign

The left-hand target of an **assign . . . deassign** procedural continuous assignment must be a register or a concatenation of registers. The **assign** statement overrides all procedural assignments to a register. The **deassign** statement terminates the continuous assignment; however, the register retains the value set by the **assign** statement until another assignment occurs.

Example 8.35   Although the **assign . . . deassign** procedural continuous assignment is used primarily in test benches, it is used here to model a *D* flip-flop. In the behavioral module of Figure 8.141, the second **always** statement has a sensitivity list of *rst_n*; thus, when reset occurs, the **assign** statements reset the flip-flop. When reset is deasserted, the **deassign** statement is executed and the flip-flop retains the value that was set with the **assign** statement until another assignment occurs. The test bench is shown in Figure 8.142. The outputs and waveforms are shown in Figure 8.143 and Figure 8.144, respectively.

```
//behavioral d flip-flop
//to illustrate assign . . . deassign
module assign_deassign2 (rst_n, clk, d, q, q_n);

input rst_n, clk, d;
output q, q_n;

wire rst_n, clk, d;
reg q, q_n;

always @ (posedge clk)
begin
   q <= d;
   q_n <= ~d;
end

always @ (rst_n)
begin
   if (rst_n == 0)
      begin
         assign q = 1'b0;
         assign q_n = 1'b1;
      end

   else
      begin
         deassign q;
         deassign q_n;
      end
end
endmodule
```

Figure 8.141   Behavioral module to demonstrate the use of the **assign . . . deassign** procedural continuous assignment.

```
//test bench for assign . . . deassign d ff
module assign_deassign2_tb;

reg rst_n, clk, d;
wire q;

initial
$monitor ("rst_n=%b, clk= b, d= b, q=%b", rst_n, clk, d, q);
//continued on next page
```

Figure 8.142   Test bench to illustrate the use of the **assign . . . deassign** statements.

```verilog
//define clock

initial
begin
   clk = 1'b0;
   forever
      #10clk = ~clk;
end


//apply inputs

initial
begin
   #0     rst_n = 1'b1;
   #5     d = 1'b0;

   #10    d = 1'b1;
   #10    d = 1'b1;
   #10    d = 1'b0;
   #10    d = 1'b0;
   #10    d = 1'b1;
   #10    rst_n = 1'b0;
   #10    d = 1'b0;
   #10    rst_n = 1'b1;
   #10    d = 1'b1;
   #10    d = 1'b1;
   #10    d = 1'b1;
   #10    d = 1'b1;

   #10    $stop;

end


//instantiate the module into the test bench
assign_deassign2 inst1 (
   .rst_n(rst_n),
   .clk(clk),
   .d(d),
   .q(q)
   );

endmodule
```

Figure 8.142   (Continued)

```
rst_n = 1, clk = 0, d = x, q = x    rst_n = 0, clk = 0, d = 1, q = 0
rst_n = 1, clk = 0, d = 0, q = x    rst_n = 0, clk = 1, d = 1, q = 0
rst_n = 1, clk = 1, d = 0, q = 0    rst_n = 0, clk = 1, d = 0, q = 0
rst_n = 1, clk = 1, d = 1, q = 0    rst_n = 0, clk = 0, d = 0, q = 0
rst_n = 1, clk = 0, d = 1, q = 0    rst_n = 1, clk = 0, d = 0, q = 0
rst_n = 1, clk = 1, d = 1, q = 1    rst_n = 1, clk = 1, d = 0, q = 0
rst_n = 1, clk = 1, d = 0, q = 1    rst_n = 1, clk = 1, d = 1, q = 0
rst_n = 1, clk = 0, d = 0, q = 1    rst_n = 1, clk = 0, d = 1, q = 0
rst_n = 1, clk = 1, d = 0, q = 0    rst_n = 1, clk = 1, d = 1, q = 1
rst_n = 1, clk = 1, d = 1, q = 0    rst_n = 1, clk = 0, d = 1, q = 1
rst_n = 1, clk = 0, d = 1, q = 0    rst_n = 1, clk = 1, d = 1, q = 1
```

Figure 8.143   Outputs for the module of Figure 8.141.

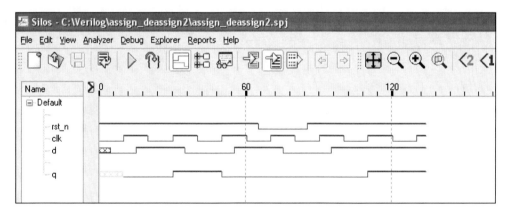

Figure 8.144   Waveforms for the module of Figure 8.141.

8.7.2 Force . . . Release

The **force** . . . **release** procedural continuous assignment is similar to the **assign** . . . **deassign** construct, but applies to both nets and registers. When the **force** assignment is applied to a net, the right-hand expression overrides any continuous assignment until the **release** is executed. The net will immediately return to its normal value when released. This is analogous to connecting a net to a high or low voltage level.

When the **force** statement is applied to a register, any procedural assignment or procedural continuous assignment is overridden and the current value is replaced by the right-hand expression. When the **release** statement is applied, the register retains the value that was forced until a procedural assignment sets a new value in the register, unless a procedural continuous assignment was already in effect when the **force** statement was applied. In that case, the register is set to the value before **force** was applied.

The **force** and **release** statements are typically used in the testing process to eliminate errors in a module by forcing nets and registers to known values. The effect on other nets and registers is then observed. The **force** and **release** statements should be used in test benches only, not in design modules.

Example 8.36 The **force** and **release** procedural continuous assignments will be used in the test bench of a *D* flip-flop that is designed in a behavioral module. The output *q* will be forced to a high level even though the *d* input is at a low level. The design module, test bench, and waveforms are shown in Figure 8.145, Figure 8.146, and Figure 8.147, respectively. In the waveforms of Figure 8.147, **release** is applied to the flip-flop at time unit 155. However, the value was not retained because a procedural continuous assignment was already in effect when the **force** statement was applied.

```
//d flip-flop behavioral
module d_ff_bh (rst_n, clk, d, q, q_n);

input rst_n, clk, d;
output q, q_n;

wire rst_n, clk, d;
reg q;

assign q_n = ~q;

always @ (rst_n or posedge clk)
begin
   if (rst_n == 0)
        q <= 1'b0;
   else q <= d;
end
endmodule
```

Figure 8.145   Behavioral module for a *D* flip-flop to illustrate the **force** and **release** statements.

```
//d_ff_bh test bench
module d_ff_bh_tb;

reg rst_n, clk, d;
wire q;

//continued on next page
```

Figure 8.146   Test bench for a *D* flip-flop illustrating the **force** and **release** statements.

```
//display variables
initial
$monitor ("rst_n = %b, clk = %b, d = %b, q = %b",
            rst_n, clk, d, q);


//define clock
initial
begin
   clk = 1'b0;
   forever
      #10clk = ~clk;
end


//apply input vectors
initial
begin
   #0    d = 1'b0;    rst_n = 1'b0;
   #5    d = 1'b1;    rst_n = 1'b1;
   #10   d = 1'b1;    rst_n = 1'b1;
   #10   d = 1'b1;    rst_n = 1'b1;
   #10   d = 1'b1;    rst_n = 1'b1;
   #10   d = 1'b0;    rst_n = 1'b1;
   #10   d = 1'b0;    rst_n = 1'b1;
   #50   force q = 1'b1;
   #50   release q;
   #10   d = 1'b0;    rst_n = 1'b1;
   #10   d = 1'b1;    rst_n = 1'b1;
   #10   d = 1'b0;    rst_n = 1'b1;

   #10   $stop;
end


//instantiate the module into the test bench
d_ff_bh inst1 (
   .rst_n(rst_n),
   .clk(clk),
   .d(d),
   .q(q)
   );

endmodule
```

Figure 8.146   (Continued)

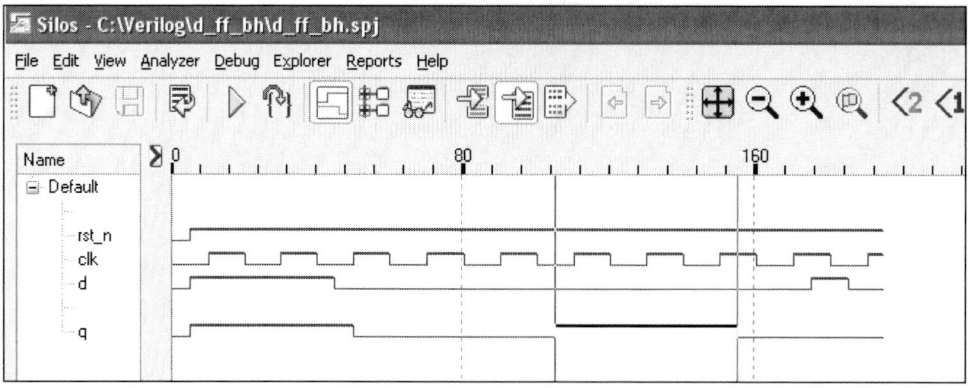

Figure 8.147 Waveforms for the *D* flip-flop illustrating the **force** and **release** statements.

8.8 Problems

8.1 Use behavioral modeling with an **always** statement to design a 2-input exclusive-OR circuit. The inputs are x_1 and x_2; the output is z_1. Obtain the design module, test bench, outputs, and waveforms for all combinations of the inputs. Use the **$monitor** system task to display the inputs and outputs.

8.2 Design a full adder using the **always** statement. The are three scalar inputs: the augend *a*, the addend *b*, and the carry-in *cin*. There are two outputs: *sum* and *cout*. Use blocking statements for *sum* and *cout* with a delay of five time units. Obtain the design module, test bench module, outputs, and waveforms. Test for all combinations of the inputs.

8.3 Using behavioral modeling with continuous assignment, design a 4-bit counter (*q[3:0]*) that counts sequentially from 0000 to 1111, where *q[0]* is the low-order bit. Connect a NAND gate, a NOR gate, and an exclusive-NOR gate to counter outputs *q[0]* and *q[1]* to generate outputs z_1, z_2, and z_3, respectively. Obtain the design module, test bench, outputs, and waveforms for the counter and for z_1, z_2, and z_3.

8.4 Design a modulo-11 counter using behavioral modeling. There are two inputs: *clk* and an active-low asynchronous reset *rst_n*; there is one output *count*. Obtain the design module, test bench, outputs, and waveforms.

8.5   Design a 4-bit counter with a mode control line x_1 that operates according to the following specifications:

(a) If $x_1 = 0$, then the counter generates the sequence 0000, 1000, 1100, 1110, 1111, 0111, 0011, 0001, 0000, ...

(b) If $x_1 = 1$, then the counter generates the sequence 0000, 1000, 0001, 0100, 0010, 1100, 0011, 1110, 0111, 1111, 0000, ...

Obtain the behavioral module using the **case** construct. Obtain the test bench, outputs, and waveforms.

8.6   Use behavioral modeling with the **case** statement to design a 5-function arithmetic and logic unit for the following five functions: add, subtract, multiply, divide, and modulus. The operands are 4-bit vectors: $a[3:0]$ and $b[3:0]$. Obtain the behavioral module, test bench, outputs, and waveforms for one input vector for each operation.

8.7   Design a 3-bit serial-in, parallel-out shift register using conditional statements only. The register shifts right one bit position for each clock pulse. Obtain the behavioral module, the test bench for several different input sequences, the outputs, and the waveforms.

8.8   Design a modulo-8 counter using behavioral modeling with conditional statements and a **case** statement. Obtain the design module, test bench, outputs, and waveforms.

8.9   Design a modulo-11 counter using behavioral modeling with conditional statements and the **case** statement. Obtain the design module, test bench, outputs, and waveforms.

8.10   Design a 5-bit Gray code counter using behavioral modeling with conditional and **case** statements. Obtain the design module, test bench, outputs, and waveforms.

8.11   Design a Moore synchronous sequential machine using behavioral modeling with conditional statements and the **case** statement. The machine will generate odd parity for each 5-bit word that is transmitted on a serial data line x_1. The format for the words is as follows:

$$\left| b_1 b_2 b_3 b_4 b_5 z_1 \;\middle|\; b_1 b_2 b_3 b_4 b_5 z_1 \;\middle|\; b_1 b_2 b_3 b_4 b_5 z_1 \;\middle|\; \cdots \right.$$

where $b_i = 0$ or 1, and z_1 is the odd parity bit. There is no space between words. Obtain the state diagram and assign state codes. Then obtain the behavioral module, the test bench for several input sequences to demonstrate odd parity generation, the outputs, and the waveforms.

8.12   Design an asynchronous sequential machine using behavioral modeling. The machine has two inputs: x_1 and x_2 and one output z_1. Input x_1 will always be asserted whenever x_2 is asserted; that is, there will never be a situation where x_1 is deasserted and x_2 is asserted. Output z_1 is asserted coincident with every third x_2 pulse and remains active for the duration of x_2.

Obtain a representative timing diagram and a reduced primitive flow table. Assign state codes so that there are no transient states that could cause an erroneous output on z_1. Obtain the behavioral module, the test bench, and the waveforms showing a complete sequence to assert output z_1.

8.13   Use the **while** construct to execute a loop if $reg\_a \neq reg\_b$ and increment a variable *count* if the registers are not equal. Register reg_b will be incremented by one until $reg\_a = reg\_b$. The variable *count* will increment by one for each increment of reg_b. Set the two registers to known unequal values. Obtain the module and outputs.

8.14   Using behavioral modeling, design a negative-edge-triggered D flip-flop to illustrate the use of the **assign** . . . **deassign** construct. The reset is active high. Obtain the design module, test bench, outputs, and waveforms.

9

Structural Modeling

Structural modeling consists of instantiation of one or more of the following design objects:

- Built-in primitives
- User-defined primitives (UDPs)
- Design modules

Instantiation means to use one or more lower-level modules — including logic primitives — that are interconnected in the construction of a higher-level structural module. A module can be a logic gate, an adder, a multiplexer, a counter, or some other logical function. The objects that are instantiated are called *instances*. Structural modeling is described by the interconnection of these lower-level logic primitives or modules. The interconnections are made by wires that connect primitive terminals or module ports.

9.1 Module Instantiation

Design modules were instantiated into every test bench module in previous chapters. The ports of the design module were instantiated by name and connected to the corresponding net names of the test bench. Each named instantiation was of the form

.design_module_port_name (test_bench_module_net_name)

Design module ports can be instantiated by name explicitly or by position. Instantiation by position is not recommended when a large number of ports are involved. Instantiation by name precludes the possibility of making errors in the instantiation process. Modules cannot be nested, but they can be instantiated into other modules. Structural modeling is analogous to placing the instances on a logic diagram and then connecting them by wires. When instantiating built-in primitives, an instance name is optional; however, when instantiating a module, an instance name must be used. Instances that are instantiated into a structural module are connected by nets of type **wire**.

A structural module may contain behavioral statements (**always**), continuous assignment statements (**assign**), built-in primitives (**and, or, nand, nor**, etc.), UDPs (*mux4*, *half_adder*, *adder4*, etc.), design modules, or any combination of these objects. Design modules can be instantiated into a higher-level structural module in order to achieve a hierarchical design.

Each module in Verilog is either a top-level (higher-level) module or an instantiated module. There is only one top-level module and it is not instantiated anywhere else in the design project. Instantiated primitives or modules, however, can be instantiated many times into a top-level module and each instance of a module is unique.

9.2 Ports

Ports provide a means for the module to communicate with its external environment. Ports, also referred to as terminals, can be declared as **input**, **output**, or **inout**. A port is a net by default; however, it can be declared explicitly as a net. A module contains an optional list of ports, as shown below for a full adder.

module full_adder (a, b, cin, sum, cout);

Ports *a*, *b*, and *cin* are input ports; ports *sum* and *cout* are output ports. The test bench for the full adder contains no ports as shown below because it does not communicate with the external environment.

module full_adder_tb;

Input ports   Input ports are those that allow signals to enter the module from external sources. The width of the input port is declared within the module. The size of the input port can be declared as either a scalar such as *a*, *b*, *cin* or as a vector such as *[3:0] a, b*, where *a* and *b* are the augend and addend inputs, respectively, to a 4-bit adder. The format of the declarations shown below is the same for both behavioral and structural modeling.

> **input** [3:0] a, b;   //declared as 4-bit vectors
> **input** cin;   //declared as a scalar

Output ports   Output ports are those that allow signals to exit the module to external destinations. The width of the output port is declared within the module. For behavioral modeling, the output is declared as type **reg** with a specified width, either scalar or vector. For structural modeling, the output port is declared as type **wire** with a specified width. The format for output ports is shown below.

> **output** [3:0] sum;   //declared as 4-bit vectors
> **output** cout;   //declared as a scalar
>
> **reg** [3:0] sum;   //for behavioral modeling
> **reg** cout;   //for behavioral modeling
>
> **wire** [3:0] sum;   //for structural modeling
> **wire** cout;   //for structural modeling

Inout ports   An **inout** port is bidirectional — it transfers signals to and from the module depending on the value of a direction control signal. A port of this type is also called a *three-state* device or buffer. A three-state device can logically disconnect its output from the bus to which it is physically connected. These devices were presented in Section 3.3.20.

All ports in a module must be declared depending on the direction of the signal. For example, in the *full_adder* module shown in Figure 9.1, ports *a*, *b*, and *cin* are declared as **input**; ports *sum* and *cout* are declared as **output**. Since port declarations are implicitly declared as type **wire**, it is not necessary to explicitly declare a port as **wire** as was done in Figure 9.1. The test bench module of Figure 9.2 is a top-level module and has no port list.

If an output port retains a value, then it is declared as type **reg** as shown in Figure 9.3 for outputs *q* and *q_n* of a *D* flip-flop. For both net and register declarations, the size of the variable must comply with that in the port declaration. An output can also be redeclared as a **reg** type variable if it is used within an **always** statement or an **initial** statement.

```
//dataflow full adder
module full_adder (a, b, cin, sum, cout);

input a, b, cin;                //list all inputs and outputs
output sum, cout;

wire a, b, cin;                 //define wires
wire sum, cout;

assign sum = (a ^ b) ^ cin;    //continuous assignment
assign cout = cin & (a ^ b) | (a & b);

endmodule
```

Figure 9.1   Module for a full adder showing the port list.

```
//full adder test bench
module full_adder_tb;

reg a, b, cin;                  //inputs are reg for a test bench
wire sum, cout;

initial                         //apply input vectors
begin: apply_stimulus
   reg [3:0] invect;
   for (invect = 0; invect < 8; invect = invect + 1)
     begin
       {a, b, cin} = invect [3:0];
       #10 $display ("a b cin = %b, sum = %b, cout = %b",
                     {a, b, cin}, sum, cout);
     end
end

//instantiate the module into the test bench
full_adder inst1 (
   .a(a),
   .b(b),
   .cin(cin),
   .sum(sum),
   .cout(cout)
   );
endmodule
```

Figure 9.2   Test bench for the full adder module of Figure 9.1.

```
//behavioral D flip-flop
module d_ff (d, clk, q, q_n, set_n, rst_n);

input d, clk, set_n, rst_n;
output q, q_n;

wire d, clk, set_n, rst_n;
reg q, q_n;

always @ (posedge clk or negedge rst_n or negedge set_n)
begin
    if (rst_n == 0)
        begin
            q <= 1'b0;
            q_n <= 1'b1;
        end

    else if (set_n == 0)
        begin
            q <= 1'b1;
            q_n <= 1'b0;
        end

    else
        begin
            q <= d;
            q_n <= ~d;
        end
end

endmodule
```

Figure 9.3   Behavioral module for a *D* flip-flop, where outputs *q* and *q_n* are declared as type **reg**.

9.2.1 Unconnected Ports

Ports can be left unconnected in an instantiation by leaving the port name blank as shown in the following for both instantiation by name and instantiation by position. If instantiating by name, the port corresponding to x_2 is left blank, indicating no connection. If instantiating by position, no port name is specified for x_3, indicating no connection. Input ports that are unconnected are assigned a value of **z**; output ports that are unconnected are unused.

```
//instantiate by name              //instantiate by position

xor_xnor inst1 (                   xor_xnor inst1 (
      .x1(x1),                           x1,
      .x2(),                             x2,
      .x3(x3),                           ,
      .x4(x4),                           x4,
      .xor_out(xor_out),                 xor_out,
      .xnor_out(xnor_out)                xnor_out
      );                                 );
```

9.2.2 Port Connection Rules

A port is an entry into a module from an external source. It connects the external unit
to the internal logic of the module. When a module is instantiated within another module, certain rules apply. An error message is indicated if the port connection rules are
not followed. Figure 9.4 illustrates the rules for port connections.

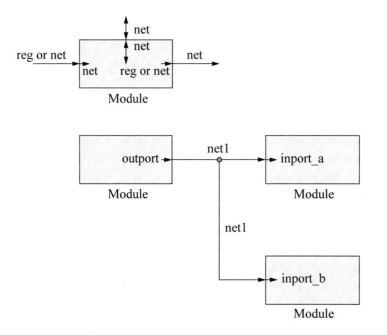

Figure 9.4   Diagram illustrating port connection rules.

Input ports must always be of type net (**wire**) internally except for test benches; externally, input ports can be **reg** or **wire**. Output ports can be of type **reg** or **wire** internally; externally, output ports must always be connected to a **wire**. The input port names can be different, but the net (**wire**) names connecting the input ports must be the same as shown in Figure 9.4.

When making intermodule port connections, it is permissible to connect ports of different widths. Port width matching occurs by right justification or truncation. Figure 9.5 shows an example of connecting ports of different widths. The bit positions that are not connected are assigned a value of **z**.

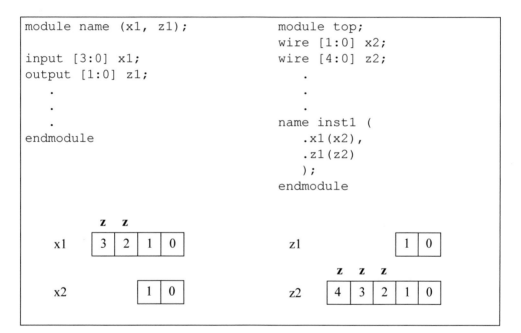

Figure 9.5   Figure to illustrate connecting ports of different widths.

9.3 Design Examples

Several examples will be presented that illustrate the structural modeling technique. These examples include code converters, counters of different moduli, adders and subtractors, a 4-function arithmetic and logic unit (ALU), a binary adder with high-speed shifter, an array multiplier, Moore and Mealy synchronous sequential machines, asynchronous sequential machines, and pulse-mode asynchronous sequential machines. Each example will be completely designed in detail and will include appropriate theory where applicable.

9.3.1 Gray-to-Binary Code Converter

The algorithm to convert from Gray code to binary code is shown in Equation 9.1. This section will convert a 4-bit Gray code to the corresponding binary code using structural modeling. The 4-bit binary numbers are generated using Equation 9.2. The structural module is shown in Figure 9.6. The test bench and outputs are shown in Figure 9.7 and Figure 9.8, respectively.

$$b_{n-1} = g_{n-1}$$
$$b_i = b_{i+1} \oplus g_i \tag{9.1}$$

$$b_3 = g_3$$
$$b_2 = b_3 \oplus g_2$$
$$b_1 = b_2 \oplus g_1$$
$$b_0 = b_1 \oplus g_0 \tag{9.2}$$

```
//structural gray-to-binary converter
module gray_to_bin_struc (g3, g2, g1, g0, b3, b2, b1, b0);

input g3, g2, g1, g0;
output b3, b2, b1, b0;

wire g3, g2, g1, g0;
wire b3, b2, b1, b0;

assign b3 = g3;

//instantiate the xor gates
xor (b2, b3, g2);
xor (b1, b2, g1);
xor (b0, b1, g0);

endmodule
```

Figure 9.6   Structural module to convert a 4-bit Gray code to the corresponding binary code.

```
//test bench for structural gray_to_bin converter
module gray_to_bin_struc_tb;

reg g3, g2, g1, g0;
wire b3, b2, b1, b0;

//apply input vectors and display variables
initial
begin: apply_stimulus
   reg [4:0] invect;
   for (invect=0; invect<16; invect=invect+1)
      begin
         {g3, g2, g1, g0} = invect [4:0];
         #10 $display ("{g3g2g1g0} = %b, {b3b2b1b0} = %b",
                        {g3, g2, g1, g0}, {b3, b2, b1, b0});
      end
end

//instantiate the module into the test bench
gray_to_bin_struc inst1 (
   .g3(g3),
   .g2(g2),
   .g1(g1),
   .g0(g0),
   .b3(b3),
   .b2(b2),
   .b1(b1),
   .b0(b0)
   );

endmodule
```

Figure 9.7   Test bench for the Gray-to-binary code converter of Figure 9.6.

```
g3g2g1g0=0000, b3b2b1b0=0000    g3g2g1g0=1000, b3b2b1b0=1111
g3g2g1g0=0001, b3b2b1b0=0001    g3g2g1g0=1001, b3b2b1b0=1110
g3g2g1g0=0010, b3b2b1b0=0011    g3g2g1g0=1010, b3b2b1b0=1100
g3g2g1g0=0011, b3b2b1b0=0010    g3g2g1g0=1011, b3b2b1b0=1101
g3g2g1g0=0100, b3b2b1b0=0111    g3g2g1g0=1100, b3b2b1b0=1000
g3g2g1g0=0101, b3b2b1b0=0110    g3g2g1g0=1101, b3b2b1b0=1001
g3g2g1g0=0110, b3b2b1b0=0100    g3g2g1g0=1110, b3b2b1b0=1011
g3g2g1g0=0111, b3b2b1b0=0101    g3g2g1g0=1111, b3b2b1b0=1010
```

Figure 9.8   Outputs for the Gray-to-binary code converter of Figure 9.6.

9.3.2 Binary-Coded Decimal (BCD)-to-Decimal Decoder

This section will design a BCD-to-decimal decoder. All inputs are asserted high and all outputs from the decoder are asserted low. There are three parts to this example:

1. Design a BCD-to-decimal decoder.

2. Use the decoder to implement the following two functions:

$$f_1 (a, b, c, d) = \Sigma_m(1, 2, 4)$$
$$f_2 (a, b, c, d) = \Sigma_m(4, 7, 9)$$

The outputs for f_1 and f_2 are to be asserted high.

3. Use the decoder to implement a BCD-to-excess-3 code converter. The outputs for the excess-3 code are to be asserted high.

The truth table for the BCD-to-decimal conversion is shown in Table 9.1, where BCD inputs 1010 through 1111 are invalid and provide inactive high outputs. The logic diagram is shown in Figure 9.9.

Table 9.1 BCD-to-Decimal Conversion

BCD Input				Decimal Output									
a	b	c	d	z_0	z_1	z_2	z_3	z_4	z_5	z_6	z_7	z_8	z_9
0	0	0	0	0	1	1	1	1	1	1	1	1	1
0	0	0	1	1	0	1	1	1	1	1	1	1	1
0	0	1	0	1	1	0	1	1	1	1	1	1	1
0	0	1	1	1	1	1	0	1	1	1	1	1	1
0	1	0	0	1	1	1	1	0	1	1	1	1	1
0	1	0	1	1	1	1	1	1	0	1	1	1	1
0	1	1	0	1	1	1	1	1	1	0	1	1	1
0	1	1	1	1	1	1	1	1	1	1	0	1	1
1	0	0	0	1	1	1	1	1	1	1	1	0	1
1	0	0	1	1	1	1	1	1	1	1	1	1	0
1	0	1	0	1	1	1	1	1	1	1	1	1	1
1	0	1	1	1	1	1	1	1	1	1	1	1	1
1	1	0	0	1	1	1	1	1	1	1	1	1	1
1	1	0	1	1	1	1	1	1	1	1	1	1	1
1	1	1	0	1	1	1	1	1	1	1	1	1	1
1	1	1	1	1	1	1	1	1	1	1	1	1	1

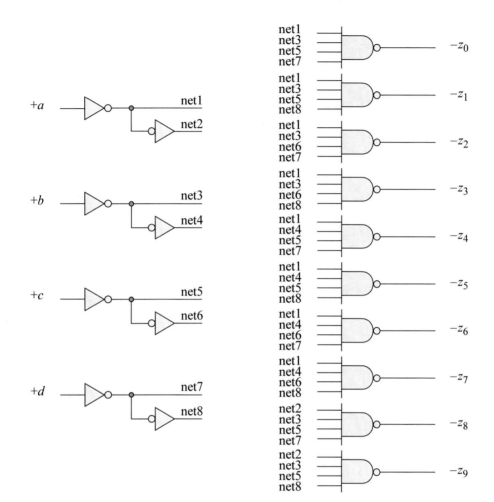

Figure 9.9   Logic diagram for a BCD-to-decimal decoder.

The logic diagram to implement the functions f_1 and f_2 is shown in Figure 9.10 using the decoder and logic primitives. The truth table for the BCD-to-excess-3 conversion is shown in Table 9.2 and the Karnaugh maps for the excess-3 outputs are shown in Figure 9.11. The equations for the excess-3 outputs are shown in Equation 9.3 as obtained from the Karnaugh maps. The logic diagram for BCD-to-excess-3 conversion is shown in Figure 9.12.

The module for the decoder is shown in Figure 9.13 which is then instantiated into the structural module of Figure 9.14 to implement the functions for f_1 and f_2 and the BCD-to-excess-3 code converter. The test bench and outputs are shown in Figure 9.15 and Figure 9.16, respectively. Note that all outputs are deasserted for invalid BCD values 1010 through 1111.

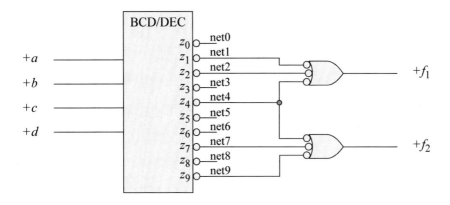

Figure 9.10   Logic diagram to implement the functions for f_1 and f_2.

Table 9.2 BCD-to-Excess-3 Conversion

BCD code				Excess-3 code			
a	b	c	d	w	x	y	z
0	0	0	0	0	0	1	1
0	0	0	1	0	1	0	0
0	0	1	0	0	1	0	1
0	0	1	1	0	1	1	0
0	1	0	0	0	1	1	1
0	1	0	1	1	0	0	0
0	1	1	0	1	0	0	1
0	1	1	1	1	0	1	0
1	0	0	0	1	0	1	1
1	0	0	1	1	1	0	0

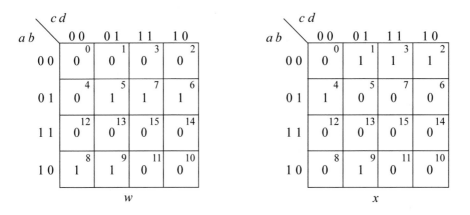

Figure 9.11   Karnaugh maps for the BCD-to-excess-3 code.

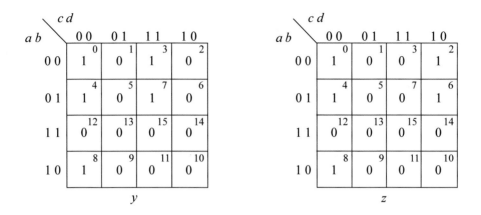

Figure 9.11   (Continued)

$$w = a'bd + a'bc + ab'c'$$

$$x = a'b'd + a'b'c + b'c'd + a'bc'd'$$

$$y = a'c'd' + a'cd + b'c'd'$$

$$z = a'c'd' + a'cd' + b'c'd' \qquad (9.3)$$

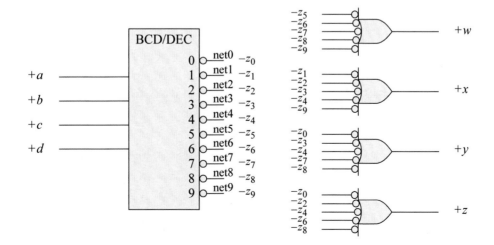

Figure 9.12   Logic diagram for BCD-to-excess-3 conversion.

```
//bcd-to-decimal decoder
module bcd_to_dec_struc (a, b, c, d,
                         z0, z1, z2, z3, z4, z5, z6, z7, z8, z9);

input a, b, c, d;
output z0, z1, z2, z3, z4, z5, z6, z7, z8, z9;

not    (net1, a);
not    (net2, net1);

not    (net3, b);
not    (net4, net3);

not    (net5, c);
not    (net6, net5);

not    (net7, d);
not    (net8, net7);

nand   (z0, net1, net3, net5, net7);
nand   (z1, net1, net3, net5, net8);
nand   (z2, net1, net3, net6, net7);
nand   (z3, net1, net3, net6, net8);
nand   (z4, net1, net4, net5, net7);
nand   (z5, net1, net4, net5, net8);
nand   (z6, net1, net4, net6, net7);
nand   (z7, net1, net4, net6, net8);
nand   (z8, net2, net3, net5, net7);
nand   (z9, net2, net3, net5, net8);

endmodule
```

Figure 9.13   Module for a BCD-to-decimal decoder.

```
//structural module to use a bcd-to-dec decoder
//to implement two functions f1 and f2
//and a bcd-to-excess-3 decoder
module bcd_to_dec_apps (a, b, c, d, f1, f2, w, x, y, z);

input a, b, c, d;
output f1, f2, w, x, y, z;

wire a, b, c, d;            //continued on next page
```

Figure 9.14   Structural module for the BCD-to-decimal decoder with applications.

```
//define internal nets
wire net0, net1, net2, net3, net4, net5, net6,
          net7, net8, net9;

//instantiate the bcd-to-dec decoder
bcd_to_dec_struc inst1 (
   .a(a),
   .b(b),
   .c(c),
   .d(d),
   .z0(net0),
   .z1(net1),
   .z2(net2),
   .z3(net3),
   .z4(net4),
   .z5(net5),
   .z6(net6),
   .z7(net7),
   .z8(net8),
   .z9(net9)
   );

//instantiate the nand gates for functions f1 and f2
nand (f1, net1, net2, net4);
nand (f2, net4, net7, net9);

//instantiate the nand gates for bcd-to-excess-3
nand (w, net5, net6, net7, net8, net9);
nand (x, net1, net2, net3, net4, net9);
nand (y, net0, net3, net4, net7, net8);
nand (z, net0, net2, net4, net6, net8);

endmodule
```

Figure 9.14 (Continued)

```
//test bench for bcd-to-dec decoder with applications
module bcd_to_dec_apps_tb;

reg a, b, c, d;
wire f1, f2, w, x, y, z;

//continued on next page
```

Figure 9.15 Test bench for the BCD-to-decimal decoder with applications.

```
initial                    //apply stimulus and display variables
begin: apply_stimulus
   reg [4:0] invect;
   for (invect=0; invect<16; invect=invect+1)
      begin
         {a, b, c, d} = invect [4:0];
         #10 $display ("abcd = %b, f1f2 = %b, wxyz = %b",
                      {a, b, c, d}, {f1, f2}, {w, x, y, z});
      end
end

bcd_to_dec_apps inst1 ( //instantiate the module
   .a(a),
   .b(b),
   .c(c),
   .d(d),
   .f1(f1),
   .f2(f2),
   .w(w),
   .x(x),
   .y(y),
   .z(z)
   );
endmodule
```

Figure 9.15   (Continued)

```
abcd = 0000, f1f2 = 00, wxyz = 0011
abcd = 0001, f1f2 = 10, wxyz = 0100
abcd = 0010, f1f2 = 10, wxyz = 0101
abcd = 0011, f1f2 = 00, wxyz = 0110
abcd = 0100, f1f2 = 11, wxyz = 0111
abcd = 0101, f1f2 = 00, wxyz = 1000
abcd = 0110, f1f2 = 00, wxyz = 1001
abcd = 0111, f1f2 = 01, wxyz = 1010
abcd = 1000, f1f2 = 00, wxyz = 1011
abcd = 1001, f1f2 = 01, wxyz = 1100
abcd = 1010, f1f2 = 00, wxyz = 0000
abcd = 1011, f1f2 = 00, wxyz = 0000
abcd = 1100, f1f2 = 00, wxyz = 0000
abcd = 1101, f1f2 = 00, wxyz = 0000
abcd = 1110, f1f2 = 00, wxyz = 0000
abcd = 1111, f1f2 = 00, wxyz = 0000
```

Figure 9.16   Outputs for the BCD-to-decimal decoder with applications.

Refer to the outputs of Figure 9.16. Function f_1 is asserted for minterms 0001, 0010, and 0100 as specified in the equation $f_1\,(a,\,b,\,c,\,d) = \Sigma_m(1, 2, 4)$. Similarly, function f_2 is asserted for minterms 0100, 0111, and 1001 as specified in the equation $f_2\,(a, b, c, d) = \Sigma_m(4, 7, 9)$.

9.3.3 Modulo-10 Counter

This section presents the design of a synchronous modulo-10 counter using *JK* flip-flops. A *JK* flip-flop will be designed and then instantiated four times into a structural module. Ports that are not required for the design will not be instantiated.

The truth table for the modulo-10 counter is shown in Table 9.3 and the excitation table for a *JK* flip-flop is shown in Table 9.4. The Karnaugh maps, obtained from the truth table and the excitation table, are shown in Figure 9.17. The *JK* input equations are shown in Equation 9.4.

Table 9.3 Truth Table for a Modulo-10 Counter

y_3	y_2	y_1	y_0
0	0	0	0
0	0	0	1
0	0	1	0
0	0	1	1
0	1	0	0
0	1	0	1
0	1	1	0
0	1	1	1
1	0	0	0
1	0	0	1
0	0	0	0

Table 9.4 Excitation Table for a *JK* Flip-Flop

Present State $Y_{(t)}$	Next State $Y_{(t+1)}$	Data Inputs J K
0 →	0	0 –
0 →	1	1 –
1 →	0	– 1
1 →	1	– 0

The behavioral module for the *JK* flip-flop is shown in Figure 9.18. The structural module for the modulo-10 counter is shown in Figure 9.19. The test bench, outputs, and waveforms are shown in Figure 9.20, Figure 9.21, and Figure 9.22, respectively.

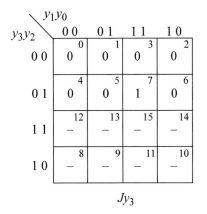

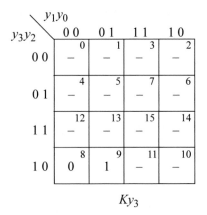

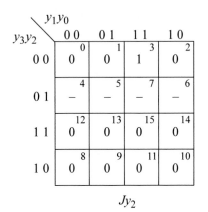

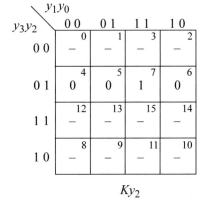

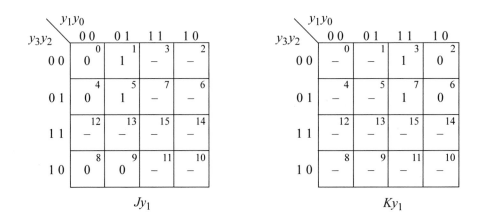

Figure 9.17   Karnaugh maps for modulo-10 counter using *JK* flip-flops.

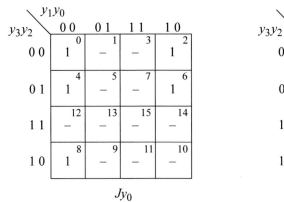 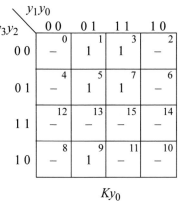

Jy_0 Ky_0

Figure 9.17   (Continued)

$$Jy_3 = y_2 y_1 y_0 \quad \text{(net1)} \qquad\qquad Ky_3 = y_0$$

$$Jy_2 = y_3' y_1 y_0 \quad \text{(net2)} \qquad\qquad Ky_2 = y_1 y_0 \quad \text{(net3)}$$

$$Jy_1 = y_3' y_0 \quad\quad \text{(net4)} \qquad\qquad Ky_1 = y_0$$

$$Jy_0 = 1 \qquad\qquad\qquad\qquad\qquad Ky_0 = 1 \qquad\qquad\qquad (9.4)$$

```
//behavioral jkff
module jkff (clk, j, k, set_n, rst_n, q, q_n);

input clk, j, k, set_n, rst_n;
output q, q_n;

wire clk, j, k, set_n, rst_n;
reg q, q_n;

//continued on next page
```

Figure 9.18   Behavioral module for a *JK* flip-flop.

```verilog
always @ (posedge clk or negedge set_n or negedge rst_n)
begin
   if (~rst_n)
      begin
         q   <= 1'b0;
         q_n <= 1'b1;
      end

   else if (~set_n)
      begin
         q   <= 1'b1;
         q_n <= 1'b0;
      end

   else if (j==1'b0 && k==1'b1)
      begin
         q   <= 1'b0;
         q_n <= 1'b1;
      end

   else if (j==1'b1 && k==1'b0)
      begin
         q   <= 1'b1;
         q_n <= 1'b0;
      end

   else if (j==1'b1 && k==1'b1)
      begin
         q   <= q_n;
         q_n <= q;
      end

//The following else statement is not necessary, since the state
//of the flip-flop will not change if all of the above
//conditions are false; that is, j==1'b0 && k==1'b0 is not
//necessary.  However, it is inserted here for completeness.

   else
      begin
         q   <= q;
         q_n <= q_n;
      end

end

endmodule
```

Figure 9.18   (Continued)

```
//structural module for a modulo-10 counter using JK flip-flops
module ctr_mod_10_jk_struc (set_n, rst_n, clk, y3, y2, y1, y0);

input set_n, rst_n, clk;
output y3, y2, y1, y0;

//define internal nets
wire set_n, rst_n, clk;
wire net1, net2, net3, net4;

//instantiate the logic primitive and the JK flip-flop for y3
and (net1, y2, y1, y0);

jkff inst3 (
   .clk(clk),
   .j(net1),
   .k(y0),
   .set_n(set_n),
   .rst_n(rst_n),
   .q(y3)
   );

//instantiate the logic primitives and the JK flip-flop for y2
and (net2, y0, y1, ~y3);
and (net3, y0, y1);

jkff inst2 (
   .clk(clk),
   .j(net2),
   .k(net3),
   .set_n(set_n),
   .rst_n(rst_n),
   .q(y2)
   );

//instantiate the logic primitive and the JK flip-flop for y1
and (net4, y0, ~y3);

jkff inst1 (
   .clk(clk),
   .j(net4),
   .k(y0),
   .set_n(set_n),
   .rst_n(rst_n),
   .q(y1)
   );
//continued on next page
```

Figure 9.19   Structural module for a modulo-10 counter using *JK* flip-flops.

```
//instantiate the logic primitive and the JK flip-flop for y0
jkff inst0 (
   .clk(clk),
   .j(1'b1),
   .k(1'b1),
   .set_n(set_n),
   .rst_n(rst_n),
   .q(y0)
   );

endmodule
```

Figure 9.19   (Continued)

```
//test bench for the structural modulo-10 counter
//using JK flip-flops
module ctr_mod_10_jk_struc_tb;

reg set_n, rst_n, clk;
wire y3, y2, y1, y0;

//display variables
initial
$monitor ("y3y2y1y0 = %b", {y3, y2, y1, y0});

//generate reset
initial
begin
   #0    set_n = 1'b1;
         rst_n = 1'b0;
   #5    rst_n = 1'b1;
end

//generate clock
initial
begin
   clk = 1'b0;
   forever
      #10   clk = ~clk;
end

//continued on next page
```

Figure 9.20   Test bench for the modulo-10 counter using *JK* flip-flops.

```
//determine length of simulation
initial
begin
   repeat (11) @ (posedge clk);
   $stop;
end

ctr_mod_10_jk_struc inst1 (    //instantiate the module
   .set_n(set_n),
   .rst_n(rst_n),
   .clk(clk),
   .y3(y3),
   .y2(y2),
   .y1(y1),
   .y0(y0)
   );
endmodule
```

Figure 9.20 (Continued)

```
y3y2y1y0 = 0000          y3y2y1y0 = 0110
y3y2y1y0 = 0001          y3y2y1y0 = 0111
y3y2y1y0 = 0010          y3y2y1y0 = 1000
y3y2y1y0 = 0011          y3y2y1y0 = 1001
y3y2y1y0 = 0100          y3y2y1y0 = 0000
y3y2y1y0 = 0101
```

Figure 9.21 Outputs for the modulo-10 counter using *JK* flip-flops.

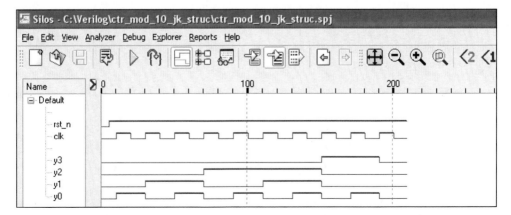

Figure 9.22 Waveforms for the modulo-10 counter using *JK* flip-flops.

9.3.4 Adder/Subtractor

This section presents the design of a 4-bit fixed-point ripple adder/subtractor. The design of a carry lookahead adder/subtractor is similar except that the carry logic uses the carry lookahead technique as described in Section 7.1.8. It is desirable to have the adder unit perform both addition and subtraction; there is no advantage to having a separate adder and subtractor. An adder similar to that of Section 7.1.7 will be modified so that it can perform subtraction while still maintaining the ability to add.

Subtraction is accomplished by adding the 2s complement of the subtrahend to the minuend as shown below for minuend A and subtrahend B, where B' is the 1s complement of B. The 2s complement of a number is obtained by adding one to the 1s complement of the number.

$$A - B = A + (B' + 1)$$

An inverter could be used to invert each subtrahend bit, but this would not allow the noninverted subtrahend bits to be used for addition. The logic that inverts the subtrahend bits should also allow for addition. The exclusive-OR operation for two variables is defined as

$$a \oplus b = ab' + a'b$$

Thus, $a \oplus b = 1$ only if $a \neq b$. The truth table for the exclusive-OR function is shown in Table 9.5. Note that when $m = 1$, b is inverted; when $m = 0$, b is noninverted. Therefore, the variable m can be used as a *mode control* input to determine whether the operation is addition or subtraction. If the mode control line is zero, then the operation is addition; if the mode control line is one, then the operation is subtraction.

Table 9.5 Rules for Exclusive-OR Operation

		b	
$\oplus$		0	1
	0	0	1
m	1	1	0

The logic diagram for the 4-bit adder/subtractor is shown in Figure 9.23 in which four full adders are used together with the requisite exclusive-OR functions. The augend/minuend is a 4-bit vector $a[3:0]$ and the addend/subtrahend is a 4-bit vector $b[3:0]$. Note that the mode control input m is connected to each exclusive-OR function and also to the carry-in of the low-order stage. The result is a 4-bit vector $rslt[3:0]$; the carry-out of each stage is also a 4-bit vector $cout[3:0]$.

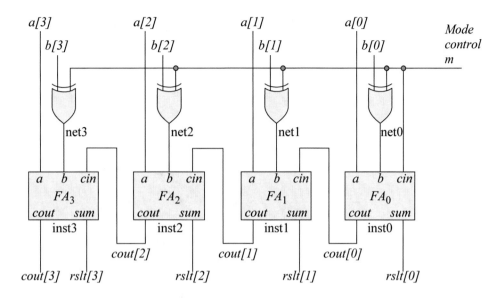

Figure 9.23   Logic diagram for a 4-bit adder/subtractor.

Examples of subtraction are shown below for 4-bit operands in 2s complement representation. The term *true addition* means that the result is the sum of the two operands, ignoring the sign bit; whereas *true subtraction* means that the result is the difference of the two operands, ignoring the sign bit. To obtain the 2s complement (negation) of an operand, the low-order zeros and the first 1 bit are unchanged; then the remaining higher-order bits are complemented.

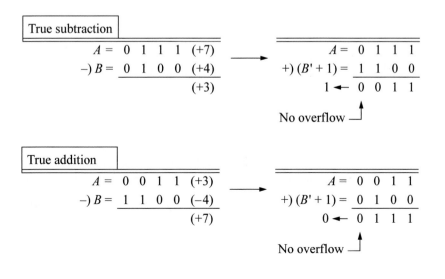

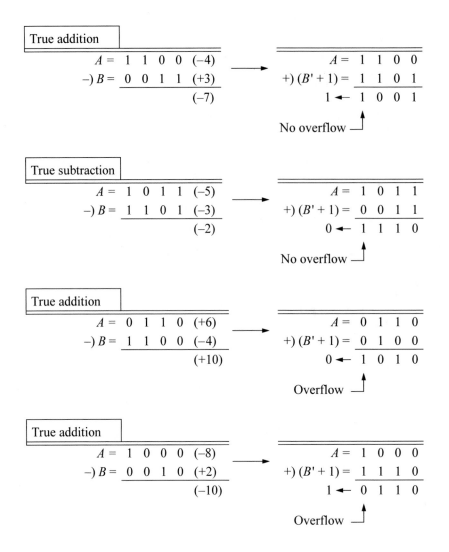

In the last two examples above, an overflow occurred because the result was too large to fit in the word size of the operands. The range for numbers in 2s complement representation is

$$-2^{n-1} \text{ to } +2^{n-1} - 1$$

where n is the number of bits in the operands. Thus, for $n = 4$, the range is from -8 to $+7$. The result of the last two examples is $+10$ and -10, both of which exceed the range for four bits. For n-bit operands

$$a_{n-1} a_{n-2} \cdots a_1 a_0$$
$$b_{n-1} b_{n-2} \cdots b_1 b_0$$

there are two ways to detect overflow:

$$\text{Overflow} = cout_{n-1} \oplus cout_{n-2}$$
$$\text{Overflow} = a_{n-1}\, b_{n-1}\, rslt_{n-1}' + a_{n-1}'\, b_{n-1}'\, rslt_{n-1}$$

The module for a full adder is shown in Figure 9.24 using dataflow modeling. This will be instantiated four times into the structural module of Figure 9.25 to implement the 4-bit adder/subtractor. The test bench is shown in Figure 9.26. The outputs and waveforms are shown in Figure 9.27 and Figure 9.28, respectively.

```
//dataflow full adder
module full_adder (a, b, cin, sum, cout);

//list all inputs and outputs
input a, b, cin;
output sum, cout;

//define wires
wire a, b, cin;
wire sum, cout;

//continuous assign
assign sum = (a ^ b) ^ cin;
assign cout = cin & (a ^ b) | (a & b);

endmodule
```

Figure 9.24   Full adder to be instantiated into a structural module to implement a 4-bit adder/subtractor.

```
//structural module for an adder/subtractor
module adder_subtr_struc (a, b, m, rslt, cout, ovfl);

input [3:0] a, b;
input m;
output [3:0] rslt, cout;
output ovfl;

//define internal nets
wire net0, net1, net2, net3;
//continued on next page
```

Figure 9.25   Structural module for a 4-bit adder/subtractor.

```verilog
//define overflow
xor (ovfl, cout[3], cout[2]);


//instantiate the xor and the full adder for FA0
xor (net0, b[0], m);
full_adder inst0 (
   .a(a[0]),
   .b(net0),
   .cin(m),
   .sum(rslt[0]),
   .cout(cout[0])
   );


//instantiate the xor and the full adder for FA1
xor (net1, b[1], m);
full_adder inst1 (
   .a(a[1]),
   .b(net1),
   .cin(cout[0]),
   .sum(rslt[1]),
   .cout(cout[1])
   );


//instantiate the xor and the full adder for FA2
xor (net2, b[2], m);
full_adder inst2 (
   .a(a[2]),
   .b(net2),
   .cin(cout[1]),
   .sum(rslt[2]),
   .cout(cout[2])
   );


//instantiate the xor and the full adder for FA3
xor (net3, b[3], m);
full_adder inst3 (
   .a(a[3]),
   .b(net3),
   .cin(cout[2]),
   .sum(rslt[3]),
   .cout(cout[3])
   );

endmodule
```

Figure 9.25   (Continued)

```
//test bench for structural adder/subtractor
module adder_subtr_struc_tb;

reg [3:0] a, b;
reg m;
wire [3:0] rslt, cout;
wire  ovfl;

//display variables
initial
$monitor ("a=%b, b=%b, m=%b, rslt=%b, cout[3]=%b, cout[2]=%b,
           ovfl=%b", a, b, m, rslt, cout[3], cout[2], ovfl);

//apply input vectors
initial
begin
//addition
    #0    a = 4'b0000;   b = 4'b0001;   m = 1'b0;
    #10   a = 4'b0010;   b = 4'b0101;   m = 1'b0;
    #10   a = 4'b0110;   b = 4'b0001;   m = 1'b0;
    #10   a = 4'b0101;   b = 4'b0001;   m = 1'b0;

//subtraction
    #10   a = 4'b0111;   b = 4'b0101;   m = 1'b1;
    #10   a = 4'b0101;   b = 4'b0100;   m = 1'b1;
    #10   a = 4'b0110;   b = 4'b0011;   m = 1'b1;
    #10   a = 4'b0110;   b = 4'b0010;   m = 1'b1;

//overflow
    #10   a = 4'b0111;   b = 4'b0101;   m = 1'b0;
    #10   a = 4'b1000;   b = 4'b1011;   m = 1'b0;
    #10   a = 4'b0110;   b = 4'b1100;   m = 1'b1;
    #10   a = 4'b1000;   b = 4'b0010;   m = 1'b1;

    #10   $stop;
end

//instantiate the module into the test bench
adder_subtr_struc inst1 (
    .a(a),
    .b(b),
    .m(m),
    .rslt(rslt),
    .cout(cout),
    .ovfl(ovfl)
    );
endmodule
```

Figure 9.26   Test bench for the structural adder/subtractor of Figure 9.25.

```
a=0000, b=0001, m=0, rslt=0001, cout[3]=0, cout[2]=0, ovfl=0
a=0010, b=0101, m=0, rslt=0111, cout[3]=0, cout[2]=0, ovfl=0
a=0110, b=0001, m=0, rslt=0111, cout[3]=0, cout[2]=0, ovfl=0
a=0101, b=0001, m=0, rslt=0110, cout[3]=0, cout[2]=0, ovfl=0
a=0111, b=0101, m=1, rslt=0010, cout[3]=1, cout[2]=1, ovfl=0
a=0101, b=0100, m=1, rslt=0001, cout[3]=1, cout[2]=1, ovfl=0
a=0110, b=0011, m=1, rslt=0011, cout[3]=1, cout[2]=1, ovfl=0
a=0110, b=0010, m=1, rslt=0100, cout[3]=1, cout[2]=1, ovfl=0
a=0111, b=0101, m=0, rslt=1100, cout[3]=0, cout[2]=1, ovfl=1
a=1000, b=1011, m=0, rslt=0011, cout[3]=1, cout[2]=0, ovfl=1
a=0110, b=1100, m=1, rslt=1010, cout[3]=0, cout[2]=1, ovfl=1
a=1000, b=0010, m=1, rslt=0110, cout[3]=1, cout[2]=0, ovfl=1
```

Figure 9.27   Outputs for the structural adder/subtractor module of Figure 9.25.

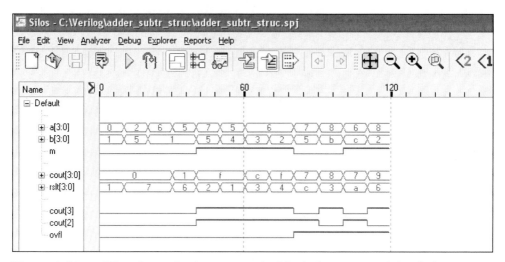

Figure 9.28   Waveforms for the structural adder/subtractor module of Figure 9.25.

9.3.5 Four-Function Arithmetic and Logic Unit (ALU)

This section will present the design of a 4-function ALU for two 4-bit operands $A[3:0]$ and $B[3:0]$. The operation codes are shown in Table 9.6, where c_1 and c_0 are two control inputs with c_0 being the low-order bit. The design will be a structural design instantiating the following modules: a 4-bit ripple adder, a 4:1 multiplexer, a 2-input AND gate, a 2-input OR gate, and an exclusive-OR circuit.

Table 9.6 Operation Codes for the 4-Function ALU

c[1]	c[0]	Operation
0	0	Add
0	1	Subtract
1	0	And
1	1	Or

The logic diagram is shown in Figure 9.29 and represents only one of several methods of designing the ALU. This method illustrates the instantiation of several different modules into a structural top-level module. Figure 9.29 also shows the instantiation names and net names.

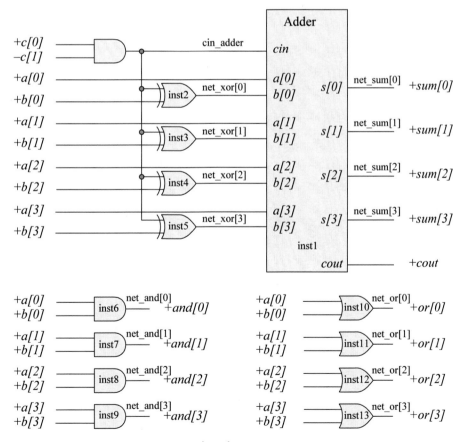

continued on next page

Figure 9.29   Logic diagram for a 4-function ALU.

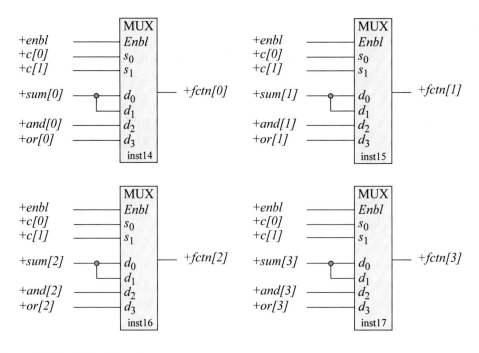

Figure 9.29   (Continued)

The structural module that instantiates dataflow modules for the adder, exclusive-OR circuits, AND gates, OR gates, and multiplexers is shown in Figure 9.30. The test bench, outputs, and waveforms are shown in Figure 9.31, Figure 9.32, and Figure 9.33, respectively.

```
//structural 4-function alu
module alu_4function (a, b, ctrl, enbl, cout, fctn);

input [3:0] a, b;
input [1:0] ctrl;
input enbl;
output [3:0] fctn;
output cout;

wire [3:0] a, b;
wire [1:0] ctrl;
wire enbl;
wire [3:0] fctn;
wire cout;
//continued on next page
```

Figure 9.30   Structural module for a 4-function ALU.

```
//define internal nets
wire cin_adder;
wire [3:0] net_xor, net_sum, net_and, net_or;

assign cin_adder = (~ctrl [1] & ctrl [0]);

//instantiate the adder
adder4_df inst1 (
    .a(a),
    .b(net_xor),
    .cin(cin_adder),
    .sum(net_sum),
    .cout(cout)
    );

//instantiate the xor circuits
xor2_df inst2 (
    .x1(cin_adder),
    .x2(b[0]),
    .z1(net_xor[0])
    );

xor2_df inst3 (
    .x1(cin_adder),
    .x2(b[1]),
    .z1(net_xor[1])
    );

xor2_df inst4 (
    .x1(cin_adder),
    .x2(b[2]),
    .z1(net_xor[2])
    );

xor2_df inst5 (
    .x1(cin_adder),
    .x2(b[3]),
    .z1(net_xor[3])
    );

//instantiate the and gates
and2_df inst6 (
    .x1(a[0]),
    .x2(b[0]),
    .z1(net_and[0])
    );
//continued on next page
```

Figure 9.30   (Continued)

```
and2_df inst7 (
    .x1(a[1]),
    .x2(b[1]),
    .z1(net_and[1])
    );

and2_df inst8 (
    .x1(a[2]),
    .x2(b[2]),
    .z1(net_and[2])
    );

and2_df inst9 (
    .x1(a[3]),
    .x2(b[3]),
    .z1(net_and[3])
    );

//instantiate the or gates
or2_df inst10 (
    .x1(a[0]),
    .x2(b[0]),
    .z1(net_or[0])
    );

or2_df inst11 (
    .x1(a[1]),
    .x2(b[1]),
    .z1(net_or[1])
    );

or2_df inst12 (
    .x1(a[2]),
    .x2(b[2]),
    .z1(net_or[2])
    );

or2_df inst13 (
    .x1(a[3]),
    .x2(b[3]),
    .z1(net_or[3])
    );

//continued on next page
```

Figure 9.30   (Continued)

```
mux4_df inst14 (                    //instantiate the multiplexers
   .s(ctrl),
   .d({net_or[0], net_and[0], net_sum[0], net_sum[0]}),
   .enbl(enbl),
   .z1(fctn[0])
   );

mux4_df inst15 (
   .s(ctrl),
   .d({net_or[1], net_and[1], net_sum[1], net_sum[1]}),
   .enbl(enbl),
   .z1(fctn[1])
   );

mux4_df inst16 (
   .s(ctrl),
   .d({net_or[2], net_and[2], net_sum[2], net_sum[2]}),
   .enbl(enbl),
   .z1(fctn[2])
   );

mux4_df inst17 (
   .s(ctrl),
   .d({net_or[3], net_and[3], net_sum[3], net_sum[3]}),
   .enbl(enbl),
   .z1(fctn[3])
   );
endmodule
```

Figure 9.30   (Continued)

```
//structural 4-function alu test bench
module alu_4function_tb;

reg [3:0] a, b;
reg [1:0] ctrl;
reg enbl;
wire [3:0] fctn;
wire cout;

initial
$monitor ("ctrl = %b, a = %b, b = %b, cout=%b, fctn = %b",
        ctrl, a, b, cout, fctn);
//continued on next page
```

Figure 9.31   Test bench for the structural module of Figure 9.30.

```
initial
begin
//add operation
  #0  ctrl=2'b00;  a=4'b0000; b=4'b0000; enbl=1'b1;//sum=0000
  #10 ctrl=2'b00;  a=4'b0001; b=4'b0011; enbl=1'b1;//sum=0100
  #10 ctrl=2'b00;  a=4'b0111; b=4'b0011; enbl=1'b1;//sum=1010
  #10 ctrl=2'b00;  a=4'b1101; b=4'b0110; enbl=1'b1;//sum=10011
  #10 ctrl=2'b00;  a=4'b0011; b=4'b1111; enbl=1'b1;//sum=10010


//subtract operation
  #10 ctrl=2'b01;  a=4'b0111; b=4'b0011; enbl=1'b1;//diff=0100
  #10 ctrl=2'b01;  a=4'b1101; b=4'b0011; enbl=1'b1;//diff=1010
  #10 ctrl=2'b01;  a=4'b1111; b=4'b0011; enbl=1'b1;//diff=1100
  #10 ctrl=2'b01;  a=4'b1111; b=4'b0001; enbl=1'b1;//diff=1110
  #10 ctrl=2'b01;  a=4'b1100; b=4'b0111; enbl=1'b1;//diff=0101


//and operation
  #10 ctrl=2'b10;  a=4'b1100; b=4'b0111; enbl=1'b1;//and=0100
  #10 ctrl=2'b10;  a=4'b0101; b=4'b1010; enbl=1'b1;//and=0000
  #10 ctrl=2'b10;  a=4'b1110; b=4'b0011; enbl=1'b1;//and=0110
  #10 ctrl=2'b10;  a=4'b1110; b=4'b1111; enbl=1'b1;//and=1110
  #10 ctrl=2'b10;  a=4'b1111; b=4'b0111; enbl=1'b1;//and=0111


//or operation
  #10 ctrl=2'b11;  a=4'b1100; b=4'b0111; enbl=1'b1;//or=1111
  #10 ctrl=2'b11;  a=4'b1100; b=4'b0100; enbl=1'b1;//or=1100
  #10 ctrl=2'b11;  a=4'b1000; b=4'b0001; enbl=1'b1;//or=1001

  #10 $stop;
end


//instantiate the module into the test bench
alu_4function inst1 (
  .a(a),
  .b(b),
  .ctrl(ctrl),
  .enbl(enbl),
  .fctn(fctn),
  .cout(cout)
  );

endmodule
```

Figure 9.31   (Continued)

```
ctrl = 00, a = 0000, b = 0000, cout=0, fctn = 0000
ctrl = 00, a = 0001, b = 0011, cout=0, fctn = 0100
ctrl = 00, a = 0111, b = 0011, cout=0, fctn = 1010
ctrl = 00, a = 1101, b = 0110, cout=1, fctn = 0011
ctrl = 00, a = 0011, b = 1111, cout=1, fctn = 0010
ctrl = 01, a = 0111, b = 0011, cout=1, fctn = 0100
ctrl = 01, a = 1101, b = 0011, cout=1, fctn = 1010
ctrl = 01, a = 1111, b = 0011, cout=1, fctn = 1100
ctrl = 01, a = 1111, b = 0001, cout=1, fctn = 1110
ctrl = 01, a = 1100, b = 0111, cout=1, fctn = 0101
ctrl = 10, a = 1100, b = 0111, cout=1, fctn = 0100
ctrl = 10, a = 0101, b = 1010, cout=0, fctn = 0000
ctrl = 10, a = 1110, b = 0111, cout=1, fctn = 0110
ctrl = 10, a = 1110, b = 1111, cout=1, fctn = 1110
ctrl = 10, a = 1111, b = 0111, cout=1, fctn = 0111
ctrl = 11, a = 1100, b = 0111, cout=1, fctn = 1111
ctrl = 11, a = 1100, b = 0100, cout=1, fctn = 1100
ctrl = 11, a = 1000, b = 0001, cout=0, fctn = 1001
```

Figure 9.32   Outputs for the structural module of Figure 9.30.

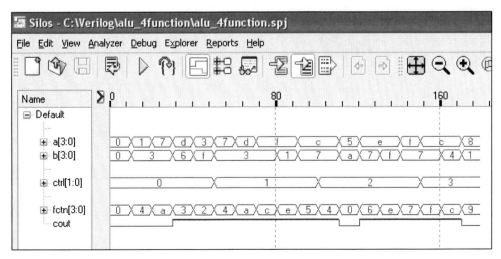

Figure 9.33   Waveforms for the structural module of Figure 9.30.

9.3.6 Adder and High-Speed Shifter

In this example, an adder and high-speed shifter will be designed by instantiating a 4-bit adder and eight 4:1 multiplexers into a structural module. The multiplexers will be used as a combinational shifter. Designing a shifter in this manner results in a shifting unit that is faster than a sequential shift register that shifts one bit per clock cycle because all the shift amounts are prewired. In order to shift the operand a specified number of bits, the shift amount is simply selected and the operand is shifted the requisite number of bits — the speed is a function only of the inertial delay of the multiplexer gates.

Four multiplexers are used for shifting left and four multiplexers are used for shifting right. Shifting a binary number left one bit position corresponds to multiplying the number by two. Shifting a binary number right one bit position corresponds to dividing the number by two.

This is a *logical* shifter; that is, zeros are shifted into the vacated low-order bit positions for a shift left operation and zeros are shifted into the vacated high-order bit positions for a shift right operation. The carry-out of the adder is not used in this design, although it could easily be gated into the low- or high-order positions of the multiplexers. Alternatively, a larger multiplexer could be used to accommodate the carry-out.

There are two operands: the augend $A[3:0] = a_3a_2a_1a_0$ and the addend $B[3:0] = b_3b_2b_1b_0$, where a_0 and b_0 are the low-order bits of A and B, respectively. There are also two shift direction inputs: *shiftleft* and *shiftright* and one input vector *shiftcount[1:0]* to determine the number of bits that the sum is to be shifted. In addition, there are two output vectors: *slmux* and *srmux* that are the outputs of the shift left and shift right multiplexers, respectively. Table 9.7 lists the control variables for shifting the sum.

Table 9.7 Shift Control Variables

Shift Direction	Shiftcount [1]	Shiftcount [0]	Shift Amount
Shiftleft	0	0	0
	0	1	1
	1	0	2
	1	1	3
Shiftright	0	0	0
	0	1	1
	1	0	2
	1	1	3

Figure 9.34 pictorially depicts the shift left and shift right operations. The logic diagram is shown in Figure 9.35 displaying the instantiation names and the net names.

Shift Left				
Adder output	Σ_3	Σ_2	Σ_1	Σ_0
No shift	Σ_3	Σ_2	Σ_1	Σ_0
Shift left 1	Σ_2	Σ_1	Σ_0	0
Shift left 2	Σ_1	Σ_0	0	0
Shift left 3	Σ_0	0	0	0

Shift Right				
Adder output	Σ_3	Σ_2	Σ_1	Σ_0
No shift	Σ_3	Σ_2	Σ_1	Σ_0
Shift right 1	0	Σ_3	Σ_2	Σ_1
Shift right 2	0	0	Σ_3	Σ_2
Shift right 3	0	0	0	Σ_3

Figure 9.34   Pictorial representation of the shift left and shift right operations.

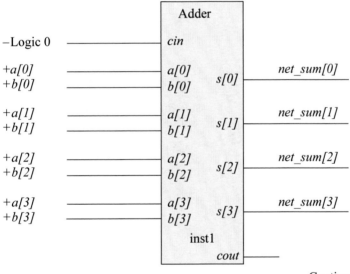

Continued on next page

Figure 9.35   Logic diagram for the adder and high-speed shifter.

The structural module is shown in Figure 9.36. The instantiated multiplexer was designed as a dataflow module with the following declarations: *input [1:0] s; input [3:0] d; input enbl; and output z1*. Therefore, when the multiplexer is instantiated, the data inputs must be in the order as specified; that is, *d[3]*, *d[2]*, *d[1]*, and *d[0]*. The test bench, outputs, and waveforms are shown in Figure 9.37, Figure 9.38, and Figure 9.39, respectively.

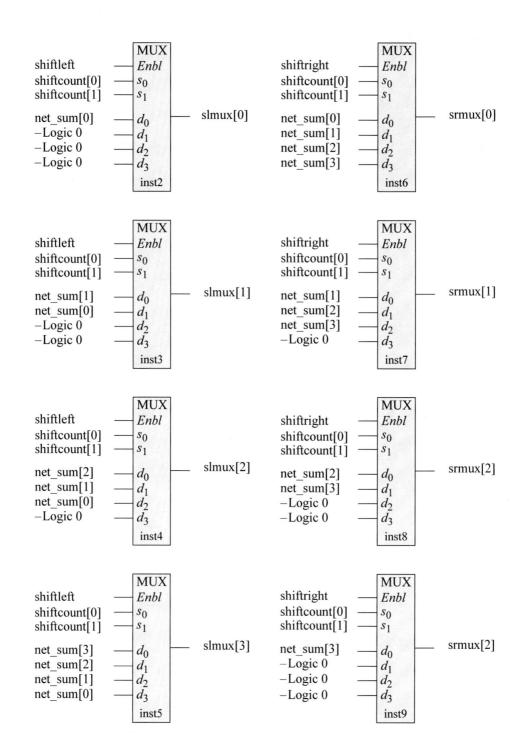

Figure 9.35   (Continued)

```
//structural adder and high-speed shifter
module adder_shifter (a, b, shiftcount, shiftleft,
                          shiftright, slmux, srmux);

input [3:0] a, b;
input [1:0] shiftcount;
input shiftleft, shiftright;

output [3:0] slmux, srmux;


//define internal nets
wire [3:0] net_sum;


//instantiate the adder
adder4_df inst1 (
    .a(a),
    .b(b),
    .cin(1'b0),
    .sum(net_sum)
    );


//instantiate the multiplexers for shifting left
mux4_df inst2 (
    .s(shiftcount),
    .d({1'b0, 1'b0, 1'b0, net_sum[0]}),
    .enbl(shiftleft),
    .z1(slmux[0])
    );


mux4_df inst3 (
    .s(shiftcount),
    .d({1'b0, 1'b0, net_sum[0], net_sum[1]}),
    .enbl(shiftleft),
    .z1(slmux[1])
    );


mux4_df inst4 (
    .s(shiftcount),
    .d({1'b0, net_sum[0], net_sum[1], net_sum[2]}),
    .enbl(shiftleft),
    .z1(slmux[2])
    );

//continued on next page
```

Figure 9.36   Structural module for the adder and high-speed shifter.

```
mux4_df inst5 (
   .s(shiftcount),
   .d({net_sum[0], net_sum[1], net_sum[2], net_sum[3]}),
   .enbl(shiftleft),
   .z1(slmux[3])
   );


//instantiate the multiplexers for shifting right
mux4_df inst6 (
   .s(shiftcount),
   .d({net_sum[3], net_sum[2], net_sum[1], net_sum[0]}),
   .enbl(shiftright),
   .z1(srmux[0])
   );


mux4_df inst7 (
   .s(shiftcount),
   .d({1'b0, net_sum[3], net_sum[2], net_sum[1]}),
   .enbl(shiftright),
   .z1(srmux[1])
   );


mux4_df inst8 (
   .s(shiftcount),
   .d({1'b0, 1'b0, net_sum[3], net_sum[2]}),
   .enbl(shiftright),
   .z1(srmux[2])
   );


mux4_df inst9 (
   .s(shiftcount),
   .d({1'b0, 1'b0, 1'b0, net_sum[3]}),
   .enbl(shiftright),
   .z1(srmux[3])
   );

endmodule
```

Figure 9.36   (Continued)

```verilog
//test bench for the adder and high-speed shifter
module adder_shifter_tb;

reg [3:0] a, b;
reg [1:0] shiftcount;
reg enbl;
reg shiftleft, shiftright;

wire [3:0] slmux, srmux;

initial
$monitor ("a=%b, b=%b, shiftcount=%b, shiftleft=%b,
          shiftright=%b, slmux=%b, srmux=%b",
      a, b, shiftcount, shiftleft, shiftright, slmux, srmux);


//apply input vectors
initial
begin
//no shift
   #0     a=4'b0011;
          b=4'b0001;        //sum=0100
          shiftcount=2'b00; //no shift
          shiftleft=1'b1;
          shiftright=1'b0;


//shift left
   #10    a=4'b0111;
          b=4'b0011;        //sum=1010
          shiftcount=2'b01; //shift one
          shiftleft=1'b1;   //shift left
          shiftright=1'b0;

   #10    a=4'b1100;
          b=4'b0011;        //sum=1111
          shiftcount=2'b10; //shift two
          shiftleft=1'b1;   //shift left
          shiftright=1'b0;

   #10    a=4'b1100;
          b=4'b0011;        //sum=1111
          shiftcount=2'b11; //shift three
          shiftleft=1'b1;   //shift left
          shiftright=1'b0;

//continued on  next page
```

Figure 9.37   Test bench for the adder and high-speed shifter.

```
//shift right
   #10    a=4'b1100;
          b=4'b0011;          //sum=1111
          shiftcount=2'b01; //shift one
          shiftleft=1'b0;
          shiftright=1'b1;   //shift right

   #10    a=4'b0110;
          b=4'b0111;          //sum=1101
          shiftcount=2'b10; //shift two
          shiftleft=1'b0;
          shiftright=1'b1;   //shift right

   #10    a=4'b0110;
          b=4'b1001;          //sum=1111
          shiftcount=2'b11; //shift three
          shiftleft=1'b0;
          shiftright=1'b1;   //shift right
   #10    $stop;
end

//instantiate the module into the test bench
adder_shifter inst1 (
   .a(a),
   .b(b),
   .shiftcount(shiftcount),
   .shiftleft(shiftleft),
   .shiftright(shiftright),
   .slmux(slmux),
   .srmux(srmux)
   );
endmodule
```

Figure 9.37   (Continued)

```
a=0011, b=0001, shiftcount=00, shiftleft=1, shiftright=0,
               slmux=0100, srmux=0000

a=0111, b=0011, shiftcount=01, shiftleft=1, shiftright=0,
               slmux=0100, srmux=0000

a=1100, b=0011, shiftcount=10, shiftleft=1, shiftright=0,
               slmux=1100, srmux=0000

//continued on next page
```

Figure 9.38   Outputs for the adder and high-speed shifter.

```
a=1100, b=0011, shiftcount=11, shiftleft=1, shiftright=0,
             slmux=1000, srmux=0000

a=1100, b=0011, shiftcount=01, shiftleft=0, shiftright=1,
             slmux=0000, srmux=0111

a=0110, b=0111, shiftcount=10, shiftleft=0, shiftright=1,
             slmux=0000, srmux=0011

a=0110, b=1001, shiftcount=11, shiftleft=0, shiftright=1,
             slmux=0000, srmux=0001
```

Figure 9.38   (Continued)

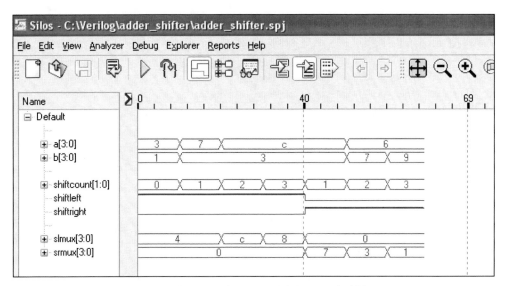

Figure 9.39   Waveforms for the adder and high-speed shifter.

9.3.7 Array Multiplier

This section presents a method for achieving high-speed multiplication using a planar array. The sequential add-shift technique requires less hardware, but is relatively slow when compared to the array multiplier method. Multiplication involves two operands; the *multiplicand* and the *multiplier*. The product of two n-bit numbers can be accommodated in $2n$ bits. Multiplication of the multiplicand by a 1 bit in the multiplier simply copies the multiplicand. If the multiplier bit is a 1, then the multiplicand is entered in the appropriately shifted position as a partial product to be added to other partial products to form the product. If the multiplier bit is 0, then 0s are entered as a partial product.

Although the array multiplier method is applicable to any size operands, this example assume two 3-bit operands as shown in Figure 9.40. The multiplicand is $A\,[2:0]$ = $a_2a_1a_0$ and the multiplier is $B = b_2b_1b_0$, where a_0 and b_0 are the low-order bits of A and B, respectively. Each bit in the multiplicand is multiplied by the low-order bit b_0 of the multiplier. This is equivalent to the AND function and generates the first of three partial products.

Each bit in the multiplicand is then multiplied by bit b_1 of the multiplier. The resulting partial product is shifted one bit position to the left. The process is repeated for bit b_2 of the multiplier. The partial products are then added together to form the product. A carry-out of any column is added to the next higher-order column.

$$
\begin{array}{ccccccc}
 & & & & a_2 & a_1 & a_0 \\
 & & \times\,) & & b_2 & b_1 & b_0 \\
\hline
\text{Partial product 0} & & & & a_2b_0 & a_1b_0 & a_0b_0 \\
\text{Partial product 1} & & & a_2b_1 & a_1b_1 & a_0b_1 & \\
\text{Partial product 2} & & a_2b_2 & a_1b_2 & a_0b_2 & & \\
\hline
 & 2^5 & 2^4 & 2^3 & 2^2 & 2^1 & 2^0
\end{array}
$$

Figure 9.40   General array multiply algorithm for two 3-bit operands.

A block diagram of the array multiplier is shown in Figure 9.41 together with a full adder to be used as the planar array elements as shown in the array multiplier of Figure 9.42.

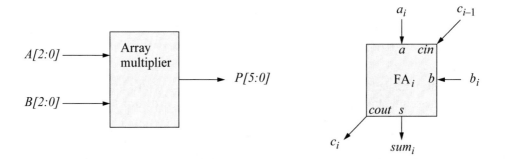

Figure 9.41   Array multiplier block diagram and full adder block diagram.

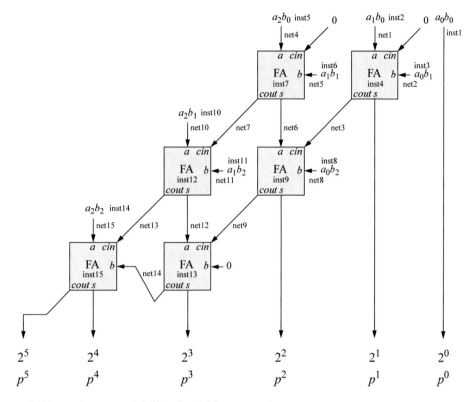

Figure 9.42   Array multiplier for 3-bit operands.

The structural module is shown in Figure 9.43 in which the net names and instantiation names correspond to those shown in Figure 9.42. A full adder is instantiated six times and a 2-input AND gate is instantiated nine times. The test bench is shown in Figure 9.44 using all combinations of the three multiplicand bits and the three multiplier bits. The outputs and waveforms are shown in Figure 9.45 and Figure 9.46, respectively.

```
//structural array multiplier
module array_mul3 (a, b, p);

input [2:0] a, b;
output [5:0] p;

wire [2:0] a, b;            //continued on next page
```

Figure 9.43   Structural module for an array multiplier for 3-bit operands.

```
//declare internal nets
wire net1, net2, net3, net4, net5, net6, net7, net8;
wire net9, net10, net11, net12, net13, net14, net15;

wire [5:0] p;//product is six bits


//instantiate the logic for product p[0]
and2_df inst1 (
    .x1(a[0]),      //AND gate input x1 connected to a[0]
    .x2(b[0]),      //AND gate input x2 connected to b[0]
    .z1(p[0])       //AND gate output z1 connected to p[0]
    );


//instantiate the logic for product p[1]
and2_df inst2 (
    .x1(a[1]),
    .x2(b[0]),
    .z1(net1)
    );

and2_df inst3 (
    .x1(a[0]),
    .x2(b[1]),
    .z1(net2)
    );

full_adder inst4 (
    .a(net1),
    .b(net2),
    .cin(1'b0),
    .sum(p[1]),
    .cout(net3)
    );


//instantiate the logic for product p[2]
and2_df inst5 (
    .x1(a[2]),
    .x2(b[0]),
    .z1(net4)
    );

//continued on next page
```

Figure 9.43   (Continued)

```
and2_df inst6 (
   .x1(a[1]),
   .x2(b[1]),
   .z1(net5)
   );

full_adder inst7 (
   .a(net4),
   .b(net5),
   .cin(1'b0),
   .sum(net6),
   .cout(net7)
   );

and2_df inst8 (
   .x1(a[0]),
   .x2(b[2]),
   .z1(net8)
   );

full_adder inst9 (
   .a(net6),
   .b(net8),
   .cin(net3),
   .sum(p[2]),
   .cout(net9)
   );


//instantiate the logic for product p[3]
and2_df inst10 (
   .x1(a[2]),
   .x2(b[1]),
   .z1(net10)
   );

and2_df inst11 (
   .x1(a[1]),
   .x2(b[2]),
   .z1(net11)
   );

//continued on next page
```

Figure 9.43   (Continued)

```
full_adder inst12 (
   .a(net10),
   .b(net11),
   .cin(net7),
   .sum(net12),
   .cout(net13)
   );


full_adder inst13 (
   .a(net12),
   .b(1'b0),
   .cin(net9),
   .sum(p[3]),
   .cout(net14)
   );



//instantiate the logic for product p[4] and p[5]
and2_df inst14 (
   .x1(a[2]),
   .x2(b[2]),
   .z1(net15)
   );


full_adder inst15 (
   .a(net15),
   .b(net14),
   .cin(net13),
   .sum(p[4]),
   .cout(p[5])
   );
endmodule
```

Figure 9.43   (Continued)

```
//test bench for structural array multiplier
module array_mul3_tb;

reg [2:0] a, b;
wire [5:0] p;

//continued on next page
```

Figure 9.44   Test bench for the array multiplier of Figure 9.43 for 3-bit operands.

```
//apply stimulus and display variables
initial
begin: apply_stimulus
    reg [6:0] invect;
        for (invect=0; invect<64; invect=invect+1)
        begin
            {a, b} = invect [6:0];
            #10 $display ("a=%d, b=%d, p=%d", a, b, p);
        end
end

//instantiate the module into the test bench
array_mul3 inst1 (
    .a(a),
    .b(b),
    .p(p)
    );
endmodule
```

Figure 9.44   (Continued)

a=0, b=0, p= 0	a=2, b=6, p=12	a=5, b=4, p=20
a=0, b=1, p= 0	a=2, b=7, p=14	a=5, b=5, p=25
a=0, b=2, p= 0	a=3, b=0, p= 0	a=5, b=6, p=30
a=0, b=3, p= 0	a=3, b=1, p= 3	a=5, b=7, p=35
a=0, b=4, p= 0	a=3, b=2, p= 6	a=6, b=0, p= 0
a=0, b=5, p= 0	a=3, b=3, p= 9	a=6, b=1, p= 6
a=0, b=6, p= 0	a=3, b=4, p=12	a=6, b=2, p=12
a=0, b=7, p= 0	a=3, b=5, p=15	a=6, b=3, p=18
a=1, b=0, p= 0	a=3, b=6, p=18	a=6, b=4, p=24
a=1, b=1, p= 1	a=3, b=7, p=21	a=6, b=5, p=30
a=1, b=2, p= 2	a=4, b=0, p= 0	a=6, b=6, p=36
a=1, b=3, p= 3	a=4, b=1, p= 4	a=6, b=7, p=42
a=1, b=4, p= 4	a=4, b=2, p= 8	a=7, b=0, p= 0
a=1, b=5, p= 5	a=4, b=3, p=12	a=7, b=1, p= 7
a=1, b=6, p= 6	a=4, b=4, p=16	a=7, b=2, p=14
a=1, b=7, p= 7	a=4, b=5, p=20	a=7, b=3, p=21
a=2, b=0, p= 0	a=4, b=6, p=24	a=7, b=4, p=28
a=2, b=1, p= 2	a=4, b=7, p=28	a=7, b=5, p=35
a=2, b=2, p= 4	a=5, b=0, p= 0	a=7, b=6, p=42
a=2, b=3, p= 6	a=5, b=1, p= 5	a=7, b=7, p=49
a=2, b=4, p= 8	a=5, b=2, p=10	
a=2, b=5, p=10	a=5, b=3, p=15	

Figure 9.45   Outputs for the array multiplier of Figure 9.43 for two 3-bit operands.

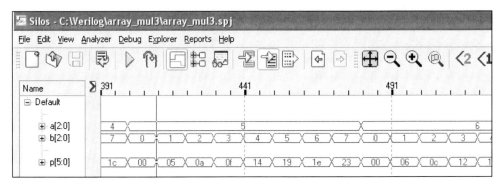

Figure 9.46   Waveforms for the array multiplier of Figure 9.43 for two 3-bit operands.

Figure 9.46   (Continued)

9.3.8 Moore-Mealy Synchronous Sequential Machine

In this section, a synchronous sequential machine will be designed containing both Moore- and Mealy-type outputs. The machine will use linear-select multiplexers for the δ next-state logic. A general block diagram for a synchronous sequential machine using multiplexers for the δ next-state logic is shown in Figure 9.47. The combinational logic that connects to the input of the multiplexer array is either very elementary or nonexistent. This will become evident in the design example.

In this method, one multiplexer is needed for each state flip-flop and each multiplexer has p select inputs, where p represents the number of storage elements. Therefore, this technique requires $p(2^p{:}1)$ multiplexers. A machine with 12 states, requiring four storage elements, would have four 16:1 multiplexers. Since most multiplexers have a single output, a design of this type is most easily implemented with D flip-flops.

The active-high output of each state flip-flop is connected to a corresponding select input line; that is, $y_1, y_2, \ldots , y_p$ connect to $s_{p-1}, \ldots , s_1, s_0$, respectively, where y_p is the low-order flip-flop and s_0 is the low-order select input. Thus, if $p = 3$, then the following state flip-flop-to-select-input connections are necessary: y_3 connects to s_0, y_2 connects to s_1, and y_1 connects to s_2. Since the flip-flop outputs connect to the multiplexer select inputs in a one-to-one mapping, this type of connection can be referred to as *linear selection*.

The state diagram for this machine is shown in Figure 9.48. Inputs x_1 and x_2 will be used as map-entered variables. The input maps for state flip-flops y_1 and y_2 are shown in Figure 9.49 and the logic diagram is shown in Figure 9.50. Output z_1 is asserted during the last half of the clock cycle; output z_2 is asserted for the entire clock cycle. Structural modeling will be used in this design by instantiating a 4:1 multiplexer, a D flip-flop, and a 3-input AND gate.

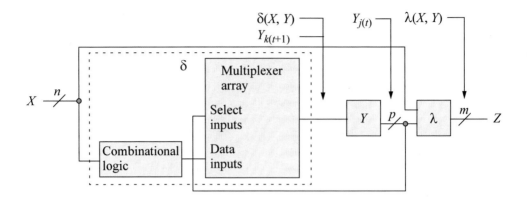

Figure 9.47   General block diagram of a synchronous sequential machine using multiplexers for the δ next-state logic.

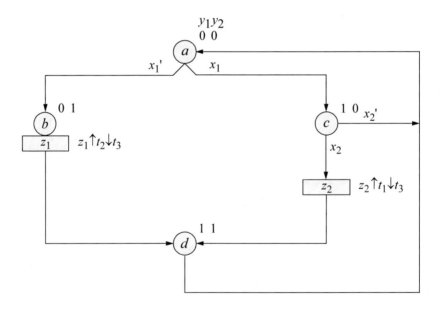

Figure 9.48   State diagram for the synchronous sequential machine of Section 9.38 with Moore- and Mealy-type outputs.

The structural module in Figure 9.51 shows the instance names and net names. Because the data inputs to the multiplexer were declared as $d[3:0]$, the instantiation of the multiplexer must have the same positional orientation for the data inputs. That is, the multiplexer data inputs for flip-flop y_1 are labeled $.d(\{1'b0, x2, 1'b1, x1\})$. The test

bench which takes the machine through all transitions in the state diagram is shown in
Figure 9.52. The outputs and waveforms are shown in Figure 9.53 and Figure 9.54, respectively.

Figure 9.49   Input maps for state flip-flops y_1 and y_2.

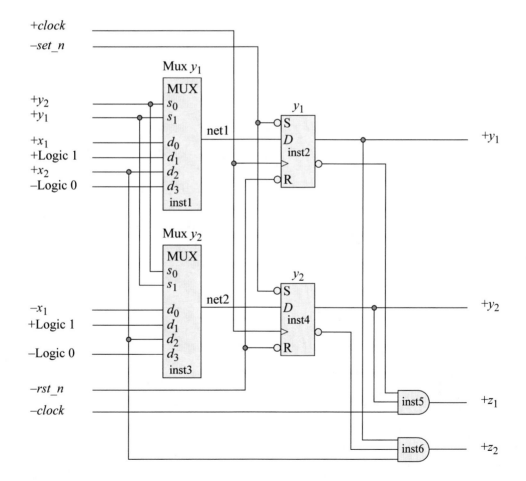

Figure 9.50   Logic diagram for the synchronous sequential machine of Figure
9.48.

```
//structural moore-mealy ssm using linear-select mux
module moore_mealy_ssm (x1, x2, clk, set_n, rst_n, y, z1, z2);

input x1, x2, clk, set_n, rst_n;
output [1:2] y;
output z1, z2;

wire x1, x2, clk, set_n, rst_n;
wire net1, net2;
wire [1:2] y;
wire z1, z2;

//instantiate the input logic for flip-flop y[1]
mux4_df inst1 (
   .s({y[1], y[2]}),
   .d({1'b0, x2, 1'b1, x1}),
   .enbl(1'b1),
   .z1(net1)
   );

d_ff inst2 (
   .d(net1),
   .clk(clk),
   .q(y[1]),
   .set_n(set_n),
   .rst_n(rst_n)
   );

//instantiate the input logic for flip-flop y[2]
mux4_df inst3 (
   .s({y[1], y[2]}),
   .d({1'b0, x2, 1'b1, ~x1}),
   .enbl(1'b1),
   .z1(net2)
   );

d_ff inst4 (
   .d(net2),
   .clk(clk),
   .q(y[2]),
   .set_n(set_n),
   .rst_n(rst_n)
   );

//continued on next page
```

Figure 9.51   Structural module for the synchronous sequential machine of Figure 9.48.

```
//instantiate the logic for output z1 and z2
and3_df inst5 (
   .x1(~y[1]),
   .x2(y[2]),
   .x3(~clk),
   .z1(z1)
   );

and3_df inst6 (
   .x1(y[1]),
   .x2(~y[2]),
   .x3(x2),
   .z1(z2)
   );

endmodule
```

Figure 9.51   (Continued)

```
//test bench for the mealy-moore ssm
module moore_mealy_ssm_tb;

reg x1, x2;
reg clk, set_n, rst_n;
wire [1:2] y;
wire z1, z2;

//display inputs and outputs
initial
$monitor ("x1x2 = %b, state = %b, z1z2 = %b",
          {x1, x2}, y, {z1, z2});

//define clock
initial
begin
   clk = 1'b0;
   forever
      #10clk = ~clk;
end

//continued on next page
```

Figure 9.52   Test bench for the synchronous sequential machine of Figure 9.48.

```
initial                     //define input sequence
begin
   #0 set_n = 1'b1;
      rst_n = 1'b0;         //reset to state_a (00)
   #5 rst_n = 1'b1;

   x1 = 1'b0;  x2 = 1'b0;
   @ (posedge clk)          //go to state_b (01)
                            //and assert z1 (t2--t3)
   x1 = 1'b0;
   @ (posedge clk)          //go to state_d (11)

   x1 = 1'b0;
   @ (posedge clk)          //go to state_a (00)

   x1 = 1'b1;
   @ (posedge clk)          //go to state_c (10)

   x2 = 1'b1;               //assert z2 with x2
   @ (posedge clk)          //then go to state_d (11)

   x1 = 1'b0;  x2=1'b0; //prevent possible glitch from d to a
   @ (posedge clk)          //go to state_a (00)

   x1 = 1'b1;
   @ (posedge clk)          //go to state_c (10)

   x2 = 1'b0;
   @ (posedge clk)          //go to state_a (00)

   #10    $stop;
end

//instantiate the module into the test bench
moore_mealy_ssm inst1 (
   .x1(x1),
   .x2(x2),
   .clk(clk),
   .set_n(set_n),
   .rst_n(rst_n),
   .y(y),
   .z1(z1),
   .z2(z2)
   );

endmodule
```

Figure 9.52   (Continued)

```
x1x2 = xx,  state = 00,  z1z2 = 00
x1x2 = 00,  state = 00,  z1z2 = 00
x1x2 = 00,  state = 01,  z1z2 = 00
x1x2 = 00,  state = 01,  z1z2 = 10
x1x2 = 00,  state = 11,  z1z2 = 00
x1x2 = 10,  state = 00,  z1z2 = 00
x1x2 = 11,  state = 10,  z1z2 = 01
x1x2 = 00,  state = 11,  z1z2 = 00
x1x2 = 10,  state = 00,  z1z2 = 00
x1x2 = 10,  state = 10,  z1z2 = 00
x1x2 = 10,  state = 00,  z1z2 = 00
```

Figure 9.53   Outputs for the synchronous sequential machine of Figure 9.48.

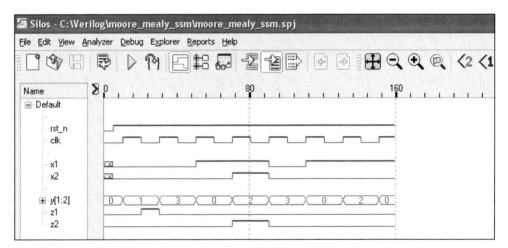

Figure 9.54   Waveforms for the synchronous sequential machine of Figure 9.48.

9.3.9 Moore Synchronous Sequential Machine

This example is slightly different than previous synchronous sequential machine designs. The storage elements consist of JK flip-flops and there are three inputs, all of which will be used as map-entered variables. There are two outputs that will be asserted according to the timing diagram shown in Figure 9.55. The time period from t_1 to t_3 represents the clock cycle for an active-low clock. Output z_1 is asserted at time t_2

and deasserted at time t_3, which is the last half of the clock cycle whereas, output z_2 is asserted at time t_2 and deasserted at time t_4, which is the last half of the clock cycle plus the first half of the following clock cycle.

The state diagram is shown in Figure 9.56 and the Karnaugh maps are shown in Figure 9.57 using x_1, x_2, and x_3 as map-entered variables. It is not intuitively obvious what the map entries should be for J and K in order to proceed to the next state from states b (001), a (000), and c (010). Therefore, a small next-state table is utilized to facilitate the determination of the JK values. The JK values are also shown for all possible state transitions. The output maps are presented in Figure 9.58.

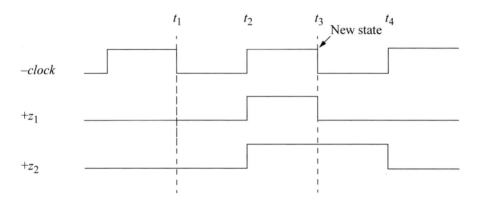

Figure 9.55   Timing diagram to illustrate the assertion/deassertion of outputs z_1 and z_2.

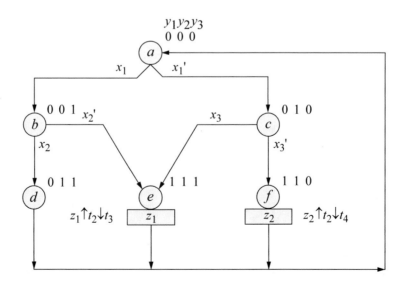

Figure 9.56   State diagram for the Moore machine of Section 9.39. Unused states: 100, 101.

Present state	Next state	JK
0	0	0 –
0	1	1 –
1	0	– 1
1	1	– 0

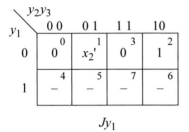

$$Jy_1$$

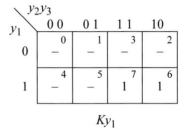

$$Ky_1$$

Present state		Next state		
$y_1y_2y_3$	x_2	y_1	Jy_1	Ky_1
0 0 1	0	1	1	–
0 0 1	1	0	0	–

$$Jy_1 = y_2{'}y_3x_2{'} + y_2y_3{'} \qquad Ky_1 = 1$$

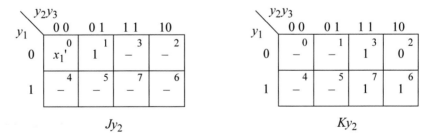

$$Jy_2 \qquad\qquad Ky_2$$

Present state		Next state		
$y_1y_2y_3$	x_1	y_2	Jy_2	Ky_2
0 0 0	0	1	1	–
0 0 0	1	0	0	–

$$Jy_2 = x_1{'} + y_3 \qquad\qquad Ky_2 = y_1 + y_3$$

Figure 9.57   Karnaugh maps for state variables y_1, y_2, and y_3.

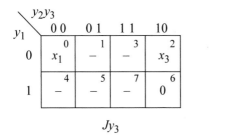

$$J_{y_3}$$ $$K_{y_3}$$

Present state	Next state			
$y_1 y_2 y_3$	x_1	y_3	J_{y_3}	K_{y_3}
0 0 0	0	0	0	−
0 0 0	1	1	1	−

Present state	Next state			
$y_1 y_2 y_3$	x_3	y_3	J_{y_3}	K_{y_3}
0 1 0	0	0	0	−
0 1 0	1	1	1	−

$$J_{y_3} = y_2' x_1 + y_1' y_2 x_3 \qquad\qquad K_{y_3} = y_2$$

Figure 9.57   (Continued)

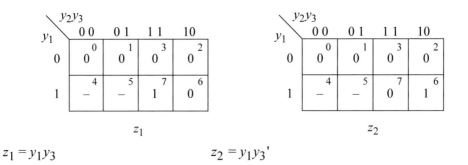

$$z_1$$ $$z_2$$

$$z_1 = y_1 y_3 \qquad\qquad\qquad z_2 = y_1 y_3'$$

Figure 9.58   Output maps for z_1 and z_2.

The logic diagram is shown in Figure 9.59 using JK flip-flops for the storage elements. In order to have output z_2 be asserted at time t_2 and deasserted at time t_4, a D flip-flop is used to maintain the asserted state until the next active clock transition. A new state is then entered and the D input is a logic 0 which deasserts z_2 at the next positive transition of the clock.

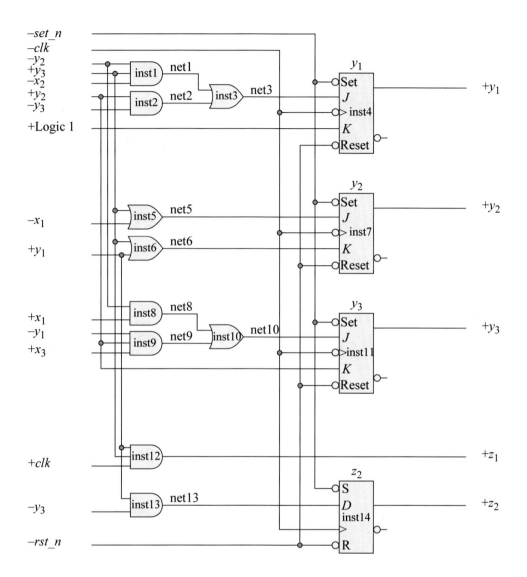

Figure 9.59   Logic diagram for the Moore synchronous sequential machine of Section 9.39.

The structural module is shown in Figure 9.60 and correlates the instantiation names and the net names with the logic diagram. The test bench is shown in Figure 9.61 and the waveforms are shown in Figure 9.62. Notice that output z_1 is asserted during the last half of the clock cycle in state e ($y_1y_2y_3 = 111$) and that output z_2 is asserted during the last half of the clock cycle in state f($y_1y_2y_3 = 110$) and the first half of the clock cycle in state a ($y_1y_2y_3 = 000$) as stated in the machine specifications.

```
//structural Moore synchronous sequential machine
module moore_ssm5 (clk, set_n, rst_n, x1, x2, x3, y1, y2, y3,
                    z1, z2);

input clk, set_n, rst_n, x1, x2, x3;
output y1, y2, y3, z1, z2;

//define internal nets
wire net1, net2, net3, net5, net6, net8, net9, net10, net13;

//instantiate the logic for flip-flop y1
and3_df inst1 (
   .x1(~y2),
   .x2(y3),
   .x3(~x2),
   .z1(net1)
   );

and2_df inst2 (
   .x1(y2),
   .x2(~y3),
   .z1(net2)
   );

or2_df inst3 (
   .x1(net1),
   .x2(net2),
   .z1(net3)
   );

jkff_neg_clk inst4 (
   .clk(clk),
   .j(net3),
   .k(1'b1),
   .set_n(set_n),
   .rst_n(rst_n),
   .q(y1)
   );

//instantiate the logic for flip-flop y2
or2_df inst5 (
   .x1(~x1),
   .x2(y3),
   .z1(net5)
   );
//continued on next page
```

Figure 9.60   Structural module for the Moore machine of Figure 9.59.

```
or2_df inst6 (
   .x1(y1),
   .x2(y3),
   .z1(net6)
   );

jkff_neg_clk inst7 (
   .clk(clk),
   .j(net5),
   .k(net6),
   .set_n(set_n),
   .rst_n(rst_n),
   .q(y2)
   );

//instantiate the logic for flip-flop y3
and2_df inst8 (
   .x1(~y2),
   .x2(x1),
   .z1(net8)
   );

and3_df inst9 (
   .x1(~y1),
   .x2(y2),
   .x3(x3),
   .z1(net9)
   );

or2_df inst10 (
   .x1(net8),
   .x2(net9),
   .z1(net10)
   );

jkff_neg_clk inst11 (
   .clk(clk),
   .j(net10),
   .k(y2),
   .set_n(set_n),
   .rst_n(rst_n),
   .q(y3)
   );

//continued on next page
```

Figure 9.60   (Continued)

```
//instantiate the logic for output z1
and3_df inst12 (
    .x1(y1),
    .x2(y3),
    .x3(clk),
    .z1(z1)
    );

//instantiate the logic for output z2
and2_df inst13 (
    .x1(y1),
    .x2(~y3),
    .z1(net13)
    );

d_ff inst14 (
    .d(net13),
    .clk(clk),
    .q(z2),
    .set_n(set_n),
    .rst_n(rst_n)
    );

endmodule
```

Figure 9.60   (Continued)

```
//test bench for the Moore ssm
module moore_ssm5_tb;

reg clk, set_n, rst_n;
reg x1, x2, x3;
wire y1, y2, y3, z1, z2;

//define clock
initial
begin
    clk = 1'b1;
    forever
        #10clk = ~clk;
end

//continued on next page
```

Figure 9.61   Test bench for the structural module of Figure 9.60.

```
//define input vectors
initial
begin
   #0      x1=1'b0;
           x2=1'b0;
           x3=1'b0;
           set_n=1'b1;
           rst_n=1'b0;          //reset to state_a (000)
   #5      rst_n=1'b1;

   @   (negedge clk)
       //if x1=0 in state_a, go to state_c (010)

   x1=1'b0; x2=1'b0; x3=1'b0; @ (negedge clk)
       //if x3=0 in state_c, go to state_f (110) and assert z2

   x1=1'b0; x2=1'b0; x3=1'b0; @ (negedge clk)
       //go to state_a (000)

   x1=1'b1; x2=1'b0; x3=1'b0; @ (negedge clk)
       //if x1=1 in state_a, go to state_b (001)

   x1=1'b0; x2=1'b1; x3=1'b0; @ (negedge clk)
       //if x2=1 in state_b, go to state_d (011)

   x1=1'b0; x2=1'b0; x3=1'b0; @ (negedge clk)
       //go to state_a (000)

   x1=1'b0; x2=1'b0; x3=1'b0; @ (negedge clk)
       //if x1=0 in state_a, go to state_c (010)

   x1=1'b0; x2=1'b0; x3=1'b1; @ (negedge clk)
       //if x3=1 in state_c, go to state_e (111) and assert z1

   x1=1'b0; x2=1'b0; x3=1'b0; @ (negedge clk)
       //go to state_a (000)

   x1=1'b1; x2=1'b0; x3=1'b0; @ (negedge clk)
       //if x1=1 in state_a, go to state_b (001)

   x1=1'b0; x2=1'b0; x3=1'b0; @ (negedge clk)
       //if x2=0 in state_b, go to state_e and assert z1

//continued on next page
```

Figure 9.61   (Continued)

```
    x1=1'b0; x2=1'b0; x3=1'b0; @ (negedge clk)
       //go to state_a (000)

    #130   $stop;
end

//instantiate the module into the test bench
moore_ssm5 inst1 (
    .clk(clk),
    .set_n(set_n),
    .rst_n(rst_n),
    .x1(x1),
    .x2(x2),
    .x3(x3),
    .y1(y1),
    .y2(y2),
    .y3(y3),
    .z1(z1),
    .z2(z2)
    );

endmodule
```

Figure 9.61   (Continued)

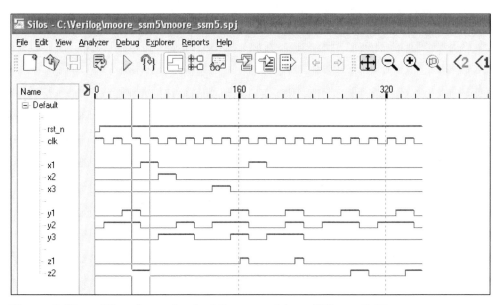

Figure 9.62   Waveforms for the Moore machine of Figure 9.60.

9.3.10 Moore Asynchronous Sequential Machine

An asynchronous sequential machine has two inputs x_1 and x_2 and one output z_1. Input x_1 will always be asserted whenever x_2 is asserted; that is, there will never be a situation where x_1 is deasserted and x_2 is asserted. Output z_1 is asserted coincident with every third x_2 pulse and remains active for the duration of x_2.

The primitive flow table is shown in Figure 9.63 in which there are no equivalent states according to the rules for equivalence which are restated here from Section 7.1.9. The primitive flow table may have an inordinate number of rows. The number of rows can be reduced by finding equivalent states and then eliminating redundant states. If the machine's operation is indistinguishable whether commencing in state Y_i or state Y_j, then one of the states is redundant and can be eliminated. The flow table thus obtained is a *reduced primitive flow table*. In order for two stable states to be equivalent, all three of the following conditions must be met:

1. The same input vector.
2. The same output value.
3. The same, or equivalent, next state for all valid input sequences.

x_1x_2 0 0	0 1	1 1	1 0	z_1
(a)	–	–	b	0
a	–	c	(b)	0
–	–	(c)	d	0
a	–	e	(d)	0
–	–	(e)	f	0
a	–	g	(f)	0
–	–	(g)	b	1

Figure 9.63 Primitive flow table for the asynchronous sequential machine of Section 9.3.10.

The merger diagram is shown in Figure 9.64. The rules for merging are restated here from Section 7.1.9. The merger diagram graphically portrays the result of the merging process in which an attempt is made to combine two or more rows of the reduced primitive flow table into a single row. The result of the merging technique is analogous to that of finding equivalent states; that is, the merging process can also reduce the number of rows in the table and hence, reduce the number of feedback variables that are required. Fewer feedback variables will result in a machine with less logic and therefore, less cost. Two rows can merge into a single row if the entries in the same column of each row satisfy one of the following three merging rules:

1.   Identical state entries, either stable or unstable.
2.   A state entry and a "don't care."
3.   Two "don't care" entries.

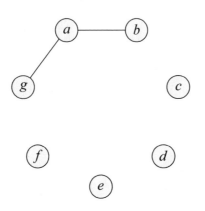

Figure 9.64   Merger diagram for the asynchronous sequential machine of Section 9.3.10.

The merged flow table is constructed from the merger diagram and is shown in Figure 9.65. The table represents the culmination of the merging process in which two or more rows of a primitive flow table are replaced by a single equivalent row that contains one stable state for each merged row.

Merging is a process of combining two or more rows of a reduced primitive flow table into a single row. The merging process reduces the number of rows in the flow table, and thus, may reduce the number of feedback variables. A reduction of feedback variables decreases the number of storage elements in the machine. The stable states in the rows that are merged are entered in the same location in the single merged row. When merging, the outputs associated with each row are disregarded. Thus, two rows of a reduced primitive flow table can merge, regardless of the output values of the rows under consideration.

The merged flow table specifies the operational characteristics of the machine in a manner analogous to that of the primitive flow table and the reduced primitive flow table, but in a more compact form. Each row in a merged flow table represents a set of maximal compatible rows.

Most unspecified entries in the reduced primitive flow table are replaced with either a stable or an unstable state entry in the merged flow table. Since more than one stable state is usually present in each row of a merged flow table, many state transition sequences do not cause a change to the feedback variables. Thus, faster operational speed is realized.

The merged flow table is derived from the merger diagram in conjunction with the reduced primitive flow table. The partition of sets of maximal compatible rows obtained from the merger diagram dictates the minimal number of rows in the merged flow table.

		x_1x_2 00	01	11	10
1	$\textcircled{a}$, $\textcircled{b}$	$\textcircled{a}$	–	c	$\textcircled{b}$
2	$\textcircled{c}$	–	–	$\textcircled{c}$	d
3	$\textcircled{d}$	a	–	e	$\textcircled{d}$
4	$\textcircled{e}$	–	–	$\textcircled{e}$	f
5	$\textcircled{f}$	a	–	g	$\textcircled{f}$
6	$\textcircled{g}$	–	–	$\textcircled{g}$	b
7		–	–	–	–
8		–	–	–	–

Figure 9.65   Merged flow table for the asynchronous sequential machine of Section 9.3.10.

If a single change to the input vector causes two or more excitation variables to change state, then multiple paths exist from the source stable state to the destination stable state. This is called a *race* condition. Inspection of column $x_1x_2 = 11$ indicates that the feedback values assigned to row 1 must be adjacent to those assigned to row 2 to provide a race-free cycle from stable state $\textcircled{b}$ through unstable state c to state $\textcircled{c}$. The same adjacency requirement is observed in column $x_1x_2 = 10$ for the sequence from state $\textcircled{c}$ through unstable state d to state $\textcircled{d}$. After observing all transitions, the following adjacency requirements for race-free operation are summarized as follows:

Column $x_1 x_2 = 11$: Rows 1 and 2 must be adjacent.
Column $x_1 x_2 = 10$: Rows 2 and 3 must be adjacent.
Column $x_1 x_2 = 00$: Rows 1 and 3 must be adjacent.
Column $x_1 x_2 = 11$: Rows 3 and 4 must be adjacent.
Column $x_1 x_2 = 00$: Rows 5 and 1 must be adjacent.
Column $x_1 x_2 = 11$: Rows 5 and 6 must be adjacent.
Column $x_1 x_2 = 10$: Rows 6 and 1 must be adjacent.

The preceding requirements listed for each column specify the adjacencies that are needed to establish race-free operation for the indicated transitions. Since it is not possible to assign adjacency to all rows requiring adjacent state codes, a *transition diagram* is used to assist in state code assignment. State codes are assigned to each row and then an appropriate path is taken which provides only one change of variable for each segment of the path. The transition diagram of Figure 9.66 graphically portrays this information. For example, the sequence from state $\textcircled{d}$ [row 3 (101)] to state $\textcircled{a}$ [row 1 (000)] can pass through the intermediate transient state in row 7 (100). This path has only one change of variable for each segment of the path.

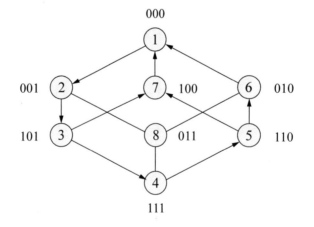

Figure 9.66   Transition diagram for the asynchronous sequential machine of Section 9.3.10.

The next step is to generate the combined excitation map for the three excitation variables Y_{1e}, Y_{2e}, and Y_{2e} as shown in Figure 9.67. The rows of the merged flow table are rearranged according to the Gray code listing of the feedback variables y_{1f}, y_{2f}, and y_{2f}, which correlate to the transition diagram. From the combined excitation map, the individual excitation maps are derived as shown in Figure 9.68. These yield the excitation equations of Equation 9.5 from which the structural module is designed.

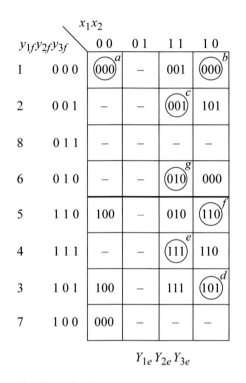

Figure 9.67 Combined excitation map.

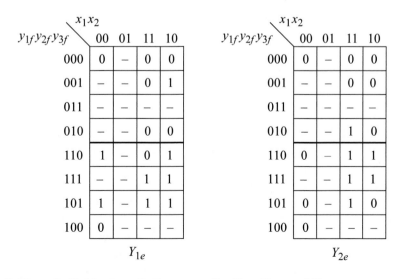

Figure 9.68 Individual excitation maps for Y_{1e}, Y_{2e}, and Y_{3e}.

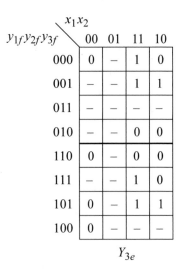

Figure 9.68   (Continued)

$$Y_{1e} = y_{3f}x_2' + y_{1f}y_{3f} + y_{1f}y_{2f}x_2'$$

$$Y_{2e} = y_{1f}x_2 + y_{2f}x_2 + y_{1f}y_{2f}x_1$$

$$Y_{3e} = y_{2f}'x_2 + y_{3f}x_2 + y_{2f}'y_{3f}x_1 \qquad (9.5)$$

The logic diagram, using NAND gates, is shown in Figure 9.69 as obtained from the excitation equations of Equation 9.5. Each AND function of each latch must be connected to the reset input to assure that the machine is initialized to a known state.

The state alphabet Y contains three storage elements in the form of SR latches. The outputs of the latches represent the excitation variables Y_{1e}, Y_{2e}, and Y_{3e}. After a delay of Δt, these same outputs represent the feedback variables y_{1f}, y_{2f}, and y_{3f} and connect to the NAND gates of the SR latches. Thus, when the feedback variables become equal to the excitation variables, the machine enters a stable state where $y_{if} = Y_{ie}$. The output alphabet Z is represented by z_1.

The structural module is shown in Figure 9.70 using NAND gates for the three latches and an AND gate for the output logic. The test bench is shown in Figure 9.71. The outputs and waveforms are shown in Figure 9.72 and Figure 9.73, respectively.

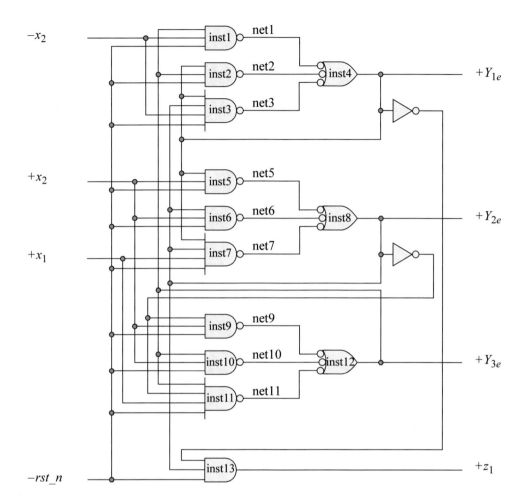

Figure 9.69   Logic diagram for the asynchronous sequential machine of Section 9.3.10.

```
//structural module for an asynchronous sequential machine
module asm10 (rst_n, x1, x2, y1e, y2e, y3e, z1);

input rst_n, x1, x2;
output y1e, y2e, y3e, z1;

//define internal nets
wire net1, net2, net3, net5, net6, net7, net9, net10, net11;
//continued on next page
```

Figure 9.70   Structural module for the asynchronous sequential machine of Figure 9.69.

```
//design for latch Y1e
nand3_df inst1 (
   .x1(y3e),
   .x2(~x2),
   .x3(rst_n),
   .z1(net1)
   );

nand3_df inst2 (
   .x1(y1e),
   .x2(y3e),
   .x3(rst_n),
   .z1(net2)
   );

nand4_df inst3 (
   .x1(y1e),
   .x2(y2e),
   .x3(~x2),
   .x4(rst_n),
   .z1(net3)
   );

nand3_df inst4 (
   .x1(net1),
   .x2(net2),
   .x3(net3),
   .z1(y1e)
   );

//design for latch Y2e
nand3_df inst5 (
   .x1(y1e),
   .x2(x2),
   .x3(rst_n),
   .z1(net5)
   );

nand3_df inst6 (
   .x1(y2e),
   .x2(x2),
   .x3(rst_n),
   .z1(net6)
   );
```

```
nand4_df inst7 (
   .x1(y1e),
   .x2(y2e),
   .x3(x1),
   .x4(rst_n),
   .z1(net7)
   );

nand3_df inst8 (
   .x1(net5),
   .x2(net6),
   .x3(net7),
   .z1(y2e)
   );

//design for latch Y3e
nand3_df inst9 (
   .x1(~y2e),
   .x2(x2),
   .x3(rst_n),
   .z1(net9)
   );

nand3_df inst10 (
   .x1(y3e),
   .x2(x2),
   .x3(rst_n),
   .z1(net10)
   );

nand4_df inst11 (
   .x1(~y2e),
   .x2(y3e),
   .x3(x1),
   .x4(rst_n),
   .z1(net11)
   );

nand3_df inst12 (
   .x1(net9),
   .x2(net10),
   .x3(net11),
   .z1(y3e)
   );

//continued on next page
```

Figure 9.70   (Continued)

```
//design for output z1
and3_df inst13 (
    .x1(~y1e),
    .x2(y2e),
    .x3(rst_n),
    .z1(z1)
    );
endmodule
```

Figure 9.70   (Continued)

```
//test bench for the asynchronous sequential machine
module asm10_tb;

reg rst_n, x1, x2;
wire y1e, y2e, y3e, z1;

//display variables
initial
$monitor ("x1=%b, x2=%b, state=%b, z1=%b",
          x1, x2, {y1e, y2e, y3e}, z1);

//apply stimulus
initial
begin
    #0      rst_n=1'b0;
            x1=1'b0;
            x2=1'b0;              //reset to state_a
    #5      rst_n=1'b1;

    #10     x1=1'b1; x2=1'b0;     //go to state_b
    #10     x1=1'b1; x2=1'b1;     //go to state_c
    #10     x1=1'b1; x2=1'b0;     //go to state_d
    #10     x1=1'b1; x2=1'b1;     //go to state_e
    #10     x1=1'b1; x2=1'b0;     //go to state_f
    #10     x1=1'b1; x2=1'b1;     //go to state_g, assert z1
    #10     x1=1'b1; x2=1'b0;     //go to state_b
    #10     x1=1'b0; x2=1'b0;     //go to state_a

    #10     $stop;
end

//continued on next page
```

Figure 9.71   Test bench for the asynchronous sequential machine module of Figure 9.70.

```
//instantiate the module into the test bench
asm10 inst1 (
    .rst_n(rst_n),
    .x1(x1),
    .x2(x2),
    .y1e(y1e),
    .y2e(y2e),
    .y3e(y3e),
    .z1(z1)
    );

endmodule
```

Figure 9.71   (Continued)

```
x1=0, x2=0, state=000, z1=0
x1=1, x2=0, state=000, z1=0
x1=1, x2=1, state=001, z1=0
x1=1, x2=0, state=101, z1=0
x1=1, x2=1, state=111, z1=0
x1=1, x2=0, state=110, z1=0
x1=1, x2=1, state=010, z1=1
x1=1, x2=0, state=000, z1=0
x1=0, x2=0, state=000, z1=0
```

Figure 9.72   Outputs for the asynchronous sequential machine module of Figure 9.70.

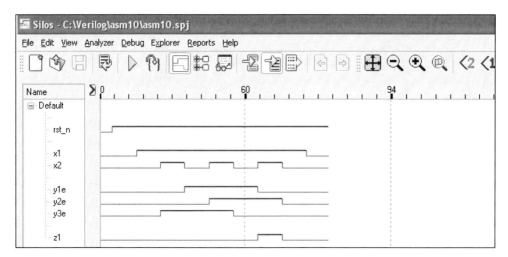

Figure 9.73   Waveforms for the asynchronous sequential machine module of Figure 9.70.

9.3.11 Moore Pulse-Mode Asynchronous Sequential Machine

In this example, a Moore pulse-mode asynchronous sequential machine will be designed using the state diagram of Figure 9.74. The machine will use SR latches and D flip-flops in a master-slave configuration. As in previous pulse-mode machines, the D flip-flops will not be triggered until the input pulse is deasserted. The pulse width restrictions that are dominant in pulse-mode sequential machines can thereby be eliminated by including D flip-flops in the feedback path from the SR latches to the δ next-state logic. Providing edge-triggered D flip-flops as a constituent part of the implementation negates the requirement of precisely controlled input pulse durations.

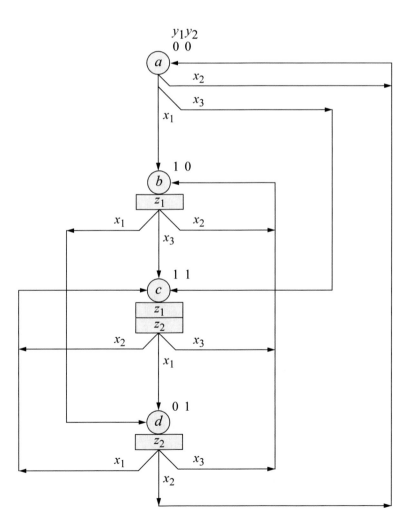

Figure 9.74   State diagram for the Moore pulse-mode asynchronous sequential machine of Section 9.3.11.

The input maps for the latches that connect to the D inputs of the flip-flops are shown in Figure 9.75. Each latch requires three input maps, one each for x_1, x_2, and x_3. The maps are arranged such that the maps corresponding to each latch are in the same row, and each column of maps corresponds to a unique input. The map entries are defined as follows:

> S indicates that the latch will be set.
> s indicates that the latch will remain set.
> R indicates that the latch will be reset.
> r indicates that the latch will remain reset.

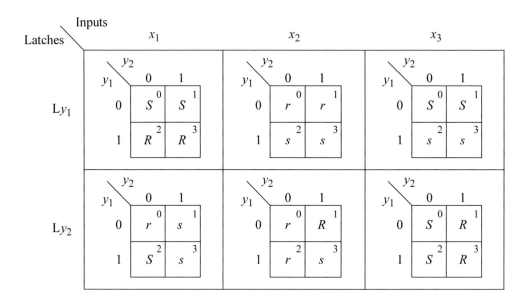

Figure 9.75   Input maps for latches Ly_1 and Ly_2 for the Moore pulse-mode machine of Section 9.3.11.

The input map entries are obtained from the state diagram. Consider state a ($y_1 y_2 = 00$) for latch Ly_1, column x_1. If input x_1 is pulsed, then the machine sequences to state b ($y_1 y_2 = 10$) and flip-flop y_1 is set. This occurs when x_1 is deasserted because the D input for flip-flop y_1 is connected to the output of latch Ly_1. Therefore, the letter S is inserted in minterm location 0 of the latch for Ly_1 in column x_1. In state b, output z_1 is asserted.

Now consider state b ($y_1 y_2 = 10$) for latch Ly_1. If x_1 is pulsed, the machine sequences to state d ($y_1 y_2 = 01$) and y_1 is reset, causing the letter R to be inserted in minterm location 2 of the map for latch Ly_1.

Consider two more entries. First, state c $(y_1 y_2 = 11)$ for latch Ly_2. If x_3 is pulsed, then the machine proceeds to state b $(y_1 y_2 = 10)$ and flip-flop y_2 is reset necessitating an entry of R in minterm location 3 of the map for Ly_2, column x_3. Next, consider state b $(y_1 y_2 = 10)$ for flip-flop y_2. If x_2 is pulsed, then the machine sequences to state b and y_2 remains reset, requiring an entry of r in minterm location 2 of the map for latch Ly_2, column x_2.

The equations for setting and resetting latches Ly_1 and Ly_2 are shown in Equation 9.6. When obtaining the equations, the upper case letters must be considered — the lower case letters are used only to minimize the equations. Thus, the equation for setting latch Ly_1 will use minterm locations 0 and 1 in the map for latch Ly_1, column x_1 and minterm locations 0, 1, 2, and 3 in the map for latch Ly_1, column x_3. This yields the set equation for Ly_1 as shown below. The other equations are obtained in a similar manner. The output equations are shown in Equation 9.7.

$$SLy_1 = y_1'x_1 + x_3$$

$$RLy_1 = y_1 x_1$$

$$SLy_2 = y_1 x_1 + y_2'x_3$$

$$RLy_2 = y_1'x_2 + y_2 x_3 \tag{9.6}$$

$$z_1 = y_1$$

$$z_2 = y_2 \tag{9.7}$$

The logic diagram is shown in Figure 9.76 and is obtained from Equation 9.6 and Equation 9.7. The structural module is shown in Figure 9.77 by instantiating the following dataflow modules: *nor3_df, nor2_df, and2_df, nand2_df,* and the behavioral module *d_ff*. The test bench shown in Figure 9.78 takes the machine through several paths in the state diagram. The outputs and waveforms are shown in Figure 9.79 and Figure 9.80, respectively. The waveforms correctly depict the state transition sequences as seen in the state diagram for the input vectors applied in the test bench.

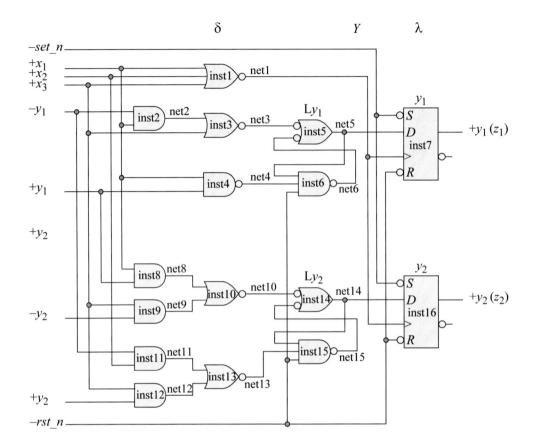

Figure 9.76 Logic diagram for the Moore pulse-mode machine of Figure 9.74.

```
//Moore pulse-mode asm
module pm_asm5 (set_n, rst_n, x1, x2, x3, y1, y2, z1, z2);

input set_n, rst_n;
input x1, x2, x3;
output y1, y2;
output z1, z2;

//define internal nets
wire net1, net2, net3, net4, net5, net6, net8, net9,
     net10, net11, net12, net13, net14, net15;

//continued on next page
```

Figure 9.77 Structural module for the Moore pulse-mode machine of Figure 9.74.

```
//design for clock input          //design for D flip-flop y1
nor3_df inst1 (                    d_ff inst7 (
   .x1(x1),                           .d(net5),
   .x2(x2),                           .clk(net1),
   .x3(x3),                           .q(y1),
   .z1(net1)                          .set_n(set_n),
   );                                 .rst_n(rst_n)
                                      );
//design for latch Ly1
and2_df inst2 (                    //design for latch Ly2
   .x1(~y1),                       and2_df inst8 (
   .x2(x1),                           .x1(x1),
   .z1(net2)                          .x2(y1),
   );                                 .z1(net8)
                                      );
nor2_df inst3 (
   .x1(net2),                      and2_df inst9 (
   .x2(x3),                           .x1(x3),
   .z1(net3)                          .x2(~y2),
   );                                 .z1(net9)
                                      );
nand2_df inst4 (
   .x1(x1),                        nor2_df inst10 (
   .x2(y1),                           .x1(net8),
   .z1(net4)                          .x2(net9),
   );                                 .z1(net10)
                                      );
nand2_df inst5 (
   .x1(net3),                      and2_df inst11 (
   .x2(net6),                         .x1(~y1),
   .z1(net5)                          .x2(x2),
   );                                 .z1(net11)
                                      );
nand3_df inst6 (
   .x1(net5),                      and2_df inst12 (
   .x2(net4),                         .x1(x3),
   .x3(rst_n),                        .x2(y2),
   .z1(net6)                          .z1(net12)
   );                                 );

                                   //continued on next page
```

Figure 9.77 (Continued)

```
nor2_df inst13 (                    //design for D flip-flop y2
   .x1(net11),                      d_ff inst16 (
   .x2(net12),                         .d(net14),
   .z1(net13)                          .clk(net1),
   );                                  .q(y2),
                                       .set_n(set_n),
nand2_df inst14 (                      .rst_n(rst_n)
   .x1(net10),                         );
   .x2(net15),
   .z1(net14)                       //design for z1 and z2
   );                               assign z1 = y1;
                                    assign z2 = y2;
nand3_df inst15 (
   .x1(net14),                      endmodule
   .x2(net13),
   .x3(rst_n),
   .z1(net15)
   );
```

Figure 9.77   (Continued)

```
//test bench for pulse-mode asm
module pm_asm5_tb;

reg x1, x2, x3, set_n, rst_n;
wire y1, y2, z1, z2;

//display inputs and outputs
initial
$monitor ("x1x2x3=%b, state=%b, z1=%b, z2=%b",
          {x1, x2, x3}, {y1, y2}, z1, z2);

//define input sequence
initial
begin
   #0   set_n = 1'b1;
        rst_n = 1'b0;   //reset to state_a (00); no output
        x1 = 1'b0;
        x2 = 1'b0;
        x3 = 1'b0;
   #5   rst_n = 1'b1;   //deassert reset

//continued on next page
```

Figure 9.78   Test bench for the Moore pulse-mode machine of Figure 9.76.

```
  #10   x1=1'b1;  x2=1'b0;  x3=1'b0;  //go to b(10); assert z1
  #10   x1=1'b0;  x2=1'b0;  x3=1'b0;
  #10   x1=1'b0;  x2=1'b0;  x3=1'b1;  //go to c(11); assert z1, z2
  #10   x1=1'b0;  x2=1'b0;  x3=1'b0;
  #10   x1=1'b1;  x2=1'b0;  x3=1'b0;  //go to d(01); assert z2
  #10   x1=1'b0;  x2=1'b0;  x3=1'b0;
  #10   x1=1'b0;  x2=1'b1;  x3=1'b0;  //go to a(00);
  #10   x1=1'b0;  x2=1'b0;  x3=1'b0;
  #10   x1=1'b0;  x2=1'b0;  x3=1'b1;  //go to c(11); assert z1, z2
  #10   x1=1'b0;  x2=1'b0;  x3=1'b0;
  #10   x1=1'b0;  x2=1'b1;  x3=1'b0;  //go to c(11); assert z1, z2
  #10   x1=1'b0;  x2=1'b0;  x3=1'b0;
  #10   x1=1'b0;  x2=1'b0;  x3=1'b1;  //go to b(10); assert z1
  #10   x1=1'b0;  x2=1'b0;  x3=1'b0;
  #10   x1=1'b1;  x2=1'b0;  x3=1'b0;  //go to d(01); assert z2
  #10   x1=1'b0;  x2=1'b0;  x3=1'b0;
  #10   x1=1'b0;  x2=1'b1;  x3=1'b0;  //go to a(00);
  #10   $stop;
end

pm_asm5 inst1 (                       //instantiate the module
  .set_n(set_n),
  .rst_n(rst_n),
  .x1(x1),
  .x2(x2),
  .x3(x3),
  .y1(y1),
  .y2(y2),
  .z1(z1),
  .z2(z2)
  );
endmodule
```

Figure 9.78   (Continued)

```
x1x2x3=000,  state=00,  z1=0,  z2=0
x1x2x3=100,  state=00,  z1=0,  z2=0
x1x2x3=000,  state=10,  z1=1,  z2=0
x1x2x3=001,  state=10,  z1=1,  z2=0
x1x2x3=000,  state=11,  z1=1,  z2=1
x1x2x3=100,  state=11,  z1=1,  z2=1
x1x2x3=000,  state=01,  z1=0,  z2=1
x1x2x3=010,  state=01,  z1=0,  z2=1
x1x2x3=000,  state=00,  z1=0,  z2=0     //continued on next page
```

Figure 9.79   Outputs for the Moore pulse-mode machine of Figure 9.76.

```
x1x2x3=001, state=00, z1=0, z2=0
x1x2x3=000, state=11, z1=1, z2=1
x1x2x3=010, state=11, z1=1, z2=1
x1x2x3=000, state=11, z1=1, z2=1
x1x2x3=001, state=11, z1=1, z2=1
x1x2x3=000, state=10, z1=1, z2=0
x1x2x3=100, state=10, z1=1, z2=0
x1x2x3=000, state=01, z1=0, z2=1
x1x2x3=010, state=01, z1=0, z2=1
```

Figure 9.79   (Continued)

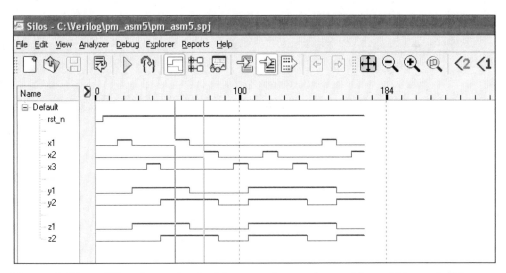

Figure 9.80   Waveforms for the Moore pulse-mode machine of Figure 9.76.

9.4 Problems

9.1   Use structural modeling to design a sum-of-products circuit that operates according to the following equation:

$$z_1 = x_1 x_2 + x_3 x_4 + x_2' x_3'$$

Design a 2-input AND gate and a 3-input OR gate to be instantiated into the structural module. Design the test bench and obtain the outputs.

9.2   Use structural modeling to design a full adder. Design the following dataflow modules to be instantiated into the structural module: *and2_df*, *or2_df*, and *xor2_df*. Obtain the structural module, test bench, outputs, and waveforms.

9.3   Use structural modeling to design a 4-bit binary-to-excess-3 code converter, where the binary inputs are *b[3:0]* and the excess-3 outputs are *x[3:0]*. Design dataflow modules for the following functions: *and2_df*, *and3_df*, *xnor2_df*, *or3_df*, and *or4_df* that will be instantiated into the structural module. The equations for the excess-3 code are shown below. Obtain the structural module, test bench, and outputs.

$$x[3] = b[3] \; b[2]' + b[3] \; b[1]' b[0]' + b[3]' \; b[2] \; b[0] + b[3]' \; b[2] \; b[1]$$
$$x[2] = b[2]' \; b[1] + b[2]' \; b[0] + b[2] \; b[1]' \; b[0]'$$
$$x[1] = b[1]' \; b[0]' + b[1] \; b[0]$$
$$x[0] = b[0]'$$

9.4   Use structural modeling to design a 3-bit comparator for the following operands:

$$A = a_2 \; a_1 \; a_0$$
$$B = b_2 \; b_1 \; b_0$$

where a_0 and b_0 are the low-order bits of A and B, respectively. Obtain the design module, test bench module, outputs, and waveforms. Test the module with inputs that demonstrate the relative magnitude of the two operands. Use the following outputs:

a_lt_b indicating $A < B$
a_eq_b indicating $A = B$
a_gt_b indicating $A > B$

The equations for the comparator are shown below.

$$(A < B) = a_2' \; b_2 + (a_2 \oplus b_2)' \; a_1' \; b_1 + (a_2 \oplus b_2)' \; (a_1 \oplus b_1)' \; a_0' \; b_0$$
$$(A = B) = (a_2 \oplus b_2)' \; (a_1 \oplus b_1)' \; (a_0 \oplus b_0)'$$
$$(A > B) = a_2 \; b_2' + (a_2 \oplus b_2)' \; a_1 \; b_1' + (a_2 \oplus b_2)' \; (a_1 \oplus b_1)' \; a_0 \; b_0'$$

9.5   Design a logic circuit that will generate a high output z_1 if a 4-bit binary input *x[3:0]* has a value less than or equal to five or greater than nine. Obtain the structural module, the test bench, and the outputs.

9.6   Use structural modeling to design a modulo-11 counter. Instantiate the following dataflow modules: *and2_df, and3_df, or2_df, or3_df*, and a behavioral module for a *D* flip-flop *d_ff*. Obtain the design module, the test bench, the outputs, and the waveforms. Display the simulation time.

9.7   Given the input map shown below for a *D* flip-flop with output q, design the δ next-state logic using a linear-select multiplexer. To review a linear-select multiplexer, refer to Section 9.3.8. Instantiate the necessary logic gates as dataflow modules, a multiplexer designed using the **case** statement, and a behavioral *D* flip-flop. Obtain the structural module, test bench, outputs, and waveforms.

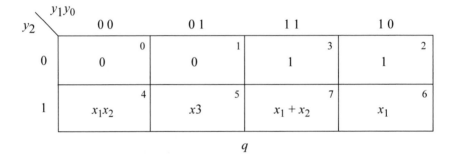

9.8   Using structural modeling with no built-in primitives, design a synchronous machine using a 3-bit counter, a 3-bit serial-in, parallel-out shift register, and logic gates. When the counter reaches a count of 7, the ~*clock* signal will be applied to the shift register clock input causing it to shift right one bit position. The shift register will operate in the two modes described below, which are controlled by two pattern select inputs: *sngl_1* and *dbl_1*.

(a) When the *sngl_1* input is active, the following pattern will be shifted through the shift register: 000, 100, 010, 001, 000, . . .

(b) When the *dbl_1* input is active, the following pattern will be shifted through the shift register: 000, 100, 110, 011, 001, 000 . . .

Obtain the structural module, the test bench, and waveforms.

9.9   A serial data line x_1 consists of 3-bit words with 1 bit space between words as shown below

$$x_1 = \left| \begin{array}{ccc} b_1 & b_2 & b_3 \end{array} \right| \quad \left| \begin{array}{ccc} b_1 & b_2 & b_3 \end{array} \right| \cdots$$

where $b_i = 0$ or 1. If there are exactly two 1s in any 3-bit word, then output z_1 is asserted during the last half of the clock cycle in the bit space between words. Since z_1 is asserted at time t_2 and deasserted at time t_3, there is no need to be concerned with output glitches because the machine will have stabilized before z_1 is asserted.

Derive the state diagram, Karnaugh maps for the flip-flops, and the input equations. Design the machine using structural modeling with D flip-flops and logic gates. Instantiate all logic gates as dataflow modules. Obtain the structural module, the test bench module, and the waveforms.

9.10   Given the state diagram shown below for a Moore synchronous sequential machine, design the machine using structural modeling. The assigned state codes preclude the possibility of glitches on output z_1 because no state transition sequence will pass through state c ($y_1 y_2 y_3 = 011$). Therefore, z_1 can be asserted during the first half of the clock cycle; that is, $z_1 \uparrow t_1 \downarrow t_2$.

There are three inputs x_1, x_2, and x_3, which allow the machine to sequence through the various states in the state diagram. The machine is initialized to a state code of $y_1 y_2 y_3 = 101$. Obtain the structural module using D flip-flops, the test bench, and the waveforms. Use dataflow modeling for the logic gates and behavioral modeling for the D flip-flop.

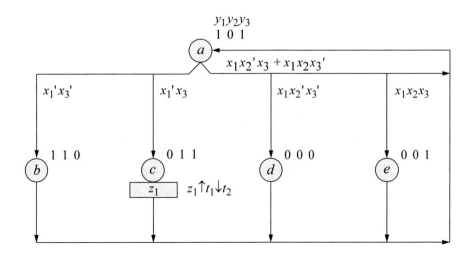

9.11   Use structural modeling to design an asynchronous sequential machine which operates according to the reduced primitive flow table shown below. Generate the merger diagram, the merged flow table, the transition diagram, the combined excitation maps, the individual excitation maps, and the output map. There must be no race conditions, no static hazards, and no momentary false outputs. Then obtain the structural module, test bench, outputs, and waveforms.

$x_1 x_2$				
0 0	0 1	1 1	1 0	z_1
$\textcircled{a}$	–	–	b	0
a	–	c	$\textcircled{b}$	0
–	–	$\textcircled{c}$	d	0
a	–	e	$\textcircled{d}$	0
–	–	$\textcircled{e}$	b	1

9.12   Design a Mealy pulse-mode asynchronous sequential machine that has three parallel inputs $x_1, x_2,$ and x_3 and one output z_1 that is asserted coincident with x_3 whenever the sequence $x_1 x_2 x_3 = 100, 010, 001$ occurs. The storage elements will consist of SR latches and positive-edge-triggered D flip-flops. The latches are to be designed using NOR gates; therefore, the reset input will be a high logic level. The D flip-flops will be reset using a low logic level.

A representative timing diagram displaying valid input sequences and corresponding outputs is shown on the next page together with the state diagram. State code assignment is arbitrary because input pulses trigger all state transitions and the machine does not begin to sequence to the next state until the input pulse, which initiated the transition, has been deasserted. Thus, output z_1 will not glitch. In order to allow the machine to stabilize between input vectors, a vector of $x_1 x_2 x_3 = 000$ is inserted between each valid input.

Obtain the structural module, the test bench for all state transitions, the outputs, and the waveforms.

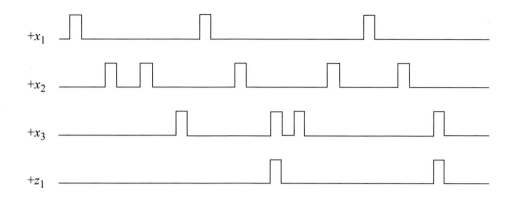

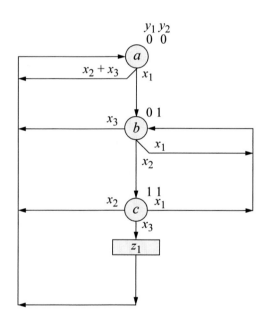

10

Tasks and Functions

Verilog provides tasks and functions that are similar to procedures or subroutines found in other programming languages. These constructs allow a behavioral module to be partitioned into smaller segments. Tasks and functions permit modules to execute common code segments that are written once then called when required, thus reducing the amount of code needed. They enhance the readability and maintainability of the Verilog modules.

Tasks and functions are defined within a module and are local to the module. They can be invoked only from within a behavior in the module. That is, they are called from an **always** block, an **initial** block, or from other tasks or functions. A function can invoke another function, but not a task. A function must have at least one **input** argument, but does not have **output** or **inout** arguments. The task and function arguments can be considered as the ports of the constructs.

10.1 Tasks

Tasks can be invoked only from within a behavior in the module. That is, they are called from an **always** block, an **initial** block, or from other tasks or functions. A task cannot be invoked from a continuous assignment statement and does not return values to an expression, but places the values on the **output** or **inout** ports. Tasks can contain delays, timing, or event control statements and can execute in nonzero simulation time when event control is applied. A task can invoke other tasks and functions and can have arguments of type **input**, **output**, or **inout**.

10.1.1 Task Declaration

A task is delimited by the keywords **task** and **endtask**. The syntax for a task declaration is as follows:

> **task** task_name
>   **input** arguments
>   **output** arguments
>   **inout** arguments
>   task declarations
>   local variable declarations
>   **begin**
>   statements
>   **end**
> **endtask**

Arguments (or parameters) that are of type **input** or **inout** are processed by the task statements; arguments that are of type **output** or **inout**, resulting from the task construct, are passed back to the task invocation statement. The keywords **input**, **output**, and **inout** are not ports of the module, they are ports used to pass values between the task invocation statement and the task construct. Additional local variables can be declared within a task, if necessary. Since tasks cannot be synthesized, they are used only in test benches. When a task completes execution, control is passed to the next statement in the module.

10.1.2 Task Invocation

A task can be invoked (or called) from a procedural statement; therefore, it must appear within an **always** or an **initial** block. A task can call itself or be invoked by tasks that it has called. The syntax for a task invocation is as follows, where the expressions are parameters passed to the task:

> task_name (expression1, expression2, . . . , expressionN);

Values for arguments of type **output** and **inout** are passed back to the variables in the task invocation statement upon completion of the task. The list of arguments in the task invocation must match the order of **input**, **output**, and **inout** variables in the task declaration. The **output** and **inout** arguments must be of type **reg** because a task invocation is a procedural statement.

Example 10.1 A task module will be generated that performs both arithmetic and logical operations. There are three inputs: *a[7:0]*, *b[7:0]*, and *c[7:0]*, where *a[0]*, *b[0]*, and *c[0]* are the low-order bits of *a*, *b*, and *c*, respectively. There are four outputs: z_1, z_2, z_3, and z_4 that perform the operations shown below.

$$z_1 = (a + b) \& (c)$$
$$z_2 = (a + b) \,|\, (c)$$
$$z_3 = (a \& b) + (c)$$
$$z_4 = (a \,|\, b) + (c)$$

Figure 10.1 is a block diagram of the module in which the task is embedded. Notice that there are no ports in the module to the external environment. The only ports are in the task which passes variables into the task declaration and results back to the task invocation.

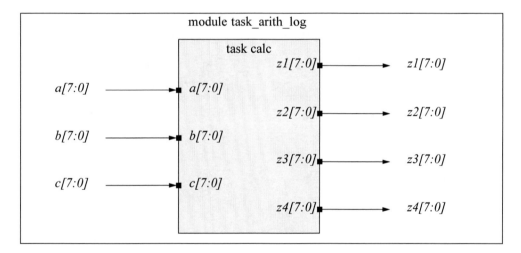

Figure 10.1   Block diagram of the task module of Example 10.1.

The task module is shown in Figure 10.2 in which no ports are listed in the module definition. The first set of variables passed to the task declaration by the task invocation are shown below.

$a = 8\text{'b0000\_1111};$ $b = 8\text{'b0011\_1100};$ $c = 8\text{'b0101\_0101};$
calc $(a, b, c, z_1, z_2, z_3, z_4);$

The variables a, b, and c are passed to the task. The variables z_1, z_2, z_3, and z_4 are the results that are passed back to the task invocation.

```
//module to illustrate a task
module task_arith_log;


reg [7:0] a, b, c;
reg [7:0] z1, z2, z3, z4;


initial
begin
      a=8'b0000_1111;   b=8'b0011_1100;   c=8'b0101_0101;
   calc (a, b, c, z1, z2, z3, z4);   //invoke task

      a=8'b1111_1111;   b=8'b0000_1100;   c=8'b1011_1101;
   calc (a, b, c, z1, z2, z3, z4);   //invoke task

      a=8'b0011_1100;   b=8'b0001_1101;   c=8'b1010_0101;
   calc (a, b, c, z1, z2, z3, z4);   //invoke task

      a=8'b1100_1001;   b=8'b1011_1101;   c=8'b0111_0111;
   calc (a, b, c, z1, z2, z3, z4);   //invoke task
end


task calc;
   input [7:0] a, b, c;
   output [7:0] z1, z2, z3, z4;

   begin
      z1 = (a + b) & (c);
      z2 = (a + b) | (c);
      z3 = (a & b) + (c);
      z4 = (a | b) + (c);
      $display ("a=%b, b=%b, c=%b, z1=%b, z2=%b, z3=%b, z4=%b",
                a, b, c, z1, z2, z3, z4);
   end
endtask

endmodule
```

Figure 10.2   Module for the task of Example 10.1.

The outputs from the task are shown in Figure 10.3. The module declares 8-bit register vectors $[7:0]$ a, $[7:0]$ b, and $[7:0]$ c. These are redeclared in the task. Output z_1 adds operands a and b and then performs a bitwise logical AND operation on the sum with operand c.

```
a=00001111, b=00111100, c=01010101,
z1=01000001, z2=01011111, z3=01100001, z4=10010100

a=11111111, b=00001100, c=10111101,
z1=00001001, z2=10111111, z3=11001001, z4=10111100

a=00111100, b=00011101, c=10100101,
z1=00000001, z2=11111101, z3=11000001, z4=11100010

a=11001001, b=10111101, c=01110111,
z1=00000110, z2=11110111, z3=00000000, z4=01110100
```

Figure 10.3   Outputs for the task module of Figure 10.2.

Example 10.2   This example illustrates a module that contains a task to perform logical operations on two 8-bit vectors $a[7:0]$ and $b[7:0]$. The logical operations are: AND, NAND, OR, NOR, exclusive-OR, and exclusive-NOR. Variables a and b are passed to the task *logical*, which returns the results as *a_and_b*, *a_nand_b*, *a_or_b*, *a_nor_b*, *a_xor_b*, and *a_xnor_b*. Figure 10.4 shows a block diagram of the task *logical* embedded in the task module *task_logical*. The task module is shown in Figure 10.5; the outputs are shown in Figure 10.6.

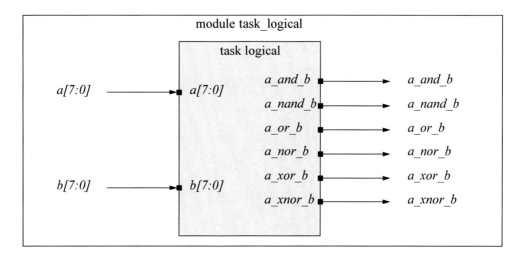

Figure 10.4   Block diagram for the task of Example 10.2.

```verilog
//module to illustrate a task for logical operations
module task_logical;


reg [7:0] a, b;
reg [7:0] a_and_b, a_nand_b, a_or_b, a_nor_b,
          a_xor_b, a_xnor_b;


initial
begin
     a=8'b1010_1010;   b=8'b1100_1100;
  logical (a, b, a_and_b, a_nand_b, a_or_b, a_nor_b,
          a_xor_b, a_xnor_b);      //invoke the task

     a=8'b1110_0111;   b=8'b1110_0111;
  logical (a, b, a_and_b, a_nand_b, a_or_b, a_nor_b,
          a_xor_b, a_xnor_b);      //invoke the task

     a=8'b0000_0111;   b=8'b0000_0111;
  logical (a, b, a_and_b, a_nand_b, a_or_b, a_nor_b,
          a_xor_b, a_xnor_b);      //invoke the task

     a=8'b0101_0101;   b=8'b1010_1010;
  logical (a, b, a_and_b, a_nand_b, a_or_b, a_nor_b,
          a_xor_b, a_xnor_b);      //invoke the task
end


task logical;
   input [7:0] a, b;
   output [7:0] a_and_b, a_nand_b, a_or_b, a_nor_b,
          a_xor_b, a_xnor_b;
   begin
     a_and_b = a & b;
     a_nand_b = ~(a & b);
     a_or_b = a | b;
     a_nor_b = ~(a | b);
     a_xor_b = a ^ b;
     a_xnor_b = ~(a ^ b);

     $display ("a=%b, b=%b, a_and_b=%b, a_nand_b=%b,
          a_or_b=%b, a_nor_b=%b, a_xor_b=%b, a_xnor_b=%b",
          a, b, a_and_b, a_nand_b, a_or_b, a_nor_b,
          a_xor_b, a_xnor_b);
end
endtask
endmodule
```

Figure 10.5   Module for the task of Example 10.2.

```
a=10101010, b=11001100,
a_and_b=10001000, a_nand_b=01110111,
a_or_b=11101110, a_nor_b=00010001,
a_xor_b=01100110, a_xnor_b=10011001

a=11100111, b=11100111,
a_and_b=11100111, a_nand_b=00011000,
a_or_b=11100111, a_nor_b=00011000,
a_xor_b=00000000, a_xnor_b=11111111

a=00000111, b=00000111,
a_and_b=00000111, a_nand_b=11111000,
a_or_b=00000111, a_nor_b=11111000,
a_xor_b=00000000, a_xnor_b=11111111

a=01010101, b=10101010,
a_and_b=00000000, a_nand_b=11111111,
a_or_b=11111111, a_nor_b=00000000,
a_xor_b=11111111, a_xnor_b=00000000
```

Figure 10.6   Outputs for the task module of Figure 10.5.

Example 10.3 This example utilizes a task to perform addition on two operands *a* and *b* with carry-in *cin*. The module is labeled *task1_adder4* and the task is labeled *add*. The operands *a*, *b*, *cin*, and *sum* are declared as registers of type **integer**. The task declaration lists the inputs and output which are ports of the task, and declares them as integers. The **$display** system task displays the variables using decimal notation. The block diagram is shown in Figure 10.7, the module is shown in Figure 10.8, and the outputs are shown in Figure 10.9.

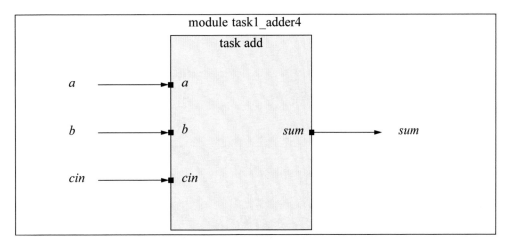

Figure 10.7   Block diagram for the task of Example 10.3.

```verilog
//module to illustrate the use of a task
module task1_adder4;

integer a, b, cin, sum;

initial
begin
      a=2;  b=5;  cin=0;
   add (a, b, cin, sum);    //invoke the task

      a=3;  b=2;  cin=1;
   add (a, b, cin, sum);    //invoke the task

      a=4;  b=6;  cin=1;
   add (a, b, cin, sum);    //invoke the task

      a=14; b=63; cin=1;
   add (a, b, cin, sum);    //invoke the task

      a=150; b=225; cin=0;
   add (a, b, cin, sum);    //invoke the task
end

task add;
   input a;
   input b;
   input cin;
   output sum;

   integer a, b, cin, sum;

   begin
      sum = a + b + cin;
      $display ("a=%d, b=%d, cin=%d, sum=%d", a, b, cin, sum);
   end
endtask
endmodule
```

Figure 10.8   Module for the task of Example 10.3.

```
a= 2,    b= 5,    cin= 0, sum= 7
a= 3,    b= 2,    cin= 1, sum= 6
a= 4,    b= 6,    cin= 1, sum= 11
a= 14,   b= 63,   cin= 1, sum= 78
a= 150, b= 225, cin= 0, sum= 375
```

Figure 10.9   Outputs for the task module of Figure 10.8.

10.2 Functions

Functions are similar to tasks, except that functions return only a single value to the expression from which they are called. Like tasks, functions provide the ability to execute common procedures from within a module. A function can be invoked from a continuous assignment statement or from within a procedural statement and is represented by an operand in an expression.

Functions cannot contain delays, timing, or event control statements and execute in zero simulation time. Although functions can invoke other functions, they are not recursive. Functions cannot invoke a task. Functions must have at least one **input** argument, but cannot have **output** or **inout** arguments.

10.2.1 Function Declaration

The syntax for a function declaration is shown below. If the optional *range or type* is omitted, the value returned to the function invocation is a scalar of type **reg**. Functions are delimited by the keywords **function** and **endfunction** and are used to implement combinational logic; therefore, functions cannot contain event controls or timing controls.

> **function** [range or type] function name
> > **input** declaration
> > other declarations
> > **begin**
> > > statement
> > **end**
> **endfunction**

10.2.2 Function Invocation

A function is invoked from an expression. The function is invoked by specifying the function name together with the input parameters. The syntax is shown below.

> function name (expression1, expression2, . . . , expressionN);

All local registers that are declared within a function are static; that is, they retain their values between invocations of the function. When the function execution is finished, the return value is positioned at the location where the function was invoked.

Example 10.4   This example calculates the parity of a 16-bit address and returns 1 bit indicating whether the parity is even (1) or odd (0). If the parity is even, then *parity is* *even* is printed; if parity is odd, then *parity is odd* is printed. The function module, like tasks, has no ports to communicate with the external environment. The only ports are input ports that receive parameters from the function invocation.

Figure 10.10 shows the block diagram of the module *fctn_parity* with the function *calc_parity* embedded in the module. The module is shown in Figure 10.11. The variable *addr* is declared as a 16-bit register; the variable *parity* is a scalar register. The statement

$$parity = calc\_parity \ (16'b1111\_0000\_1111\_0000);$$

invokes the function *calc_parity* and passes the address (1111_0000_1111_0000) to the function input port *[15:0] address*. Then the function uses the reduction exclusive-OR operator to determine the parity of the address. The parity of the address is returned to the left-hand side of the *parity* statement, then the parity is displayed. The outputs of the module are shown in Figure 10.12.

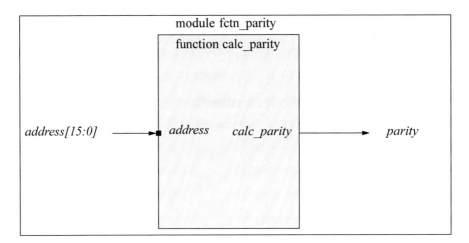

Figure 10.10   Block diagram for the function of Example 10.4.

```
//module to illustrate a function
module fctn_parity;

reg [15:0] addr;
reg parity;

//continued on next page
```

Figure 10.11   Module for the function of Example 10.4.

```
initial
begin
   parity = calc_parity (16'b1111_0000_1111_0000);
      if (parity ==1)
         $display ("parity is even");
      else
         $display ("parity is odd");


   parity = calc_parity (16'b1111_0000_1111_0001);
      if (parity ==1)
         $display ("parity is even");
      else
         $display ("parity is odd");
end


function calc_parity;
   input [15:0] address;
   begin
      calc_parity = ^address;
   end
endfunction

endmodule
```

Figure 10.11   (Continued)

```
parity is even
parity is odd
```

Figure 10.12   Outputs for the function module of Figure 10.11.

Example 10.5   A module will be designed using the function construct to count the number of 1s in an 8-bit word. The module is called *fctn_count1s* and the function is called *count1s*. Since there can be up to eight 1s in the word, a global register is declared as *[3:0] count* to contain up to four bits. The code shown below invokes the functon *count1s* and passes an 8-bit argument (*8'b1100_0011*) to the function. The returned value is placed on the right-hand side of the expression, then the number of 1s in the variable *count* is displayed by the **$display** system task.

$$count = count1s \ (8'b1100\_0011);$$
$$\textbf{\$display} \ (``number \ of \ 1s = \%d", count);$$

The function module is shown in Figure 10.13 and the outputs are shown in Figure 10.14. A local variable *[3:0] cnt* is declared within the function. The returned value is assigned to the global variable *[3:0] count*. The **while** loop executes as long as *reg_a* contains at least one 1 bit. When the function has ended execution, the return value is placed where the function was invoked [*count = count1s (8'b1100_0011);*]. Notice that the return (*count1s = cnt*) to the invocation is placed outside the **while** loop.

```
//module to illustrate a function
module fctn_count1s;
reg [3:0] count;

//invoke the function
initial
begin
   count = count1s (8'b1100_0011);
      $display ("number of 1s = %d", count);


   count = count1s (8'b1101_0011);
      $display ("number of 1s = %d", count);


   count = count1s (8'b1111_1011);
      $display ("number of 1s = %d", count);


   count = count1s (8'b1010_1010);
      $display ("number of 1s = %d", count);


   count = count1s (8'b0000_0000);
      $display ("number of 1s = %d", count);


   count = count1s (8'b0000_0010);
      $display ("number of 1s = %d", count);


   count = count1s (8'b1111_1111);
      $display ("number of 1s = %d", count);


   count = count1s (8'b0110_0100);
      $display ("number of 1s = %d", count);
end
//continue on next page
```

Figure 10.13   Module for a function to count the number of 1s in a word.

```
function [3:0] count1s;
input [7:0] reg_a;
reg [3:0] cnt;
begin
   cnt = 0;
   while (reg_a)
      begin
         cnt = cnt + reg_a[0];
         reg_a = reg_a >> 1;
      end
   count1s = cnt;
end
endfunction

endmodule
```

Figure 10.13   (Continued)

```
number of 1s = 4
number of 1s = 5
number of 1s = 7
number of 1s = 4
number of 1s = 0
number of 1s = 1
number of 1s = 8
number of 1s = 3
```

Figure 10.14   Outputs for a function to count the number of 1s in a word.

Refer to the **while** loop in Figure 10.13. If *reg_a* contains at least one 1 bit, then the **while** loop returns a value of 1, indicating a true condition. The loop is repeated and *reg_a* is shifted right one bit position with a zero filling the leftmost vacated position. The process repeats until *reg_a* contains all zeroes at which time control passes out of the loop to the next statement.

Example 10.6   This example will use a function to implement a full adder using the **case** statement. Figure 10.15 shows the block diagram for the full adder module *fctn_full_adder* which contains the full adder function *full_add*. The only ports are those in the function — there are no ports in the module to the external environment. Operands *a*, *b*, and *cin* are scalar inputs to the function; *sum* is a 3-bit vector of type **reg** that is returned to function invocation *full_add* where it is assigned to the variable *sum*, which is then displayed.

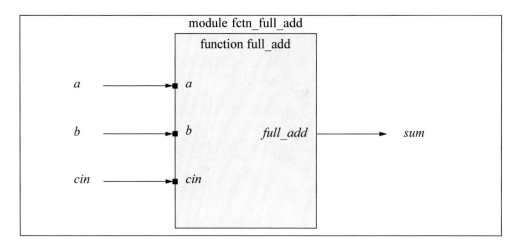

Figure 10.15   Block diagram for the full adder function module.

The function module is shown in Figure 10.16 where *a*, *b*, *cin*, and *sum* are declared as registers. The function invocation is *full_add* which passes scalar variables *a*, *b*, and *cin* to the function. The variable *sum* is the result of the **case** statement and is declared as type **reg**, which is passed back to the function call, assigned to *sum*, and displayed.

Since functions can be synthesized, using the **case** construct to perform the add operation will cause the synthesis tool to construct a multiplexer instead of a full adder. The functionality, however, is identical. The outputs are shown in Figure 10.17.

```
//module for a full adder using a function
module fctn_full_add;

reg a, b, cin;
reg [1:0] sum;

initial
begin
   sum = full_add (1'b0, 1'b0, 1'b0);   //invoke the function
      $display ("abcin=000, cout, sum = %b", sum);

   sum = full_add (1'b0, 1'b0, 1'b1);
      $display ("abcin=001, cout, sum = %b", sum);

//continued on next page
```

Figure 10.16   Module for the full adder function of Example 10.6.

```verilog
      sum = full_add (1'b0, 1'b1, 1'b0);
         $display ("abcin=010, cout, sum = %b", sum);


      sum = full_add (1'b0, 1'b1, 1'b1);
         $display ("abcin=011, cout, sum = %b", sum);


      sum = full_add (1'b1, 1'b0, 1'b0);
         $display ("abcin=100, cout, sum = %b", sum);


      sum = full_add (1'b1, 1'b0, 1'b1);
         $display ("abcin=101, cout, sum = %b", sum);


      sum = full_add (1'b1, 1'b1, 1'b0);
         $display ("abcin=110, cout, sum = %b", sum);


      sum = full_add (1'b1, 1'b1, 1'b1);
         $display ("abcin=111, cout, sum = %b", sum);

end


function [2:0] full_add;
input a, b, cin;
reg [1:0] sum;

begin
   case ({a,b,cin})
      3'b000: sum = 2'b00;
      3'b001: sum = 2'b01;
      3'b010: sum = 2'b01;
      3'b011: sum = 2'b10;
      3'b100: sum = 2'b01;
      3'b101: sum = 2'b10;
      3'b110: sum = 2'b10;
      3'b111: sum = 2'b11;
      default:sum = 2'bxx;
   endcase

      full_add = sum;
end
endfunction


endmodule
```

Figure 10.16   (Continued)

```
abcin=000, cout, sum = 00
abcin=001, cout, sum = 01
abcin=010, cout, sum = 01
abcin=011, cout, sum = 10
abcin=100, cout, sum = 01
abcin=101, cout, sum = 10
abcin=110, cout, sum = 10
abcin=111, cout, sum = 11
```

Figure 10.17   Outputs for the full adder function module of Figure 10.16.

Example 10.7   This example designs a function to perform a shift operation on an 8-bit word under control of a scalar shift direction variable *ctrl*. If *ctrl* = 0, then the word is shifted left one bit position; if *ctrl* = 1, then the word is shifted right one bit position. The variables *[7:0] left_word* and *[7:0] right_word* contain the words that are to be shifted and are declared as type **reg**. They invoke the function *shift*.

There are two inputs to the function: *[7:0] word*, which contains the word to be shifted, and *ctrl*, which determines the shift direction. The conditional operator (**? :**) is used to evaluate the shift control variable. The shifted word is returned to the task invocation *shift* and is assigned to *left_word* or *right_word*, and then displayed. The function module is shown in Figure 10.18 and the outputs are shown in Figure 10.19.

```
//module for a shifter using a function
module fctn_shifter;

reg [7:0] left_word, right_word;
reg ctrl;

initial
begin
   left_word = shift (8'b1010_1010, 1'b0);//invoke function
      $display ("word = %b", left_word);

   right_word = shift (8'b1100_0011, 1'b1);
      $display ("word = %b", right_word);

   left_word = shift (8'b0000_1111, 1'b0);
      $display ("word = %b", left_word);

//continued on next page
```

Figure 10.18   Module for a function to shift a word left or right 1 bit position.

```
    right_word = shift (8'b0000_1111, 1'b1);
       $display ("word = %b", right_word);

   left_word = shift (8'b0010_0000, 1'b0);
       $display ("word = %b", left_word);

   right_word = shift (8'b0010_0000, 1'b1);
       $display ("word = %b", right_word);
end

function [7:0] shift;
input [7:0] word;
input ctrl;
   begin
       shift = (ctrl == 1'b0) ? (word << 1) : (word >> 1);
   end
endfunction

endmodule
```

Figure 10.18   (Continued)

```
word = 01010100
word = 01100001
word = 00011110
word = 00000111
word = 01000000
word = 00010000
```

Figure 10.19   Outputs for the shift function module of Figure 10.18.

Example 10.8   This example uses a function to design a simple 4-operation arithmetic and logic unit (ALU). The operations are: add, subtract, multiply, and divide. The module is called *fctn_alu* and the function is called *alu*. There are two 8-bit operands a and b and one result *rslt*. The result is 16 bits to accommodate the product of two 8-bit operands because the product is always $2n$ bits for two n-bit operands. There is also a 2-bit mode control input *mode* for the function to specify the operation to be performed.

The block diagram is shown in Figure 10.20, the module is shown in Figure 10.21, and the outputs are shown in Figure 10.22. In the second subtract operation, $a = 15$ and $b = 16$. Therefore, the result is -1, which is 1111_1111_1111_1111 in 2s complement representation. In the second divide operation, $a = 55$ and $b = 7$. The quotient is 7.857142857, which is truncated to 7. The function returns the result to the task invocation *alu* and assigns it to *rslt*, which is then displayed.

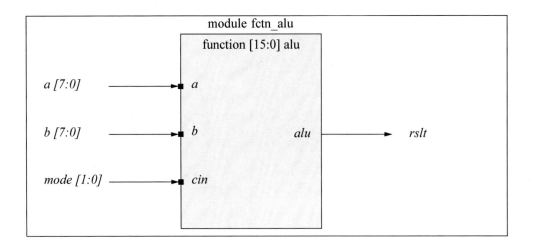

Figure 10.20   Block diagram for the 4-operation ALU of Example 10.8.

```
//module for an arithmetic and logic unit using a function
module fctn_alu;

reg [7:0] a, b;
reg [1:0] mode;
reg [15:0] rslt;

initial
begin
   rslt = alu (8'b0000_1111, 8'b0000_0011, 2'b00);
   $display ("add, a=0000_1111, b=0000_0011, rslt = %b", rslt);

   rslt = alu (8'b0011_1101, 8'b0010_0011, 2'b00);
   $display ("add, a=0011_1101, b=0010_0011, rslt = %b", rslt);

   rslt = alu (8'b0000_1111, 8'b0000_0011, 2'b01);
   $display ("sub, a=0000_1111, b=0000_0011, rslt = %b", rslt);

   rslt = alu (8'b0000_1111, 8'b0001_0000, 2'b01);
   $display ("sub, a=0000_1111, b=0001_0000, rslt = %b", rslt);

   rslt = alu (8'b0000_1111, 8'b0000_0011, 2'b10);
   $display ("mul, a=0000_1111, b=0000_0011, rslt = %b", rslt);

//continued on next page
```

Figure 10.21   Module to implement a function for a 4-operation ALU.

```
      rslt = alu (8'b0000_1111, 8'b0001_1001, 2'b10);
      $display ("mul, a=0000_1111, b=0001_1001, rslt = %b", rslt);


      rslt = alu (8'b0000_1111, 8'b0000_0011, 2'b11);
      $display ("div, a=0000_1111, b=0000_0011, rslt = %b", rslt);


      rslt = alu (8'b0011_0111, 8'b0000_0111, 2'b11);
      $display ("div, a=0011_0111, b=0000_0111, rslt = %b", rslt);

end

function [15:0] alu;
input [7:0] a, b;
input [1:0] mode;
reg [15:0] rslt;

begin
   case (mode)
      2'b00:   rslt = a + b;
      2'b01:   rslt = a - b;
      2'b10:   rslt = a * b;
      2'b11:   rslt = a / b;
      default: rslt = 16'bxxxx_xxxx_xxxx_xxxx;
   endcase

   alu = rslt;
end
endfunction

endmodule
```

Figure 10.21   (Continued)

```
add, a=0000_1111, b=0000_0011, rslt = 0000000000010010
add, a=0011_1101, b=0010_0011, rslt = 0000000001100000
sub, a=0000_1111, b=0000_0011, rslt = 0000000000001100
sub, a=0000_1111, b=0001_0000, rslt = 1111111111111111
mul, a=0000_1111, b=0000_0011, rslt = 0000000000101101
mul, a=0000_1111, b=0001_1001, rslt = 0000000101110111
div, a=0000_1111, b=0000_0011, rslt = 0000000000000101
div, a=0011_0111, b=0000_0111, rslt = 0000000000000111
```

Figure 10.22   Outputs for Figure 10.21, which implements a 4-operation ALU.

10.3   Problems

10.1   Design a module that contains a task to detect even or odd parity in a 16-bit register *reg_a*. The task returns a value of 0 if the parity is even and a value of 1 if the parity is odd. The task declaration returns the parity value to an integer called *parity*. Obtain the outputs from the module.

10.2   Design a module that contains a task to count the number of 1s in an 8-bit register *reg_a*. The task returns the number of 1s to a 4-bit register *count*. Obtain the outputs from the module.

10.3   Design a module that contains a task to add two operands *a* and *b*, and a carry-in *cin*. Obtain the binary outputs from the module.

10.4   Design a module that contains a task to obtain the sum, difference, product, and quotient according to the equations shown below. There is an integer *int* that is initialized to a value of 1 and a variable *sum* that is initialized to a value of 0 — both are initialized prior to the task invocation. Assign numbers *num* of 1, 2, 4, and 8 that are passed to the task declaration by four task invocations. Obtain the outputs of the module as decimal numbers.

> sum = sum + int;
> diff = int − sum;
> prod = sum * int;
> quot = prod / sum;
> int = in + num;

10.5   Design a function for a half adder that adds operands *a* and *b* and produces a result *sum*. The **case** statement can be used in the function. Display the results.

10.6   Design a module that contains a function to perform the logical operations of AND, NAND, OR, NOR, exclusive-OR, and exclusive-NOR. Display the outputs. A 3-bit mode control is passed to the function, together with the two operands, to specify the function to be performed.

10.7   Design a function to convert from the binary code to the excess-3 code. Display the outputs.

11

Additional Design Examples

This chapter will present several design examples utilizing the modeling methods presented in previous chapters. The examples will include a Johnson counter using structural modeling; a counter-shifter using structural modeling which shifts different patterns; a universal shift register using mixed-design modeling; a Hamming code error detection and correction module using dataflow modeling; a module to illustrate the Booth multiply algorithm using mixed-design modeling; various Moore and Mealy machines, including a Mealy one-hot synchronous sequential machine using structural modeling; a binary-coded decimal (BCD) adder/subtractor, and a pipelined reduced instruction set computer (RISC) processor.

11.1 Johnson Counter

A 3-bit Johnson counter will be designed using *D* flip-flops as the storage elements. The counter of Figure 11.1 represents a *Johnson* counter in which any two contiguous state codes (or code words) differ by only one variable. It is similar, in this respect, to a Gray code counter.

The Johnson counter is also referred to as a Möbius counter after August F. Möbius, a German mathematician who discovered a one-sided surface constructed from a rectangle by holding one end fixed, rotating the opposite end through 180 degrees, and applying it to the first end. The last stage output of a Johnson counter is inverted and fed back to the first stage.

Using the design procedure described in previous chapters, the input maps are obtained using D flip-flops, as shown in Figure 11.2. The maps can be derived directly from the state diagram without the necessity of generating a next-state table. For example, from state b ($y_1y_2y_3 = 100$), the machine sequences to state c ($y_1y_2y_3 = 110$) where the next state for flip-flop y_1 is 1. Thus, a 1 is entered in minterm location $y_1y_2y_3 = 100$ for flip-flop y_1. Likewise, from state c the machine proceeds to state d where the next state for y_1 is 1; therefore, a 1 is entered in minterm location $y_1y_2y_3 = 110$ for flip-flop y_1. In a similar manner, the remaining entries are obtained for the input map for y_1, as well as for the input maps for y_2 and y_3. The input equations are listed in Equation 11.1.

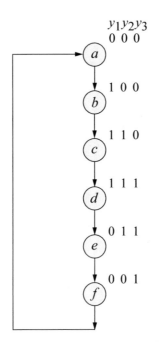

Figure 11.1   State diagram for a Johnson counter with a nonsequential counting sequence. There are two unused states, $y_1y_2y_3 = 010$ and 101.

The logic diagram is shown in Figure 11.3. The counting sequence is easily verified by asserting the appropriate input logic levels to the flip-flop D inputs for each state of the counter and then applying the active clock transition.

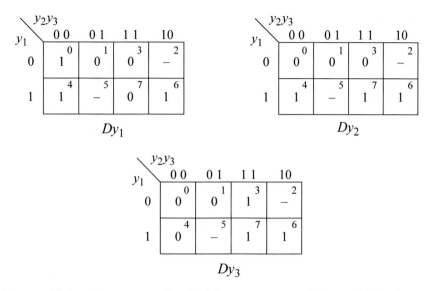

Figure 11.2 Input maps for the Johnson counter of Figure 11.1 using D flip-flops. The unused states are $y_1 y_2 y_3 = 010$ and 101.

$$Dy_1 = y_3'$$
$$Dy_2 = y_1$$
$$Dy_3 = y_2 \qquad\qquad (11.1)$$

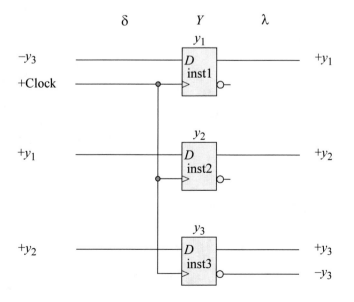

Figure 11.3 Logic diagram for the Johnson counter of Figure 11.1 using D flip-flops.

The structural module is shown in Figure 11.4 in which the behavioral module for a *D* flip-flop is instantiated three times. The instance names correlate to those in the logic diagram. The test bench is shown in Figure 11.5. The outputs are shown in Figure 11.6, which also displays the simulation time. The waveforms are shown in Figure 11.7.

```verilog
//structural 3-bit johnson counter
module ctr_johnson3 (clk, rst_n, y1, y2, y3, y3_n);

input clk, rst_n;
output y1, y2, y3, y3_n;

wire y1, y2, y3, y1_n, y2_n, y3_n;

//instantiate D flip-flop for y1
d_ff inst1 (
   .d(~y3),
   .clk(clk),
   .q(y1),
   .q_n(),
   .set_n(set_n),
   .rst_n(rst_n)
   );

//instantiate D flip-flop for y2
d_ff inst2 (
   .d(y1),
   .clk(clk),
   .q(y2),
   .q_n(),
   .set_n(set_n),
   .rst_n(rst_n)
   );

//instantiate D flip-flop for y3
d_ff inst3 (
   .d(y2),
   .clk(clk),
   .q(y3),
   .q_n(y3_n),
   .set_n(set_n),
   .rst_n(rst_n)
   );

endmodule
```

Figure 11.4   Structural module for the 3-bit Johnson counter of Figure 11.3.

```
//3-bit Johnson counter test bench
module ctr_johnson3_tb;


reg clk, rst_n;
wire y1, y2, y3, y3_n;


//display outputs at simulation time
initial
$monitor ($time, "Count = %b", {y1, y2, y3});


//define reset
initial
begin
   #0  rst_n = 1'b0;
   #5  rst_n = 1'b1;
end


//define clk
initial
begin
   clk = 1'b0;
   forever
      #10 clk = ~clk;
end

//finish simulation at time 200
initial
begin
   #200 $finish;
end

//instantiate the module into the test bench
ctr_johnson3 inst1 (
   .clk(clk),
   .rst_n(rst_n),
   .y1(y1),
   .y2(y2),
   .y3(y3),
   .y3_n(y3_n)
   );

endmodule
```

Figure 11.5   Test bench for the 3-bit Johnson counter of Figure 11.4.

```
0    Count = 000
10   Count = 100
30   Count = 110
50   Count = 111
70   Count = 011
90   Count = 001
110  Count = 000
130  Count = 100
150  Count = 110
170  Count = 111
190  Count = 011
```

Figure 11.6   Outputs for the 3-bit Johnson counter of Figure 11.4.

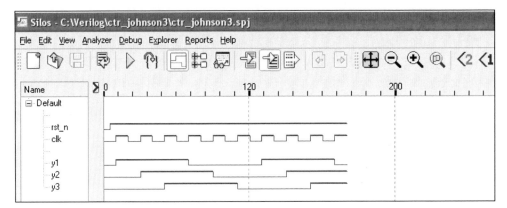

Figure 11.7   Waveforms for the 3-bit Johnson counter of Figure 11.4.

11.2 Counter-Shifter

This section presents a counter-shifter application using structural modeling in which different patterns are shifted through the shift register depending on pattern select inputs. When the counter reaches a count of seven, a pulse is applied to the shift register clock input and one of two patterns is directed to the serial input of the shift register. If the *sngl_1* input is active, then the following pattern is shifted: 000, 100, 010, 001, 000, ... If the *dbl_1* input is active, then the following pattern is shifted: 000, 100, 110, 011, 001, 000, ...

   The logic diagram is shown in Figure. 11.8. A 3-bit counter and a 3-bit shift register will be designed for instantiation into the structural module. These modules are shown in Figure 11.9 and Figure 11.10, respectively. The structural module for the counter-shifter application is shown in Figure 11.11, the test bench in Figure 11.12, and the outputs are shown in Figure 11.13.

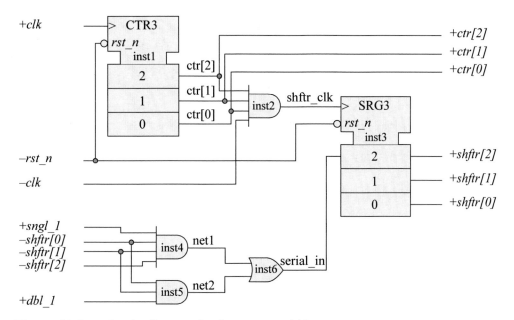

Figure 11.8   Logic diagram for the counter-shifter.

```
//behavioral 3-bit binary counter
module ctr_bin3 (clk, rst_n, ctr);

input clk, rst_n;
output [2:0] ctr;

wire clk, rst_n;           //inputs are wire
reg [2:0] ctr, next_cnt;   //outputs are reg

//latch next count
always @ (posedge clk or negedge rst_n)
begin
    if (rst_n == 0)
        ctr <= 3'b000;
    else
        ctr <= ctr + 1;
end
endmodule
```

Figure 11.9   Module for a 3-bit binary counter.

```
//behavioral 3-bit shifter
module shftr3 (clk, rst_n, serial_in, shftr);

input clk, rst_n, serial_in;
output [2:0] shftr;

wire clk, rst_n, serial_in;
reg [2:0] shftr;

//establish reset and shift operation
always @ (posedge clk or rst_n)
begin
   if (rst_n == 0)
         shftr <= 3'b000;
   else
         shftr <= {serial_in, shftr [2:1]};
end
endmodule
```

Figure 11.10   Module for a 3-bit shift register.

```
//structural counter shifter
module ctr_shftr_struct (clk, rst_n, sngl_1, dbl_1,
                              ctr, shftr);

input clk, rst_n, sngl_1, dbl_1;
output [2:0] ctr, shftr;

//The following statements are not necessary because
//signals are wire by default.  They have been added
//as a reminder that input/output signals are wires in
//structural modeling.
wire clk, rst_n, sngl_1, dbl_1;
wire [2:0] ctr, shftr;

//instantiate the counter
ctr_bin3 inst1 (
   .clk(clk),
   .rst_n(rst_n),
   .ctr(ctr)
   );

//continued on next page
```

Figure 11.11   Structural module for the counter-shifter of Section 11.2.

```
//instantiate the shifter clk logic
and4_df inst2 (
   .x1(ctr[0]),
   .x2(ctr[1]),
   .x3(ctr[2]),
   .x4(~clk),
   .z1(shftr_clk)
   );


//instantiate the pattern select logic
and4_df inst4 (
   .x1(sngl_1),
   .x2(~shftr[0]),
   .x3(~shftr[1]),
   .x4(~shftr[2]),
   .z1(net1)
   );


and3_df inst5 (
   .x1(~shftr[0]),
   .x2(~shftr[1]),
   .x3(dbl_1),
   .z1(net2)
   );


or2_df inst6 (
   .x1(net1),
   .x2(net2),
   .z1(serial_in)
   );


//instantiate the shifter
shftr3 inst3 (
   .clk(shftr_clk),
   .rst_n(rst_n),
   .serial_in(serial_in),
   .shftr(shftr)
   );

endmodule
```

Figure 11.11   (Continued)

```verilog
//test bench for the counter shifter
module ctr_shftr_struct_tb;


reg clk, rst_n, sngl_1, dbl_1, shftr_clk;
wire [2:0] ctr, shftr;


initial
$monitor ("ctr=%b, sngl_1=%b, dbl_1=%b, shftr=%b",
          ctr, sngl_1, dbl_1, shftr);

//define clk
initial
begin
   clk = 1'b0;
   forever
      #10 clk = ~clk;
end


//define reset and pattern
initial
begin
   #0     rst_n  = 1'b0;
          sngl_1 = 1'b1;
          dbl_1  = 1'b0;

   #5     rst_n  = 1'b1;

   #640   sngl_1 = 1'b0;
          dbl_1  = 1'b1;

   #900   $stop;
end


//instantiate the module into the test bench
ctr_shftr_struct inst1 (
   .clk(clk),
   .rst_n(rst_n),
   .ctr(ctr),
   .sngl_1(sngl_1),
   .dbl_1(dbl_1),
   .shftr(shftr)
   );

endmodule
```

Figure 11.12   Test bench for the structural module of Figure 11.11.

```
SINGLE 1                                 DOUBLE 1

ctr=000, sngl_1=1, dbl_1=0, shftr=000
ctr=001, sngl_1=1, dbl_1=0, shftr=000    ctr=000, sngl_1=0, dbl_1=1, shftr=000
ctr=010, sngl_1=1, dbl_1=0, shftr=000    . . .
ctr=011, sngl_1=1, dbl_1=0, shftr=000    ctr=111, sngl_1=0, dbl_1=1, shftr=100
ctr=100, sngl_1=1, dbl_1=0, shftr=000    ------------------------------------
ctr=101, sngl_1=1, dbl_1=0, shftr=000    ctr=000, sngl_1=0, dbl_1=1, shftr=100
ctr=110, sngl_1=1, dbl_1=0, shftr=000    . . .
ctr=111, sngl_1=1, dbl_1=0, shftr=100    ctr=111, sngl_1=0, dbl_1=1, shftr=110
------------------------------------     ------------------------------------
ctr=000, sngl_1=1, dbl_1=0, shftr=100    ctr=000, sngl_1=0, dbl_1=1, shftr=110
. . .                                    . . .
ctr=111, sngl_1=1, dbl_1=0, shftr=010    ctr=111, sngl_1=0, dbl_1=1, shftr=011
------------------------------------     ------------------------------------
ctr=000, sngl_1=1, dbl_1=0, shftr=010    ctr=000, sngl_1=0, dbl_1=1, shftr=011
. . .                                    . . .
ctr=111, sngl_1=1, dbl_1=0, shftr=001    ctr=111, sngl_1=0, dbl_1=1, shftr=001
------------------------------------     ------------------------------------
ctr=000, sngl_1=1, dbl_1=0, shftr=001    ctr=000, sngl_1=0, dbl_1=1, shftr=001
. . .                                    . . .
ctr=111, sngl_1=1, dbl_1=0, shftr=000    ctr=111, sngl_1=0, dbl_1=1, shftr=000
------------------------------------     ------------------------------------
```

Figure 11.13   Outputs for the counter-shifter module of Figure 11.11.

11.3 Universal Shift Register

A universal shift register will be designed using behavioral modeling. The shift register will execute the four functions listed in Table 11.1. The shift right and shift left operations are logical shifts; that is, zeros fill the high-order vacated positions during a right shift and fill the low-order vacated positions during a left shift. The block diagram is shown in Figure 11.14(a) and a more detailed diagram is shown in Figure 11.14(b). Since behavioral modeling is used, the logic design details are left to the synthesis tool.

Table 11.1 Functions for the Universal Shift Register

Function	Function Code	Shift Amount (right/left)
NOP (No operation)	00	00
Shift right	01	01
Shift left	10	10
Load	11	11

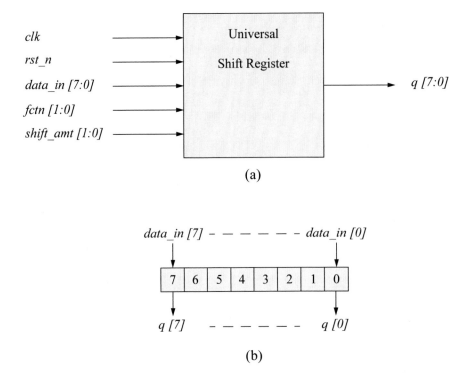

(a)

(b)

Figure 11.14   Universal shift register.

The mixed-design module is shown in Figure 11.15 using the **parameter** keyword to define the function codes and the **case** statement to perform the operations. The test bench, which applies the four functions to the shift register together with various data vectors and shift amounts, is shown in Figure 11.16. The waveforms are shown in Figure 11.17.

```
//mixed-design universal shift register
module shift_reg1 (clk, rst_n, data_in, fctn, shift_amt, q);

input clk, rst_n;
input [7:0] data_in;
input [1:0] fctn, shift_amt;

output [7:0] q;

reg [7:0] q;
wire [7:0] shift_l_data, shift_r_data;
reg [7:0] d;                          //continued on next page
```

Figure 11.15   Mixed-design module for the universal shift register of Figure 11.14.

```
parameter    nop = 2'b00;
parameter    shr = 2'b01;
parameter    shl = 2'b10;
parameter    ld  = 2'b11;

assign shift_l_data = q << shift_amt;
assign shift_r_data = q >> shift_amt;

always @ (shift_l_data or shift_r_data or fctn or data_in or q)
begin
   case (fctn)
      nop: d = q;
      shr: d = shift_r_data;
      shl: d = shift_l_data;
      ld : d = data_in;
      default: d = 8'h00;
   endcase
end

always @ (posedge clk or negedge rst_n)
begin
   if (~rst_n)
      q <= 8'h00;
   else
      q <= d;
end
endmodule
```

Figure 11.15   (Continued)

```
//test bench for the universal shift register
module shift_reg1_tb;
reg clk, rst_n;
reg [7:0] data_in;
reg [1:0] fctn, shift_amt;
wire [7:0] q;

initial      //display variables
$monitor ("data_in=%b, fctn=%b, shift_amt=%b, q=%b",
          data_in, fctn, shift_amt, q);
initial      //define clock
begin
   clk = 1'b0;
   forever
      #10 clk = ~clk;
end          //continued on next page
```

Figure 11.16   Test bench for the universal shift register of Figure 11.15.

```
initial              //define operand, fctn, and shift amount
begin
   #0 rst_n = 1'b0;
      data_in = 8'b0000_0000;
      fctn = 2'b00;
      shift_amt = 2'b00;
   #10 rst_n = 1'b1;
//shift right 0, 1, 2, and 3*******************************
   data_in = 8'b1111_0000;
   fctn = 2'b11;
   @ (posedge clk)          //load q

   fctn = 2'b01;
   shift_amt = 2'b00;
   @ (posedge clk)          //shift right 0
//------------------------------------------------------------
   data_in = 8'b1111_0000;
   fctn = 2'b11;
   @ (posedge clk)          //load q

   fctn = 2'b01;
   shift_amt = 2'b01;
   @ (posedge clk)          //shift right 1
//------------------------------------------------------------
   data_in = 8'b1111_0000;
   fctn = 2'b11;
   @ (posedge clk)          //load q

   fctn = 2'b01;
   shift_amt = 2'b10;
   @ (posedge clk)          //shift right 2
//------------------------------------------------------------
   data_in = 8'b1111_0000;
   fctn = 2'b11;
   @ (posedge clk)          //load q

   fctn = 2'b01;
   shift_amt = 2'b11;
   @ (posedge clk)          //shift right 3
//------------------------------------------------------------
   data_in = 8'b1111_1111;
   fctn = 2'b11;
   @ (posedge clk)          //load q

   fctn = 2'b01;
   shift_amt = 2'b01;
   @ (posedge clk)       //shift right 1 (0111_1111), next pg
```

Figure 11.16   (Continued)

```
   fctn = 2'b01;          //use previous q
   shift_amt = 2'b10;
   @ (posedge clk)        //shift right 2 (0001_1111)

   fctn = 2'b01;          //use previous q
   shift_amt = 2'b11;
   @ (posedge clk)        //shift right 3 (0000_0011)
//shift left 0, 1, 2, and 3 *********************************
   data_in = 8'b0000_1111;
   fctn = 2'b11;
   @ (posedge clk)        //load q

   fctn = 2'b10;
   shift_amt = 2'b00;
   @ (posedge clk)        //shift left 0
//-------------------------------------------------------
   data_in = 8'b0000_1111;
   fctn = 2'b11;
   @ (posedge clk)        //load q

   fctn = 2'b10;
   shift_amt = 2'b01;
   @ (posedge clk)        //shift left 1
//-------------------------------------------------------
   data_in = 8'b0000_1111;
   fctn = 2'b11;
   @ (posedge clk)        //load q

   fctn = 2'b10;
   shift_amt = 2'b10;
   @ (posedge clk)        //shift left 2
//-------------------------------------------------------
   data_in = 8'b0000_1111;
   fctn = 2'b11;
   @ (posedge clk)        //load q

   fctn = 2'b10;
   shift_amt = 2'b11;
   @ (posedge clk)        //shift left 3
//-------------------------------------------------------
   data_in = 8'b1111_1111;
   fctn = 2'b11;
   @ (posedge clk)        //load q

   fctn = 2'b10;
   shift_amt = 2'b01;
   @ (posedge clk)        //shift left 1 (1111_1110).  next pg
```

Figure 11.16   (Continued)

```
   fctn = 2'b10;
   shift_amt = 2'b10;    //use previous q
   @ (posedge clk)       //shift left 2 (1111_1000)

   fctn = 2'b10;
   shift_amt = 2'b11;    //use previous q
   @ (posedge clk)       //shift left 3 (1100_0000)

   #40 $stop;

end


////instantiate the module into the test bench
shift_reg1 inst1 (
   .clk(clk),
   .rst_n(rst_n),
   .data_in(data_in),
   .q(q),
   .fctn(fctn),
   .shift_amt(shift_amt)
   );

endmodule
```

Figure 11.16   (Continued)

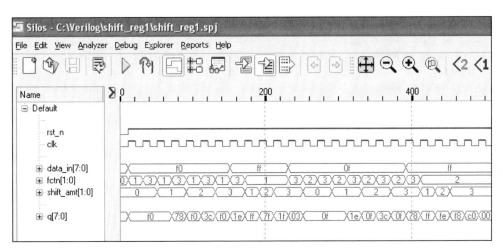

Figure 11.17   Waveforms for the universal shift register of Figure 11.15.

11.4 Hamming Code Error Detection and Correction

A common error detection technique is to add a parity bit to the message being transmitted. Thus, single-bit errors — and an odd number of errors — can be detected in the received message. The error bits, however, cannot be corrected because their location in the message is unknown. For example, assume that the message shown below was transmitted using odd parity.

Bit position	7	6	5	4	3	2	1	0	Parity bit
Message	0	1	1	0	0	1	0	0	0

Assume that the message shown below is the received message. The parity of the message is even; therefore, the received message has an error. However, the location of the bit (or bits) in error is unknown. Therefore, in this example, bit 5 cannot be corrected.

Bit position	7	6	5	4	3	2	1	0	Parity bit
Message	0	1	0	0	0	1	0	0	0

Richard W. Hamming developed a code that resolves this problem. The basic Hamming code can detect single or double errors and can correct a single error. The mathematics to develop the Hamming code is beyond the scope of this book. An entire chapter could be utilized in the presentation of linear codes and matrix algebra from which the Hamming code is derived.

Assume a 2-element code as shown below; that is, there are only two code words in the code. If a single error occurred in code word X, then the assumed message is X = 000. Similarly, if a single error occurred in code word Y, then the assumed message is Y = 111. Therefore, detection and correction is possible because the code words *differ in three bit positions*.

Code word X =	0 0 0
	0 0 1
	0 1 0
	1 0 0

Code word Y =	1 1 1
	0 1 1
	1 0 1
	1 1 0

A *code word* contains n bits consisting of m message bits plus k parity check bits as shown in Figure 11.18. The m bits represent the information or message part of the code word; the k bits are used for detecting and correcting errors, where $k = n - m$. The

Hamming distance of two code words X and Y is the number of bits in which the two words differ in their corresponding columns. For example, the Hamming distance is three for code words X and Y as shown Figure 11.19.

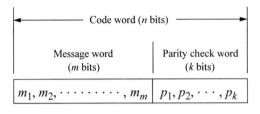

Code word $X = x_1, x_2, \cdots\cdots\cdots, x_m, \quad x_{m+1}, \cdots, x_n$

Figure 11.18   Code word of n bits containing m message bits and k parity check bits.

$$
\begin{array}{lcccccccc}
X = & 1 & 0 & 1 & 1 & 1 & 1 & 0 & 1 \\
Y = & 1 & 0 & 1 & 0 & 1 & 1 & 1 & 0
\end{array}
$$

Figure 11.19   Two code words to illustrate a Hamming distance of three.

Let $X = x_1, x_2, \cdots x_n$ and $Y = y_1, y_2, \cdots y_n$ be the ordered n-tuples that represent two code words, where $x_i, y_i \in \{0, 1\}$ for all i. The Hamming distance is the number of x_i and y_i that are different and can be defined mathematically as

$$
H_d(x_i, y_i) = \sum_{i=1}^{n}(x_i \oplus y_i)
$$

The Hamming distance is characterized by the following mathematical properties:

1. $H_d(X, Y) = H_d(Y, X)$
2. $H_d(X, Y) = 0$ if and only if $x_i = y_i$ for $i = 1, 2, \cdots, n$
3. $H_d(X, Y) > 0$ if at least one $x_i \neq y_i$ for $i = 1, 2, \cdots, n$
4. $H_d(X, Y) + H_d(Y, Z) \geq H_d(X, Z)$

Since there can be an error in *any* bit position, including the parity check bits, there must be a sufficient number of k parity check bits to identify any of the $m + k$ bit positions. The parity check bits are normally embedded in the code word and are positioned in columns with column numbers that are a power of two, as shown below for a code word containing four message bits (m_3, m_5, m_6, m_7) and three parity bits (p_1, p_2, p_4).

Column number	1	2	3	4	5	6	7
Code word =	p_1	p_2	m_3	p_4	m_5	m_6	m_7

Each parity bit maintains odd parity over a unique group of bits as shown below for a code word of four message bits.

$E_1 =$	p_1	m_3	m_5	m_7
$E_2 =$	p_2	m_3	m_6	m_7
$E_4 =$	p_4	m_5	m_6	m_7

The placement of the parity bits in certain columns is not arbitrary. Each of the variables in group E_1 contain a 1 in column 1 (2^0) of the binary representation of the column number as shown below.

	8	4	2	1
Group E_1	2^3	2^2	2^1	2^0
p_1	0	0	0	1
m_3	0	0	1	1
m_5	0	1	0	1
m_7	0	1	1	1
. . .				

Since p_1 has only a single 1 in the binary representation of column 1, p_1 can therefore be used as a parity check bit for a message bit in *any* column in which the binary representation of the column number has a 1 in column 1 (2^0). Thus, group E_1 can be expanded to include other message bits, as shown below.

$$p_1, m_3, m_5, m_7, m_9, m_{11}, m_{13}, m_{15}, m_{17}, \cdots$$

Each of the variables in group E_2 contain a 1 in column 2 (2^1) of the binary representation of the column number as shown below.

	8	4	2	1
Group E_2	2^3	2^2	2^1	2^0
p_2	0	0	1	0
m_3	0	0	1	1
m_6	0	1	1	0
m_7	0	1	1	1
$\cdots$				

Since p_2 has only a single 1 in the binary representation of column 2, p_2 can therefore be used as a parity check bit for a message bit in *any* column in which the binary representation of the column number has a 1 in column 2 (2^1). Thus, group E_2 can be expanded to include other message bits, as shown below.

$$p_2, m_3, m_6, m_7, m_{10}, m_{11}, m_{14}, m_{15}, m_{18}, \cdots$$

Each of the variables in group E_4 contain a 1 in column 4 (2^2) of the binary representation of the column number as shown below.

	8	4	2	1
Group E_4	2^3	2^2	2^1	2^0
p_4	0	1	0	0
m_5	0	1	0	1
m_6	0	1	1	0
m_7	0	1	1	1
$\cdots$				

Since p_4 has only a single 1 in the binary representation of column 4, p_4 can therefore be used as a parity check bit for a message bit in *any* column in which the binary representation of the column number has a 1 in column 4 (2^2). Thus, group E_4 can be expanded to include other message bits, as shown below.

$$p_4, m_5, m_6, m_7, m_{12}, m_{13}, m_{14}, m_{15}, m_{20}, \cdots$$

The format for embedding parity bits in the code word can be extended easily to any size message. For example, the code word for an 8-bit message is encoded as follows:

$$p_1, p_2, m_3, p_4, m_5, m_6, m_7, p_8, m_9, m_{10}, m_{11}, m_{12}$$

where $m_3, m_5, m_6, m_7, m_9, m_{10}, m_{11}, m_{12}$ are the message bits and p_1, p_2, p_4, p_8 are the parity check bits for groups E_1, E_2, E_4, E_8, respectively, as shown below.

Group $E_1 =$	p_1	m_3	m_5	m_7	m_9	m_{11}
Group $E_2 =$	p_2	m_3	m_6	m_7	m_{10}	m_{11}
Group $E_4 =$	p_4	m_5	m_6	m_7	m_{12}	
Group $E_8 =$	p_8	m_9	m_{10}	m_{11}	m_{12}	

For messages, the bit with the highest numbered subscript is the low-order bit. Thus, the low-order message bit is m_{12} for a byte of data that is encoded using the Hamming code. A 32-bit message requires six parity check bits:

$$p_1, p_2, p_4, p_8, p_{16}, p_{32}$$

There is only one parity bit in each group. The parity bits are independent and no parity bit checks any other parity bit. Consider the following code word for a 4-bit message:

$$p_1, p_2, m_3, p_4, m_5, m_6, m_7$$

The parity bits are generated so that there are an odd number of 1s in the following groups:

$E_1 =$	p_1	m_3	m_5	m_7
$E_2 =$	p_2	m_3	m_6	m_7
$E_4 =$	p_4	m_5	m_6	m_7

For example, the parity bits are generated as follows:

$$p_1 = (m_3 \oplus m_5 \oplus m_7)'$$
$$p_2 = (m_3 \oplus m_6 \oplus m_7)'$$
$$p_4 = (m_5 \oplus m_6 \oplus m_7)'$$

Example 11.1 A 4-bit message (0110) will be encoded using the Hamming code then transmitted. The message, transmitted code word, and received code word are shown below.

	p_1	p_2	m_3	p_4	m_5	m_6	m_7
Message to be sent			0		1	1	0
Code word sent	0	0	0	1	1	1	0
Code word received	0	0	0	1	0	1	0

From the received code word, it is seen that bit 5 is in error. When the code word is received, the parity of each group is checked using the bits assigned to that group, as shown below.

Group $E_1 =$	$p_1\, m_3\, m_5\, m_7 =$	$0\ 0\ 0\ 0 =$	Error $=$	1
Group $E_2 =$	$p_2\, m_3\, m_6\, m_7 =$	$0\ 0\ 1\ 0 =$	No error $=$	0
Group $E_4 =$	$p_4\, m_5\, m_6\, m_7 =$	$1\ 0\ 1\ 0 =$	Error $=$	1

A parity error is assigned a value of 1; no parity error is assigned a value of 0. The groups are then listed according to their binary weight. The resulting binary number is called the *syndrome word* and indicates the bit in error; in this case, bit 5. The bit in error is then complemented to yield a correct message of 0110.

	2^2	2^1	2^0
Groups	E_4	E_2	E_1
Syndrome word	1	0	1

Since there are three groups in this example, there are eight combinations. Each combination indicates the column of the bit in error, including the parity bits. A syndrome word of $E_4 E_2 E_1 = 000$ indicates no error in the received code word.

Double error detection and single error correction can be achieved by adding a parity bit for the entire code word. The format is shown below, where p_{cw} is the parity bit for the code word.

$$\text{Code word} = p_1\, p_2\, m_3\, p_4\, m_5\, m_6\, m_7\, p_{cw}$$

Several examples using the code word parity bit are shown in Table 11.2. The examples include both single error correction and double error detection. In the final example, the syndrome word is 000, indicating no error. However, the code word parity is incorrect; therefore, there must be no single error and also no double error — the error occurred in the code word parity bit. The logic diagram is shown in Figure 11.20.

Table 11.2 Examples of Single Error Correction and Double Error Detection

	Code Word Format								Syndrome	Code Word Parity	Single Error	Double Error
	p_1	p_2	m_3	p_4	m_5	m_6	m_7	p_{cw}	$E_4E_2E_1$			
Sent	0	0	1	1	0	0	0	1				
Received	0	0	0	1	0	0	0	1	0 1 1	Bad	Yes	No
Sent	0	0	0	1	1	1	0	0				
Received	0	1	0	1	1	1	0	0	0 1 0	Bad	Yes	No
Sent	1	0	0	1	1	0	1	1				
Received	1	1	0	1	0	0	1	1	1 1 1	Good	No	Yes
Sent	0	0	1	1	0	0	0	1				
Received	0	0	1	1	0	1	1	1	0 1 1	Good	No	Yes
Sent	0	0	1	1	0	0	0	1				
Received	0	0	1	1	0	0	0	0	0 0 0	Bad	No	No

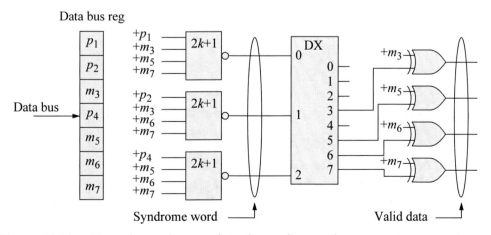

Figure 11.20   Hamming code error detection and correction.

The bit in error can be complemented (inverted) by using an exclusive-OR circuit. Any bit that is exclusive-ORed with a logic 1 is inverted as shown below. The inputs of the exclusive-OR circuits connect to the output of the decoder — which indicates the bit in error — and the error bit is obtained from the data register. If only the correct message is required, then the parity bits do not have to be corrected.

$$0 \oplus 0 = 0$$
$$0 \oplus 1 = 1$$
$$1 \oplus 0 = 1$$
$$1 \oplus 1 = 0$$

Example 11.2 This example encodes an 8-bit message in Hamming code for transmission. At the receiver, the message is checked for errors and any single-bit error is corrected. The code word, including the eight message bits and the four parity bits, is as follows:

Code word = p_1 p_2 m_3 p_4 m_5 m_6 m_7 p_8 m_9 m_{10} m_{11} m_{12}

The code word is partitioned into the following four groups that will be used to generate the syndrome word: E_1, E_2, E_4, and E_8. The four groups and their respective code word bits are shown below. The parity of each group is odd.

Group $E_1 = $ p_1 m_3 m_5 m_7 m_9 m_{11}
Group $E_2 = $ p_2 m_3 m_6 m_7 m_{10} m_{11}
Group $E_4 = $ p_4 m_5 m_6 m_7 m_{12}
Group $E_8 = $ p_8 m_9 m_{10} m_{11} m_{12}

The logic diagram for this example is shown in Figure 11.21. The m_i bits are the message bits that are being transmitted and are used to generate the parity bits p_i to maintain odd parity over each group. The parity bits for each group are generated in the test bench by a task labeled *task pbit_generate*. The mr_i bits represent the received message bits that may contain errors. Errors are injected into the received message by a task in the test bench labeled *task error_inject*. The el_err, e2_err, e4_err, and e8_err outputs of the exclusive-NOR functions represent the syndrome word which connects to the inputs of the 4:16 decoder.

The decoder outputs specify the message bit of the received code word in error. The active decoder output is then exclusive-ORed with the corresponding received message bit to correct the bit in error. The valid message is represented by the mv_i bits.

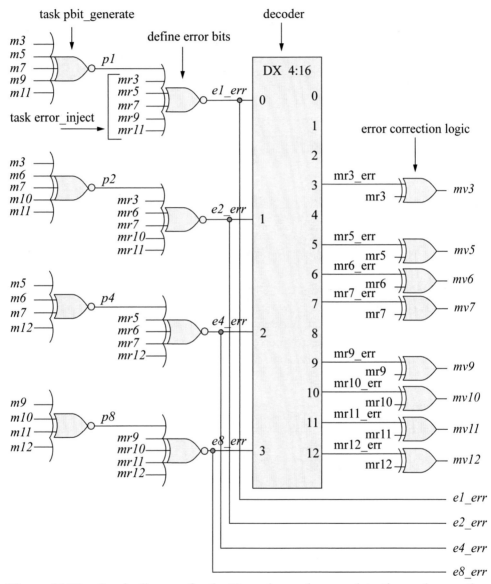

Figure 11.21 Logic diagram for the Hamming code error detection and correction circuit of Examlple 11.2.

The dataflow module, test bench module, and outputs are shown in Figure 11.22, Figure 11.23, and Figure 11.24, respectively. In the test bench, parity bits $p1, p2, p4$, and $p8$ are generated for each corresponding group by the task *task pbit_generate*. Single-bit errors are inserted in the received mr_i bits in the test bench by the task *task error_inject*. The outputs of Figure 11.24 indicate the message that was sent, the received message containing the error, the column number in which the error occurred, and the corrected message containing valid data. Since parity bits are not corrected, the byte of data specifies the message bits only. Recall that the column numbers for the message bits are m_3, m_5, m_6, m_7, m_9, m_{10}, m_{11}, and m_{12}. Thus, the high-order message bit (m_3) is the leftmost bit listed and the low-order message bit (m_{12}) is the rightmost bit listed.

Consider the case where an error is inserted into message bit *m3*. The code word, shown in the test bench, is formatted as follows:

Bit position =	11	10	9	8	7	6	5	4	3	2	1	0
Data =	m3	m5	m6	m7	m9	m10	m11	m12	p1	p2	p4	p8

The statement *error_inject (11)* passes the constant 11 to the task as the *bit_number*. Then the statement

$$bit\_position = 1'b1 << bit\_number$$

shifts a 1 bit eleven bit positions to the left to location *m3*. The message bit *m3* is exclusive-ORed with the 1 bit as shown below, thereby inverting *m3*. The message containing the error is then passed back to the task invocation as the received message, which is then corrected by the error correction logic. In a similar manner, errors are injected into message bits *m7*, *m9*, and *m12*.

Data =	m3	m5	m6	m7	m9	m10	m11	m12	p1	p2	p4	p8
XOR =	1	0	0	0	0	0	0	0	0	0	0	0
Received message =	m3'	m5	m6	m7	m9	m10	m11	m12	p1	p2	p4	p8

```
//dataflow module for Hamming code to encode an 8-bit message
module hamming_code (m3, m5, m6, m7, m9, m10, m11, m12,
                p1, p2, p4,p8,
                mv3, mv5, mv6, mv7, mv9, mv10, mv11, mv12,
                e1_err, e2_err, e4_err, e8_err);

input m3, m5, m6, m7, m9, m10, m11, m12;
input p1, p2, p4, p8;
//continued on next  page
```

Figure 11.22   Dataflow module to illustrate Hamming code error detection and correction.

```
output mv3, mv5, mv6, mv7, mv9, mv10, mv11, mv12;
output e1_err, e2_err, e4_err, e8_err;

wire mr3_err, mr5_err, mr6_err, mr7_err;
wire mr9_err, mr10_err, mr11_err, mr12_err;

//define the error bits
assign   e1_err = ~(p1 ^ m3 ^ m5 ^ m7 ^ m9 ^ m11),
         e2_err = ~(p2 ^ m3 ^ m6 ^ m7 ^ m10 ^ m11),
         e4_err = ~(p4 ^ m5 ^ m6 ^ m7 ^ m12),
         e8_err = ~(p8 ^ m9 ^ m10 ^ m11 ^ m12);

//design the decoder
assign   mr3_err= (~e8_err) & (~e4_err) & (e2_err)  & (e1_err),
      mr5_err= (~e8_err) & (e4_err)  & (~e2_err) & (e1_err),
      mr6_err= (~e8_err) & (e4_err)  & (e2_err)  & (~e1_err),
      mr7_err= (~e8_err) & (e4_err)  & (e2_err)  & (e1_err),
      mr9_err= (e8_err)  & (~e4_err) & (~e2_err) & (e1_err),
      mr10_err = (e8_err) & (~e4_err) & (e2_err)  & (~e1_err),
      mr11_err = (e8_err) & (~e4_err) & (e2_err)  & (e1_err),
      mr12_err = (e8_err) & (e4_err)  & (~e2_err) & (~e1_err);

//design the correction logic
assign   mv3 = (mr3_err)   ^ (m3),
         mv5 = (mr5_err)   ^ (m5),
         mv6 = (mr6_err)   ^ (m6),
         mv7 = (mr7_err)   ^ (m7),
         mv9 = (mr9_err)   ^ (m9),
         mv10= (mr10_err) ^ (m10),
         mv11= (mr11_err) ^ (m11),
         mv12= (mr12_err) ^ (m12);
endmodule
```

Figure 11.22   (Continued)

```
//test bench for the Hamming code module
module hamming_code_tb;

reg m3, m5, m6, m7, m9, m10, m11, m12;
reg ms3, ms5, ms6, ms7, ms9, ms10, ms11, ms12;
reg mr3, mr5, mr6, mr7, mr9, mr10, mr11, mr12;
wire mv3, mv5, mv6, mv7, mv9, mv10, mv11, mv12;
wire e1_err, e2_err, e4_err, e8_err;
//continued on next page
```

Figure 11.23   Test bench for the Hamming code module of Figure 11.22.

```
reg p1, p2, p4, p8;


initial
$display ("bit_order = m3, m5, m6, m7, m9, m10, m11, m12");


initial
$monitor ("sent=%b, rcvd=%b, error=%b, valid=%b",
          {ms3, ms5, ms6, ms7, ms9, ms10, ms11, ms12},
          {mr3, mr5, mr6, mr7, mr9, mr10, mr11, mr12},
          {e8_err, e4_err, e2_err, e1_err},
          {mv3, mv5, mv6, mv7, mv9, mv10, mv11, mv12});


initial
begin
//------------------------------------------------------
    #0    {ms3,ms5,ms6,ms7,ms9,ms10,ms11,ms12}=8'b1010_1010;
          {m3,m5,m6,m7,m9,m10,m11,m12} =
          {ms3,ms5,ms6,ms7,ms9,ms10,ms11,ms12};
    pbit_generate(m3,m5,m6,m7,m9,m10,m11,m12);   //invoke task

          //no error injected
          {mr3, mr5, mr6, mr7, mr9, mr10, mr11, mr12} =
          {m3, m5, m6, m7, m9, m10, m11, m12};
//------------------------------------------------------
    #10   {ms3,ms5,ms6,ms7,ms9,ms10,ms11,ms12}=8'b1010_1010;
          {m3,m5,m6,m7,m9,m10,m11,m12} =
          {ms3,ms5,ms6,ms7,ms9,ms10,ms11,ms12};
    pbit_generate(m3,m5,m6,m7,m9,m10,m11,m12);   //invoke task

          //inject error into m3
    error_inject(11);                            //invoke task
          {mr3,mr5,mr6,mr7,mr9,mr10,mr11,mr12} =
          {m3,m5,m6,m7,m9,m10,m11,m12};
//------------------------------------------------------
    #10   {ms3,ms5,ms6,ms7,ms9,ms10,ms11,ms12}=8'b0101_0101;
          {m3,m5,m6,m7,m9,m10,m11,m12} =
          {ms3,ms5,ms6,ms7,ms9,ms10,ms11,ms12};
    pbit_generate(m3,m5,m6,m7,m9,m10,m11,m12);   //invoke task

          //inject error into m7
    error_inject(8);                             //invoke task
          {mr3,mr5,mr6,mr7,mr9,mr10,mr11,mr12} =
          {m3,m5,m6,m7,m9,m10,m11,m12};
//------------------------------------------------------
//continued on next page
```

Figure 11.23   (Continued)

```
//-----------------------------------------------------------
   #10   {ms3,ms5,ms6,ms7,ms9,ms10,ms11,ms12}=8'b1111_0000;
         {m3,m5,m6,m7,m9,m10,m11,m12} =
         {ms3,ms5,ms6,ms7,ms9,ms10,ms11,ms12};
   pbit_generate(m3,m5,m6,m7,m9,m10,m11,m12);   //invoke task

         //inject error into m9
   error_inject(7);                             //invoke task
         {mr3,mr5,mr6,mr7,mr9,mr10,mr11,mr12} =
         {m3,m5,m6,m7,m9,m10,m11,m12};
//-----------------------------------------------------------
   #10   {ms3,ms5,ms6,ms7,ms9,ms10,ms11,ms12}=8'b0110_1101;
         {m3,m5,m6,m7,m9,m10,m11,m12} =
         {ms3,ms5,ms6,ms7,ms9,ms10,ms11,ms12};
   pbit_generate(m3,m5,m6,m7,m9,m10,m11,m12);   //invoke task

         //inject error into m12
   error_inject(4);                             //invoke task
         {mr3,mr5,mr6,mr7,mr9,mr10,mr11,mr12} =
         {m3,m5,m6,m7,m9,m10,m11,m12};
//-----------------------------------------------------------
   #10   $stop;
end


task pbit_generate;
input m3, m5, m6, m7, m9, m10, m11, m12;
begin
   p1 = ~(m3 ^ m5 ^ m7 ^ m9 ^ m11);
   p2 = ~(m3 ^ m6 ^ m7 ^ m10 ^ m11);
   p4 = ~(m5 ^ m6 ^ m7 ^ m12);
   p8 = ~(m9 ^ m10 ^ m11 ^ m12);
end
endtask


task error_inject;
   input [3:0] bit_number;
   reg [11:0] bit_position;
   reg [11:0] data;
   begin
      bit_position = 1'b1 << bit_number;

      data = {m3,m5,m6,m7,m9,m10,m11,m12,p1,p2,p4,p8};
      {m3,m5,m6,m7,m9,m10,m11,m12,p1,p2,p4,p8} =
         data ^ bit_position;
end
endtask
//continued on next page
```

Figure 11.23   (Continued)

```
//instantiate the module into the test bench
hamming_code inst1 (
    .m3(m3),
    .m5(m5),
    .m6(m6),
    .m7(m7),
    .m9(m9),
    .m10(m10),
    .m11(m11),
    .m12(m12),
    .p1(p1),
    .p2(p2),
    .p4(p4),
    .p8(p8),


    .mv3(mv3),
    .mv5(mv5),
    .mv6(mv6),
    .mv7(mv7),
    .mv9(mv9),
    .mv10(mv10),
    .mv11(mv11),
    .mv12(mv12),


    .e1_err(e1_err),
    .e2_err(e2_err),
    .e4_err(e4_err),
    .e8_err(e8_err)
    );

endmodule
```

Figure 11.23   (Continued)

```
bit_order = m3, m5, m6, m7, m9, m10, m11, m12

sent=10101010, rcvd=10101010, error=0000, valid=10101010
sent=10101010, rcvd=00101010, error=0011, valid=10101010
sent=10101010, rcvd=10111010, error=0111, valid=10101010
sent=11110000, rcvd=11111000, error=1001, valid=11110000
sent=01101101, rcvd=01101100, error=1100, valid=01101101
```

Figure 11.24   Outputs for the Hamming code test bench of Figure 11.23.

11.5 Booth Algorithm

The Booth algorithm is an effective technique for 2s complement multiplication. Unlike the sequential add-shift or the planar array methods, it treats both positive and negative numbers uniformly. In the sequential add-shift method, each multiplier bit generates a rendering of the multiplicand that is added to the partial product. For large operands, the delay to obtain the product can be substantial. The Booth algorithm reduces the number of partial products by shifting over strings of zeros in a recoded version of the multiplier. This method is referred to as *skipping over zeros*. The increase in speed is proportional to the number of zeros in the recoded version of the multiplier.

Consider a string of k consecutive 1s in the multiplier as shown below. The multiplier will be recoded so that the k consecutive 1s will be transformed into $k - 1$ consecutive 0s. The multiplier hardware will then shift over the $k - 1$ consecutive 0s without having to generate partial products.

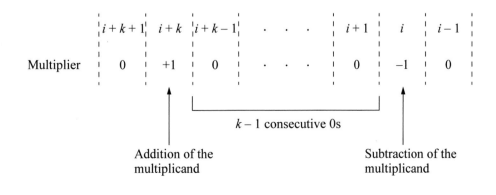

In the sequential add-shift method, the multiplicand would be added k times to the shifted partial product. The number of additions can be reduced by the following property of binary strings:

$$2^{i + k} - 2^i = 2^{i + k - 1} + 2^{i + k - 2} + \cdots 2^{i + 1} + 2^i \tag{11.2}$$

The right-hand side of the equation is a binary string that can be replaced by the difference of two numbers on the left-hand side of the equation. Thus, the k consecutive 1s can be replaced by the following string:

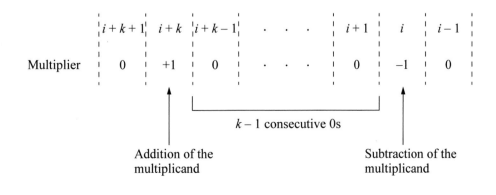

632   Chapter 11   Additional Design Examples

An example will help to clarify the procedure. Let the multiplier be +30 as shown below.

	2^{i+k}							
	2^{i+4}					2^i		
	2^5	2^4	2^3	2^2	2^1	2^0		
Multiplier	0	1	1	1	1	0	+30

$$k = 4$$

The validity of Equation 11.2 can be verified using the multiplier of +30 shown above.

2^{i+k}	2^i	=	2^{i+k-1}			
2^5	-2^1	=	2^4	$+2^3$	$+2^2$	$+2^1$
32	-2	=	16	$+8$	$+4$	$+2$
	30	=	30			

Thus, the multiplier 011110 (+30) can be regarded as the difference of two numbers: $32 - 2$, as shown below.

```
     0  1  0  0  0  0  0   (32)
 -)  0  0  0  0  0  1  0   (2)
     0  0  1  1  1  1  0   (30)
```

The product can be generated by one subtraction in column 2^i and one addition in column 2^{i+k}, as shown below; that is, by adding 32 and subtracting 2. In this case, adding 2^5 times the multiplicand and subtracting 2^1 times the multiplicand yields the appropriate result.

	2^6	2^5	2^4	2^3	2^2	2^1	2^0
Standard multiplier	0	0	+1	+1	+1	+1	0
			$\leftarrow$	$k = 4$	$\rightarrow$		
Recoded multiplier	0	+1	0	0	0	-1	0
			$\leftarrow k - 1 = 3 \rightarrow$				

Note that -1 times the left-shifted multiplicand occurs at $0 \to 1$ boundaries and $+1$ times the left-shifted multiplicand occurs at $1 \to 0$ boundaries as the multiplier is scanned from right to left. The Booth algorithm can be applied to any number of groups of 1s in a multiplier, including the case where the group consists of a single 1 bit. Table 11.3 shows the multiplier recoding table for two consecutive bits that specify which version of the multiplicand will be added to the shifted partial product.

Table 11.3 Booth Multiplier Recoding Table

Multiplier Bit i	Bit $i-1$	Version of Multiplicand
0	0	$0 \times$ multiplicand
0	1	$+1 \times$ multiplicand
1	0	$-1 \times$ multiplicand
1	1	$0 \times$ multiplicand

The Booth algorithm converts both positive and negative multipliers into a form that generates versions of the multiplicand to be added to the shifted partial products. Since the increase in speed is a function of the bit configuration of the multiplier, the efficiency of the Booth algorithm is data dependent. Several examples will now be presented to illustrate the Booth algorithm before proceeding to the Verilog design.

Example 11.3 An 8-bit positive multiplicand (+53) will be multiplied by an 8-bit positive multiplier (+30) — first using the standard sequential add-shift technique, then using the Booth algorithm to show the reduced number of partial products. In the sequential add-shift technique, there are six additions; using the Booth algorithm, there are only two additions. The second partial product in the Booth algorithm is the 2s complement of the multiplicand shifted left. The third partial product is the result of shifting left over three 0s.

Standard sequential add-shift																	
Multiplicand					0	0	1	1	0	1	0	1					+53
Multiplier			×)		0	0	0	1	1	1	1	0					+30
					0	0	0	0	0	0	0	0					
				0	0	1	1	0	1	0	1						
			0	0	1	1	0	1	0	1							
		0	0	1	1	0	1	0	1								
	0	0	1	1	0	1	0	1									
	0	0	0	0	0	0	0	0									
0	0	0	0	0	0	0	0										
0	0	0	0	0	0	0	0										
0	0	0	0	1	1	0	0	0	1	1	0	1	1	0			+1590

Booth algorithm																
Multiplicand								0	0	1	1	0	1	0	1	+53
Recoded multiplier							×)	0	0	+1	0	0	0	−1	0	
0	0	0	0	0	0	0	0	0	0	0	0	0	0	0	0	
1	1	1	1	1	1	1	1	1	0	0	1	0	1	1		
0	0	0	0	0	1	1	0	1	0	1						
0	0	0	0	0	1	1	0	0	0	1	1	0	1	1	0	+1590

Example 11.4 This is an example of a 5-bit positive multiplicand (+13) and a 5-bit negative multiplier (−12).

Multiplicand		0	1	1	0	1	+13
Multiplier	×)	1	0	1	0	0	−12
							−156

Booth algorithm										
Multiplicand				0	1	1	0	1		+13
Recoded multiplier			×)	−1	+1	−1	0	0		
	0	0	0	0	0	0	0	0	0	
	1	1	1	1	0	0	1	1		
	0	0	0	1	1	0	1			
	1	1	0	0	1	1				
	1	0	1	1	0	0	1	0	0	−156

Example 11.5 If the multiplier has a low-order bit of 1, then an implied 0 is placed to the right of the low-order multiplier bit. This provides a boundary of $0 \rightarrow 1$ (−1 times the multiplicand) for the low-order bit as the multiplier is scanned from right to left. This is shown in the example below, which multiplies a 5-bit positive multiplicand by a 5-bit positive multiplier.

Multiplicand		0	1	1	1	1		+15
Multiplier	×)	0	0	0	1	1	0	+3
								+45

Booth algorithm											
Multiplicand						0	1	1	1	1	+15
Recoded multiplier ×)						0	0	+1	0	−1	
	1	1	1	1	1	1	0	0	0	1	
	0	0	0	0	1	1	1	1			
		0	0	0	1	0	1	1	0	1	+45

Example 11.6   This is an example of a 6-bit negative multiplicand (−19) and a 6-bit positive multiplier (+14).

Multiplicand	1	0	1	1	0	1	−19
Multiplier ×)	0	0	1	1	1	0	+14
							−266

Booth algorithm											
Multiplicand					1	0	1	1	0	1	−19
Recoded multiplier ×)					0	+1	0	0	−1	0	
	0	0	0	0	0	0	0	0	0	0	
	0	0	0	0	1	0	0	1	1		
	1	0	1	1	0	1					
		0	1	1	1	1	0	1	1	0	−266

Example 11.7   This is an example of a 5-bit negative multiplicand (−13) and a 5-bit negative multiplier (−7). There is an implied 0 to the right of the low-order multiplier bit to provide a boundary of $0 \rightarrow 1$ (−1 times the multiplicand) for the low-order bit.

Multiplicand	1	0	0	1	1		−13
Multiplier ×)	1	1	0	0	1	0	−7
							+91

Booth algorithm										
Multiplicand						1	0	0	1	1	−13
Recoded multiplier				×)	0	−1	0	+1	−1	
0	0	0	0	0	0	1	1	0	1	
1	1	1	1	1	0	0	1	1		
0	0	0	1	1	0	1				
	0	0	1	0	1	1	0	1	1	+91

Example 11.8   A final example shows a multiplier that has very little advantage over the sequential add-shift method because of alternating 1s and 0s. The multiplicand is a positive 6-bit operand (+19) and the multiplier is a negative 6-bit operand (−11). There is an implied 0 to the right of the low-order multiplier bit.

Multiplicand		0	1	0	0	1	1	+19
Multiplier	×)	1	1	0	1	0	1	0	−11
								−209

Booth algorithm												
Multiplicand					0	1	0	0	1	1		+19
Recoded multiplier				×)	0	−1	+1	−1	+1	−1		
1	1	1	1	1	0	1	1	0	1		
0	0	0	0	1	0	0	1	1			
1	1	1	0	1	1	0	1				
0	0	1	0	0	1	1					
1	0	1	1	0	1						
	1	0	0	1	0	1	1	1	1		−209

The mixed-design module for the Booth algorithm is shown in Figure 11.25. The operands are 4-bit vectors a[3:0] and b[3:0]; the product is an 8-bit result rslt[7:0]. The following internal wires are defined: a_ext_pos[7:0], which is operand a with sign extended, and a_ext_neg[7:0], which is the negation (2s complement) of operand a with sign extended. The following internal registers are defined: a_neg[3:0], which is the negation of operand a, and pp1[7:0], pp2[7:0], pp3[7:0], and pp4[7:0], which are the partial products to be added together to obtain the product *rslt[7:0]*. The example below illustrates the use of the internal registers. The right-most four bits of partial product 1 (*pp1*) are the negation [*a_neg* (2s complement)] of operand *a*, which is generated as a result of the –1 times operand *a* operation. The entire row of partial product 1 (*pp1*) corresponds to *a_neg* with the sign bit extended; that is, *a_ext_neg*.

Multiplicand		0	1	1	1		+7
Multiplier	×)	0	1	0	1	0	+5
							+35

	Booth algorithm										
	Multiplicand					0	1	1	1	+7	
	Recoded multiplier				×)	+1	–1	+1	–1		
a_ext_neg →	*pp1*	1	1	1	1	1	0	0	1		← *a_neg*
a_ext_pos →	*pp2*	0	0	0	0	1	1	1	0		
	pp3	1	1	1	0	0	1	0	0		
	pp4	0	0	1	1	1	0	0	0		
		0	0	1	0	0	0	1	1	+35	

```
//mixed-design for the Booth multiply algorithm
module booth2 (a, b, rslt);

input [3:0] a, b;
output [7:0] rslt;

wire [3:0] a, b;
wire [7:0] rslt;

wire [3:0] a_bar;

//define internal wires and registers
wire [7:0] a_ext_pos;
wire [7:0] a_ext_neg;
reg [3:0] a_neg;
reg [7:0] pp1, pp2, pp3, pp4;
//continued on next page
```

Figure 11.25 Mixed-design module to implement the Booth algorithm.

```verilog
//test b[1:0] -------------------------------------

assign a_bar = ~a;

//the following will cause synthesis of a single adder
//rather than multiple adders in the case statement

always @ (a_bar)
    a_neg = a_bar + 1;

assign a_ext_pos = {{4{a[3]}}, a};
assign a_ext_neg = {{4{a_neg[3]}}, a_neg};


always @ (b, a_ext_neg)
begin
    case (b[1:0])
        2'b00 :
            begin
                pp1 = 8'h00;
                pp2 = 8'h00;
            end

        2'b01 :
            begin
                pp1 = a_ext_neg;
                pp2 = {{3{a[3]}}, a[3:0], 1'b0};
            end

        2'b10 :
            begin
                pp1 = 8'h00;
                pp2 = {a_ext_neg[6:0], 1'b0};
            end

        2'b11 :
            begin
                pp1 = a_ext_neg;
                pp2 = 8'h00;
            end
    endcase

end

//continued on next page
```

Figure 11.25   (Continued)

```
//test b[2:1] -------------------------------------
always @ (b, a_ext_pos, a_ext_neg)

begin
   case (b[2:1])
      2'b00: pp3 = 8'h00;

      2'b01: pp3 = {a_ext_pos[5:0], 2'b0};

      2'b10: pp3 = {a_ext_neg[5:0], 2'b00};

      2'b11: pp3 = 8'h00;
   endcase
end


//test b[3:2] -------------------------------------
always @ (b, a_ext_pos, a_ext_neg)

begin
   case (b[3:2])
      2'b00: pp4 = 8'h00;

      2'b01: pp4 = {a_ext_pos[4:0], 3'b000};

      2'b10: pp4 = {a_ext_neg[4:0], 3'b000};

      2'b11: pp4 = 8'h00;

   endcase
end

assign rslt = pp1 + pp2 + pp3 + pp4;

endmodule
```

Figure 11.25   (Continued)

The test bench is shown in Figure 11.26, which tests all pairs of bits; that is, *b[1:0]*, *b[2:1]*, and *b[3:2]*. The outputs are shown in Figure 11.27 with binary inputs for operands *a* and *b* that include both positive and negative multiplicands and multipliers. The product *rslt* is in hexadecimal notation. The waveforms are shown in Figure 11.28.

```
//test bench for booth algorithm
module booth2_tb;

reg [3:0] a, b;
wire [7:0] rslt;

//display operands a, b, and rslt
initial
$monitor ("a = %b, b = %b, rslt = %h", a, b, rslt);

//apply input vectors
initial
begin
//test b[1:0] ---------------------------------------
   #0    a = 4'b0111;
         b = 4'b1000;

   #10   a = 4'b0110;
         b = 4'b0101;

   #10   a = 4'b1110;
         b = 4'b0110;

   #10   a = 4'b1011;
         b = 4'b1011;

//test b[2:1] ---------------------------------------
   #10   a = 4'b0001;
         b = 4'b1000;

   #10   a = 4'b0111;
         b = 4'b1011;

   #10   a = 4'b1011;
         b = 4'b1100;

   #10   a = 4'b0111;
         b = 4'b0111;

//test b[3:2] ---------------------------------------
   #10   a = 4'b0111;
         b = 4'b0000;

   #10   a = 4'b1111;
         b = 4'b0101;

//continue on next page
```

Figure 11.26   Test bench for the Booth algorithm module.

```
    #10    a = 4'b0101;
           b = 4'b1010;

    #10    a = 4'b1101;
           b = 4'b1100;

    #10    $stop;
end

//instantiate the module into the test bench
booth2 inst1 (
    .a(a),
    .b(b),
    .rslt(rslt)
    );

endmodule
```

Figure 11.26   (Continued)

```
a = 0111, b = 1000, rslt = c8  │  a = 1011, b = 1100, rslt = 14
a = 0110, b = 0101, rslt = 1e  │  a = 0111, b = 0111, rslt = 31
a = 1110, b = 0110, rslt = f4  │  a = 0111, b = 0000, rslt = 00
a = 1011, b = 1011, rslt = 19  │  a = 1111, b = 0101, rslt = fb
a = 0001, b = 1000, rslt = f8  │  a = 0101, b = 1010, rslt = e2
a = 0111, b = 1011, rslt = dd  │  a = 1101, b = 1100, rslt = 0c
```

Figure 11.27   Outputs for the Booth algorithm module of Figure 11.25.

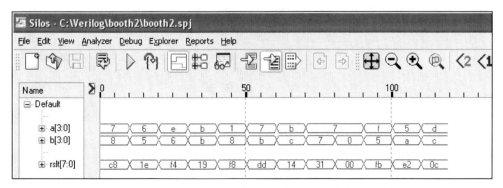

Figure 11.28   Waveforms for the Booth algorithm module of Figure 11.25.

11.6 Moore Synchronous Sequential Machine

A Moore synchronous sequential machine will be designed using structural modeling according to the state diagram shown in Figure 11.29. Since the outputs are to be asserted for the entire clock period ($z_1 \uparrow t_1 \downarrow t_3$, $z_2 \uparrow t_1 \downarrow t_3$), state codes are chosen so that no state transition sequence will pass through state d ($y_1 y_2 y_3 = 000$) or state e ($y_1 y_2 y_3 = 111$) for a transition that does not involve state d or state e.

The storage elements will consist of D flip-flops. Because the machine is initialized to $y_1 y_2 y_3 = 001$, there will be separate sets and resets for each flip-flop as shown in the block diagram of Figure 11.30 and the logic diagram of Figure 11.31.

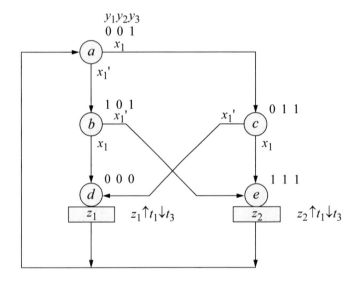

Figure 11.29   State diagram for the Moore machine of Section 11.6.

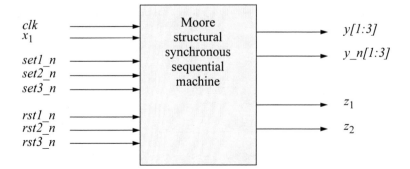

Figure 11.30   Block diagram for the Moore machine of Section 11.6.

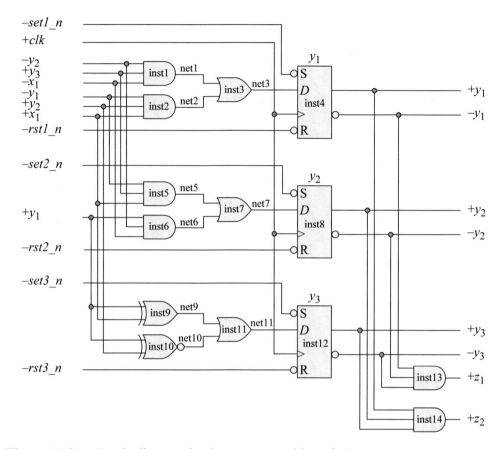

Figure 11.31 Logic diagram for the Moore machine of Figure 11.29.

The Verilog structural module is shown in Figure 11.32 and the test bench is shown in Figure 11.33. The outputs are shown in Figure 11.34 and the waveforms are shown in Figure 11.35.

```
//structural Moore ssm
module moore_ssm8 (x1, clk, set1_n, set2_n, set3_n, rst1_n,
                   rst2_n, rst3_n, y, y_n, z1, z2);

input x1, clk;
input set1_n, set2_n, set3_n;
input rst1_n, rst2_n, rst3_n;
output [1:3] y, y_n;
output z1, z2;
//continued on next page
```

Figure 11.32 Structural module for the Moore machine of Figure 11.31.

```verilog
wire x1, clk;
wire set1_n, set2_n, set3_n;
wire rst1_n, rst2_n, rst3_n;
wire net1, net2, net3, net5, net6, net7, net9, net10, net11;
wire [1:3] y, y_n;
wire z1, z2;

//instantiate the input logic for flip-flop y[1] -----------
and3_df inst1 (
   .x1(y_n[2]),
   .x2(y[3]),
   .x3(~x1),
   .z1(net1)
   );

and3_df inst2 (
   .x1(y_n[1]),
   .x2(y[2]),
   .x3(x1),
   .z1(net2)
   );

or2_df inst3 (
   .x1(net1),
   .x2(net2),
   .z1(net3)
   );

d_ff inst4 (
   .d(net3),
   .clk(clk),
   .q(y[1]),
   .q_n(y_n[1]),
   .set_n(set1_n),
   .rst_n(rst1_n)
   );

//instantiate the input logic for flip-flop y[2] -----------
and3_df inst5 (
   .x1(y_n[1]),
   .x2(y[3]),
   .x3(x1),
   .z1(net5)
   );

//continued on next page
```

Figure 11.32   (Continued)

```
and3_df inst6 (
   .x1(y[1]),
   .x2(y_n[2]),
   .x3(~x1),
   .z1(net6)
   );


or2_df inst7 (
   .x1(net5),
   .x2(net6),
   .z1(net7)
   );


d_ff inst8 (
   .d(net7),
   .clk(clk),
   .q(y[2]),
   .q_n(y_n[2]),
   .set_n(set2_n),
   .rst_n(rst2_n)
   );



//instantiate the input logic for flip-flop y[3] -----------
xor2_df inst9 (
   .x1(y[1]),
   .x2(x1),
   .z1(net9)
   );


xnor2_df inst10 (
   .x1(y[1]),
   .x2(y[2]),
   .z1(net10)
   );


or2_df inst11 (
   .x1(net9),
   .x2(net10),
   .z1(net11)
   );

//continued on next page
```

Figure 11.32   (Continued)

```
d_ff inst12 (
   .d(net11),
   .clk(clk),
   .q(y[3]),
   .q_n(y_n[3]),
   .set_n(set3_n),
   .rst_n(rst3_n)
   );

//instantiate the logic for outputs z1 and z2 ----------
and3_df inst13 (
   .x1(y_n[1]),
   .x2(y_n[2]),
   .x3(y_n[3]),
   .z1(z1)
   );

and3_df inst14 (
   .x1(y[1]),
   .x2(y[2]),
   .x3(y[3]),
   .z1(z2)
   );

endmodule
```

Figure 11.32   (Continued)

```
//test bench for Moore ssm
module moore_ssm8_tb;

reg x1, clk;
reg set1_n, set2_n, set3_n;
reg rst1_n, rst2_n, rst3_n;
wire [1:3] y, y_n;
wire z1, z2;

initial      //display inputs and outputs
$monitor ("x1 = %b, state = %b, z1z2 = %b", x1, y, {z1, z2});


//continued on next page
```

Figure 11.33   Test bench for the Moore machine of Figure 11.31.

```
//define clock
initial
begin
   clk = 1'b0;
   forever
      #10   clk = ~clk;
end

//define input sequence
initial
begin
   #0 set1_n = 1'b1;
      set2_n = 1'b1;
      set3_n = 1'b0;

      rst1_n = 1'b0;
      rst2_n = 1'b0;
      rst3_n = 1'b1;
      x1 = 1'b0;

   #5 set1_n = 1'b1;
      set2_n = 1'b1;
      set3_n = 1'b1;

      rst1_n = 1'b1;
      rst2_n = 1'b1;
      rst3_n = 1'b1;

   x1 = 1'b0;
   @ (posedge clk)    //go to state_b (101)

   x1 = 1'b1;
   @ (posedge clk)    //go to state_d (000)
                      //and assert z1 (t1 -- t3)

   x1 = 1'b0;
   @ (posedge clk)    //go to state_a (001)

   x1 = 1'b1;
   @ (posedge clk)    //go to state_c (011)

   x1 = 1'b1;
   @ (posedge clk)    //go to state_e (111)
                      //and assert z2 (t1 -- t3)

//continued on next page
```

Figure 11.33   (Continued)

```
   x1 = 1'b0;
   @ (posedge clk)     //go to state_a (001)

   x1 = 1'b0;
   @ (posedge clk)     //go to state_b (101)

   x1 = 1'b0;
   @ (posedge clk)     //go to state_e (111)
                       //and assert z2 (t1 -- t3)

   x1 = 1'b0;
   @ (posedge clk)     //go to state_a (001)

   x1 = 1'b1;
   @ (posedge clk)     //go to state_c (011)

   x1 = 1'b0;
   @ (posedge clk)     //go to state_d (000)
                       //and assert z1 (t1 -- t3)

   x1 = 1'b0;
   @ (posedge clk)     //go to state_a (001)

   #10     $stop;

end


//instantiate the module into the test bench
moore_ssm8 inst1 (
   .x1(x1),
   .clk(clk),
   .set1_n(set1_n),
   .set2_n(set2_n),
   .set3_n(set3_n),
   .rst1_n(rst1_n),
   .rst2_n(rst2_n),
   .rst3_n(rst3_n),
   .y(y),
   .y_n(y_n),
   .z1(z1),
   .z2(z2)
   );

endmodule
```

Figure 11.33   (Continued)

```
x1 = 0, state = 001, z1z2 = 00
x1 = 1, state = 101, z1z2 = 00
x1 = 0, state = 000, z1z2 = 10
x1 = 1, state = 001, z1z2 = 00
x1 = 1, state = 011, z1z2 = 00
x1 = 0, state = 111, z1z2 = 01
x1 = 0, state = 001, z1z2 = 00
x1 = 0, state = 101, z1z2 = 00
x1 = 0, state = 111, z1z2 = 01
x1 = 1, state = 001, z1z2 = 00
x1 = 0, state = 011, z1z2 = 00
x1 = 0, state = 000, z1z2 = 10
x1 = 0, state = 001, z1z2 = 00
```

Figure 11.34   Outputs for the Moore machine of Figure 11.31.

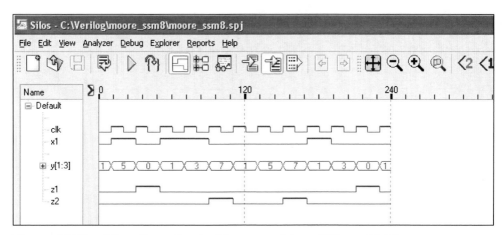

Figure 11.35   Waveforms for the Moore machine of Figure 11.31.

11.7 Mealy Pulse-Mode Asynchronous Sequential Machine

A Mealy pulse-mode asynchronous sequential machine will be designed using structural modeling that operates according to the state diagram shown in Figure 11.36. The machine will use NAND SR latches and D flip-flops in a master-slave configuration. Each latch requires two input maps, one each for x_1 and x_2, as shown in

Figure 11.37. As in previous examples, the maps are arranged such that the maps corresponding to each latch are in the same row, and each column of maps corresponds to a unique input. The map entries are defined as follows:

> S indicates that the latch will be set.
> s indicates that the latch will remain set.
> R indicates that the latch will be reset.
> r indicates that the latch will remain reset.

The time duration between successive input pulses must be sufficient to allow the machine to stabilize before the application of the next pulse. Therefore, after the deassertion of an input pulse, a vector of $x_1 x_2 = 00$ is applied before the assertion of the next input pulse. This allows the machine to operate in a deterministic manner. Since only one input pulse can be active at a time, the following input vectors are allowed: $x_1 x_2 = 00, 01, 10$. The output map is shown in Figure 11.38.

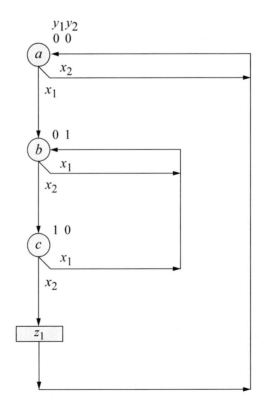

Figure 11.36   State diagram for the Mealy pulse-mode machine of Section 11.7.

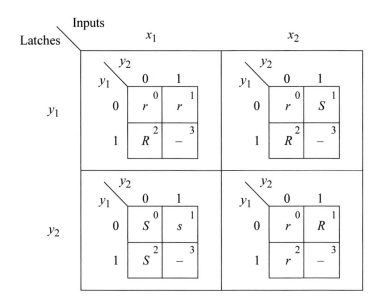

Figure 11.37   Input maps for the Mealy pulse-mode machine of Figure 11.36.

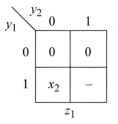

Figure 11.38   Output map for the Mealy pulse-mode machine of Figure 11.36.

The input equations for latches Ly_1 and Ly_2 are shown in Equation 11.3. Since the input equations for the latches also apply to the inputs of the corresponding D flip-flops, there is no need to have separate input equations for the flip-flops. The output equation for z_1 is shown in Equation 11.4.

$$SLy_1 = y_2 x_2$$

$$RLy_1 = x_1 + y_1 x_2$$

$$SLy_2 = x_1$$

$$RLy_2 = x_2 \tag{11.3}$$

$$z_1 = y_1 x_2 \qquad\qquad (11.4)$$

The logic diagram is shown in Figure 11.39 and the Verilog structural module is shown in Figure 11.40. The test bench, outputs, and waveforms are shown in Figure 11.41, Figure 11.42, and Figure 11.43, respectively.

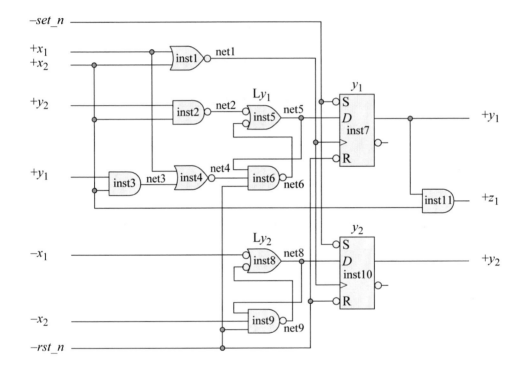

Figure 11.39   Logic diagram for the Mealy pulse-mode machine of Figure 11.36.

```
//structural Mealy pulse-mode asynchronous sequential machine
module pm_asm_mealy (set_n, rst_n, x1, x2, y1, y2, z1);

input set_n, rst_n;
input x1, x2;
output y1, y2;
output z1;
//continued on next page
```

Figure 11.40   Structural module for the Mealy pulse-mode machine of Figure 11.39.

```verilog
//define internal nets
wire net1, net2, net3, net4, net5, net6, net8, net9;


//design for clock input ------------------------
nor2_df inst1 (
   .x1(x1),
   .x2(x2),
   .z1(net1)
   );


//design for latch Ly1 --------------------------
nand2_df inst2 (
   .x1(y2),
   .x2(x2),
   .z1(net2)
   );

and2_df inst3 (
   .x1(y1),
   .x2(x2),
   .z1(net3)
   );

nor2_df inst4 (
   .x1(x1),
   .x2(net3),
   .z1(net4)
   );

nand2_df inst5 (
   .x1(net2),
   .x2(net6),
   .z1(net5)
   );

nand3_df inst6 (
   .x1(net5),
   .x2(net4),
   .x3(rst_n),
   .z1(net6)
   );

//continued on next page
```

Figure 11.40   (Continued)

```
//design for D flip-flop y1 ----------------------
d_ff inst7 (
   .d(net5),
   .clk(net1),
   .q(y1),
   .q_n(),
   .set_n(set_n),
   .rst_n(rst_n)
   );


//design for latch Ly2 ---------------------------
nand2_df inst8 (
   .x1(~x1),
   .x2(net9),
   .z1(net8)
   );

nand3_df inst9 (
   .x1(net8),
   .x2(~x2),
   .x3(rst_n),
   .z1(net9)
   );


//design for D flip-flop y2 ----------------------
d_ff inst10 (
   .d(net8),
   .clk(net1),
   .q(y2),
   .q_n(),
   .set_n(set_n),
   .rst_n(rst_n)
   );


//design for output z1 ---------------------------
and2_df inst11 (
   .x1(y1),
   .x2(x2),
   .z1(z1)
   );


endmodule
```

Figure 11.40   (Continued)

```verilog
//test bench for the Mealy pulse-mode machine
module pm_asm_mealy_tb;

reg x1, x2;
reg set_n, rst_n;
wire y1, y2;
wire z1;


//display variables
initial
$monitor ("x1x2=%b, state=%b, z1=%b", {x1, x2}, {y1, y2}, z1);


//define input sequence
initial
begin
   #0    set_n = 1'b1;
         rst_n = 1'b0;          //reset to state_a(00)
         x1 = 1'b0;
         x2 = 1'b0;

   #5    rst_n = 1'b1;          //deassert reset

   #10   x1=1'b1; x2=1'b0;      //go to state_b(01)
   #10   x1=1'b0; x2=1'b0;
   #10   x1=1'b0; x2=1'b1;      //go to state_c(10)
   #10   x1=1'b0; x2=1'b0;
   #10   x1=1'b0; x2=1'b1;      //assert z1; go to state_a(00)
   #10   x1=1'b0; x2=1'b0;
   #10   x1=1'b0; x2=1'b1;      //go to state_a(00)
   #10   x1=1'b0; x2=1'b0;
   #10   x1=1'b1; x2=1'b0;      //go to state_b(01)
   #10   x1=1'b0; x2=1'b0;
   #10   x1=1'b1; x2=1'b0;      //go to state_b(01)
   #10   x1=1'b0; x2=1'b0;
   #10   x1=1'b0; x2=1'b1;      //go to state_c(10)
   #10   x1=1'b0; x2=1'b0;
   #10   x1=1'b0; x2=1'b1;      //assert z1; go to state_a(00)
   #10   x1=1'b0; x2=1'b0;
   #10   x1=1'b0; x2=1'b1;      //go to state_a(00)
   #10   x1=1'b0; x2=1'b0;

   #10   $stop;
end

//continued on next page
```

Figure 11.41   Test bench for the Mealy pulse-mode machine of Figure 11.39.

```
//instantiate the module into the test bench
pm_asm_mealy inst1 (
    .set_n(set_n),
    .rst_n(rst_n),
    .x1(x1),
    .x2(x2),
    .y1(y1),
    .y2(y2),
    .z1(z1)
    );
endmodule
```

Figure 11.41   (Continued)

```
x1x2=00, state=00, z1=0          x1x2=00, state=01, z1=0
x1x2=10, state=00, z1=0          x1x2=10, state=01, z1=0
x1x2=00, state=01, z1=0          x1x2=00, state=01, z1=0
x1x2=01, state=01, z1=0          x1x2=01, state=01, z1=0
x1x2=00, state=10, z1=0          x1x2=00, state=10, z1=0
x1x2=01, state=10, z1=1          x1x2=01, state=10, z1=1
x1x2=00, state=00, z1=0          x1x2=00, state=00, z1=0
x1x2=01, state=00, z1=0          x1x2=01, state=00, z1=0
x1x2=00, state=00, z1=0          x1x2=00, state=00, z1=0
x1x2=10, state=00, z1=0
```

Figure 11.42   Outputs for the Mealy pulse-mode machine of Figure 11.39.

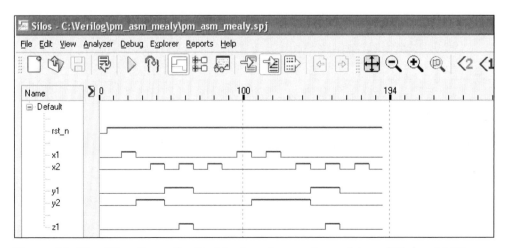

Figure 11.43   Waveforms for the Mealy pulse-mode machine of Figure 11.39.

11.8 Mealy One-Hot Machine

Using unspecified entries to provide intermediate unstable states may result in cycles with different transition times. If equal cycle times are required for all state transitions, then this can be achieved by utilizing only one asserted feedback variable for each row of the merged flow table. Thus, the assignment of state codes is straightforward, thereby eliminating the sometimes tedious task of finding an optimal set of intermediate unstable states. Since only one feedback variable is active in each row, the codes are classified as *one-hot codes*.

In general, the feedback variables are denoted as $y_{1f} y_{2f} y_{3f} \ldots y_{pf}$. In row i of the merged flow table, feedback variable $y_{if} = 1$, while all other variables are equal to 0. When a transition is required from row i to row j, y_{jf} is set to a value of 1 such that $y_{if} y_{jf} = 11$. Then y_{if} is reset to 0. Thus, each transition between two rows passes through one intermediate transient state in an appended row. This allows the duration of all state transition sequences to be identical; that is, each transition from a stable state in row i to a stable state in row j requires two state transition durations of Δt_i and Δt_j. This method uses a maximal number of flip-flops instead of a minimal number of flip-flops for a sequential machine.

Example 11.9 Consider the state diagram for the Mealy one-hot code synchronous sequential machine of Figure 11.44 in which each state code has only one asserted value. All state transitions pass through only unused states; therefore, there will be no glitches on output z_1. Since there are three states, three flip-flops are required. If the machine of Figure 11.44 did not use the one-hot code technique, only two flip-flops would be required.

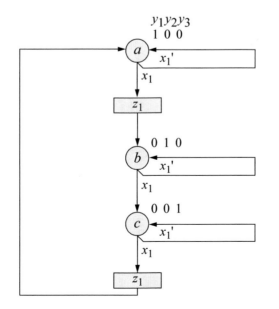

Figure 11.44   State diagram for the Mealy one-hot machine of Section 11.8.

An inherent characteristic of one-hot code machines is that the input maps can be derived directly from the state diagram. The input maps and output map are shown in Figure 11.45(a) and Figure 11.45(b), respectively. The logic diagram using D flip-flops is shown in Figure 11.46 using separate set and reset inputs for flip-flop y_1.

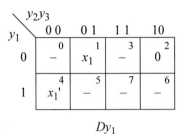

$$Dy_1 = y_3 x_1 + y_1 x_1'$$

$$Dy_2 = y_1 x_1 + y_2 x_1'$$

$$Dy_3 = y_2 x_1 + y_3 x_1'$$

(a)

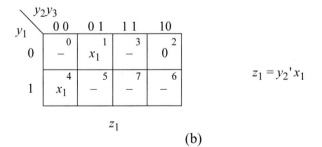

$$z_1 = y_2' x_1$$

(b)

Figure 11.45   (a) Input maps and (b) output map for the Mealy one-hot machine of Figure 11.44.

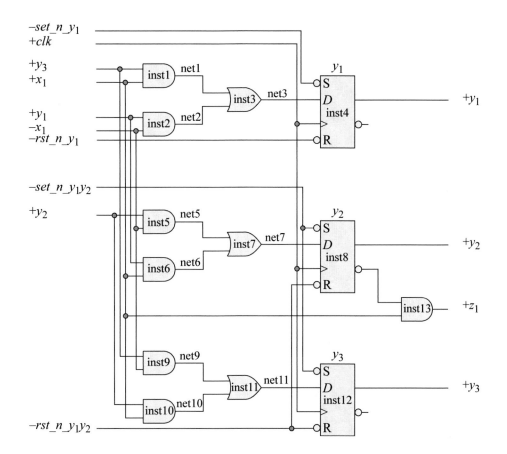

Figure 11.46   Logic diagram for the Mealy one-hot machine of Figure 11.44.

The structural module is shown in Figure 11.47 in which the instantiation names and net names correspond to those in the logic diagram. The test bench is shown in Figure 11.48, the outputs are shown in Figure 11.49, and the waveforms are shown in Figure 11.50.

```
//structural module for a Mealy one-hot machine
module mealy_one_hot_struc (set_n_y1, set_n_y2y3, rst_n_y1,
              rst_n_y2y3, clk, x1, y1, y1_n,
              y2, y2_n, y3, y3_n, z1);

input set_n_y1, set_n_y2y3, rst_n_y1, rst_n_y2y3, clk, x1;
output y1, y1_n, y2, y2_n, y3, y3_n, z1;
//continued on next page
```

Figure 11.47   Structural module for the Mealy one-hot machine of Figure 11.46.

```verilog
//define internal nets
wire net1, net2, net3, net5, net6, net7, net9, net10, net11;

//design the logic for flip-flop y1
and2_df inst1 (
    .x1(y3),
    .x2(x1),
    .z1(net1)
    );

and2_df inst2 (
    .x1(y1),
    .x2(~x1),
    .z1(net2)
    );

or2_df inst3 (
    .x1(net1),
    .x2(net2),
    .z1(net3)
    );

d_ff inst4 (
    .d(net3),
    .clk(clk),
    .q(y1),
    .q_n(y1_n),
    .set_n(set_n_y1),
    .rst_n(rst_n_y1)
    );

//design the logic for flip-flop y2
and2_df inst5 (
    .x1(y2),
    .x2(~x1),
    .z1(net5)
    );

and2_df inst6 (
    .x1(y1),
    .x2(x1),
    .z1(net6)
    );

//continued on next page
```

Figure 11.47   (Continued)

```
or2_df inst7 (
   .x1(net5),
   .x2(net6),
   .z1(net7)
   );

d_ff inst8 (
   .d(net7),
   .clk(clk),
   .q(y2),
   .q_n(y2_n),
   .set_n(set_n_y2y3),
   .rst_n(rst_n_y2y3)
   );

//design the logic for flip-flop y3
and2_df inst9 (
   .x1(y3),
   .x2(~x1),
   .z1(net9)
   );

and2_df inst10 (
   .x1(y2),
   .x2(x1),
   .z1(net10)
   );

or2_df inst11 (
   .x1(net9),
   .x2(net10),
   .z1(net11)
   );

d_ff inst12 (
   .d(net11),
   .clk(clk),
   .q(y3),
   .q_n(y3_n),
   .set_n(set_n_y2y3),
   .rst_n(rst_n_y2y3)
   );


//continued on next page
```

Figure 11.47   (Continued)

```
//design the logic for output z1
and2_df inst13 (
   .x1(y2_n),
   .x2(x1),
   .z1(z1)
   );

endmodule
```

Figure 11.47   (Continued)

```
//test bench for the Mealy one-hot machine
module mealy_one_hot_struc_tb;

reg set_n_y1, set_n_y2y3, rst_n_y1, rst_n_y2y3, clk, x1;
wire y1, y1_n, y2, y2_n, y3, y3_n, z1;

//display variables
initial
$monitor ("x1=%b, y1 y2 y3=%b, z1=%b", x1, {y1, y2, y3}, z1);

//define clock
initial
begin
   clk = 1'b0;
   forever
      #10clk = ~clk;
end

//define input sequence
initial
begin
   #0      set_n_y1 = 1'b0;
           rst_n_y1 = 1'b1;
           set_n_y2y3 = 1'b1;
           rst_n_y2y3 = 1'b0;
           x1 = 1'b0;

   #10     rst_n_y2y3 = 1'b1;
           set_n_y1 = 1'b1;

   x1 = 1'b1;               //assert z1
   @ (posedge clk)         //and go to state 010
//continued on next page
```

Figure 11.48   Test bench for the Mealy one-hot machine of Figure 11.46.

```
   x1 = 1'b1;
   @ (posedge clk)          //go to state 001

   x1 = 1'b1;               //assert z1
   @ (posedge clk)          //and go to state 100

   x1 = 1'b0;
   @ (posedge clk)          //remain in state 100

   x1 = 1'b1;               //assert z1
   @ (posedge clk)          //and go to state 010

   x1 = 1'b0;
   @ (posedge clk)          //remain in state 010

   x1 = 1'b1;
   @ (posedge clk)          //go to state 001

   x1 = 1'b0;
   @ (posedge clk)          //remain in state 001

   x1 = 1'b1;               //assert z1
   @ (posedge clk)          //and go to state 100

   x1 = 1'b0;
   @ (posedge clk)          //remain in state 100

   #10    $stop;
end

//instantiate the module into the test bench
mealy_one_hot_struc inst1 (
   .set_n_y1(set_n_y1),
   .set_n_y2y3(set_n_y2y3),
   .rst_n_y1(rst_n_y1),
   .rst_n_y2y3(rst_n_y2y3),
   .clk(clk),
   .x1(x1),
   .y1(y1),
   .y1_n(y1_n),
   .y2(y2),
   .y2_n(y2_n),
   .y3(y3),
   .y3_n(y3_n),
   .z1(z1)
   );
endmodule
```

Figure 11.48   (Continued)

```
x1=0,  y1 y2 y3=100,  z1=0
x1=1,  y1 y2 y3=100,  z1=1
x1=1,  y1 y2 y3=010,  z1=0
x1=0,  y1 y2 y3=001,  z1=0
x1=1,  y1 y2 y3=001,  z1=1
x1=0,  y1 y2 y3=100,  z1=0
x1=1,  y1 y2 y3=100,  z1=1
x1=0,  y1 y2 y3=010,  z1=0
x1=1,  y1 y2 y3=010,  z1=0
x1=0,  y1 y2 y3=001,  z1=0
```

Figure 11.49   Outputs for the Mealy one-hot machine of Figure 11.46.

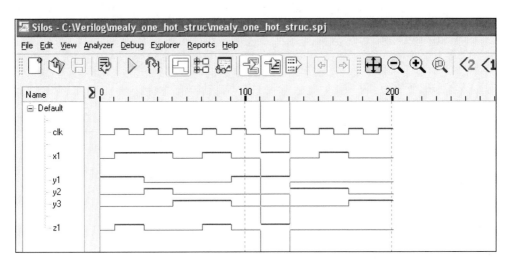

Figure 11.50   Waveforms for the Mealy one-hot machine of Figure 11.46.

Example 11.10   To illustrate the simplicity of designing one-hot machines, the same Mealy one-hot synchronous sequential machine will be designed using an alternative approach. There is no need for Karnaugh maps, equations, or even a logic diagram — the machine can be designed directly from the state diagram. The state diagram is reproduced in Figure 11.51 for convenience. This design will be in a behavioral modeling style using the **case** statement. In this method, the functional operation of the machine is described in an algorithmic style. The synthesis tool determines the necessary logic based on area and timing budget requirements.

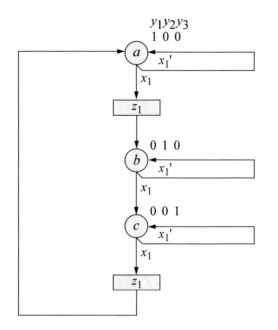

Figure 11.51   State diagram for a Mealy one-hot sequential machine.

Figure 11.52 shows the behavioral module for the Mealy one-hot sequential machine using the **case** statement. The state of the machine is represented by the *state* variable, not by individual flip-flops as in the previous design. The state transitions can be followed easily in conjunction with the state diagram. The test bench is shown in Figure 11.53, the outputs are shown in Figure 11.54, and the waveforms are shown in Figure 11.55.

```
//behavioral mealy one-hot state machine
module mealy_one_hot2 (clk, rst_n, x1, state, z1);

input clk, rst_n, x1;
output z1;
output [2:0] state;

wire clk, rst_n, x1;
reg z1;
reg [2:0] state;
//continued on next page
```

Figure 11.52   Behavioral module for the Mealy one-hot machine of Figure 11.51.

```verilog
parameter   a = 3'b100;
parameter   b = 3'b010;
parameter   c = 3'b001;

//define internal registers
reg [2:0] next_state;

always @ (negedge rst_n or posedge clk)
   if (rst_n == 0)
      state = a;
   else
      state = next_state;

//define next state and output
always @ (state or x1)
begin
   case (state)
      a :    if (x1)
                begin
                   z1 = 1'b1;
                   next_state = b;
                end
             else
                begin
                   z1 = 1'b0;
                   next_state = a;
                end

      b :    if (x1)
                begin
                   z1 = 1'b0;
                   next_state = c;
                end
             else
                begin
                   z1 = 1'b0;
                   next_state = b;
                end

      c :    if (x1)
                begin
                   z1 = 1'b1;
                   next_state = a;
                end

//continued on next page
```

Figure 11.52   (Continued)

```
        else
            begin
                z1 = 1'b0;
                next_state = c;
            end
    default: begin
                next_state = a;
                z1 = 1'b0;
            end
    endcase
end
endmodule
```

Figure 11.52 (Continued)

```
//test bench for the Mealy one-hot machine
module mealy_one_hot2_tb;

reg clk, rst_n, x1;
wire [2:0] state;
wire z1;

//display variables
initial
$monitor ("x1 = %b, state = %b, z1 = %b", x1, state, z1);

//define clock
initial
begin
    clk = 1'b0;
    forever
        #10clk = ~clk;
end

//define input sequence
initial
begin
    #0      rst_n = 1'b0;
            x1 = 1'b0;
    #10     rst_n = 1'b1;

//continued on next page
```

Figure 11.53 Test bench for the Mealy one-hot machine of Figure 11.51.

```
      x1 = 1'b1;              //assert z1
      @ (posedge clk)         //and go to state_b (010)

      x1 = 1'b1;
      @ (posedge clk)         //go to state_c (001)

      x1 = 1'b1;              //assert z1
      @ (posedge clk)         //and go to state_a (100)

      x1 = 1'b0;
      @ (posedge clk)         //remain in state_a (100)

      x1 = 1'b1;              //assert z1
      @ (posedge clk)         //and go to state_b (010)

      x1 = 1'b0;
      @ (posedge clk)         //remain in state_b (010)

      x1 = 1'b1;
      @ (posedge clk)         //go to state_c (001)

      x1 = 1'b0;
      @ (posedge clk)         //remain in state_c (001)

      x1 = 1'b1;              //assert z1
      @ (posedge clk)         //and go to state_a (100)

      x1 = 1'b0;
      @ (posedge clk)         //remain in state_a (100)

      #10     $stop;
end


//instantiate the module into the test bench
mealy_one_hot2 inst1 (
   .clk(clk),
   .rst_n(rst_n),
   .x1(x1),
   .state(state),
   .z1(z1)
   );

endmodule
```

Figure 11.53   (Continued)

```
x1 = 0, state = 100, z1 = 0
x1 = 1, state = 010, z1 = 0
x1 = 1, state = 001, z1 = 1
x1 = 0, state = 100, z1 = 0
x1 = 1, state = 100, z1 = 1
x1 = 0, state = 010, z1 = 0
x1 = 1, state = 010, z1 = 0
x1 = 0, state = 001, z1 = 0
x1 = 1, state = 001, z1 = 1
x1 = 0, state = 100, z1 = 0
```

Figure 11.54   Outputs for the Mealy one-hot machine of Figure 11.51.

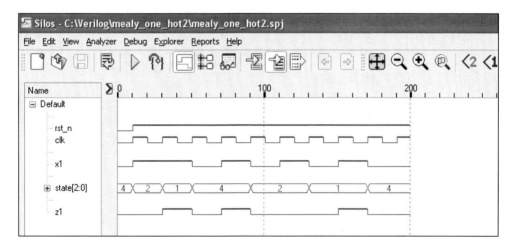

Figure 11.55   Waveforms for the Mealy one-hot machine of Figure 11.51.

11.9 Binary-Coded Decimal (BCD) Adder/ Subtractor

A decimal arithmetic element has two 4-bit BCD operands as inputs and one carry-in; there is one 4-bit output that represents a valid BCD sum and one carry-out, as shown in Figure 11.56. The most commonly used code in BCD arithmetic is the 8421 code, which allows binary fixed-point arithmetic to be used. In fixed-point arithmetic, each bit is treated as a digit, whereas in decimal arithmetic a digit consists of four bits encoded in the 8421 code. Before proceeding with the Verilog design, BCD addition and subtraction will be reviewed.

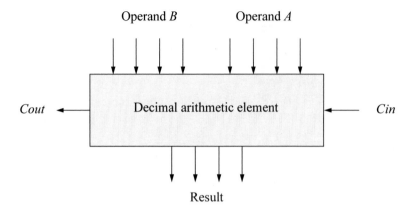

Figure 11.56   Decimal arithmetic element.

11.9.1 BCD Addition

BCD instructions operate on decimal numbers that are encoded as 4-bit binary numbers. For example, the decimal number 15 is encoded in BCD as 0001 0101. The BCD add operation to add decimal numbers 11 and 33 is shown below.

$$
\begin{array}{rr}
11 & 0001 \quad 0001 \\
+) \ \ 33 & +) \ \ 0011 \quad 0011 \\
\hline
44 & 0100 \quad 0100
\end{array}
$$

BCD numbers have a range of 0 to 9; therefore, any number greater than 9 must be adjusted by adding a value of 6 to the number to yield a valid BCD number. For example, if the result of an operation is 1010, then 0110 is added to the result as shown below. This yields a value of 0000 with a carry of 1, or 0001 0000 in BCD which is 10_{10} in decimal.

$$
\begin{array}{r}
1010 \\
+) \ \ 0110 \\
\hline
1 \leftarrow \ 0000 \\
\downarrow \quad \ \downarrow \\
0001 \ 0000
\end{array}
$$

The reason 6 is added is because $10_{10} + 6_{10} = 16_{10}$. However, 16_{10} modulo-16 equals 0 with a carry of 1, or 0001 0000.

If the result of an operation is 1110 (14_{10}), then 0110 is added to the result as shown below. This yields a value of 0100 with a carry of 1, or 0001 0100 in BCD which is 14_{10} in decimal.

$$
\begin{array}{r}
1110 \\
+) \ \underline{0110} \\
1 \leftarrow 0100 \\
\downarrow \quad \downarrow \\
0001 \quad 0100
\end{array}
$$

The reason 6 is added is because $14_{10} + 6_{10} = 20_{10}$. However, 20_{10} modulo-16 equals 4 with a carry of 1, or 0001 0100.

The process of adding 6 to the intermediate result is shown in Table 11.4, which enumerates the valid BCD numbers (0 to 9) and the invalid BCD numbers (10 to 15) that are adjusted to obtain a valid BCD number. For any binary number that is invalid for BCD, count six numbers beyond the invalid number to obtain the correct BCD number accompanied by a carry.

Table 11.4 Binary-Coded Decimal Sum Digits

BCD Result 8421			Carry	8421
0000	Valid			
0001	Valid			
0010	Valid			
0011	Valid			
0100	Valid			
0101	Valid			
0110	Valid			
1110	Valid			
1000	Valid			
1001	Valid			
1010	Invalid	+ 0110 =	1	0000
1011	Invalid	+ 0110 =	1	0001
1100	Invalid	+ 0110 =	1	0010
1101	Invalid	+ 0110 =	1	0011
1110	Invalid	+ 0110 =	1	0100
1111	Invalid	+ 0110 =	1	0101
1 0000				

Example 11.11 The numbers 68_{10} and 35_{10} will be added in BCD to yield a result of 103_{10} as shown below. Both intermediate sums (1101 and 1010) are invalid for BCD; therefore, 0110 must be added to the intermediate sums. Any carry that results from adding six to the intermediate sum is ignored because it provides no new information. The result of the BCD add operation is 0001 0000 0011. The carry produced from the low-order decade is referred to as the *auxiliary carry* flag.

$$
\begin{array}{rcc}
68 & 0110 & 1000 \\
+) \ \ 35 & 0011 & 0101 \\
\hline
103 & 1 \leftarrow & 1101 \\
& 1 \leftarrow 1010 & 0110 \\
& 0110 & 0011 \\
\cline{2-2}
& 0000 &
\end{array}
$$

Example 11.12 A final example of BCD addition is shown below, in which the intermediate sums are valid BCD numbers; however, there is a carry-out of the high-order decade. Whenever the unadjusted sum produces a carry-out, the intermediate sum must be corrected by adding six.

$$
\begin{array}{rcc}
96 & 1001 & 0110 \\
+) \ \ 93 & 1001 & 0011 \\
\hline
189 & 1 \leftarrow 0010 & 1001 \\
& 0110 & \\
\cline{2-2}
& 1000 &
\end{array}
$$

The condition for a correction (adjustment) that also produces a carry-out is shown in Equation 11.5. This specifies that a carry-out will be generated whenever bit position b_8 is a 1 in both decades, when bit positions b_8 and b_4 are both 1s, or when bit positions b_8 and b_2 are both 1s.

$$\text{Carry} = c_8 + b_8 b_4 + b_8 b_2 \tag{11.5}$$

11.9.2 BCD Subtraction

In fixed-point binary arithmetic, subtraction is performed by adding the rs complement of the subtrahend (B) to the minuend (A); that is, by adding the 2s complement of the subtrahend as shown below.

$$A - B = A + (B' + 1)$$

where B' is the 1s complement ($r - 1$) and 1 is added to the low-order bit position to form the 2s complement. The same rationale is used in decimal arithmetic; however, the rs complement is obtained by adding 1 to the 9s complement ($r - 1$) of the subtrahend to form the 10s complement.

Example 11.13 The number 42_{10} will be subtracted from the number 76_{10}. This yields a result of $+34_{10}$. The 10s complement of the subtrahend is obtained as follows using radix 10 numbers: $9 - 4 = 5$; $9 - 2 = 7 + 1 = 8$, where 5 and 7 are the 9s complement of 4 and 2, respectively. A carry-out of the high-order decade indicates that the result is a positive number in BCD.

76	0111	0110
$-$) 42	$+$) 0101	1000
34	1 $\leftarrow$	1110
	1 $\leftarrow$ 1101	0110
	0110	0100
	0011	

Example 11.14 The number 87_{10} will be subtracted from the number 76_{10}. This yields a difference of -11_{10}, which is 1000 1001 represented as a negative BCD number in 10s complement. The 10s complement of the subtrahend is obtained as follows using radix 10 numbers: $9 - 8 = 1$; $9 - 7 = 2 + 1 = 3$, where 1 and 2 are the 9s complement of 8 and 7, respectively. A carry-out of 0 from the high-order decade indicates that the result is a negative BCD number in 10s complement. To obtain the result in radix 10, form the 10s complement of 89_{10}, which will yield 11_{10}. This is interpreted as a negative number.

76	0111	0110
$-$) 87	$+$) 0001	0011
-11 (89)	0 $\leftarrow$ 1000	1001

Before presenting the organization for the BCD adder/subtractor, a 9s complementer will be designed which will be used in the adder/subtractor module together
with a carry-in to form the 10s complement of the subtrahend. A 9s complementer is
required because BCD is not a self-complementing code; that is, it cannot form the $r -$
1 complement by inverting the bits. The truth table for a 9s complementer is shown in
Table 11.5. A mode control input m will be used to determine whether operand $b[3:0]$
will be added to operand $a[3:0]$ or subtracted from operand $a[3:0]$ as shown in the
block diagram of the 9s complementer of Figure 11.57. The function $f[3:0]$ is the 9s
complement of the subtrahend $b[3:0]$.

Table 11.5 Nines Complementer

Subtrahend				9s Complement			
b_3	b_2	b_1	b_0	f_3	f_2	f_1	f_0
0	0	0	0	1	0	0	1
0	0	0	1	1	0	0	0
0	0	1	0	0	1	1	1
0	0	1	1	0	1	1	0
0	1	0	0	0	1	0	1
0	1	0	1	0	1	0	0
0	1	1	0	0	0	1	1
0	1	1	1	0	0	1	0
1	0	0	0	0	0	0	1
1	0	0	1	0	0	0	0

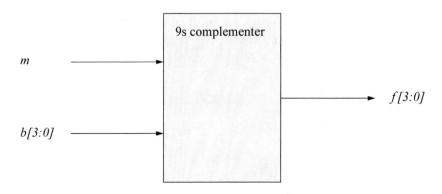

Figure 11.57   Block diagram for a 9s complementer.

The equations for the output $f[3:0]$ of the 9s complementer in conjunction with the mode control input m are shown in Equation 11.6. The equations are obtained directly from Table 11.5. Note that if the operation is add ($m = 0$), then b_0 is passed through the 9s complementer unchanged; if the operation is subtract ($m = 1$), then b_0 is inverted. The b_1 bit is unchanged.

$$f_0 = b_0 \oplus m$$

$$f_1 = b_1$$

$$f_2 = m' b_2 + m(b_2 \oplus b_1)$$

$$f_3 = m' b_3 + m b_3' b_2' b_1' \qquad (11.6)$$

The logic diagram for the 9s complementer is shown in Figure 11.58. The mixed-design module, test bench, and outputs for the 9s complementer are shown in Figure 11.59, Figure 11.60, and Figure 11.61, respectively.

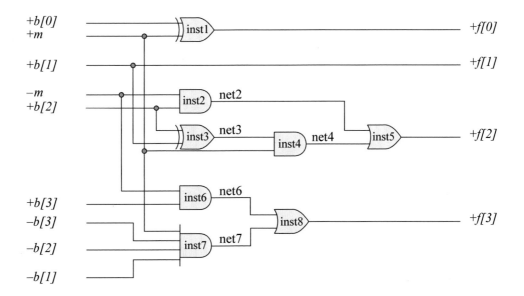

Figure 11.58   Logic diagram for a 9s complementer.

```
//mixed-design module for 9s complementer
module nines_compl (m, b, f);

input m;
input [3:0] b;
output [3:0] f;


//define internal nets
wire net2, net3, net4, net6, net7;


//instantiate the logic gates for the 9s complementer
xor2_df inst1 (
    .x1(b[0]),
    .x2(m),
    .z1(f[0])
    );

assign f[1] = b[1];

and2_df inst2 (
    .x1(~m),
    .x2(b[2]),
    .z1(net2)
    );

xor2_df inst3 (
    .x1(b[2]),
    .x2(b[1]),
    .z1(net3)
    );

and2_df inst4 (
    .x1(net3),
    .x2(m),
    .z1(net4)
    );

or2_df inst5 (
    .x1(net2),
    .x2(net4),
    .z1(f[2])
    );

//continued on next page
```

Figure 11.59   Structural module for a 9s complementer.

```
and2_df inst6 (
   .x1(~m),
   .x2(b[3]),
   .z1(net6)
   );

and4_df inst7 (
   .x1(m),
   .x2(~b[3]),
   .x3(~b[2]),
   .x4(~b[1]),
   .z1(net7)
   );

or2_df inst8 (
   .x1(net6),
   .x2(net7),
   .z1(f[3])
   );

endmodule
```

Figure 11.59   (Continued)

```
//test bench for 9s complementer
module nines_compl_tb;

reg m;
reg [3:0] b;
wire [3:0] f;

//display variables
initial
$monitor ("m=%b, b=%b, f=%b", m, b, f);

//apply input vectors
initial
begin
//add -- do not complement
   #0     m = 1'b0;    b = 4'b0000;
   #10    m = 1'b0;    b = 4'b0001;
   #10    m = 1'b0;    b = 4'b0010;
   #10    m = 1'b0;    b = 4'b0011;
//continued on next page
```

Figure 11.60   Test bench for the 9s complementer of Figure 11.59.

```
    #10    m = 1'b0;    b = 4'b0100;
    #10    m = 1'b0;    b = 4'b0101;
    #10    m = 1'b0;    b = 4'b0110;
    #10    m = 1'b0;    b = 4'b0111;
    #10    m = 1'b0;    b = 4'b1000;
    #10    m = 1'b0;    b = 4'b1001;

//subtract -- complement
    #10    m = 1'b1;    b = 4'b0000;
    #10    m = 1'b1;    b = 4'b0001;
    #10    m = 1'b1;    b = 4'b0010;
    #10    m = 1'b1;    b = 4'b0011;
    #10    m = 1'b1;    b = 4'b0100;
    #10    m = 1'b1;    b = 4'b0101;
    #10    m = 1'b1;    b = 4'b0110;
    #10    m = 1'b1;    b = 4'b0111;
    #10    m = 1'b1;    b = 4'b1000;
    #10    m = 1'b1;    b = 4'b1001;

    #10    $stop;
end

//instantiate the module into the test bench
nines_compl inst1 (
    .m(m),
    .b(b),
    .f(f)
    );

endmodule
```

Figure 11.60   (Continued)

Add	Subtract
m=0, b=0000, f=0000	m=1, b=0000, f=1001
m=0, b=0001, f=0001	m=1, b=0001, f=1000
m=0, b=0010, f=0010	m=1, b=0010, f=0111
m=0, b=0011, f=0011	m=1, b=0011, f=0110
m=0, b=0100, f=0100	m=1, b=0100, f=0101
m=0, b=0101, f=0101	m=1, b=0101, f=0100
m=0, b=0110, f=0110	m=1, b=0110, f=0011
m=0, b=0111, f=0111	m=1, b=0111, f=0010
m=0, b=1000, f=1000	m=1, b=1000, f=0001
m=0, b=1001, f=1001	m=1, b=1001, f=0000

Figure 11.61   Outputs for the 9s complementer of Figure 11.59.

The logic diagram for a BCD adder/subtractor is shown in Figure 11.62. Operand $a[3:0]$ is connected directly to the A inputs of a fixed-point adder; operand $b[3:0]$ is connected to the inputs of a 9s complementer whose outputs $f[3:0]$ connect to the B inputs of the adder. There is also a mode control input m to the 9s complementer which specifies either an add operation ($m = 0$) or a subtract operation ($m = 1$). The mode control is also connected to the carry-in of the low-order adder.

If the operation is addition, then the mode control adds 0 to the uncomplemented version of operand $b[3:0]$. If the operation is subtraction, then the mode control adds a 1 to the 9s complement of operand $b[3:0]$ to form the 10s complement. The outputs of the adder are a 4-bit intermediate sum $sum[3:0]$ and a carry-out labeled $cout3$. The intermediate sum is connected to a second fixed-point adder, which will add 0000 to the intermediate sum if no adjustment is required or add 0110 to the intermediate sum if adjustment is required.

The carry-out of the low-order BCD stage is specified as an auxiliary carry aux_cy and is defined by Equation 11.7. The auxiliary carry determines whether 0000 or 0110 is added to the intermediate sum produce a valid BCD digit, as well as providing a carry-in to the high-order stage.

$$aux\_cy = cout3 + sum[3] \, sum[1] + sum[3] \, sum[2] \qquad (11.7)$$

The same rationale applies to the high-order BCD stage for operand $a[7:4]$ and operand $b[7:4]$. The 9s complementer produces outputs $f[7:4]$ which are connected to the B inputs of the adder. The mode control input is also connected to the input of the 9s complementer. The fixed-point adder generates an intermediate sum of $sum[7:4]$ and a carry-out $cout7$. The carry-out of the BCD adder/subtractor is labeled $cout$ and determines whether 0000 or 0110 is added to the intermediate sum. The equation for $cout$ is shown in Equation 11.8. The BCD adder/subtractor can be extended to any size operand by simply adding more stages.

$$cout = cout7 + sum[7] \, sum[5] + sum[7] \, sum[6] \qquad (11.8)$$

The module for the BCD adder/subtractor is shown in Figure 11.63 using structural modeling. There are two input operands $a[7:0]$ and $b[7:0]$ and one input mode control m. There are two outputs, $bcd[7:0]$, which represents a valid BCD number, and a carry-out $cout$. The test bench is shown in Figure 11.64 and contains operands for addition and subtraction, including numbers that result in negative differences in BCD. The outputs are shown in Figure 11.65 for both addition and subtraction and the waveforms are shown in Figure 11.66.

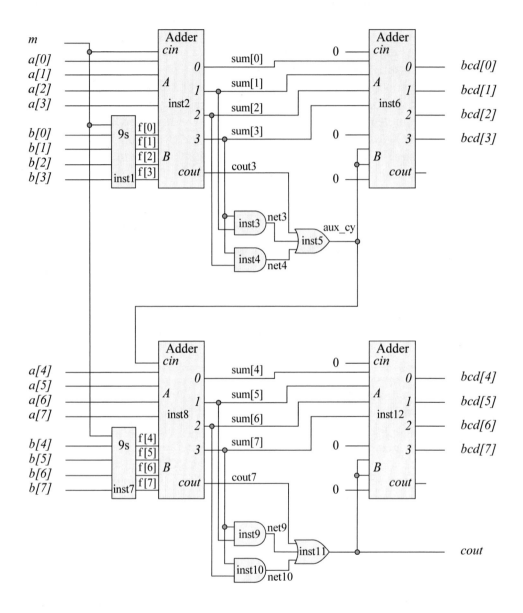

Figure 11.62   Logic diagram for a BCD adder/subtractor.

The two adders that generate valid BCD digits each have a carry-out signal *cout*; however *cout* is not used because it provides no new information. The result of the BCD adder/subtractor is two valid BCD digits, *bcd[7:4]* and *bcd[3:0]*, plus a carry-out *cout*.

```
//structural bcd adder subtractor
module add_sub_bcd (a, b, m, bcd, cout);

input [7:0] a, b;
input m;
output [7:0] bcd;
output cout;

//define internal nets
wire [7:0] f;
wire [7:0] sum;
wire cout3, aux_cy, cout7;
wire net3, net4, net9, net10;


//instantiate the logic for the low-order stage [3:0]
//instantiate the 9s complementer
nines_compl inst1 (
    .m(m),
    .b(b[3:0]),
    .f(f[3:0])
    );

//instantiate the adder for the intermediate sum
adder4 inst2 (
    .a(a[3:0]),
    .b(f[3:0]),
    .cin(m),
    .sum(sum[3:0]),
    .cout(cout3)
    );

//instantiate the logic gates
and2_df inst3 (
    .x1(sum[3]),
    .x2(sum[1]),
    .z1(net3)
    );

and2_df inst4 (
    .x1(sum[2]),
    .x2(sum[3]),
    .z1(net4)
    );

//continued on next page
```

Figure 11.63   Structural module for the BCD adder/subtractor of Figure 11.62.

```
or3_df inst5 (
   .x1(cout3),
   .x2(net3),
   .x3(net4),
   .z1(aux_cy)
   );

//instantiate the adder for the bcd sum [3:0]
adder4 inst6 (
   .a(sum[3:0]),
   .b({1'b0, aux_cy, aux_cy, 1'b0}),
   .cin(1'b0),
   .sum(bcd[3:0])
   );

//instantiate the logic for the high-order stage [7:4]
//instantiate the 9s complementer
nines_compl inst7 (
   .m(m),
   .b(b[7:4]),
   .f(f[7:4])
   );

//instantiate the adder for the intermediate sum
adder4 inst8 (
   .a(a[7:4]),
   .b(f[7:4]),
   .cin(aux_cy),
   .sum(sum[7:4]),
   .cout(cout7)
   );

//instantiate the logic gates
and2_df inst9 (
   .x1(sum[7]),
   .x2(sum[5]),
   .z1(net9)
   );

and2_df inst10 (
   .x1(sum[6]),
   .x2(sum[7]),
   .z1(net10)
   );
//continued on next page
```

Figure 11.63   (Continued)

```
or3_df inst11 (
   .x1(cout7),
   .x2(net9),
   .x3(net10),
   .z1(cout)
   );

//instantiate the adder for the bcd sum [7:0]
adder4 inst12 (
   .a(sum[7:4]),
   .b({1'b0, cout, cout, 1'b0}),
   .cin(1'b0),
   .sum(bcd[7:4])
   );

endmodule
```

Figure 11.63   (Continued)

```
//test bench for the bcd adder subtractor
module add_sub_bcd_tb;

reg [7:0] a, b;
reg m;

wire [7:0] bcd;
wire cout;


//display variables
initial
$monitor ("a=%b, b=%b, m=%b, cout=%b, bcd=%b",
          a, b, m, cout, bcd);


//apply input vectors
initial
begin
//add bcd
   #0    a = 8'b1001_1001;    b = 8'b0110_0110;    m = 1'b0;
   #10   a = 8'b0010_0110;    b = 8'b0101_1001;    m = 1'b0;
   #10   a = 8'b0001_0001;    b = 8'b0011_0011;    m = 1'b0;

//continued on next page
```

Figure 11.64   Test bench for the BCD adder/subtractor of Figure 11.63.

```
   #10    a = 8'b0000_1000;     b = 8'b0000_0101;     m = 1'b0;
   #10    a = 8'b0110_1000;     b = 8'b0011_0101;     m = 1'b0;
   #10    a = 8'b1000_1001;     b = 8'b0101_1001;     m = 1'b0;
   #10    a = 8'b1001_0110;     b = 8'b1001_0011;     m = 1'b0;
   #10    a = 8'b1001_1001;     b = 8'b0000_0001;     m = 1'b0;
   #10    a = 8'b1001_1001;     b = 8'b0110_0110;     m = 1'b0;

//subtract bcd
   #10    a = 8'b1001_1001;     b = 8'b0110_0110;     m = 1'b1;
   #10    a = 8'b1001_1001;     b = 8'b0110_0110;     m = 1'b1;
   #10    a = 8'b0011_0011;     b = 8'b0110_0110;     m = 1'b1;
   #10    a = 8'b0111_0110;     b = 8'b0100_0010;     m = 1'b1;
   #10    a = 8'b0111_0110;     b = 8'b1000_0111;     m = 1'b1;
   #10    a = 8'b0001_0001;     b = 8'b1001_1001;     m = 1'b1;
   #10    a = 8'b0001_1000;     b = 8'b0010_0110;     m = 1'b1;
   #10    a = 8'b0001_1000;     b = 8'b0010_1000;     m = 1'b1;
   #10    a = 8'b1001_0100;     b = 8'b0111_1000;     m = 1'b1;
   #10    $stop;
end

//instantiate the module into the test bench
add_sub_bcd inst1 (
   .a(a),
   .b(b),
   .m(m),
   .bcd(bcd),
   .cout(cout)
   );

endmodule
```

Figure 11.64   (Continued)

```
Addition
a=1001_1001, b=0110_0110, m=0, cout=1, bcd=0110_0101
a=0010_0110, b=0101_1001, m=0, cout=0, bcd=1000_0101
a=0001_0001, b=0011_0011, m=0, cout=0, bcd=0100_0100
a=0000_1000, b=0000_0101, m=0, cout=0, bcd=0001_0011
a=0110_1000, b=0011_0101, m=0, cout=1, bcd=0000_0011
a=1000_1001, b=0101_1001, m=0, cout=1, bcd=0100_1000
a=1001_0110, b=1001_0011, m=0, cout=1, bcd=1000_1001
a=1001_1001, b=0000_0001, m=0, cout=1, bcd=0000_0000
a=1001_1001, b=0110_0110, m=0, cout=1, bcd=0110_0101
//continued on next page
```

Figure 11.65   Outputs for the BCD adder/subtractor of Figure 11.63.

```
Subtraction
a=1001_1001, b=0110_0110, m=1, cout=1, bcd=0011_0011
a=0011_0011, b=0110_0110, m=1, cout=0, bcd=0110_0111
a=0111_0110, b=0100_0010, m=1, cout=1, bcd=0011_0100
a=0111_0110, b=1000_0111, m=1, cout=0, bcd=1000_1001
a=0001_0001, b=1001_1001, m=1, cout=0, bcd=0001_0010
a=0001_1000, b=0010_0110, m=1, cout=0, bcd=1001_0010
a=0001_1000, b=0010_1000, m=1, cout=0, bcd=1001_0000
a=1001_0100, b=0111_1000, m=1, cout=1, bcd=0001_0110
```

Figure 11.65   (Continued)

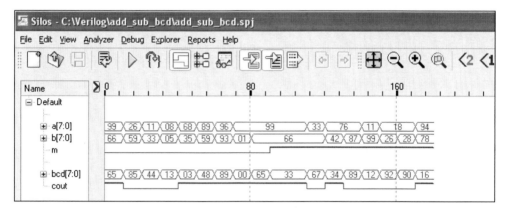

Figure 11.66   Waveforms for the BCD adder/subtractor of Figure 11.63.

11.10 Pipelined Reduced Instruction Set Computer (RISC) Processor

Before designing the project, a brief introduction to pipelined RISC processors will be presented. A pipelined computer improves CPU speed and system throughput because several instructions can be processed in parallel. The processor partitions an operation into many autonomous but related suboperations. The machine partitions the work over several concurrently operating units. The CPU can begin processing the next instruction before the current instruction is completed; that is, the system can overlap the execution of contiguous instructions.

Fetching instructions from main memory or cache is a major bottleneck due to the relatively slow access times. The slow access times can be alleviated by prefetching instructions before they are required by the processing unit. The prefetched instructions are loaded into a prefetch buffer where they are retained until needed by the processor. Prefetching can be overlapped with normal processing and can be accomplished on unused memory bus cycles.

Figure 11.67 shows one type of instruction prefetch buffer which contains two 4-byte registers *Ibufr A* and *Ibufr B* that store instructions of either two bytes or four bytes. Associated with each instruction buffer is a modulo-4 counter (*A full* and *B full*) that increments by one for each byte that is loaded into the respective instruction buffer. When a counter reaches a count of three, no additional bytes are loaded into the buffer. When a buffer is empty, the counter is reset and the corresponding buffer is again loaded from memory.

The two buffers connect to an array of 4:1 multiplexers that align the instructions so that they can be loaded into an *instruction register* (IR), with the operation code byte in the leftmost byte position. The alignment is accomplished by a modulo-4 instruction buffer program counter *IPC*.

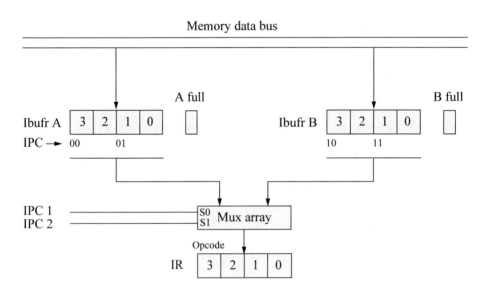

Figure 11.67   Instruction buffer for 2- and 4-byte instructions.

There are many different stages in a pipeline, with each stage performing one unique operation in the instruction processing. When the pipeline is full, a result is obtained every clock cycle. An example of a 4-stage pipeline is shown in Figure 11.68. The *Ifetch* stage fetches the instruction from memory; the *Decode* stage decodes the instruction and fetches the operands; the *Execute* stage performs the operation specified in the instruction; the *Store* stage stores the result in the destination location. Four instructions are in progress at any given clock cycle. Each stage of the pipeline performs its task in parallel with all other stages.

If the instruction required is not available in the cache, then a cache miss occurs, necessitating a fetch from main memory. This is referred to as a pipeline *stall* and delays processing the next instruction. Information is passed from one stage to the next by means of a storage buffer, as shown in Figure 11.69. There must be a register in the input of each stage (or between stages) to store information that is transmitted from the preceding stage. This prevents data being processed by one stage from interfering with the following stage during the same clock period.

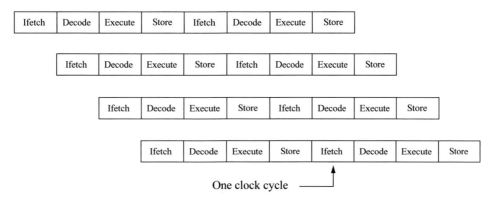

Figure 11.68 Example of a 4-stage pipeline.

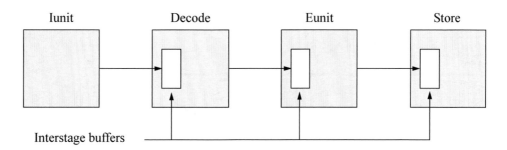

Figure 11.69 Four stages of a pipeline showing the interstage storage buffers.

RISC processors have only a small number of instructions compared to a complex instruction set computer (CISC). The instructions are also smaller in size with a smaller number of fields and usually of fixed length. Most instructions have the same format with a limited number of addressing modes and are executed by hardware. RISC processors have an instruction cache and a data cache; only load and store instructions reference memory.

The operands are loaded from register files and stored in register files. These are large register files that provide fast access, thereby eliminating many slow memory accesses. The control unit is hardwired, not microprogram controlled, for increased speed. The RISC architecture reduces the number of clock cycles required to execute an instruction; instruction execution requires only one clock cycle per pipeline stage.

The RISC architecture uses fewer, very simple instructions compared to complex instruction set computers. RISC instructions are ideal for pipelined execution. Hardwired implementation of simple instructions provides faster execution rates. A key advantage of RISC computers is that they can be used effectively by optimizing compilers. Another advantage is that they provide a smaller chip area, thus allowing more on-chip space for processor register files and cache.

This section presents the Verilog design of a pipelined RISC processor with no prefetch buffer. Although this is a small RISC processor, the same techniques can be applied to any size processor. There are 16 instructions as shown in Table 11.6. Figure 11.70 shows the various fields for each instruction. All instructions, except load and store, have the same format.

Table 11.6 Instructions for the RISC Processor of Section 11.10

Program	
LD	
LD	
LD	
LD	regfile [0:7] = dcache [0:7]
LD	
LD	
LD	
LD	
ADD	
SUB	
AND	
OR	
XOR	
INC	See instruction cache table for op code and operands
DEC	
NOT	
NEG	
SHR	
SHL	
ROR	
ROL	
Continued on next page	

Table 11.6 Instructions for the RISC Processor of Section 11.10

Program	
ST	
ST	
ST	
ST	dcache [8:15] = regfile [0:7]
ST	
ST	
ST	
ST	
NOP	
NOP	clear pipeline
NOP	

Op Code	Opnd A	Opnd B	Dst
nnnn	RRR	RRR	RRR
12 9	8 6	5 3	2 0

where RRR specifies one of eight registers
(0 – 7) for opnd A, opnd B, and Dst.

Op Code		Opnd A		Dst
1110 (Load)	0	Memory address	0	RRR
12 9	8	7 4	3	2 0

Op Code		Opnd A	Dst
1111 (Store)	00	RRR	Memory address
12 9	8 7	6 4	3 0

Figure 11.70   Instruction format for the pipelined RISC processor.

There are six units (modules) in the processor: instruction cache (*icache*), instruction unit (*iunit*), decode unit (*decode*), execution unit (*eunit*), data cache (*dcache*), and a register file (*regfile*). Each unit will be a behavioral or a mixed-design module and will be compiled and simulated using a test bench to show binary outputs and waveforms. The entire processor will then be configured using a structural module and will display the waveforms. The architecture of Figure 11.71 shows the six units and their interconnections. The structural block diagram of Figure 11.72 shows the hierarchy.

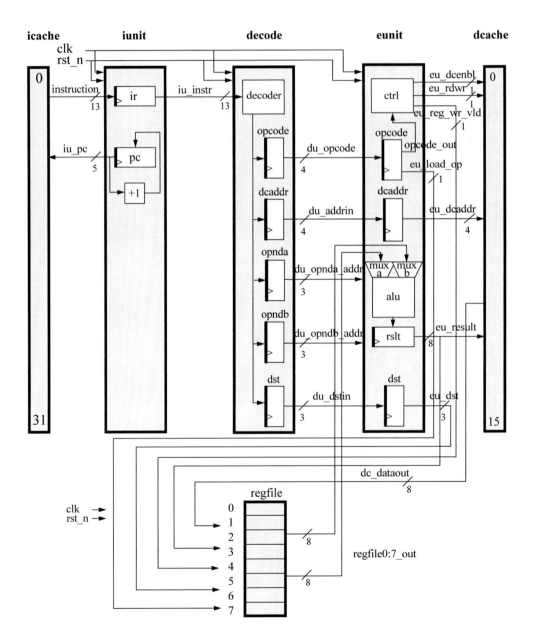

Figure 11.71   Architecture for the pipelined RISC processor of Section 11.10.

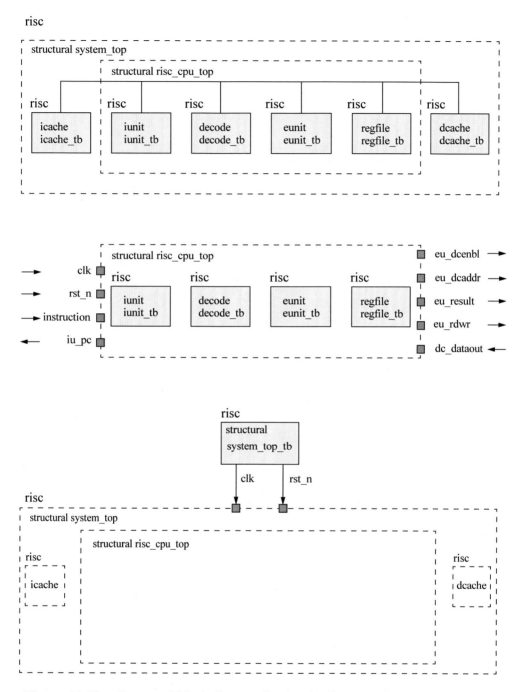

Figure 11.72   Structural block diagram for the pipelined RISC processor of Section 11.10.

The instruction cache and data cache will be preloaded in their respective modules with the instructions and data given in Table 11.7 and Table 11.8, respectively. The instruction cache contents show all the instructions and the application program. A predefined program is used so that the results can be easily verified. The operand size is eight bits.

Table 11.7 Instruction Cache Contents

Address	Operation	Op Code	Opnd A	Opnd B	Dst
00000	LD	1110	0 0000	Not used	0000
00001	LD	1110	0 0001	Not used	0001
00010	LD	1110	0 0010	Not used	0010
00011	LD	1110	0 0011	Not used	0011
00100	LD	1110	0 0100	Not used	0100
00101	LD	1110	0 0101	Not used	0101
00110	LD	1110	0 0110	Not used	0110
00111	LD	1110	0 0111	Not used	0111
01000	ADD	0001	000	001	000
01001	SUB	0010	111	110	001
01010	AND	0011	010	101	010
01011	OR	0100	011	100	011
01100	XOR	0101	100	100	100
01101	INC	0110	101	000	101
01110	DEC	0111	110	000	110
01111	NOT	1000	111	000	111
10000	NEG	1001	000	000	000
10001	SHR	1010	001	000	001
10010	SHL	1011	010	000	010
10011	ROR	1100	011	000	011
10100	ROL	1101	100	000	100
10101	ST	1111	00 000	Not used	1000
10110	ST	1111	00 001	Not used	1001
10111	ST	1111	00 010	Not used	1010
11000	ST	1111	00 011	Not used	1011
11001	ST	1111	00 100	Not used	1100
11010	ST	1111	00 101	Not used	1101
11011	ST	1111	00 110	Not used	1110
11100	ST	1111	00 111	Not used	1111
11101	NOP	0000	000	000	000
11110	NOP	0000	000	000	000
11111	NOP	0000	000	000	000

Table 11.8 Data Cache Contents

	Address	Data
SRC	0000	0000 0000
	0001	0010 0010
	0010	0100 0100
	0011	0110 0110
	0100	1000 1000
	0101	1010 1010
	0110	1100 1100
	0111	1111 1111
DST	1000	0000 0000
	1001	0000 0000
	1010	0000 0000
	1011	0000 0000
	1100	0000 0000
	1101	0000 0000
	1110	0000 0000
	1111	0000 0000

RISC instructions typically have a fixed-length instruction format. The instruction length for this design is 13 bits. Only load and store instructions can access memory. The addressing mode is register; that is, the operand is in the register file. In this design there are eight registers, which are initialized by eight load instructions that load certain data cache contents into the register file. The initial contents of the register file are shown in Table 11.9; the final contents of the register file are shown in Table 11.10.

Table 11.9 Register File Contents before Execution

Address	Data
000	0000 0000
001	0010 0010
010	0100 0100
011	0110 0110
100	1000 1000
101	1010 1010
110	1100 1100
111	1111 1111

Table 11.10 Register File Contents after Execution

Address	Data
000	1101 1110
001	0001 1001
010	0000 0000
011	0111 0111
100	0000 0000
101	1010 1011
110	1100 1011
111	0000 0000

Detailed block diagrams of the six units are shown in Figure 11.73 to Figure 11.78.

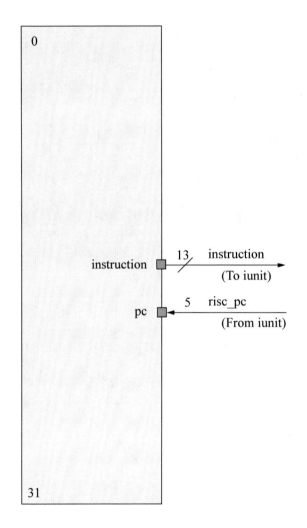

Figure 11.73   Instruction cache for the pipelined RISC processor.

The small dark squares represent the ports of the module. There are two ports in the instruction cache: *instruction*, which sends 13-bit instructions to the *iunit*, and a port for a 5-bit program counter *pc* that receives the next instruction cache address from the *iunit*.

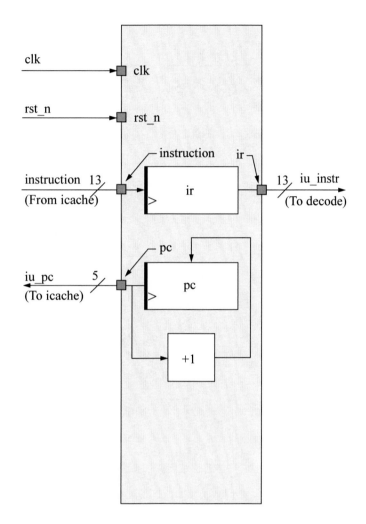

Figure 11.74   Instruction unit for the pipelined RISC processor.

The instruction unit has five ports: one each for the clock *clk* and reset *rst_n* inputs; a port to receive a 13-bit instruction from the instruction cache called *instruction[12:0]* that is loaded into the instruction register *ir* — the net name from the instruction cache is labeled *instruction[12:0]*; a port for the output of the program counter *pc[4:0]* that sends the program counter to the instruction cache on a net labeled *iu_pc[4:0]*; and a port called *ir[12:0]* that passes the 13-bit instruction to the decode unit on a net called *iu_instr[12:0]*.

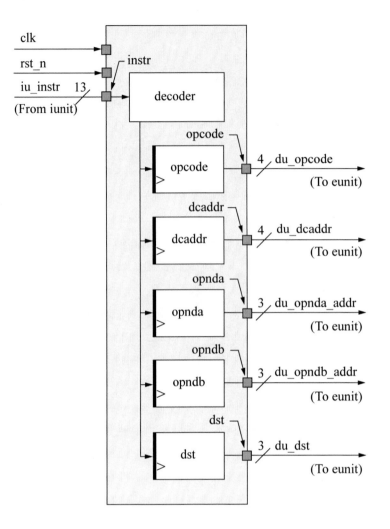

Figure 11.75   Decode unit for the pipelined RISC processor.

The decode unit has three input ports and five output ports. The input ports are *clk* and *rst_n*, plus a port *instr* to receive the instruction from the instruction unit. There is an output port *opcode[3:0]* that sends the operation code to the execution unit on a net labeled *du_opcode[3:0]*; a port *dcaddr[3:0]* that sends the data cache address to the execution unit on a net labeled *du_dcaddr[3:0]*; a port *opnda[2:0]* that sends the address of operand *A* to the execution unit on a net labeled *du_opnda_addr[2:0]*; a port *opndb[2:0]* that sends the address of operand *B* to the execution unit on a net labeled *du_opndb_addr[2:0]*; and a port *dst* that sends the destination address to the execution unit on a net labeled *du_dst[2:0]* to be used by the register file.

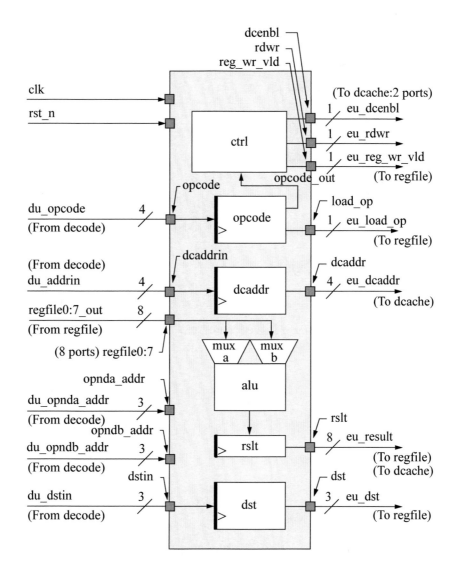

Figure 11.76   Execution unit for the pipelined RISC processor.

The execution unit is the dominant unit in the processor — all operations are performed by the execution unit. It obtains the opcode and operand addresses from the decode unit. There are eight 8-bit input ports that connect the eight registers of the register file to the ALU by means of multiplexers. The multiplexer inputs are selected by signals from the *ctrl* block. The operand addresses then select the appropriate operands to be utilized in the execution of the instruction. The result of the operation is placed in a *rslt* register — also controlled by the *ctrl* block — then sent to the register file and data cache. This will become more apparent when the execution unit module is presented.

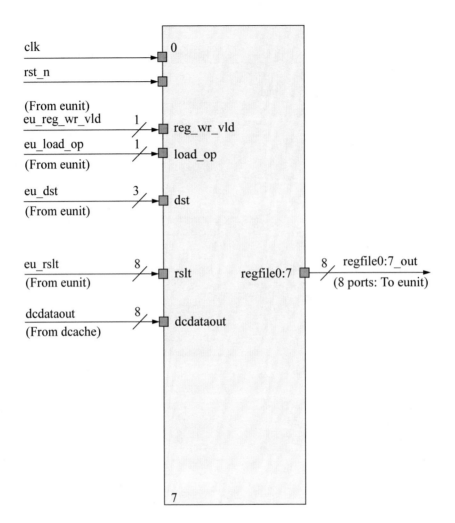

Figure 11.77   Register file for the pipelined RISC processor.

The *reg_wr_vld* port is an input from the execution unit that enables a write operation to the register file. The signal at the *load_op* port then loads the register file with the result of an instruction execution or data from the data cache. The result of an instruction execution is available at the input port *rslt*; the data from the data cache is available at the input port *dcdataout*. There are eight 8-bit output ports that provide data to the execution unit. A logic diagram for the register file will be shown when the register file module is presented.

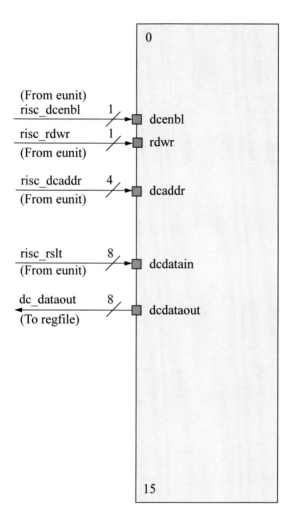

Figure 11.78   Data cache for the pipelined RISC processor.

The data cache input port *dcenbl* enables the data cache for a read or write operation. The *rdwr* port provides a read enable (*rdwr* = 1) or a write enable (*rdwr* = 0) for the data cache at the address specified by the address at port *dcaddr*. The data to be written to the data cache is available at the *dcdatain* port; the data to be read from the data cache is available at the *dcdataout* port.

The instruction cache contents, operations, and results are shown in Figure 11.79. Each instruction cache address indicates the instruction at that address. This is followed by a comment specifying the instruction, register file addresses from where the operands are obtained, and the destination address in the register file. The operands are also shown together with the result of the operation. The instruction formats are shown in Figure 11.70.

```
//load dcache into regfile; regfile [0:7] = dcache [0:7]
icache [00] = 13'h1c00;    //ld : regfile[0]=dcache[0]=00h
icache [01] = 13'h1c11;    //ld : regfile[1]=dcache[1]=22h
icache [02] = 13'h1c22;    //ld : regfile[2]=dcache[2]=44h
icache [03] = 13'h1c33;    //ld : regfile[3]=dcache[3]=66h
icache [04] = 13'h1c44;    //ld : regfile[4]=dcache[4]=88h
icache [05] = 13'h1c55;    //ld : regfile[5]=dcache[5]=aah
icache [06] = 13'h1c66;    //ld : regfile[6]=dcache[6]=cch
icache [07] = 13'h1c77;    //ld : regfile[7]=dcache[7]=ffh


//execute instruction set. All operations are on the regfile
icache [08] = 13'h0208;    //add: [0]+[1]--->[0]; 00h+22h=22h
icache [09] = 13'h05f1;    //sub: [7]-[6]--->[1]; ffh-cch=33h
icache [10] = 13'h0238;    //and: [2]&[5]--->[2]; 44h&aah=00h
icache [11] = 13'h08e3;    //or : [3]|[4]--->[3]; 66h|88h=eeh
icache [12] = 13'h0b24;    //xor: [4]^[4]--->[4]; 88h^88h=00h
icache [13] = 13'h0d45;    //inc: [5]+1  --->[5]; aah+1=abh
icache [14] = 13'h0f86;    //dec: [6]-1  --->[6]; cch-1=cbh
icache [15] = 13'h11c7;    //not: [7]not --->[7]; ffh=00h
icache [16] = 13'h1200;    //neg: [0]neg --->[0]; 22h=deh
icache [17] = 13'h1441;    //shr: [1]shr --->[1]; 33h=19h
icache [18] = 13'h1682;    //shl: [2]shl --->[2]; 00h=00h
icache [19] = 13'h18c3;    //ror: [3]ror --->[3]; eeh=77h
icache [20] = 13'h1b04;    //rol: [4]rol --->[4]; 00h=00h


//store regfile into dcache; dcache [8:15] = regfile [0:7]
icache [21] = 13'h1e08;    //st : dcache[8]=regfile[0]=deh
icache [22] = 13'h1e19;    //st : dcache[9]=regfile[1]=19h
icache [23] = 13'h1e2a;    //st : dcache[10]=regfile[2]=00h
icache [24] = 13'h1e3b;    //st : dcache[11]=regfile[3]=77h
icache [25] = 13'h1e4c;    //st : dcache[12]=regfile[4]=00h
icache [26] = 13'h1e5d;    //st : dcache[13]=regfile[5]=abh
icache [27] = 13'h1e6e;    //st : dcache[14]=regfile[6]=cbh
icache [28] = 13'h1e7f;    //st : dcache[15]=regfile[7]=00h


//execute 3 nop instructions to clear the pipeline
icache [29] = 13'h0000;    //nop: clear iunit
icache [30] = 13'h0000;    //nop: clear decode
icache [31] = 13'h0000;    //nop: clear eunit
```

Figure 11.79 Contents of the instruction cache, operations to be executed, and the results of the operations.

11.10.1 Instruction Cache

Instructions are assigned to the cache by the system task **$readmemb**, which reads and loads data from a specified text file into the instruction cache. The text file is saved in the project file as *icache.instr* with no *.v* extension. The text file is shown in Figure 11.80. The behavioral module for the instruction cache is shown in Figure 11.81. The test bench is shown in Figure 11.82, the outputs are shown in Figure 11.83, and the waveforms are shown in Figure 11.84.

```
1_1100_0000_0000
1_1100_0001_0001
1_1100_0010_0010
1_1100_0011_0011
1_1100_0100_0100        regfile[0:7] = dcache[0:7]
1_1100_0101_0101
1_1100_0110_0110
1_1100_0111_0111
0_0010_0000_1000  ◄──  ADD
0_0101_1111_0001  ◄──  SUB
0_0110_1010_1010  ◄──  AND
0_1000_1110_0011
0_1011_0010_0100
0_1101_0100_0101        execute the instruction set
0_1111_1000_0110
1_0001_1100_0111
1_0010_0000_0000
1_0100_0100_0001
1_0110_1000_0010
1_1000_1100_0011
1_1011_0000_0100
1_1110_0000_1000
1_1110_0001_1001
1_1110_0010_1010
1_1110_0011_1011        dcache[8:15] = regfile[0:7]
1_1110_0100_1100
1_1110_0101_1101
1_1110_0110_1110
1_1110_0111_1111
0_0000_0000_0000
0_0000_0000_0000        clear pipeline
0_0000_0000_0000
```

Figure 11.80   Text file for the instruction cache.

```
//behavioral icache contents
module risc_icache (pc, instruction);

//list which are inputs and which are outputs
input [4:0] pc;
output [12:0] instruction;

//list which are wire and which are reg
wire [4:0] pc;
reg [12:0] instruction;

//define memory size
reg [12:0] icache [0:31];//# of bits per reg; # of regs
                         //13 bits per reg; 31 regs
                         //icache is an array of 32 13-bit regs

//define memory contents
initial
begin
   $readmemb ("icache.instr", icache);
end

/* alternatively, the icache could have been loaded
by initializing each location separately as shown below
initial
begin
   icache [00] = 13'h1c00;
   icache [01] = 13'h1c11;
   icache [02] = 13'h1c22;
   icache [03] = 13'h1c33;
   icache [04] = 13'h1c44;
   icache [05] = 13'h1c55;
   icache [06] = 13'h1c66;
   icache [07] = 13'h1c77;
   icache [08] = 13'h0208;
   icache [09] = 13'h05f1;
   icache [10] = 13'h06aa;
   icache [11] = 13'h08e3;
   icache [12] = 13'h0b24;
   icache [13] = 13'h0d45;
   icache [14] = 13'h0f86;
   icache [15] = 13'h11c7;
   icache [16] = 13'h1200;
   icache [17] = 13'h1441;
   icache [18] = 13'h1682;
   icache [19] = 13'h18c3;
   icache [20] = 13'h1b04;           //continued on next page
```

Figure 11.81   Behavioral module for the instruction cache.

```
      icache [21] = 13'h1e08;
      icache [22] = 13'h1e19;
      icache [23] = 13'h1e2a;
      icache [24] = 13'h1e3b;
      icache [25] = 13'h1e4c;
      icache [26] = 13'h1e5d;
      icache [27] = 13'h1e6e;
      icache [28] = 13'h1e7f;
      icache [29] = 13'h0000;
      icache [30] = 13'h0000;
      icache [31] = 13'h0000;
end
*/

always @ (pc)
begin
   instruction = icache [pc];
end

endmodule
```

Figure 11.81   (Continued)

```
//icache test bench
module risc_icache_tb;

integer i;                   //used for display

reg [4:0] pc;                //inputs are reg for test bench
wire [12:0] instruction;     //outputs are wire for test bench

initial
begin
   #10    pc = 5'b00000;
   #10    pc = 5'b00000;
   #10    pc = 5'b00001;
   #10    pc = 5'b00010;
   #10    pc = 5'b00011;
   #10    pc = 5'b00100;
   #10    pc = 5'b00101;
   #10    pc = 5'b00110;
   #10    pc = 5'b00111;
   #10    pc = 5'b01000;
   #10    pc = 5'b01001;
//continued on next page
```

Figure 11.82   Test bench for the instruction cache.

```
   #10    pc = 5'b01010;
   #10    pc = 5'b01011;
   #10    pc = 5'b01100;
   #10    pc = 5'b01101;
   #10    pc = 5'b01110;
   #10    pc = 5'b01111;
   #10    pc  = 5'b10000;
   #10    pc = 5'b10001;
   #10    pc = 5'b10010;
   #10    pc = 5'b10011;
   #10    pc = 5'b10100;
   #10    pc = 5'b10101;
   #10    pc = 5'b10110;
   #10    pc = 5'b10111;
   #10    pc = 5'b11000;
   #10    pc = 5'b11001;
   #10    pc = 5'b11010;
   #10    pc = 5'b11011;
   #10    pc = 5'b11100;
   #10    pc = 5'b11101;
   #10    pc = 5'b11110;
   #10    pc = 5'b11111;

   #20    $stop;              //must not stop before display ends
end                          //otherwise not all addrs will display


initial
begin
//#20 synchs waveforms produced by the first initial with
//the display produced by the second initial

   #20    for (i=0; i<32; i=i+1)
          begin
             #10 $display ("address %h = %h", i, instruction);
          end
   #600 $stop;
end


//instantiate the behavioral module into the test bench
risc_icache inst1 (
   .pc(pc),
   .instruction(instruction)
   );

endmodule
```

Figure 11.82   (Continued)

```
address 00000000 = 1c00        address 00000010 = 1200
address 00000001 = 1c11        address 00000011 = 1441
address 00000002 = 1c22        address 00000012 = 1682
address 00000003 = 1c33        address 00000013 = 18c3
address 00000004 = 1c44        address 00000014 = 1b04
address 00000005 = 1c55        address 00000015 = 1e08
address 00000006 = 1c66        address 00000016 = 1e19
address 00000007 = 1c77        address 00000017 = 1e2a
address 00000008 = 0208        address 00000018 = 1e3b
address 00000009 = 05f1        address 00000019 = 1e4c
address 0000000a = 06aa        address 0000001a = 1e5d
address 0000000b = 08e3        address 0000001b = 1e6e
address 0000000c = 0b24        address 0000001c = 1e7f
address 0000000d = 0d45        address 0000001d = 0000
address 0000000e = 0f86        address 0000001e = 0000
address 0000000f = 11c7        address 0000001f = 0000
```

Figure 11.83   Outputs for the instruction cache.

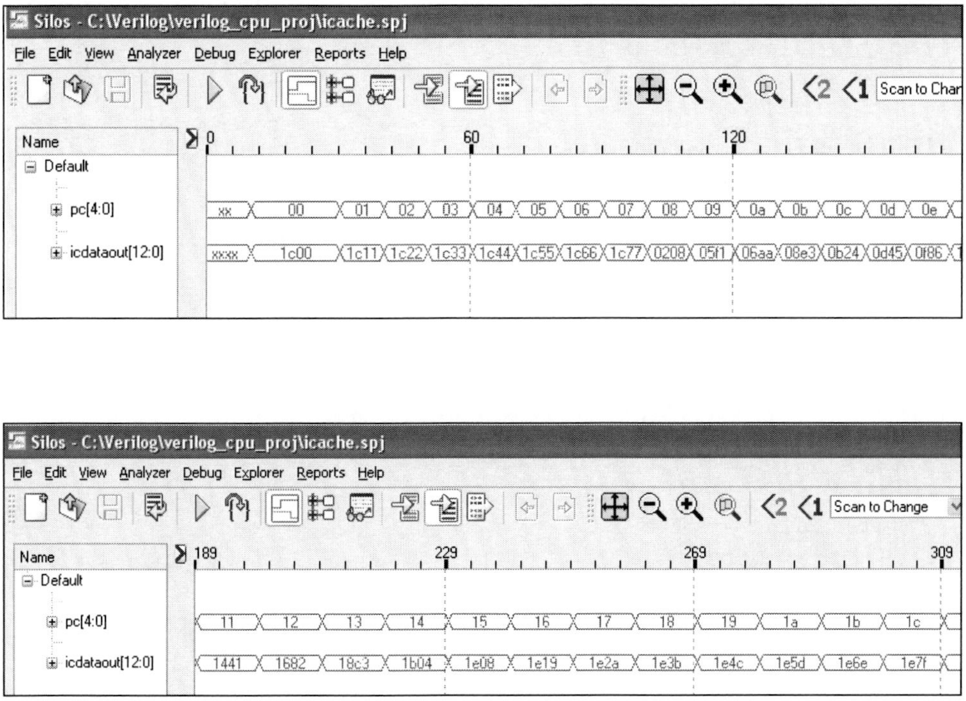

Figure 11.84   Waveforms for the instruction cache.

11.10.2 Instruction Unit

The behavioral module for the instruction unit is shown in Figure 11.85. Note that the port labeled *ir*, the statement *output [12:0] ir;*, and the statement *reg [12:0] ir;* imply a connection from the *ir* register to the *ir* port. The test bench is shown in Figure 11.86, the outputs are shown in Figure 11.87, and the waveforms are shown in Figure 11.88. The program counter places the instruction on *icdataout* and the next clock loads the instruction into the instruction register.

```
//behavioral iunit
module risc_iunit (instruction, pc, ir, clk, rst_n);

//list which are inputs and which are outputs
input [12:0] instruction;
input clk, rst_n;
output [4:0] pc;
output [12:0] ir;

//list which are wire and which are reg
wire [12:0] instruction;
wire clk, rst_n;
reg [4:0] pc, next_pc;
reg [12:0] ir;

parameter nop = 13'h0000;

always @ (posedge clk or negedge rst_n)
begin                          //must have always for each reg
   if (rst_n == 1'b0)          //== means compare
      pc = 5'b00000;           //initialize pc = 00000
   else
      pc <= pc + 1;            //determine next pc
end

always @ (posedge clk or negedge rst_n)
begin
   if (rst_n == 1'b0)
      ir = nop;
   else
      ir <= instruction;    //load ir from instruction
end

endmodule
```

Figure 11.85   Behavioral module for the instruction unit.

```
//test bench for iunit
module risc_iunit_tb;

//list input and output ports
reg [12:0] instruction;     //inputs are reg for test bench
reg clk, rst_n;
wire [4:0] pc;              //outputs are wire for test bench
wire [12:0] ir, instr;

//define clock
initial
begin
   clk = 1'b0;
   forever
      #10 clk = ~clk;
end

//define reset and simulation duration
//define instruction
initial
begin
   #0    rst_n = 1'b0;
         instruction = 13'h0208;        //pc = 00
         $display ("pc = %h, instruction = %h, ir = %h",
                     pc, instruction, ir);

   #5    rst_n = 1'b1;

   #10   instruction = 13'h05f1;        //pc = 01; sub
         $display ("pc = %h, instruction = %h, ir = %h",
                     pc, instruction, ir);

   #20   instruction = 13'h06aa;        //pc = 02; and
         $display ("pc = %h, instruction = %h, ir = %h",
                     pc, instruction, ir);

   #20   instruction = 13'h08e3;        //pc = 03; or
         $display ("pc = %h, instruction = %h, ir = %h",
                     pc, instruction, ir);

   #20   instruction = 13'h0b24;        //pc = 04; xor
         $display ("pc = %h, instruction = %h, ir = %h",
                     pc, instruction, ir);

   #20   instruction = 13'h0d45;        //pc = 05; inc
         $display ("pc = %h, instruction = %h, ir = %h",
                     pc, instruction, ir);        //next page
```

Figure 11.86   Test bench for the instruction unit.

```
  #20    instruction = 13'h0f86;          //pc = 06; dec
         $display ("pc = %h, instruction = %h, ir = %h",
                       pc, instruction, ir);

  #20    instruction = 13'h11c7;          //pc = 07; not
         $display ("pc = %h, instruction = %h, ir = %h",
                       pc, instruction, ir);

  #20    instruction = 13'h1200;          //pc = 08; neg
         $display ("pc = %h, instruction = %h, ir = %h",
                       pc, instruction, ir);

  #20    instruction = 13'h1441;          //pc = 09; shr
         $display ("pc = %h, instruction = %h, ir = %h",
                       pc, instruction, ir);

  #20    instruction = 13'h1682;          //pc = 10; shl
         $display ("pc = %h, instruction = %h, ir = %h",
                       pc, instruction, ir);

  #20    instruction = 13'h18c3;          //pc = 11; ror
         $display ("pc = %h, instruction = %h, ir = %h",
                       pc, instruction, ir);

  #20    instruction = 13'h1b04;          //pc = 12; rol
         $display ("pc = %h, instruction = %h, ir = %h",
                       pc, instruction, ir);

  #20    $stop;

end


//instantiate the behavioral module into the test bench
risc_iunit inst1 (
   .instruction(instruction),
   .clk(clk),
   .rst_n(rst_n),
   .pc(pc),
   .ir(ir)
   );

endmodule
```

Figure 11.86   (Continued)

```
pc = xx, instruction = 0208, ir = xxxx
pc = 01, instruction = 05f1, ir = 0208
pc = 02, instruction = 06aa, ir = 05f1
pc = 03, instruction = 08e3, ir = 06aa
pc = 04, instruction = 0b24, ir = 08e3
pc = 05, instruction = 0d45, ir = 0b24
pc = 06, instruction = 0f86, ir = 0d45
pc = 07, instruction = 11c7, ir = 0f86
pc = 08, instruction = 1200, ir = 11c7
pc = 09, instruction = 1441, ir = 1200
pc = 0a, instruction = 1682, ir = 1441
pc = 0b, instruction = 18c3, ir = 1682
pc = 0c, instruction = 1b04, ir = 18c3
```

Figure 11.87   Outputs for the instruction unit.

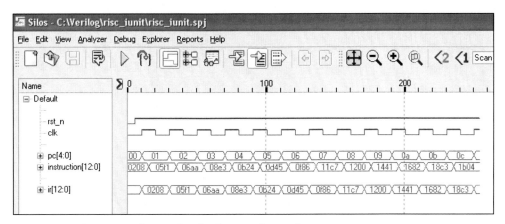

Figure 11.88   Waveforms for the instruction unit.

11.10.3 Decode Unit

The 13-bit instruction received from the instruction unit is decoded into its constituent parts: a 4-bit operation code, a 3-bit operand A address, a 3-bit operand B address, and a 3-bit destination address in the register file.

If the operation is a *load* instruction (load register file from data cache), then the data cache address is *instr[7:4]* and the destination address in the register file is *instr[2:0]*. If the operation is a *store* instruction (store register file to data cache), then register file address is *instr[6:4]* and the data cache address is *instr[3:0]*.

The mixed-design module for the decode unit is shown in Figure 11.89. The test bench is shown in Figure 11.90, the outputs are shown in Figure 11.91, and the waveforms are shown in Figure 11.92.

```verilog
//mixed-design decode unit
module risc_decode (clk, rst_n, instr, dcaddr,
                    opnda, opndb, dst, opcode);

input clk, rst_n;
input [12:0] instr;
output [3:0] dcaddr, opcode;
output [2:0] opnda, opndb, dst;

reg [3:0] dcaddr;
reg [2:0] opnda, opndb, dst;

//define internal registers and net
reg [3:0] opcode;
wire [3:0] opcode_i;
reg [3:0] dcaddr_i;
reg [2:0] opnda_i, opndb_i, dst_i;

parameter   ld = 4'b1110,
            st = 4'b1111;

assign opcode_i = instr[12:9];

always @ (opcode_i or instr)
begin
   case (opcode_i)
      ld:    begin
                dcaddr_i = instr[7:4];
                dst_i    = instr[2:0];
                opnda_i  = 3'b000;
             end

      st:    begin
                dcaddr_i  = instr [3:0];
                dst_i     = 3'b000;
                opnda_i   = instr[6:4];
                opndb_i   = 3'b000;
             end

//continued on next page
```

Figure 11.89   Mixed-design module for the decode unit.

```
        default: begin
                    dcaddr_i    = 4'b0000;
                    dst_i     = instr [2:0];
                    opnda_i   = instr [8:6];
                    opndb_i   = instr [5:3];
                end
    endcase
end

always @ (posedge clk or negedge rst_n)
begin
    if (~rst_n)
        begin
            opcode      <= 4'b0000;
            dcaddr      <= 4'b0000;
            dst         <= 3'b000;
            opnda       <= 3'b000;
            opndb       <= 3'b000;
        end

    else
        begin
            opcode      <= opcode_i;
            dcaddr      <= dcaddr_i;
            dst         <= dst_i;
            opnda       <= opnda_i;
            opndb       <= opndb_i;
        end
end
endmodule
```

Figure 11.89   (Continued)

```
//test bench for decode unit
module risc_decode_tb;

reg clk, rst_n;            //inputs are reg for test bench
reg [12:0] instr;

wire [3:0] dcaddr;         //outputs are wire for test bench
wire [2:0] opnda, opndb, dst;

//continued on next page
```

Figure 11.90   Test bench for the decode unit.

```
//define clock
initial
begin
   clk = 1'b0;
   forever
      #10 clk = ~clk;
end

//define reset
initial
begin
   #0 rst_n = 1'b0;
   #5 rst_n = 1'b1;
end

initial
begin
   #5     instr = 13'h1c00;     //ld
   #20    instr = 13'h0208;     //add
          $display ("opcode=%b, opnda=%b, opndb=%b, dst=%b",
                      instr[12:9], opnda, opndb, dst);

   #20    instr = 13'h05f1;     //sub
          $display ("opcode=%b, opnda=%b, opndb=%b, dst=%b",
                      instr[12:9], opnda, opndb, dst);

   #20    instr = 13'h06aa;     //and
          $display ("opcode=%b, opnda=%b, opndb=%b, dst=%b",
                      instr[12:9], opnda, opndb, dst);

   #20    instr = 13'h08e3;     //or
          $display ("opcode=%b, opnda=%b, opndb=%b, dst=%b",
                      instr[12:9], opnda, opndb, dst);

   #20    instr = 13'h0b1c;     //xor
          $display ("opcode=%b, opnda=%b, opndb=%b, dst=%b",
                      instr[12:9], opnda, opndb, dst);

   #20    instr = 13'h0d45;     //inc
          $display ("opcode=%b, opnda=%b, opndb=%b, dst=%b",
                      instr[12:9], opnda, opndb, dst);

   #20    instr = 13'h0f86;     //dec
          $display ("opcode=%b, opnda=%b, opndb=%b, dst=%b",
                      instr[12:9], opnda, opndb, dst);

//continued on next page
```

Figure 11.90   (Continued)

```
    #20    instr = 13'h11c7;      //not
           $display ("opcode=%b, opnda=%b, opndb=%b, dst=%b",
                        instr[12:9], opnda, opndb, dst);

    #20    instr = 13'h1200;      //neg
           $display ("opcode=%b, opnda=%b, opndb=%b, dst=%b",
                        instr[12:9], opnda, opndb, dst);

    #20    instr = 13'h1441;      //shr
           $display ("opcode=%b, opnda=%b, opndb=%b, dst=%b",
                        instr[12:9], opnda, opndb, dst);

    #20    instr = 13'h1682;      //shl
           $display ("opcode=%b, opnda=%b, opndb=%b, dst=%b",
                        instr[12:9], opnda, opndb, dst);

    #20    instr = 13'h18c3;      //ror
           $display ("opcode=%b, opnda=%b, opndb=%b, dst=%b",
                        instr[12:9], opnda, opndb, dst);

    #20    instr = 13'h1b04;      //rol
           $display ("opcode=%b, opnda=%b, opndb=%b, dst=%b",
                        instr[12:9], opnda, opndb, dst);

    #20    instr = 13'h1e00;      //st
           $display ("opcode=%b, opnda=%b, opndb=%b, dst=%b",
                        instr[12:9], opnda, opndb, dst);

    #20    $stop;
end


//instantiate the behavioral module into the test bench
risc_decode inst1 (
   .clk(clk),
   .rst_n(rst_n),
   .instr(instr),
   .dcaddr(dcaddr),
   .opnda(opnda),
   .opndb(opndb),
   .opcode(),
   .dst(dst)
   );

endmodule
```

Figure 11.90   (Continued)

```
opcode=0001, opnda=000, opndb=000, dst=000
opcode=0010, opnda=000, opndb=001, dst=000
opcode=0011, opnda=111, opndb=110, dst=001
opcode=0100, opnda=010, opndb=101, dst=010
opcode=0101, opnda=011, opndb=100, dst=011
opcode=0110, opnda=100, opndb=011, dst=100
opcode=0111, opnda=101, opndb=000, dst=101
opcode=1000, opnda=110, opndb=000, dst=110
opcode=1001, opnda=111, opndb=000, dst=111
opcode=1010, opnda=000, opndb=000, dst=000
opcode=1011, opnda=001, opndb=000, dst=001
opcode=1100, opnda=010, opndb=000, dst=010
opcode=1101, opnda=011, opndb=000, dst=011
opcode=1111, opnda=100, opndb=000, dst=100
```

Figure 11.91 Outputs for the decode unit.

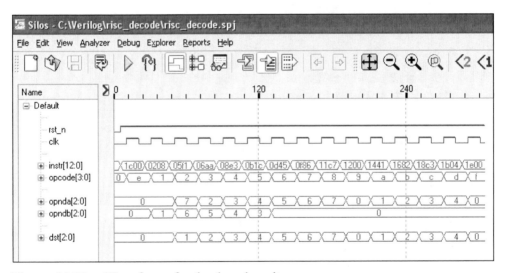

Figure 11.92 Waveforms for the decode unit.

Refer to the outputs of Figure 11.91 and the waveforms of Figure 11.92. Consider the subtract operation with an instruction format of hexadecimal 05f1. The fields in the instruction are shown on the next page. The operation code is 0010 (2); the address for operand A in the register file is 111 (7); the address for operand B in the register file is 110 (6); and the destination address in the register file is 001 (1). That is, operand B in register file 6 is subtracted from operand A in register file 7, and the difference is stored in register file 1. The same rationale applies to all instructions.

Op Code	Opnd A	Opnd B	Dst
0010 (sub)	111	110	001
12 9	8 6	5 3	2 0

In Figure 11.91, the *opnda*, *opndb*, and *dst* values are offset by one row. This is because the **$display** system task displays the variables whenever a variable changes, but the registers are not set to the new values until the next clock pulse, as shown in the waveforms.

11.10.4 Execution Unit

The instructions and their functions are listed in Table 11.11. The multiplexer logic that gates the operands to the ALU is shown in Figure 11.93. There are two sets of eight 8:1 multiplexers, one set for operand *A* and one set for operand *B*. Each multiplexer selects bits 0 through 7 of register file 0 through register file 7 for the appropriate operand, as a function of the multiplexer select inputs.

Table 11.11 Instructions for the Execution Unit

Instruction	Function
nop	No operation is performed
add	Operand *A* plus operand *B*
sub	Operand *A* minus operand *B*
and	Operand *A* AND operand *B*
or	Operand *A* OR operand *B*
xor	Operand *A* exclusive-OR operand *B*
inc	Increment operand *A* by 1
dec	Decrement operand *A* by 1
not	Form the 1s complement of operand *A*
neg	Form the 2s complement of operand *A*
shr	Shift right logical operand *A*
shl	Shift left logical operand *A*
ror	Rotate right operand *A*
rol	Rotate left operand *A*
ld	Load register file from memory
st	Store register file to memory

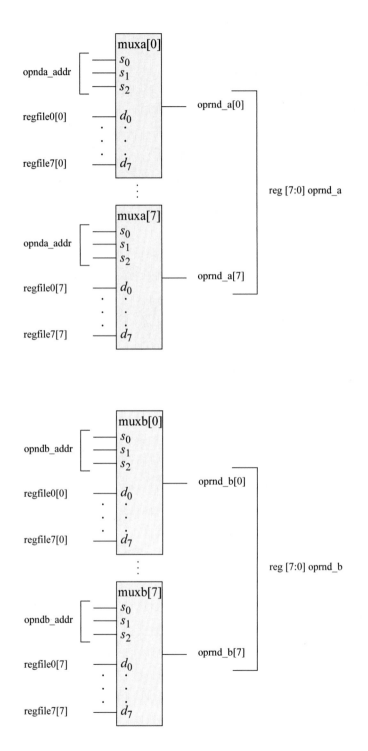

Figure 11.93   Multiplexer array input to the execution unit ALU.

The mixed-design module for the execution unit is shown in Figure 11.94. The test bench is shown in Figure 11.95 in which the register file is initialized to known values. The outputs and waveforms are shown in Figure 11.96 and Figure 11.97, respectively.

```verilog
//mixed-design eunit
module risc_eunit (clk, rst_n, opcode, dcaddrin,
      opnda_addr, opndb_addr, dstin,
      regfile0, regfile1, regfile2, regfile3,
      regfile4, regfile5, regfile6, regfile7,
      dcenbl, rdwr, dcaddr, rslt, dst, reg_wr_vld,
      dcdatain, load_op);

//list all inputs and outputs
input clk, rst_n;
input [3:0] dcaddrin, opcode;
input [7:0] regfile0, regfile1, regfile2, regfile3,
            regfile4, regfile5, regfile6, regfile7;
input [2:0] opnda_addr, opndb_addr, dstin;

output dcenbl, rdwr, reg_wr_vld, load_op;
output [3:0] dcaddr;
output [7:0] rslt, dcdatain;
output [2:0] dst;

//list all wire and reg
wire reg_wr_vld;        //wire, because used in assign stmt
wire rdwr;              //wire, because used in assign stmt
wire [7:0] adder_in_a;
wire [7:0] rslt_or;
wire [7:0] rslt_xor;
wire [7:0] rslt_and;
wire [7:0] rslt_shl;
wire [7:0] rslt_shr;
wire [7:0] rslt_ror;
wire [7:0] rslt_rol;
wire [7:0] rslt_not;
wire ci;                // carry in

reg dcenbl;             //outputs are reg in behavioral
reg [3:0] dcaddr;       //except when used in assign stmt
reg [7:0] rslt;
reg [7:0] oprnd_a, oprnd_b;
reg [7:0] rslt_i;
reg [7:0] rslt_sum;   //continued on next page
```

Figure 11.94   Mixed-design module for the execution unit.

```verilog
reg [7:0] adder_in_b;
reg adder_mode;
reg co;          // carry out

//define internal registers
reg [2:0] dst;
reg [3:0] opcode_out;

parameter add = 1'b0;
parameter sub = 1'b1;

parameter    nop_op = 4'b0000,
             add_op = 4'b0001,
             sub_op = 4'b0010,
             and_op = 4'b0011,
             or_op  = 4'b0100,
             xor_op = 4'b0101,
             inc_op = 4'b0110,
             dec_op = 4'b0111,
             not_op = 4'b1000,
             neg_op = 4'b1001,
             shr_op = 4'b1010,
             shl_op = 4'b1011,
             ror_op = 4'b1100,
             rol_op = 4'b1101,
             ld_op  = 4'b1110,
             st_op  = 4'b1111;


//use assign stmt to show mix of dataflow and behavioral
assign load_op = (opcode_out == ld_op);
assign rdwr = ~(opcode_out == st_op);
assign reg_wr_vld = (opcode_out != st_op) &&
                    (opcode_out != nop_op);
assign dcdatain = rslt;

always @ (posedge clk or negedge rst_n)
begin
   if (~rst_n)
      rslt <= 8'h00;
   else
      rslt <= rslt_i;
end

//continued on next page
```

Figure 11.94   (Continued)

```verilog
always @ (posedge clk or negedge rst_n)
begin
   if (~rst_n)
      dcaddr <= 4'h0;
   else
      dcaddr <= dcaddrin;
end


always @ (posedge clk or negedge rst_n)
begin
   if (~rst_n)
      opcode_out <= 4'h0;
   else
      opcode_out <= opcode;
end


always @ (posedge clk or negedge rst_n)
begin
   if (~rst_n)
      dcenbl <= 1'b0;
   else if (opcode == st_op || opcode == ld_op)
      dcenbl <= 1'b1;
   else
      dcenbl <= 1'b0;
end


always @ (posedge clk or negedge rst_n)
begin
   if (~rst_n)
      dst <= 3'b000;
   else
      dst <= dstin;
end


always @ (opcode)
begin
   if ((opcode == sub_op) || (opcode == dec_op))
       adder_mode = sub;
   else
       adder_mode = add;
end

//continued on next page
```

Figure 11.94   (Continued)

```verilog
//determine the b-input into the adder based on the opcode

always @ (opcode or oprnd_b)
begin
   case (opcode)
      sub_op:  adder_in_b = ~oprnd_b;
      inc_op:  adder_in_b = 8'h01;
      neg_op:  adder_in_b = 8'h01;
      dec_op:  adder_in_b = ~(8'h01);
      default: adder_in_b = oprnd_b;
   endcase
end

assign adder_in_a = (opcode ==  neg_op) ? ~oprnd_a : oprnd_a;

assign ci = adder_mode;

always @ (adder_in_a or adder_in_b or ci)
begin
   {co, rslt_sum} = adder_in_a + adder_in_b + ci;
end


assign rslt_or  = oprnd_a | oprnd_b;
assign rslt_xor = oprnd_a ^ oprnd_b;
assign rslt_and = oprnd_a & oprnd_b;
assign rslt_shl = oprnd_a << 1;
assign rslt_shr = oprnd_a >> 1;
assign rslt_ror = {oprnd_a[0], oprnd_a[7:1]};
assign rslt_rol = {oprnd_a[6:0], oprnd_a[7]};
assign rslt_not = ~oprnd_a;

// mux in the result based on opcode

always @ (opcode or oprnd_a or rslt_and or rslt_or
            or rslt_xor or rslt_shr or rslt_shl
            or rslt_rol or rslt_ror or rslt_sum )

begin
   case (opcode)
      st_op : rslt_i = oprnd_a;
      nop_op: rslt_i = 8'h00;
      and_op: rslt_i = rslt_and;
      or_op : rslt_i = rslt_or;
      xor_op: rslt_i = rslt_xor;
      shr_op: rslt_i = rslt_shr;
//continued on next page
```

Figure 11.94   (Continued)

```verilog
      shl_op: rslt_i = rslt_shl;
      ror_op: rslt_i = rslt_ror;
      rol_op: rslt_i = rslt_rol;
      not_op: rslt_i = rslt_not;
      add_op,
      sub_op,
      inc_op,
      dec_op,
      neg_op: rslt_i = rslt_sum;
      default: rslt_i = 8'h00;
   endcase
end

always @ (opnda_addr or regfile0 or regfile1 or regfile2 or
          regfile3 or regfile4 or regfile5 or regfile6 or
          regfile7)
begin
   case (opnda_addr)
          0: oprnd_a = regfile0;
          1: oprnd_a = regfile1;
          2: oprnd_a = regfile2;
          3: oprnd_a = regfile3;
          4: oprnd_a = regfile4;
          5: oprnd_a = regfile5;
          6: oprnd_a = regfile6;
          7: oprnd_a = regfile7;
          default: oprnd_a = 0;
   endcase
end

always @ (opndb_addr or regfile0 or regfile1 or regfile2 or
          regfile3 or regfile4 or regfile5 or regfile6 or
          regfile7)
begin
   case (opndb_addr)
          0: oprnd_b = regfile0;
          1: oprnd_b = regfile1;
          2: oprnd_b = regfile2;
          3: oprnd_b = regfile3;
          4: oprnd_b = regfile4;
          5: oprnd_b = regfile5;
          6: oprnd_b = regfile6;
          7: oprnd_b = regfile7;
          default: oprnd_b = 0;
   endcase
end
endmodule
```

Figure 11.94   (Continued)

```verilog
//test bench for the execution unit
module risc_eunit_tb;

//list all input and output ports
reg clk, rst_n;              //inputs are reg for test bench
reg [3:0] dcaddrin, opcode;
reg [7:0] regfile0, regfile1, regfile2, regfile3,
          regfile4, regfile5, regfile6, regfile7;
reg [2:0] opnda_addr, opndb_addr, dstin;

wire dcenbl, rdwr;           //outputs are wire for test bench
wire [3:0] dcaddr;
wire [7:0] rslt;
wire [2:0] dst;
wire reg_wr_vld;

//initialize regfile
initial
begin
   regfile0 = 8'h00;
   regfile1 = 8'h22;
   regfile2 = 8'h44;
   regfile3 = 8'h66;
   regfile4 = 8'h88;
   regfile5 = 8'haa;
   regfile6 = 8'hcc;
   regfile7 = 8'hff;
   dstin = 3'h0;
end

//define clock
initial
begin
   clk = 1'b0;
   forever
      #10 clk = ~clk;
end

initial          //define input vectors
begin
   #0 rst_n = 1'b0;
//add ----------------------------------------------------------
   @ (negedge clk)
         rst_n = 1'b1;
         opcode = 4'b0001;          //add
         opnda_addr = 3'b000;       //00h + ffh
         opndb_addr = 3'b111;       //rslt = ffh, next page
```

Figure 11.95   Test bench for the execution unit.

```
    @ (negedge clk)
      $display ("regfile0 = %h, regfile7 = %h, add, rslt = %h",
                regfile0, regfile7, rslt);

//sub ------------------------------------------------------
      opcode = 4'b0010;              //sub
      opnda_addr = 3'b001;           //22h - cch
      opndb_addr = 3'b110;           //rslt = 56h
    @ (negedge clk)
      $display ("regfile1 = %h, regfile6 = %h, sub, rslt = %h",
                regfile1, regfile6, rslt);

//and ------------------------------------------------------
      opcode = 4'b0011;              //and
      opnda_addr = 3'b010;           //44h & aah
      opndb_addr = 3'b101;           //rslt = 00h
    @ (negedge clk)
      $display ("regfile2 = %h, regfile5 = %h, and, rslt = %h",
                regfile2, regfile5, rslt);

//or ------------------------------------------------------
      opcode = 4'b0100;              //or
      opnda_addr = 3'b011;           //66h | 88h
      opndb_addr = 3'b100;           //rslt = eeh
    @ (negedge clk)
      $display ("regfile3 = %h, regfile4 = %h, or , rslt = %h",
                regfile3, regfile4, rslt);

//xor ------------------------------------------------------
      opcode = 4'b0101;              //xor
      opnda_addr = 3'b100;           //88h ^ eeh
      opndb_addr = 3'b011;           //rslt = 66h
    @ (negedge clk)
      $display ("regfile4 = %h, regfile3 = %h, xor, rslt = %h",
                regfile4, regfile3, rslt);

//inc ------------------------------------------------------
      opcode = 4'b0110;              //inc
      opnda_addr = 3'b101;           //aah + 1
      opndb_addr = 3'b000;           //rslt = abh
    @ (negedge clk)
      $display ("regfile5 = %h, regfile0 = %h, inc, rslt = %h",
                regfile5, regfile0, rslt);


//continued on next page
```

Figure 11.95   (Continued)

```
//dec ----------------------------------------------------
      opcode = 4'b0111;            //dec
      opnda_addr = 3'b110;         //cch - 1
      opndb_addr = 3'b000;         //rslt = cbh
   @ (negedge clk)
      $display ("regfile6 = %h, regfile0 = %h, dec, rslt = %h",
                regfile6, regfile0, rslt);

//not ----------------------------------------------------
      opcode = 4'b1000;            //not
      opnda_addr = 3'b111;         //ffh = 00h
      opndb_addr = 3'b000;         //rslt = 00h
   @ (negedge clk)
      $display ("regfile7 = %h, regfile0 = %h, not, rslt = %h",
                regfile7, regfile0, rslt);

//neg ----------------------------------------------------
      opcode = 4'b1001;            //neg
      opnda_addr = 3'b000;         //00h = 00h
      opndb_addr = 3'b000;         //rslt = 00h
   @ (negedge clk)
      $display ("regfile0 = %h, regfile0 = %h, neg, rslt = %h",
                regfile0, regfile0, rslt);

//shr ----------------------------------------------------
      opcode = 4'b1010;            //shr
      opnda_addr = 3'b001;         //56h = 2bh
      opndb_addr = 3'b000;         //rslt = 2bh
   @ (negedge clk)
      $display ("regfile1 = %h, regfile0 = %h, shr, rslt = %h",
                regfile1, regfile0, rslt);

//shl ----------------------------------------------------
      opcode = 4'b1011;            //shl
      opnda_addr = 3'b010;         //44h = 88h
      opndb_addr = 3'b000;         //rslt = 88h
   @ (negedge clk)
      $display ("regfile2 = %h, regfile0 = %h, shl, rslt = %h",
                regfile2, regfile0, rslt);

//ror ----------------------------------------------------
      opcode = 4'b1100;            //ror
      opnda_addr = 3'b011;         //eeh = 77h
      opndb_addr = 3'b000;         //rslt = 77h
   @ (negedge clk)
      $display ("regfile3 = %h, regfile0 = %h, ror, rslt = %h",
                regfile3, regfile0, rslt);    //next page
```

Figure 11.95   (Continued)

```
//rol -------------------------------------------------------
     opcode = 4'b1101;              //rol
     opnda_addr = 3'b100;          //66h = cch
     opndb_addr = 3'b000;          //rslt = cch
   @ (negedge clk)
     $display ("regfile4 = %h, regfile0 = %h, rol, rslt = %h",
               regfile4, regfile0, rslt);


//use either $display or nop to get the final output
     opcode = 4'b0000;             //nop


//add st and nop to show the rdwr and reg_wr_vld lines changing.
//do not need opnda_addr or opndb_addr because st and nop
//do not use the regfile

   #20    opcode = 4'b1111;        //st

   #20    opcode = 4'b0000;        //nop

   #20    $stop;

end


//instantiate the behavioral model into the test bench
risc_eunit inst1 (
   .clk(clk),
   .rst_n(rst_n),
   .opcode(opcode),
   .dcaddrin(dcaddrin),
   .opnda_addr(opnda_addr),
   .opndb_addr(opndb_addr),
   .regfile0(regfile0),
   .regfile1(regfile1),
   .regfile2(regfile2),
   .regfile3(regfile3),
   .regfile4(regfile4),
   .regfile5(regfile5),
   .regfile6(regfile6),
   .regfile7(regfile7),


//continued on next page
```

Figure 11.95   (Continued)

```
    .dcenbl(dcenbl),
    .rdwr(rdwr),
    .dcaddr(dcaddr),
    .rslt(rslt),
    .dst(dst),
    .dstin(dstin),
    .reg_wr_vld(reg_wr_vld),
    .dcdatain(),
    .load_op()
    );

endmodule
```

Figure 11.95   (Continued)

```
regfile0 = 00, regfile7 = ff, add, rslt = ff
regfile1 = 22, regfile6 = cc, sub, rslt = 56
regfile2 = 44, regfile5 = aa, and, rslt = 00
regfile3 = 66, regfile4 = 88, or , rslt = ee
regfile4 = 88, regfile3 = 66, xor, rslt = ee
regfile5 = aa, regfile0 = 00, inc, rslt = ab
regfile6 = cc, regfile0 = 00, dec, rslt = cb
regfile7 = ff, regfile0 = 00, not, rslt = 00
regfile0 = 00, regfile0 = 00, neg, rslt = 00
regfile1 = 22, regfile0 = 00, shr, rslt = 11
regfile2 = 44, regfile0 = 00, shl, rslt = 88
regfile3 = 66, regfile0 = 00, ror, rslt = 33
regfile4 = 88, regfile0 = 00, rol, rslt = 11
```

Figure 11.96   Outputs for the execution unit.

Refer to the waveforms of Figure 11.97 for the discussion that follows. Operation code 0001 is an *add* operation. The operand *A* address is register file 0, which contains the value 0000_0000. The operand *B* address is register file 7, which contains the value 1111_1111. After adding the two operands, the result is 1111_1111 (hexadecimal ff).

Now consider an *exclusive-OR* instruction with an operation code of 0101. Operand *A* is in register file 4, which contains the value 1000_1000. Operand *B* is in register file 3, which contains the value 0110_0110. The result is 1110_1110 (hexadecimal ee).

The *negate* instruction has an operation code of 1001 and obtains the 2s complement of operand *A*, which is located in register file 0 with a value of 0000_0000. When the 2s complement of 0000_0000 is obtained, the result will still be a value of 0000_0000 (hexadecimal 00) as shown in the waveforms.

The *shift right* instruction has an operation code of 1010. The operation is performed on operand *A* only, which is located in register file 1 containing a value of 0010_0010. After shifting right one bit position, the result is 0001_0001 (hexadecimal 11).

The *rotate left* instruction has an operation code of 1101 and uses operand *A* only, which is contained in register file 4 and has a value of 1000_1000. After the execution of the instruction, the result is 0001_0001 (hexadecimal 11).

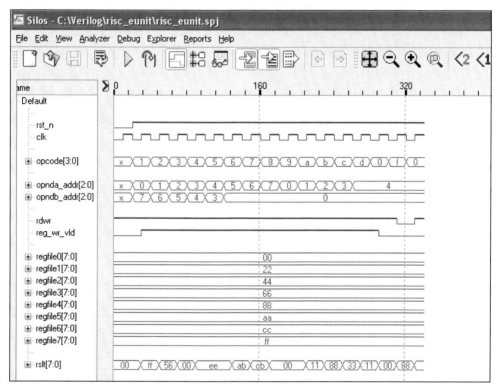

Figure 11.97   Waveforms for the execution unit.

11.10.5 Register File

The register file control logic is shown in Figure 11.98. A register file is selected by means of a 3:8 decoder with inputs that are the destination address *dst[2:0]*. The output of the decoder in conjunction with a valid write signal (*reg_wr_vld*) selects the appropriate register file. The data that is to be written to the register file comes from one of two sources: from the result of an operation *rslt[7:0]* or from the data cache *dcdataout[7:0]*.

The mixed-design module that implements the register file logic directly is shown in Figure 11.99. The test bench, outputs, and waveforms are shown in Figure 11.100, Figure 11.101, and Figure 11.102, respectively. Only selected values of *rslt* and *dcdataout* are provided.

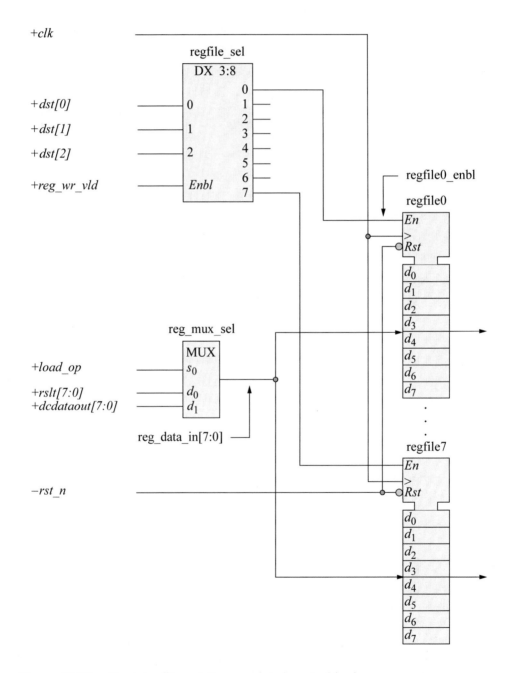

Figure 11.98   Register file and the associated control logic.

```
//mixed-design for regfile
module risc_regfile (clk, rst_n, rslt, reg_wr_vld, load_op,
        dst, dcdataout, regfile0, regfile1, regfile2,
        regfile3, regfile4, regfile5, regfile6, regfile7);


//specify which are input or output by width
input clk, rst_n, reg_wr_vld, load_op;
input [7:0] rslt, dcdataout;
input [2:0] dst;


output [7:0] regfile0, regfile1, regfile2, regfile3,
            regfile4, regfile5, regfile6, regfile7;


//specify which are wire or reg by width
//inputs are wire for behavioral
wire [7:0] rslt, dcdataout;
wire [2:0] dst;


//declare internal regfile_enbl signals
wire regfile0_enbl;
wire regfile1_enbl;
wire regfile2_enbl;
wire regfile3_enbl;
wire regfile4_enbl;
wire regfile5_enbl;
wire regfile6_enbl;
wire regfile7_enbl;


//outputs are reg for behavioral
reg [7:0]regfile0, regfile1, regfile2, regfile3,
        regfile4, regfile5, regfile6, regfile7;
wire[7:0] reg_data_in;


//generate enables for each register file
assign regfile0_enbl = (dst == 3'h0) & reg_wr_vld;
assign regfile1_enbl = (dst == 3'h1) & reg_wr_vld;
assign regfile2_enbl = (dst == 3'h2) & reg_wr_vld;
assign regfile3_enbl = (dst == 3'h3) & reg_wr_vld;
assign regfile4_enbl = (dst == 3'h4) & reg_wr_vld;
assign regfile5_enbl = (dst == 3'h5) & reg_wr_vld;
assign regfile6_enbl = (dst == 3'h6) & reg_wr_vld;
assign regfile7_enbl = (dst == 3'h7) & reg_wr_vld;
//continued on next page
```

Figure 11.99   Mixed-design module for the register file.

```
//define reg_mux outputs from the eight 2:1 muxs
assign reg_data_in = load_op ? dcdataout : rslt;

//define the operation of the regfile
always @ (posedge clk or negedge rst_n)
begin
    if (~rst_n)
        regfile0 <= 8'h00;
    else if (regfile0_enbl)
        regfile0 <= reg_data_in;
end

always @ (posedge clk or negedge rst_n)
begin
    if (~rst_n)
        regfile1 <= 8'h00;
    else if (regfile1_enbl)
        regfile1 <= reg_data_in;
end

always @ (posedge clk or negedge rst_n)
begin
    if (~rst_n)
        regfile2 <= 8'h00;
    else if (regfile2_enbl)
        regfile2 <= reg_data_in;
end

always @ (posedge clk or negedge rst_n)
begin
    if (~rst_n)
        regfile3 <= 8'h00;
    else if (regfile3_enbl)
        regfile3 <= reg_data_in;
end

always @ (posedge clk or negedge rst_n)
begin
    if (~rst_n)
        regfile4 <= 8'h00;
    else if (regfile4_enbl)
        regfile4 <= reg_data_in;
end

//continued on next page
```

Figure 11.99   (Continued)

```
always @ (posedge clk or negedge rst_n)
begin
   if (~rst_n)
      regfile5 <= 8'h00;
   else if (regfile5_enbl)
      regfile5 <= reg_data_in;
end

always @ (posedge clk or negedge rst_n)
begin
   if (~rst_n)
      regfile6 <= 8'h00;
   else if (regfile6_enbl)
      regfile6 <= reg_data_in;
end

always @ (posedge clk or negedge rst_n)
begin
   if (~rst_n)
      regfile7 <= 8'h00;
   else if (regfile7_enbl)
      regfile7 <= reg_data_in;
end
endmodule
```

Figure 11.99   (Continued)

```
//test bench for regfile
module risc_regfile_tb;

//inputs are reg for test bench
reg clk, rst_n, reg_wr_vld, load_op;
reg [2:0] dst;
reg [7:0] rslt, dcdataout;

//outputs are wire for test bench
wire [7:0]regfile0, regfile1, regfile2, regfile3,
          regfile4, regfile5, regfile6, regfile7;

//define clock
initial
begin
   clk = 1'b0;
   forever #10 clk = ~clk;
end          //continued on next page
```

Figure 11.100 Test bench for the register file.

```
//define input vectors
initial
begin
   #0 rst_n = 1'b0;

   @ (negedge clk)
      rst_n = 1'b1;

//regfile0 ---------------------------------------------
      dst = 3'b000;          //regfile0 is dst
      reg_wr_vld = 1'b1;
      load_op = 1'b0;        //rslt is used
      rslt = 8'h00;          //8'h00 ---> regfile0
      dcdataout = 8'h00;
   @ (negedge clk)
      $display ("rslt = %h, regfile0 = %h", rslt, regfile0);

//regfile2 ---------------------------------------------
      dst = 3'b010;          //regfile2 is dst
      reg_wr_vld = 1'b1;
      load_op = 1'b0;        //rslt is used
      rslt = 8'h22;          //8'h22 ---> regfile2
      dcdataout = 8'h00;
   @ (negedge clk)
      $display ("rslt = %h, regfile2 = %h", rslt, regfile2);

//regfile4 ---------------------------------------------
      dst = 3'b100;          //regfile4 is dst
      reg_wr_vld = 1'b1;
      load_op = 1'b0;        //rslt is used
      rslt = 8'h44;          //8'h44 ---> regfile4
      dcdataout = 8'h00;
   @ (negedge clk)
      $display ("rslt = %h, regfile4 = %h", rslt, regfile4);

//regfile6 ---------------------------------------------
      dst = 3'b110;          //regfile6 is dst
      reg_wr_vld = 1'b1;
      load_op = 1'b0;        //rslt is used
      rslt = 8'h66;          //8'h66 ---> regfile6
      dcdataout = 8'h00;
   @ (negedge clk)
      $display ("rslt = %h, regfile6 = %h", rslt, regfile6);

//continued on next page
```

Figure 11.100   (Continued)

```
//regfile1 -------------------------------------------------
    dst = 3'b001;          //regfile1 is dst
    reg_wr_vld = 1'b1;
    load_op = 1'b1;        //dcdataout is used
    rslt = 8'h00;
    dcdataout = 8'h11;     //8'h11 ---> regfile1
  @ (negedge clk)
    $display ("dcdataout = %h, regfile1 = %h",
                dcdataout, regfile1);


//regfile3 -------------------------------------------------
    dst = 3'b011;          //regfile3 is dst
    reg_wr_vld = 1'b1;
    load_op = 1'b1;        //dcdataout is used
    rslt = 8'h00;
    dcdataout = 8'h33;     //8'h33 ---> regfile3
  @ (negedge clk)
    $display ("dcdataout = %h, regfile3 = %h",
                dcdataout, regfile3);


//regfile5 -------------------------------------------------
    dst = 3'b101;          //regfile5 is dst
    reg_wr_vld = 1'b1;
    load_op = 1'b1;        //dcdataout is used
    rslt = 8'h00;
    dcdataout = 8'h55;     //8'h55 ---> regfile5
  @ (negedge clk)
    $display ("dcdataout = %h, regfile5 = %h",
                dcdataout, regfile5);


//regfile7 -------------------------------------------------
    dst = 3'b111;          //regfile7 is dst
    reg_wr_vld = 1'b1;
    load_op = 1'b1;        //dcdataout is used
    rslt = 8'h00;
    dcdataout = 8'h77;     //8'h77 ---> regfile7
  @ (negedge clk)
    $display ("dcdataout = %h, regfile7 = %h",
                dcdataout, regfile7);


//continued on next page
```

Figure 11.100   (Continued)

```
//regfile0 ----------------------------------------------
     dst = 3'b000;          //regfile0 is dst
     reg_wr_vld = 1'b1;
     load_op = 1'b1;        //dcdataout is used
     rslt = 8'h00;          //8'h00 ---> regfile0
     dcdataout = 8'h00;
   @ (negedge clk)
     $display ("rslt = %h, regfile0 = %h", rslt, regfile0);

   $stop;
end

//instantiate the behavioral module into the test bench
risc_regfile inst1 (
   .clk(clk),
   .rst_n(rst_n),
   .rslt(rslt),
   .reg_wr_vld(reg_wr_vld),
   .load_op(load_op),
   .dst(dst),
   .dcdataout(dcdataout),
   .regfile0(regfile0),
   .regfile1(regfile1),
   .regfile2(regfile2),
   .regfile3(regfile3),
   .regfile4(regfile4),
   .regfile5(regfile5),
   .regfile6(regfile6),
   .regfile7(regfile7)
   );

endmodule
```

Figure 11.100   (Continued)

```
      rslt= 00,  regfile0 = 00
      rslt= 22,  regfile2 = 22
      rslt= 44,  regfile4 = 44
      rslt= 66,  regfile6 = 66
dcdataout= 11,  regfile1 = 11
dcdataout= 33,  regfile3 = 33
dcdataout= 55,  regfile5 = 55
dcdataout= 77,  regfile7 = 77
      rslt= 00,  regfile0 = 00
```

Figure 11.101 Outputs for the register file.

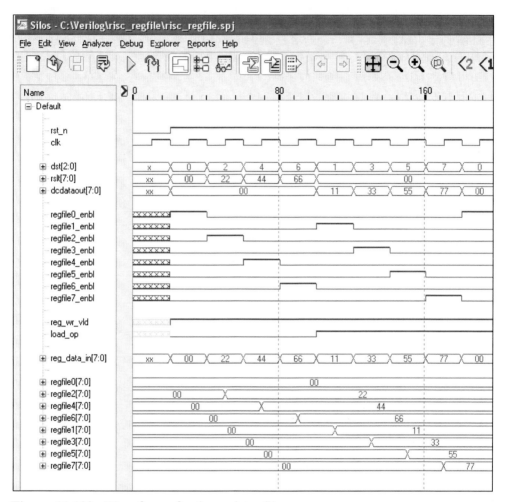

Figure 11.102   Waveforms for the register file.

11.10.6 Data Cache

The data cache is a memory of 16 8-bit words that is preloaded with specific operands so that the results of the program execution will be known. The data cache has five ports: *dcenbl* to enable the cache for a read or a write operation, *rdwr* to allow reading (*rdwr* = 1) or writing (*rdwr* = 0), *dcaddr[3:0]* selects an address in the cache to either read or write data, *dcdatain[7:0]* provides data for a write operation, and *dcdataout[7:0]* provides cache contents for a read operation.

The behavioral module is shown in Figure 11.103 in which each word of the cache is loaded with preassigned data. Alternatively, the cache could have been loaded using the system tasks **$readmemb** for binary data or **$readmenh** for hexadecimal data, which load a specified text file into the cache.

The **always** statement is used to select a cache read operation if the statement **if** *(rdwr)* returns a value of 1; otherwise, a write operation is performed. The test bench shown in Figure 11.104, reads the contents of the cache addresses *dcache[0000]* through *dcache[0111]*, then writes alternating 0000_0000 and 1111_1111 to cache addresses *dcache[1000]* through *dcache[1111]*. The outputs and waveforms are shown in Figure 11.105 and Figure 11.106, respectively.

```
//behavioral data cache
module risc_dcache (dcenbl, dcaddr, dcdatain, dcdataout, rdwr);

input [3:0] dcaddr;
input [7:0] dcdatain;
input dcenbl, rdwr;
output [7:0] dcdataout;

wire [3:0] dcaddr;
wire [7:0] dcdatain;
wire dcenbl, rdwr;

reg [7:0] dcdataout;

//define memory size: # of bits per reg; # of regs
//8 bits per reg; 16 regs
reg [7:0] dcache [0:15];

//define memory contents
initial
begin
   dcache [00] = 8'h00;
   dcache [01] = 8'h22;
   dcache [02] = 8'h44;
   dcache [03] = 8'h66;
   dcache [04] = 8'h88;
   dcache [05] = 8'haa;
   dcache [06] = 8'hcc;
   dcache [07] = 8'hff;
   dcache [08] = 8'h00;
   dcache [09] = 8'h00;
   dcache [10] = 8'h00;
   dcache [11] = 8'h00;
   dcache [12] = 8'h00;
   dcache [13] = 8'h00;
   dcache [14] = 8'h00;
   dcache [15] = 8'h00;
end                        //continued on next page
```

Figure 11.103 Behavioral module for the data cache.

```
always @ (dcenbl or dcaddr or dcdatain or rdwr)
begin
   if (rdwr)                //if true, read op (ld regfile)
      dcdataout = dcache [dcaddr];
   else                     //if false, write op (st regfile)
      dcache [dcaddr] = dcdatain;
end

endmodule
```

Figure 11.103   (Continued)

```
//test bench for data cache
module risc_dcache_tb;


reg [3:0] dcaddr;
reg [7:0] dcdatain;
reg dcenbl, rdwr;
wire [7:0] dcdataout;


//display variables
initial
$monitor ("dcaddr = %b, datain = %h, dataout = %h",
          dcaddr, dcdatain, dcdataout);


//read the contents of cache
initial
begin
   #0     dcaddr = 4'b0000; dcdatain = 8'h00;
          dcenbl = 1'b1; rdwr = 1'b1;           //read
   #10    dcaddr = 4'b0001; dcenbl = 1'b1; rdwr = 1'b1;
   #10    dcaddr = 4'b0010; dcenbl = 1'b1; rdwr = 1'b1;
   #10    dcaddr = 4'b0011; dcenbl = 1'b1; rdwr = 1'b1;
   #10    dcaddr = 4'b0100; dcenbl = 1'b1; rdwr = 1'b1;
   #10    dcaddr = 4'b0101; dcenbl = 1'b1; rdwr = 1'b1;
   #10    dcaddr = 4'b0110; dcenbl = 1'b1; rdwr = 1'b1;
   #10    dcaddr = 4'b0111; dcenbl = 1'b1; rdwr = 1'b1;


//continued on next page
```

Figure 11.104 Test bench for the data cache.

```
//write alternating all 0s and all 1s to cache
   #10   dcaddr = 4'b1000; dcdatain = 8'h00; dcenbl = 1'b1;
         rdwr = 1'b0;                        //write
   #10   dcaddr = 4'b1001; dcdatain = 8'hff; dcenbl = 1'b1;
         rdwr = 1'b0;
   #10   dcaddr = 4'b1010; dcdatain = 8'h00; dcenbl = 1'b1;
         rdwr = 1'b0;
   #10   dcaddr = 4'b1011; dcdatain = 8'hff; dcenbl = 1'b1;
         rdwr = 1'b0;
   #10   dcaddr = 4'b1100; dcdatain = 8'h00; dcenbl = 1'b1;
         rdwr = 1'b0;
   #10   dcaddr = 4'b1101; dcdatain = 8'hff; dcenbl = 1'b1;
         rdwr = 1'b0;
   #10   dcaddr = 4'b1110; dcdatain = 8'h00; dcenbl = 1'b1;
         rdwr = 1'b0;
   #10   dcaddr = 4'b1111; dcdatain = 8'hff; dcenbl = 1'b1;
         rdwr = 1'b0;

   #20   $stop;
end

//instantiate the module into the test bench
risc_dcache inst1 (
   .dcenbl(dcenbl),
   .dcaddr(dcaddr),
   .dcdatain(dcdatain),
   .dcdataout(dcdataout),
   .rdwr(rdwr)
   );
endmodule
```

Figure 11.104   (Continued)

```
dcaddr = 0000, datain = 00, dataout = 00
dcaddr = 0001, datain = 00, dataout = 22
dcaddr = 0010, datain = 00, dataout = 44
dcaddr = 0011, datain = 00, dataout = 66
dcaddr = 0100, datain = 00, dataout = 88
dcaddr = 0101, datain = 00, dataout = aa
dcaddr = 0110, datain = 00, dataout = cc
dcaddr = 0111, datain = 00, dataout = ff

//continued on next page
```

Figure 11.105 Outputs for the data cache.

```
dcaddr = 1000, datain = 00, dataout = ff
dcaddr = 1001, datain = ff, dataout = ff
dcaddr = 1010, datain = 00, dataout = ff
dcaddr = 1011, datain = ff, dataout = ff
dcaddr = 1100, datain = 00, dataout = ff
dcaddr = 1101, datain = ff, dataout = ff
dcaddr = 1110, datain = 00, dataout = ff
dcaddr = 1111, datain = ff, dataout = ff
```

Figure 11.105   (Continued)

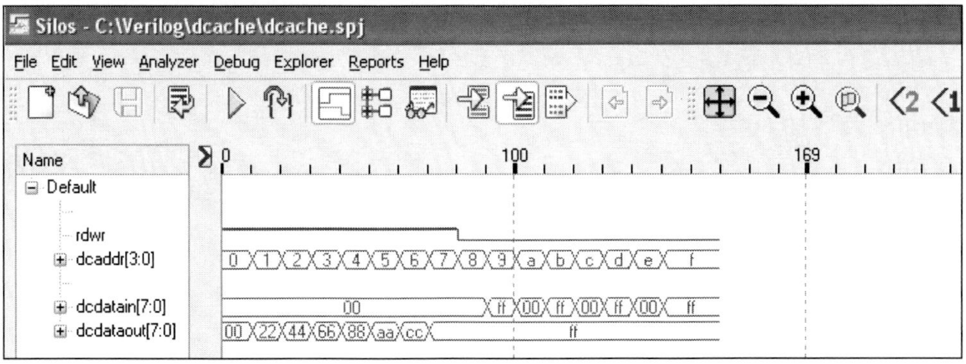

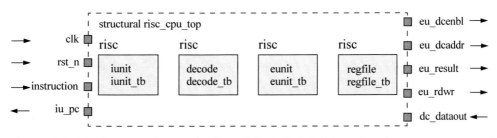

Figure 11.106   Waveforms for the data cache.

11.10.7 RISC CPU Top

The organization of the structural RISC CPU top level with interface ports to the instruction cache and the data cache is shown in Figure 11.107. The instruction unit, decode unit, execution unit, and register file will be instantiated into the top level structure of the RISC CPU labeled *rsic_cpu_top1* as shown in the structural module of Figure 11.108. The wires that interconnect the four modules are also listed in the structural module.

Figure 11.107   Structural organization of the RISC CPU top level.

```
//structural risc cpu top
module risc_cpu_top1 (clk, rst_n, instruction, pc, dcenbl,
                      dcaddr, dcdatain, dcdataout, rdwr);

input clk, rst_n;
input [12:0] instruction;
output [4:0] pc;
output dcenbl, rdwr;
output [3:0] dcaddr;
output [7:0] dcdatain, dcdataout;

wire [12:0] instruction, instr;
wire [4:0] pc;
wire [3:0] opcode, dcaddrin;
wire [2:0] opnda_addr, opndb_addr, dstin;
wire clk, rst_n;
wire [2:0] dst;
wire [7:0] regfile0_in, regfile1_in, regfile2_in, regfile3_in,
      regfile4_in, regfile5_in, regfile6_in, regfile7_in;
wire dcenbl, rdwr, reg_wr_vld, load_op;
wire [3:0] dcaddr;
wire [7:0] result;

risc_iunit inst1 (
   .instruction(instruction),
   .pc(pc),
   .clk(clk),
   .rst_n(rst_n),
   .ir(instr)
   );

risc_decode inst2 (
   .clk(clk),
   .rst_n(rst_n),
   .instr(instr),
   .opcode(opcode),
   .dcaddr(dcaddrin),
   .opnda(opnda_addr),
   .opndb(opndb_addr),
   .dst(dstin)
   );

risc_eunit inst3 (
   .clk(clk),              //inputs begin here
   .rst_n(rst_n),
   .opcode(opcode),
//continued on next page
```

Figure 11.108 Structural module for the RISC CPU top level.

```
        .dcaddrin(dcaddrin),
        .opnda_addr(opnda_addr),
        .opndb_addr(opndb_addr),
        .dstin(dstin),
        .regfile0(regfile0_in),
        .regfile1(regfile1_in),
        .regfile2(regfile2_in),
        .regfile3(regfile3_in),
        .regfile4(regfile4_in),
        .regfile5(regfile5_in),
        .regfile6(regfile6_in),
        .regfile7(regfile7_in),

        .dcenbl(dcenbl),              //outputs begin here
        .dcdatain(dcdatain),
        .rdwr(rdwr),
        .reg_wr_vld(reg_wr_vld),
        .dcaddr(dcaddr),
        .rslt(result),
        .dst(dst),
        .load_op(load_op)
        );

risc_regfile inst4 (
        .clk(clk),
        .rst_n(rst_n),
        .reg_wr_vld(reg_wr_vld),
        .load_op(load_op),
        .dst(dst),
        .rslt(result),
        .dcdataout(dcdataout),
        .regfile0(regfile0_in),
        .regfile1(regfile1_in),
        .regfile2(regfile2_in),
        .regfile3(regfile3_in),
        .regfile4(regfile4_in),
        .regfile5(regfile5_in),
        .regfile6(regfile6_in),
        .regfile7(regfile7_in)
        );

endmodule
```

Figure 11.108   (Continued)

11.10.8 System Top

The structural organization for the complete pipelined RISC processor is shown in Figure 11.109. The system top structural module is shown in Figure 11.110. This module instantiates the *risc_cpu_top* module of Figure 11.108 together with the instruction cache and the data cache. The only inputs are the reset and clock signals. The test bench is shown in Figure 11.111, which defines the reset and clock signals and the simulation duration.

The waveforms are displayed in Figure 11.112 and show the complete execution of the application program together with the values of the registers and wires for the modules. The register file contents at the end of the waveforms correctly depict the results of the program as specified in Table 11.10 and based upon the original register file contents of Table 11.9.

Although this was a comparatively simple processor design, the same rationale applies equally well to a more complex pipelined RISC processor. This section has introduced pipelining and RISC processors and provided a foundation for further study of this intriguing architecture.

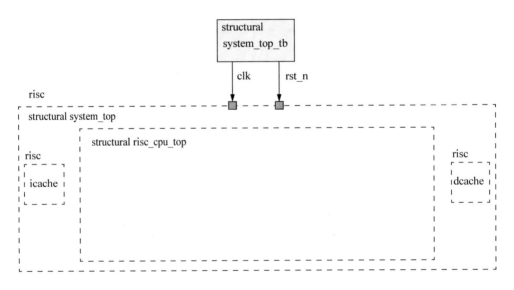

Figure 11.109 Structural organization

```
//structural system_top
module risc_system_top (clk, rst_n);

input clk, rst_n;
//continued on next page
```

Figure 11.110 Structural module for the system top pipelined RISC processor.

```
wire clk, rst_n, dcenbl, rdwr;
wire [12:0] instruction;
wire [4:0] pc;
wire [3:0] dcaddr;
wire [7:0] dcdatain, dcdataout;

risc_icache inst1 (
    .pc(pc),
    .instruction(instruction)
    );

risc_dcache inst2 (      //no clk or rst_n needed
    .dcenbl(dcenbl),
    .dcaddr(dcaddr),
    .dcdatain(dcdatain),
    .rdwr(rdwr),
    .dcdataout(dcdataout)
    );

risc_cpu_top1 inst3 (
    .clk(clk),
    .rst_n(rst_n),
    .pc(pc),
    .instruction(instruction),
    .dcenbl(dcenbl),
    .dcaddr(dcaddr),
    .dcdatain(dcdatain),
    .rdwr(rdwr),
    .dcdataout(dcdataout)
    );
endmodule
```

Figure 11.110   (Continued)

```
//structural risc_cpu_top test bench
module risc_system_top_tb;

reg clk, rst_n;

initial
begin
   clk = 1'b0;
      forever
   #10 clk = ~clk;
end                         //continued on next page
```

Figure 11.111 Test bench for the system top module of Figure 11.110.

```
initial
begin
   rst_n = 1'b0;
   #5 rst_n = 1'b1;
   #640 $stop;              //32 instr times 20 ns per clk cycle
end

risc_system_top inst1 (
   .clk(clk),
   .rst_n(rst_n)
   );
endmodule
```

Figure 11.111   (Continued)

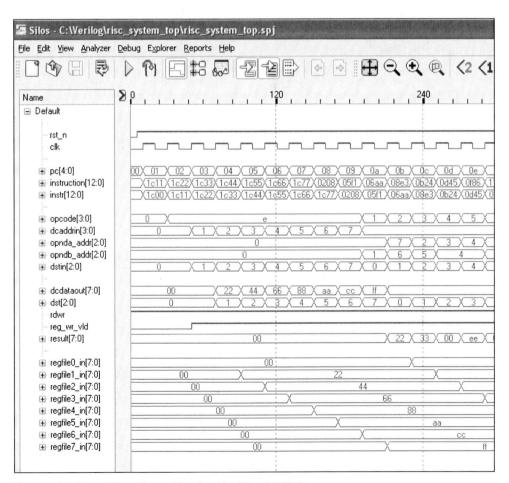

Figure 11.112   Waveforms for the pipelined RISC processor.

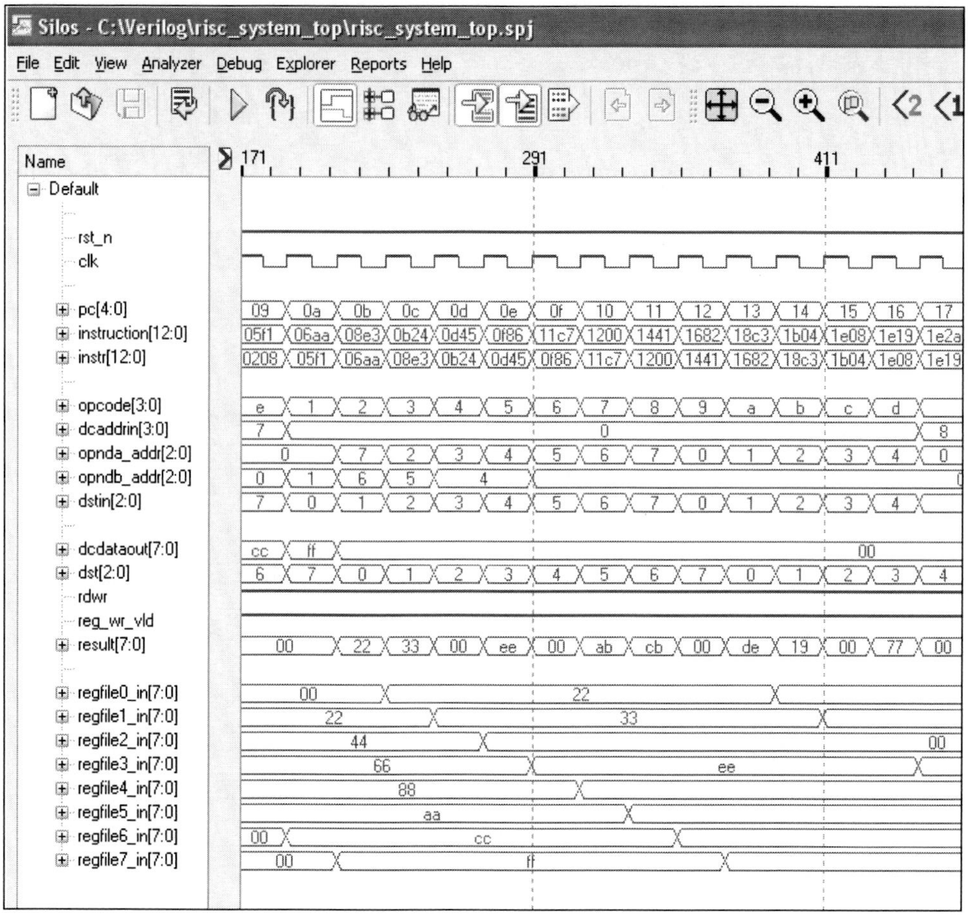

Figure 11.112   (Continued)

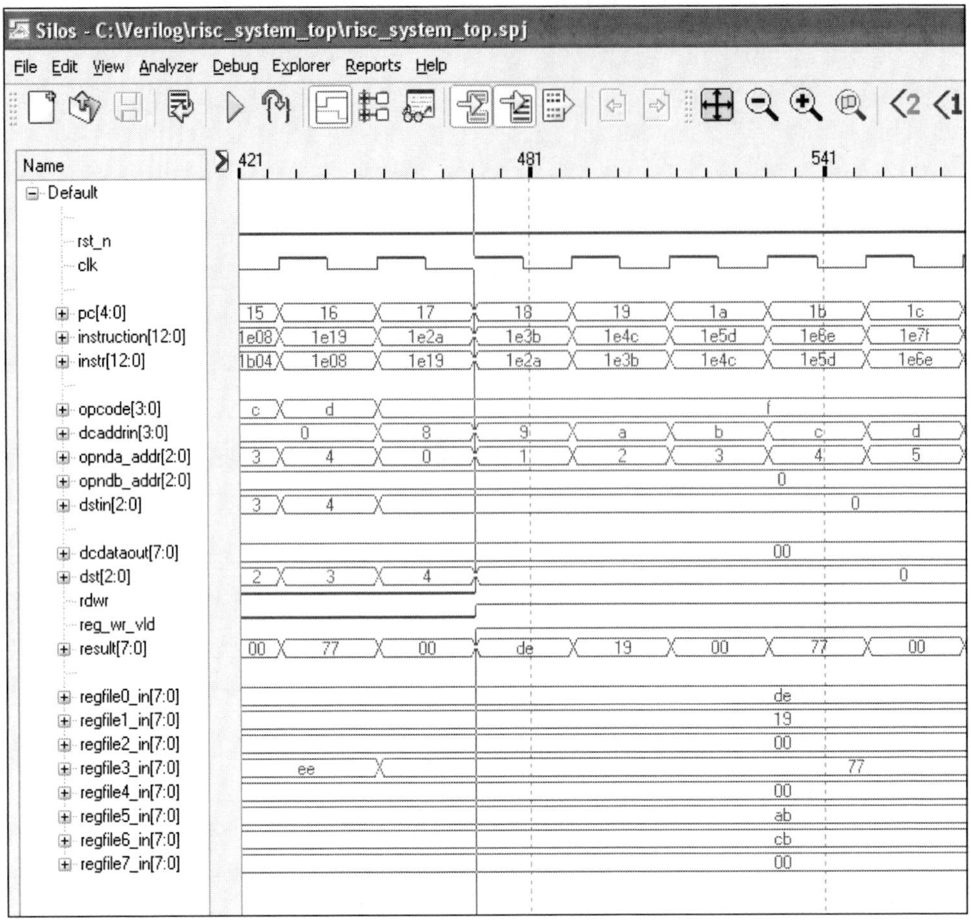

Figure 11.112 (Continued)

11.11 Problems

11.1   Design a behavioral module for an 8-bit serial-in, parallel-out shift register. Obtain the test bench, outputs, and waveforms for various input sequences.

11.2   Using structural modeling with D flip-flops, design a 4-bit binary counter that counts in the following decimal sequence: 15, 13, 11, 9, 7, 5, 3, 1, 0, 2, 4, 6, 8, 10, 12, 14, 15, ... Show the Karnaugh maps for the four flip-flops $y[3:0]$ and obtain the input equations. Instantiate the necessary logic gates and the D flip-flop. Obtain the design module, the test bench module, the outputs showing the counting sequence, and the waveforms.

11.3   Using structural modeling, design a gated D flip-flop. The D input to the flip-flop is $x_1x_2 + x_3x_4$. Then instantiate the gated D flip-flop three times to design a 3-bit Gray code counter using structural modeling. Obtain the design module, the test bench module, the outputs showing the counting sequence, and the waveforms.

11.4   Using behavioral/dataflow modeling, design a counter-shifter module that shifts two different patterns through a shift register. When the counter reaches a count of three, a shifter clock pulse is generated during the last half of the system clock. If the pattern select input is *pattern = 0*, then the shift pattern is 000, 100, 010, 001, 000, ... If the pattern select input is *pattern = 1*, then the shift pattern is 000, 100, 110, 011, 001, 000, ... Use the **case** statement to determine the next count and next pattern. Obtain the design module, test bench module, and waveforms.

11.5   Design a 4-bit Gray code counter using structural modeling that incorporates a parity bit to maintain odd parity for the concatenated 4-bit Gray code and parity bit itself. Include error detection logic that generates an error output if the five bits have even parity. Use *JK* flip-flops and any additional dataflow logic primitives. Obtain the structural module, the test bench, the outputs, and the waveforms. After the counting sequence has been verified, the test bench will force a parity error to check the parity error detection logic.

11.6   Design an 8-bit carry lookahead adder using dataflow modeling. The adder has two groups of four bits per group. The high-order group has operands *a[7:4]* and *b[7:4]* that produce a sum of *sum[7:4]*; the low-order group has operands *a[3:0]* and *b[3:0]* that produce a sum of *sum[3:0]*. The carry-in is *cin*; the carry-out is *cout*, which is the carry-out of bit 7.
    Obtain the design module, the test bench module for several different operands, the outputs in decimal notation, and the waveforms.

11.7   Use structural modeling to design a 4-bit array multiplier. Use dataflow modeling to design the logic primitives and structural modeling to design a full adder, all of which will be instantiated into the multiplier module. Obtain the design module, the test bench module, and the outputs in decimal notation for all combinations of the two operands *a[3:0]* and *b[3:0]* to yield the product *product[7:0]*.

11.8   Design a Mealy synchronous sequential machine using structural modeling to detect a sequence of 0110 on a serial input line x_1. Overlapping sequences are valid. Output z_1 will be asserted during the last half of the clock cycle in the state in which the final 0 is detected. Obtain the state diagram, the Karnaugh maps for the input equations using D flip-flops, and the logic diagram.

Use dataflow modeling for any logic primitives that are required. Use behavioral modeling for the D flip-flop. Obtain the structural module, the test bench, the outputs, and the waveforms. In the test bench, check all possible paths in the state diagram for correct functional operation and include at least two valid overlapping sequences.

11.9   Design a Moore synchronous sequential machine that operates according to the following two sequences:

1.   State a $(x_1 \oplus x_2)'$ $\rightarrow$ state b $(x_1'x_2)$ $\rightarrow$ state c $(z_1 \uparrow t_1 \downarrow t_3)$
2.   State a $(x_1 \oplus x_2)'$ $\rightarrow$ state b $(x_1 x_2')$ $\rightarrow$ state e $(z_2 \uparrow t_1 \downarrow t_3)$

where $\uparrow t_1 \downarrow t_3$ specifies assertion for the entire clock period. There are two sequences in the machine specifications: one to assert output z_1 and one to assert output z_2. Both sequences begin in state a and proceed to state b if inputs x_1 and x_2 are the same value — either both 0s or both 1s. In sequence 1, the machine proceeds to state b if the conditions are met in state a. In state b, if $x_1 x_2$ = 01, then the machine moves to state c and asserts output z_1 for one clock period.

In sequence 2, if the conditions are met in state a, then the transition is again to state b; however, a valid input sequence in state b is now $x_1 x_2$ = 10. Under these conditions, the machine proceeds to state e and generates output z_2 for one clock period. Maintain three state levels for all possible state transition sequences, including any path that does not produce an output.

Because the machine represents a Moore model and outputs z_1 and z_2 are asserted at time t_1 and deasserted at time t_3, state code assignments must be carefully chosen so that no state transition will produce a glitch on the outputs. If the outputs were asserted at time t_2, then any arbitrary state code assignment would suffice, because the assertion time occurs long after the machine has stabilized.

Obtain the state diagram, the input maps, the output maps, the logic diagram using linear-select multiplexers for the δ next-state logic, D flip-flops for the storage elements, and logic primitives for the output logic. Then generate the structural module, test bench, outputs, and waveforms.

11.10   In Chapter 9, the δ next-state logic was implemented with linear-select multiplexers. Although the machines functioned correctly according to the specifications, the designs illustrated an inefficient use of the 2^p:1 multiplexers. Smaller multiplexers with fewer data inputs could be utilized effectively with a corresponding reduction in machine cost.

If the number of unique entries in an input map for flip-flop y_i satisfies the expression in the following equation, where u is the number of unique entries and p is the number of storage elements, then at most a $(2^p \div 2){:}1$ multiplexer will satisfy the requirements of Dy_i.

$$1 < u \geq (2^p \div 2)$$

If, however, $u > 2^p \div 2$, then a $2^p{:}1$ multiplexer is necessary. The largest multiplexer with which to economically implement the input logic is a 16:1 multiplexer, and then only if the number of distinct entries in the input map warrants a multiplexer of this size.

A general block diagram for a synchronous sequential machine using *nonlinear-select multiplexers* for the δ next-state logic is shown below. In this method, one multiplexer is still required for each state flip-flop, but the number of multiplexer select inputs p' is less than p, such that $1 \leq p' < p$.

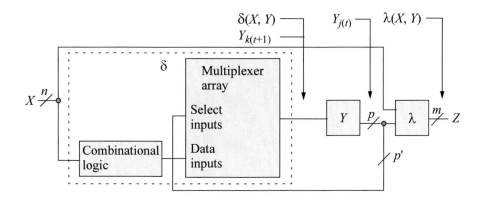

If a multiplexer has unused data inputs — corresponding to unused states in the input map — then these unused inputs can be connected to logically adjacent multiplexer inputs. The resulting linked set of inputs can be addressed by a common select variable. Thus, in a 4:1 multiplexer, if data input $d_2 = 1$ and $d_3 =$ "don't care," then d_2 and d_3 can both be connected to a logic 1. The two inputs can now be selected by $s_1 s_0 = 10$ or 11; that is, $s_1 s_0 = 1{-}$. Also, multiplexers containing the same number of data inputs should be addressed by the same select input variables, if possible. This permits the utilization of noncustom technology, where multiplexers in the same integrated circuit share common select inputs.

Given the state diagram shown on the next page for a Moore synchronous sequential machine, implement the design using nonlinear-select multiplexers for the δ next-state logic, D flip-flops for the storage elements, and a 3:8 decoder for the λ output logic. Outputs z_1, z_2, and z_3 are asserted at time t_2 and deasserted at time t_3. Recall that t_1 is the beginning of the clock cycle, t_2 is the midpoint of the clock cycle, and t_3 is the end of the clock cycle. Therefore, all three outputs are asserted during the last half of their respective clock cycle.

Generate the Karnaugh maps, the input equations, and the logic diagram. Then design the Moore machine using structural modeling. Implement a 4:1 multiplexer using dataflow modeling, a D flip-flop using behavioral modeling, and a 3:8 decoder using dataflow modeling. Any logic primitives are to be designed using dataflow modeling. Obtain the test bench that checks all paths in the state diagram, the outputs, and the waveforms.

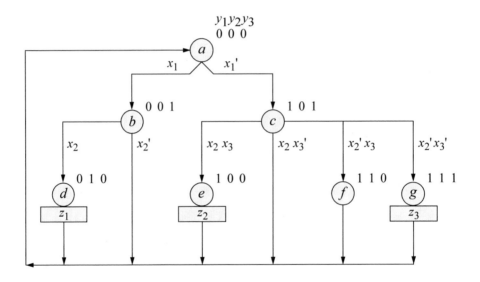

11.11   Using mixed-design modeling, design a 4-function arithmetic and logic unit to execute the following four operations: add, multiply, AND, and OR. The operands are eight bits defined as $a[7:0]$ and $b[7:0]$. The result of all operations will be 16 bits $rslt[15:0]$. Determine the state of the following flags:

The parity flag $pf = 1$ if the result has an even number of 1s; otherwise, $pf = 0$.

The sign flag sf represents the state of the leftmost result bit.

The zero flag $zf = 1$ if the result of an operation is zero; otherwise, $zf = 0$.

Obtain the design module, the test bench module using several values for each of the four operations, the outputs, and the waveforms.

11.12 Design a Mealy synchronous sequential machine using structural modeling to detect a sequence of 1001 on a serial input line x_1. Overlapping sequences are valid. Output z_1 will be asserted for the entire clock cycle in the state in which the final 1 is detected. Obtain the state diagram, the Karnaugh maps for the input equations using D flip-flops, and the logic diagram.

Use dataflow modeling for any logic primitives that are required. Use behavioral modeling for the D flip-flop. Obtain the structural module, the test bench, the outputs, and the waveforms. In the test bench, check all possible paths in the state diagram for correct functional operation and include at least two valid overlapping sequences.

11.13 Design a pipelined RISC processor using the eight instructions shown in Table 11.12. The format for the instructions is shown in Table 11.13. There are five units (modules) in the processor: instruction cache (*icache*), instruction unit (*iunit*), decode unit (*decode*), execution unit (*eunit*), and a register file (*regfile*). Each unit will be designed using behavioral modeling and will be compiled and simulated using a test bench to show the binary outputs and waveforms.

Table 11.12 Operation Codes

Instruction	Operation Code		
ADD	0	0	0
SUB	0	0	1
AND	0	1	0
OR	0	1	1
XOR	1	0	0
NOT	1	0	1
SRA	1	1	0
ROL	1	1	1

Table 11.13 Instruction Format

Operation code			Destination			Source		
8	7	6	5	4	3	2	1	0

The entire processor will then be configured using a structural module and simulated by means of a test bench to verify correct operation. The result of the simulation will be waveforms that illustrate the entire processor operation using the register file initial contents shown in Table 11.14 for the program given in Table 11.15.

**Table 11.14 Register File
Initial Contents**

Register 0		0000	0000
Register 1		0000	0001
Register 2		0000	0010
Register 3		0000	0011
Register 4		0000	1111
Register 5		1111	0000
Register 6		0101	0101
Register 7		1111	1111

Table 11.15 Program in the Instruction Cache

Icache Address	Instruction			Result		
00	ADD	R1, R2		R1 =	0000	0011
01	SUB	R4, R3		R4 =	0000	1100
02	AND	R5, R6		R5 =	0101	0000
03	OR	R6, R7		R6 =	1111	1111
04	XOR	R5, R7		R5 =	1010	1111
05	NOT	R4		R4 =	1111	0011
06	SRA	R3		R3 =	0000	0001
07	ROL	R5		R5 =	0101	1111
08	SUB	R1, R2		R1 =	0000	0001
09	ADD	R3, R2		R3 =	0000	0011
10	AND	R4, R1		R4 =	0000	0001
11	SRA	R5		R5 =	0010	1111
12	ADD	R4, R3		R4 =	0000	0100
13	OR	R1, R1		R1 =	0000	0001
14	OR	R1, R1		R1 =	0000	0001
15	OR	R1, R1		R1 =	0000	0001

The register file final contents are shown in Table 11.16 and should correspond to the register file contents as shown in the waveforms at the completion of simulation.

**Table 11.16 Register File
Final Contents**

Register 0		0000	0000
Register 1		0000	0001
Register 2		0000	0010
Register 3		0000	0011
Register 4		0000	0100
Register 5		0010	1111
Register 6		1111	1111
Register 7		1111	1111

Appendix A

Event Queue

Event management in Verilog hardware description language (HDL) is controlled by an event queue. Verilog modules generate events in the test bench, which provide stimulus to the module under test. These events can then produce new events by the modules under test. Since the Verilog HDL Language Reference Manual (LRM) does not specify a method of handling events, the simulator must provide a way to arrange and schedule these events in order to accurately model delays and obtain the correct order of execution. The manner of implementing the event queue is vendor-dependent.

Time in the event queue advances when every event that is scheduled in that time step is executed. Simulation is finished when all event queues are empty. An event at time t may schedule another event at time t or at time $t + n$.

A.1 Event Handling for Dataflow Assignments

Dataflow constructs consist of continuous assignments using the **assign** statement. The assignment occurs whenever simulation causes a change to the right-hand side expression. Unlike procedural assignments, continuous assignments are order independent — they can be placed anywhere in the module.

Consider the logic diagram shown in Figure A.1 which is represented by the two dataflow modules of Figure A.2 and Figure A.3. The test bench for both modules is shown in Figure A.4. The only difference between the two dataflow modules is the reversal of the two **assign** statements. The order in which the two statements execute is not defined by the Verilog HDL LRM; therefore, the order of execution is indeterminate.

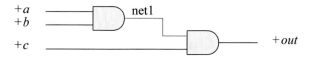

Figure A.1   Logic diagram to demonstrate event handling.

```
module dataflow (a, b, c, out);    module dataflow (a, b, c, out)

input a, b, c;                     input a, b, c;
output out;                        output out;

wire a, b, c;                      wire a, b, c;
wire out;                          wire out;

//define internal net             //define internal net
wire net1;                        wire net1;

assign net1 = a & b;              assign out = net1 & c;
assign out = net1 & c;            assign net1 = a & b;

endmodule                         endmodule
```

Figure A.2   Dataflow module 1.   **Figure A.3**   Dataflow module 2.

```
module dataflow_tb;               end
                                  //instantiate the module
reg test_a, test_b, test_c;       dataflow inst1
wire test_out;                        .a(test_a),
                                      .b(test_b),
initial                               .c(test_c),
begin                                 .out(test_out)
   test_a = 1'b1;                     );
   test_b = 1'b0;
   test_c = 1'b0;                 endmodule

   #10   test_b = 1'b1;
         test_c = 1'b1;
   #10   $stop;
```

Figure A.4   Test bench for Figure A.2 and Figure A.3.

Assume that the simulator executes the assignment order shown in Figure A.2 first. When the simulator reaches time unit #10 in the test bench, it will evaluate the right-hand side of *test_b = 1'b1;* and place its value in the event queue for an immediate scheduled assignment. Since this is a blocking statement, the next statement will not execute until the assignment has been made. Figure A.5 represents the event queue after the evaluation. The input signal *b* will assume the value of *test_b* through instantiation.

Event queue					
Scheduled event 5	Scheduled event 4	Scheduled event 3	Scheduled event 2	Scheduled event 1	Time unit
				test_b ← 1'b1 b ← 1'b1	$t = \#10$
◄────────────────────────────				Order of execution	

Figure A.5   Event queue after execution of *test_b = 1'b1;*.

After the assignment has been made, the simulator will execute the *test_c = 1'b1;* statement by evaluating the right-hand side, and then placing its value in the event queue for immediate assignment. The new event queue is shown in Figure A.6. The entry that is not shaded represents an executed assignment.

Event queue					
Scheduled event 5	Scheduled event 4	Scheduled event 3	Scheduled event 2	Scheduled event 1	Time unit
			test_c ← 1'b1 c ← 1'b1	test_b ← 1'b1 b ← 1'b1	$t = \#10$
◄────────────────────────────				Order of execution	

Figure A.6   Event queue after execution of *test_c = 1'b1;*.

When the two assignments have been made, time unit #10 will have ended in the test bench, which is the top-level module in the hierarchy. The simulator will then enter the instantiated dataflow module during this same time unit and determine that events have occurred on input signals *b* and *c* and execute the two continuous assignments. At this point, inputs *a*, *b*, and *c* will be at a logic 1 level. However, *net1* will still contain a logic 0 level as a result of the first three assignments that executed at time #0 in the test bench. Thus, the statement *assign out = net1 & c;* will evaluate to a logic 0, which will be placed in the event queue and immediately assigned to *out*, as shown in Figure A.7.

Event queue					
Scheduled event 5	Scheduled event 4	Scheduled event 3	Scheduled event 2	Scheduled event 1	Time unit
		out ← 1'b0 test_out ← 1'b0	test_c ← 1'b1 c ← 1'b1	test_b ← 1'b1 b ← 1'b1	$t = \#10$
				Order of execution	

Figure A.7   Event queue after execution of *assign out = net1 & c;*.

The simulator will then execute the *assign net1 = a & b;* statement in which the right-hand side evaluates to a logic 1 level. This will be placed on the queue and immediately assigned to *net1* as shown in Figure A.8.

Event queue					
Scheduled event 5	Scheduled event 4	Scheduled event 3	Scheduled event 2	Scheduled event 1	Time unit
	net1 ← 1'b1	out ← 1'b0 test_out ← 1'b0	test_c ← 1'b1 c ← 1'b1	test_b ← 1'b1 b ← 1'b1	$t = \#10$
				Order of execution	

Figure A.8   Event queue after execution of *assign net1 = a & b;*.

When the assignment has been made to *net1*, the simulator will recognize this as an event on *net1*, which will cause all statements that use *net1* to be reevaluated. The only statement to be reevaluated is *assign out = net1 & c;*. Since both *net1* and *c* equal a logic 1 level, the right-hand side will evaluate to a logic 1, resulting in the event queue shown in Figure A.9.

Event queue					
Scheduled event 5	Scheduled event 4	Scheduled event 3	Scheduled event 2	Scheduled event 1	Time unit
out ← 1'b1 test_out ← 1'b1	net1 ← 1'b1	out ← 1'b0 test_out ← 1'b0	test_c ← 1'b1 c ← 1'b1	test_b ← 1'b1 b ← 1'b1	t = #10
				Order of execution	

Figure A.9   Event queue after execution of *assign out = net1 & c;*.

The test bench signal *test_out* must now be updated because it is connected to *out* through instantiation. Because the signal *out* is not associated with any other statements within the module, the output from the module will now reflect the correct output. Since all statements within the dataflow module have been processed, the simulator will exit the module and return to the test bench. All events have now been processed; therefore, time unit #10 is complete and the simulator will advance the simulation time.

Since the order of executing the **assign** statements is irrelevant, processing of the dataflow events will now begin with the *assign net1 = a & b;* statement to show that the result is the same. The event queue is shown in Figure A.10.

Event queue					
Scheduled event 5	Scheduled event 4	Scheduled event 3	Scheduled event 2	Scheduled event 1	Time unit
		net1 ← 1'b1	test_c ← 1'b1 c ← 1'b1	test_b ← 1'b1 b ← 1'b1	t = #10
				Order of execution	

Figure A.10   Event queue beginning with the statement *assign net1 = a & b;*.

Once the assignment to *net1* has been made, the simulator recognizes this as a new event on *net1*. The existing event on input *c* requires the evaluation of statement *assign out = net1 & c;*. The right-hand side of the statement will evaluate to a logic 1, and will be placed on the event queue for immediate assignment, as shown in Figure A.11.

Event queue					
Sched-uled event 5	Scheduled event 4	Scheduled event 3	Scheduled event 2	Scheduled event 1	Time unit
	out ← 1'b1 test_out ← 1'b1	net1 ← 1'b1	test_c ← 1'b1 c ← 1'b1	test_b ← 1'b1 b ← 1'b1	t = #10
◄────────────────────────────────				Order of execution	

Figure A.11   Event queue after execution of *assign out = net1 & c;*.

A.2 Event Handling for Blocking Assignments

The blocking assignment operator is the equal (=) symbol. A blocking assignment evaluates the right-hand side arguments and completes the assignment to the left-hand side before executing the next statement; that is, the assignment *blocks* other assignments until the current assignment has been executed.

Example A.1   Consider the code segment shown in Figure A.12 using blocking assignments in conjunction with the event queue of Figure A.13. There are no inter-statement delays and no intrastatement delays associated with this code segment. In the first blocking assignment, the right-hand side is evaluated and the assignment is scheduled in the event queue. Program flow is blocked until the assignment is executed. This is true for all blocking assignment statements in this code segment. The assignments all occur in the same simulation time step *t*.

```
always @ (x2 or x3 or x5 or x7)
begin
    x1 = x2 | x3;
    x4 = x5;
    x6 = x7;
end
```

Figure A.12   Code segment with blocking assignments.

Event queue					
Scheduled event 5	Scheduled event 4	Scheduled event 3	Scheduled event 2	Scheduled event 1	Time unit
		x6 ← x7 (*t*)	x4 ← x5 (*t*)	x1 ← x2 | x3 (*t*)	*t*
			Order of execution		

Figure A.13   Event queue for Figure A.12.

Example A.2   The code segment shown in Figure A.14 contains an interstatement delay. Both the evaluation and the assignment are delayed by two time units. When the delay has taken place, the right-hand side is evaluated and the assignment is scheduled in the event queue as shown in Figure A.15. The program flow is blocked until the assignment is executed.

```
always @ (x2)
begin
   #2 x1 = x2;
end
```

Figure A.14   Blocking statement with interstatement delay.

Event queue					
Scheduled event 5	Scheduled event 4	Scheduled event 3	Scheduled event 2	Scheduled event 1	Time unit
					t
				x1 ← x2 (*t* + 2)	*t* + 2
			Order of execution		

Figure A.15   Event queue for Figure A.14.

Example A.3   The code segment of Figure A.16 shows three statements with interstatement delays of $t + 2$ time units. The first statement does not execute until simulation time $t + 2$ as shown in Figure A.17. The right-hand side $(x_2 \mid x_3)$ is evaluated at the current simulation time which is $t + 2$ time units, and then assigned to the left-hand side. At $t + 2$, x_1 receives the output of $x_2 \mid x_3$.

```
always @ (x2 or x3 or x5 or x7)
begin
   #2 x1 = x2 | x3;
   #2 x4 = x5;
   #2 x6 = x7;
end
```

Figure A.16   Code segment for delayed blocking assignment with interstatement delays.

		Event queue			
Scheduled event 5	Scheduled event 4	Scheduled event 3	Scheduled event 2	Scheduled event 1	Time unit
					t
				$x1 \leftarrow x2 \mid x3\ (t+2)$	$t+2$
				$x4 \leftarrow x5\ (t+4)$	$t+4$
				$x6 \leftarrow x7\ (t+6)$	$t+6$
				Order of execution	

Figure A.17   Event queue for Figure A.16.

Example A.4   The code segment in Figure A.18 shows three statements using blocking assignments with intrastatement delays. Evaluation of $x_3 = \#2\ x_4$ and $x_5 = \#2\ x_6$ is blocked until x_2 has been assigned to x_1, which occurs at $t + 2$ time units. When the second statement is reached, it is scheduled in the event queue at time $t + 2$, but the assignment to x_3 will not occur until $t + 4$ time units. The evaluation in the third statement is blocked until the assignment is made to x_3. Figure A.19 shows the event queue.

```
always @ (x2 or x4 or x6)
begin
   x1 = #2 x2;      //first statement
   x3 = #2 x4;      //second statement
   x5 = #2 x6;      //third statement
end
```

Figure A.18   Code segment using blocking assignments with interstatement delays.

Event queue					
Scheduled event 5	Scheduled event 4	Scheduled event 3	Scheduled event 2	Scheduled event 1	Time unit
					t
				$x1 \leftarrow x2\ (t)$	$t+2$
				$x3 \leftarrow x4\ (t+2)$	$t+4$
				$x5 \leftarrow x6\ (t+4)$	$t+6$
←				Order of execution	

Figure A.19   Event queue for the code segment of Figure A.18.

A.3 Event Handling for Nonblocking Assignments

Whereas blocking assignments block the sequential execution of an **always** block until the simulator performs the assignment, nonblocking statements evaluate each statement in succession and place the result in the event queue. Assignment occurs when all of the **always** blocks in the module have been processed for the current time unit. The assignment may cause new events that require further processing by the simulator for the current time unit.

Example A.5   For nonblocking statements, the right-hand side is evaluated and the assignment is scheduled at the end of the queue. The program flow continues and the assignment occurs at the end of the time step. This is shown in the code segment of Figure A.20 and the event queue of Figure A.21.

```
always @ (posedge clk)
begin
   x1 <= x2;
end
```

Figure A.20   Code segment for a nonblocking assignment.

Event queue					
Scheduled event 5	Scheduled event 4	Scheduled event 3	Scheduled event 2	Scheduled event 1	Time unit
$x1 \leftarrow x2$ (t)					t
← ———————————————— Order of execution					

Figure A.21   Event queue for Figure A.20.

Example A.6   The code segment of Figure A.22 shows a nonblocking statement with an interstatement delay. The evaluation is delayed by the timing control, and then the right-hand side expression is evaluated and assignment is scheduled at the end of the queue. Program flow continues and assignment is made at the end of the current time step as shown in the event queue of Figure A.23.

```
always @ (posedge clk)
begin
    #2 x1 <= x2;
end
```

Figure A.22   Nonblocking assignment with interstatement delay.

Event queue					
Scheduled event 5	Scheduled event 4	Scheduled event 3	Scheduled event 2	Scheduled event 1	Time unit
					t
$x1 \leftarrow x2$ $(t+2)$					$t+2$
← ———————————————— Order of execution					

Figure A.23   Event queue for Figure A.22.

Example A.7   The code segment of Figure A.24 shows a nonblocking statement with an intrastatement delay. The right-hand side expression is evaluated and assignment is delayed by the timing control and is scheduled at the end of the queue. Program flow continues and assignment is made at the end of the current time step as shown in the event queue of Figure A.25.

```
always @ (posedge clk)
begin
   x1 <= #2 x2;
end
```

Figure A.24   Nonblocking assignment with intrastatement delay.

Event queue					
Scheduled event 5	Scheduled event 4	Scheduled event 3	Scheduled event 2	Scheduled event 1	Time unit
					t
x1 ← x2 (t)					$t + 2$
				Order of execution	

Figure A.25   Event queue for Figure A.24.

Example A.8   The code segment of Figure A.26 shows nonblocking statements with intrastatement delays. The right-hand side expressions are evaluated and assignment is delayed by the timing control and is scheduled at the end of the queue. Program flow continues and assignment is made at the end of the current time step as shown in the event queue of Figure A.27.

```
always @ (posedge clk)
begin
   x1 <= #2 x2;
   x3 <= #2 x4;
   x5 <= #2 x6;
end
```

Figure A.26   Nonblocking assignments with intrastatement delays.

Event queue					
Scheduled event 5	Scheduled event 4	Scheduled event 3	Scheduled event 2	Scheduled event 1	Time unit
					t
x5 ← x6 (t)	x3 ← x4 (t)	x1 ← x2 (t)			$t+2$
← Order of execution					

Figure A.27   Event queue for Figure A.26.

Example A.9   Figure A.28 shows a code segment using nonblocking assignment with an intrastatement delay. The right-hand expression is evaluated at the current time. The assignment is scheduled, but delayed by the timing control #2. This method allows for propagation delay through a logic element; for example, a D flip-flop. The event queue is shown in Figure A.29.

```
always @ (posedge clk)
begin
   q <= #2 d;
end
```

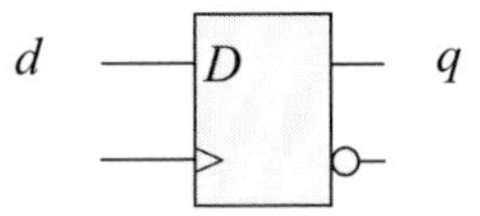

Figure A.28   Code segment using intrastatement delay with blocking assignment.

Event queue					
Scheduled event 5	Scheduled event 4	Scheduled event 3	Scheduled event 2	Scheduled event 1	Time unit
					t
				q ← d (t)	$t+2$
← Order of execution					

Figure A.29   Event queue for the code segment of Figure A.28.

A.4   Event Handling for Mixed Blocking and Nonblocking Assignments

All nonblocking assignments are placed at the end of the queue while all blocking assignments are placed at the beginning of the queue in their respective order of evaluation. Thus, for any given simulation time t, all blocking statements are evaluated and assigned first, then all nonblocking statements are evaluated.

This is the reason why combinational logic requires the use of blocking assignments while sequential logic, such as flip-flops, requires the use of nonblocking assignments. In this way, Verilog events can model real hardware in which combinational logic at the input to a flip-flop can stabilize before the clock sets the flip-flop to the state of the input logic. Therefore, blocking assignments are placed at the top of the queue to allow the input data to be stable, whereas nonblocking assignments are placed at the bottom of the queue to be executed after the input data has stabilized.

The logic diagram of Figure A.30 illustrates this concept for two multiplexers connected to the D inputs of their respective flip-flops. The multiplexers represent combinational logic; the D flip-flops represent sequential logic. The behavioral module is shown in Figure A.31 and the event queue is shown in Figure A.32.

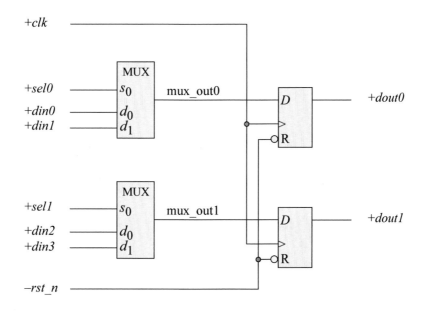

Figure A.30   Combinational logic connected to sequential logic to illustrate the use of blocking and nonblocking assignments.

Because multiplexers are combinational logic, the outputs *mux_out0* and *mux_out1* are placed at the beginning of the queue, as shown in Figure A.32. Nets *mux_out0* and *mux_out1* are in separate **always** blocks; therefore, the order in which they are placed in the queue is arbitrary and can differ with each simulator. The result, however, is the same. If *mux_out0* and *mux_out1* were placed in the same **always** block, then the order in which they are placed in the queue must be the same order as they appear in the **always** block.

Because *dout0* and *dout1* are sequential, they are placed at the end of the queue. Since they appear in separate **always** blocks, the order of their placement in the queue is irrelevant. Once the values of *mux_out0* and *mux_out1* are assigned in the queue, their values will then be used in the assignment of *dout0* and *dout1*; that is, the state of *mux_out0* and *mux_out1* will be set into the *D* flip-flops at the next positive clock transition and assigned to *dout0* and *dout1*.

```
//behavioral module with combinational and sequential logic
//to illustrate their placement in the event queue

module mux_plus_flop (clk, rst_n,
        din0, din1, sel0, dout0,
        din2, din3, sel1, dout1);

input clk, rst_n;
input din0, din1, sel0;
input din2, din3, sel1;
output dout0, dout1;

reg mux_out0, mux_out1;
reg dout0, dout1;

//combinational logic for multiplexers
always @ (din0 or din1 or sel0)
begin
   if (sel0)
      mux_out0 = din1;
   else
      mux_out0 = din0;
end

always @ (din2 or din3 or sel1)
begin
   if (sel1)
      mux_out1 = din3;
   else
      mux_out1 = din2;
end
//continued on next page
```

Figure A.31   Mixed blocking and nonblocking assignments that represent combinational and sequential logic.

```
//sequential logic for D flip-flops
always @ (posedge clk or negedge rst_n)
begin
    if (~rst_n)
        dout0 <= 1'b0;
    else
        dout0 <= mux_out0;
end

always @ (posedge clk or negedge rst_n)
begin
    if (~rst_n)
        dout1 <= 1'b0;
    else
        dout1 <= mux_out1;
end

endmodule
```

Figure A.31   (Continued)

Event queue					
Scheduled event 4	Scheduled event 3	N/A	Scheduled event 2	Scheduled event 1	Time unit
dout1 ← mux_out1 (t)	dout0 ← mux_out0 (t)		mux_out1 ← din3 (t)	mux_out0 ← din1 (t)	t
◄───────────────────────────				Order of execution	

Figure A.32   Event queue for Figure A.31.

Appendix B

Verilog Project Procedure

- **Create a folder** (Do only once)

  Windows Explorer > C > New Folder <Verilog> > Enter > Exit Windows Explorer.

- **Create a project** (Do for each project)

  Bring up Silos Simulation Environment.
  > File > Close Project. Minimize Silos.
  Windows Explorer > Verilog > File > New Folder <new folder name> Enter.
  Exit Windows Explorer. Maximize Silos.
  File > New Project.
  Create New Project. Save In: Verilog folder.
  > Click new folder name. Open.
  Create New Project. Filename: Give project name — usually same name as the folder name. Save
  Project Properties > Cancel.

- **File > New**

  .
  . Design module code goes here
  .

- **File > Save As > File name: <filename.v> > Save**

- **Compile code**

  Edit > Project Properties > Add. Select one or more files to add.
  > Click on the file > Open.
  Project Properties. The selected files are shown > OK.
  Load/Reload Input Files. This compiles the code.
  Check screen output for errors. "Simulation stopped at the end of time 0" indicates no compilation errors.

- **Test bench**

  File > New

    .
    .   Test bench module code goes here
    .

- **File > Save As > File name: < filename.v> > Save.**

- **Compile test bench**

  Edit > Project Properties > Add. Select one or more files to add.
     Click on the file > Open
  Project Properties. The selected files are shown > OK.
  Load/Reload Input Files. This compiles the code.
  Check screen output for errors. "Simulation stopped at end of time 0"
     indicates no compilation errors.

- **Binary Output and Waveforms**

  For binary output: click on the GO icon.
  For waveforms: click on the Analyzer icon.
       Click on the Explorer icon. The signals are listed in Silos Explorer.
       Click on the desired signal names.
       Right click. Add Signals to Analyzer.
       Waveforms are displayed.
       Exit Silos Explorer.

- **Change Time Scale**

  With the waveforms displayed, click on Analyzer > Timeline > Timescale
     Enter Time / div > OK

- **Exit the project**.

  Close the waveforms, module, and test bench.
  File > Close Project.

Appendix C

Answers to Select Problems

Chapter 2   Overview

2.1   Design a 4-input AND gate using dataflow modeling. Design a test bench and obtain the binary outputs and waveforms.

```
//dataflow 4-input and gate
module and4_df (x1, x2, x3, x4, z1);

//list all inputs and outputs
input x1, x2, x3, x4;
output z1;

//define signals as wire
wire x1, x2, x3, x4;
wire z1;

//continuous assign used for dataflow
assign z1 = (x1 & x2 & x3 & x4);
endmodule
```

```
//4-input and gate test bench
module and4_df_tb;

reg x1, x2, x3, x4;
wire z1;

//apply input vectors
initial
begin: apply_stimulus
   reg [5:0] invect;
   for (invect=0; invect<16; invect=invect+1)
      begin
         {x1, x2, x3, x4} = invect [5:0];
         #10 $display ("{x1x2x3x4}=%b, z1=%b",
                       {x1, x2, x3, x4},z1);
      end
end              //continued on next page
```

```
//instantiate the module into the test bench
and4_df inst1 (
    .x1(x1),
    .x2(x2),
    .x3(x3),
    .x4(x4),
    .z1(z1)
    );

endmodule
```

```
{x1x2x3x4} = 0000, z1 = 0
{x1x2x3x4} = 0001, z1 = 0
{x1x2x3x4} = 0010, z1 = 0
{x1x2x3x4} = 0011, z1 = 0
{x1x2x3x4} = 0100, z1 = 0
{x1x2x3x4} = 0101, z1 = 0
{x1x2x3x4} = 0110, z1 = 0
{x1x2x3x4} = 0111, z1 = 0
{x1x2x3x4} = 1000, z1 = 0
{x1x2x3x4} = 1001, z1 = 0
{x1x2x3x4} = 1010, z1 = 0
{x1x2x3x4} = 1011, z1 = 0
{x1x2x3x4} = 1100, z1 = 0
{x1x2x3x4} = 1101, z1 = 0
{x1x2x3x4} = 1110, z1 = 0
{x1x2x3x4} = 1111, z1 = 1
```

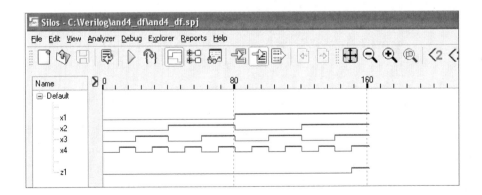

2.4   Design a 2-input NOR gate using dataflow modeling. Design a test bench and obtain the binary outputs and waveforms.

```verilog
//dataflow 2-input nor gate
module nor2_df (x1, x2, z1);

input x1, x2;
output z1;

wire x1, x2;
wire z1;

assign z1 = ~(x1 | x2);

endmodule
```

```verilog
//nor2_df test bench
module nor2_df_tb;

reg x1, x2;
wire z1;

//display the variables
initial
$monitor ("x1x2 = %b, z1 = %b",
          {x1, x2}, z1);

//apply input vectors
initial
begin
   #0   x1 = 1'b0;
        x2 = 1'b0;

   #10  x1 = 1'b0;
        x2 = 1'b1;
   #10  x1 = 1'b1;
        x2 = 1'b0;

   #10  x1 = 1'b1;
        x2 = 1'b1;

   #10  $stop;
end

//continued on next page
```

```
//instantiate the module into
//the test bench
nor2_df inst1 (
    .x1(x1),
    .x2(x2),
    .z1(z1)
    );

endmodule
```

```
x1x2 = 00, z1 = 1
x1x2 = 01, z1 = 0
x1x2 = 10, z1 = 0
x1x2 = 11, z1 = 0
```

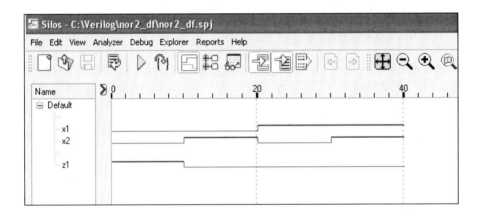

2.8   Implement the logic diagram shown below using dataflow modeling. Design a test bench and obtain the binary outputs and waveforms.

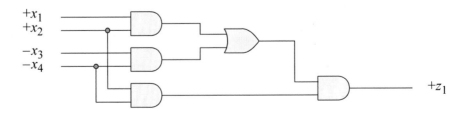

$$z_1 = x_1 x_2 x_4' + x_2 x_3' x_4'$$

```verilog
//dataflow for logic diagram
module log_diag_eqn1 (x1, x2, x3, x4, z1);

input x1, x2, x3, x4;
output z1;

wire x1, x2, x3, x4;
wire z1;

assign z1 = (x1 & x2 & ~x4) | (x2 & ~x3 & ~x4);
endmodule
```

```verilog
//logic diag eqtn test bench
module log_diag_eqn1_tb;

reg x1, x2, x3, x4;
wire z1;
//display variables
initial
$monitor ("x1x2x3x4=%b,z1=%b", {x1, x2, x3, x4}, z1);

//apply stimulus
initial
begin
   #0    x1=0; x2=0; x3=0; x4=0;
   #10   x1=0; x2=0; x3=0; x4=1;
   #10   x1=0; x2=0; x3=1; x4=0;
   #10   x1=0; x2=0; x3=1; x4=1;
   #10   x1=0; x2=1; x3=0; x4=0;
   #10   x1=0; x2=1; x3=0; x4=1;
   #10   x1=0; x2=1; x3=1; x4=0;
   #10   x1=0; x2=1; x3=1; x4=1;
   #10   x1=1; x2=0; x3=0; x4=0;
   #10   x1=1; x2=0; x3=0; x4=1;
   #10   x1=1; x2=0; x3=1; x4=0;
   #10   x1=1; x2=0; x3=1; x4=1;
   #10   x1=1; x2=1; x3=0; x4=0;
   #10   x1=1; x2=1; x3=0; x4=1;
   #10   x1=1; x2=1; x3=1; x4=0;
   #10   x1=1; x2=1; x3=1; x4=1;
   #10   $stop;
end

//continued on next page
```

```
//instantiate the module into the test bench
log_diag_eqn1 inst1 (
   .x1(x1),
   .x2(x2),
   .x3(x3),
   .x4(x4),
   .z1(z1)
   );

endmodule
```

```
x1x2x3x4 = 0000, z1 = 0
x1x2x3x4 = 0001, z1 = 0
x1x2x3x4 = 0010, z1 = 0
x1x2x3x4 = 0011, z1 = 0
x1x2x3x4 = 0100, z1 = 1
x1x2x3x4 = 0101, z1 = 0
x1x2x3x4 = 0110, z1 = 0
x1x2x3x4 = 0111, z1 = 0
x1x2x3x4 = 1000, z1 = 0
x1x2x3x4 = 1001, z1 = 0
x1x2x3x4 = 1010, z1 = 0
x1x2x3x4 = 1011, z1 = 0
x1x2x3x4 = 1100, z1 = 1
x1x2x3x4 = 1101, z1 = 0
x1x2x3x4 = 1110, z1 = 1
x1x2x3x4 = 1111, z1 = 0
```

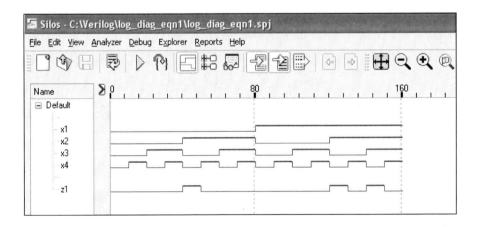

2.12   Given the Karnaugh map shown below, obtain the minimum sum-of-products expression and the minimum product-of-sums expression, then implement both expressions using behavioral modeling. Design two test benches and compare the binary outputs and waveforms.

x_1x_2 \\ x_3x_4	0 0	0 1	1 1	1 0
0 0	0	0	1	0
0 1	0	1	1	0
1 1	0	1	1	0
1 0	1	1	1	1

z_1

Minimum sum of products $= x_1x_2' + x_2x_4 + x_3x_4$

```
//behavioral logic equation
module log_eqn_sop1 (x1, x2, x3, x4, z1);

input x1, x2, x3, x4;
output z1;

wire x1, x2, x3, x4;
reg z1;

always @ (x1 or x2 or x3 or x4)
begin
    z1 = (x1 & ~x2) | (x2 & x4) | (x3 & x4);
end

endmodule
```

```
//logic equation sop test bench
module log_eqn_sop1_tb;


reg x1, x2, x3, x4;
wire z1;


initial
$monitor ("x1x2x3x4=%b, z1=%b", {x1, x2, x3, x4}, z1);


//apply input vectors
initial
begin
   #0    x1=0; x2=0; x3=0; x4=0;
   #10   x1=0; x2=0; x3=0; x4=1;
   #10   x1=0; x2=0; x3=1; x4=0;
   #10   x1=0; x2=0; x3=1; x4=1;
   #10   x1=0; x2=1; x3=0; x4=0;
   #10   x1=0; x2=1; x3=0; x4=1;
   #10   x1=0; x2=1; x3=1; x4=0;
   #10   x1=0; x2=1; x3=1; x4=1;
   #10   x1=1; x2=0; x3=0; x4=0;
   #10   x1=1; x2=0; x3=0; x4=1;
   #10   x1=1; x2=0; x3=1; x4=0;
   #10   x1=1; x2=0; x3=1; x4=1;
   #10   x1=1; x2=1; x3=0; x4=0;
   #10   x1=1; x2=1; x3=0; x4=1;
   #10   x1=1; x2=1; x3=1; x4=0;
   #10   x1=1; x2=1; x3=1; x4=1;

   #10   $stop;
end


//instantiate the module into the test bench
log_eqn_sop1 inst1 (
   .x1(x1),
   .x3(x3),
   .x4(x4),
   .z1(z1)
   );

endmodule
```

```
x1x2x3x4 = 0000, z1 = 0
x1x2x3x4 = 0001, z1 = 0
x1x2x3x4 = 0010, z1 = 0
x1x2x3x4 = 0011, z1 = 1
x1x2x3x4 = 0100, z1 = 0
x1x2x3x4 = 0101, z1 = 1
x1x2x3x4 = 0110, z1 = 0
x1x2x3x4 = 0111, z1 = 1
x1x2x3x4 = 1000, z1 = 1
x1x2x3x4 = 1001, z1 = 1
x1x2x3x4 = 1010, z1 = 1
x1x2x3x4 = 1011, z1 = 1
x1x2x3x4 = 1100, z1 = 0
x1x2x3x4 = 1101, z1 = 1
x1x2x3x4 = 1110, z1 = 0
x1x2x3x4 = 1111, z1 = 1
```

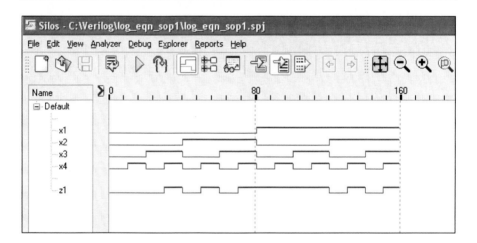

Minimum product of sums = $(x_2' + x_4)(x_1 + x_4)(x_1 + x_2 + x_3)$

```
//behavioral logic equation
module log_eqn_pos1 (x1, x2, x3, x4, z1);

input x1, x2, x3, x4;
output z1;
wire x1, x2, x3, x4;
reg z1;

always @ (x1 or x2 or x3 or x4)
begin
    z1 = (~x2 | x4) & (x1 | x4) & (x1 | x2 | x3);
end
endmodule
```

```verilog
//behavioral logic equation test bench
module log_eqn_pos1_tb;


reg x1, x2, x3, x4;
wire z1;


//display variables
initial
$monitor ("x1x2x3x4=%b,z1=%b", {x1, x2, x3, x4}, z1);


//apply input vectors
initial
begin
   #0    x1=0; x2=0; x3=0; x4=0;
   #10   x1=0; x2=0; x3=0; x4=1;
   #10   x1=0; x2=0; x3=1; x4=0;
   #10   x1=0; x2=0; x3=1; x4=1;
   #10   x1=0; x2=1; x3=0; x4=0;
   #10   x1=0; x2=1; x3=0; x4=1;
   #10   x1=0; x2=1; x3=1; x4=0;
   #10   x1=0; x2=1; x3=1; x4=1;
   #10   x1=1; x2=0; x3=0; x4=0;
   #10   x1=1; x2=0; x3=0; x4=1;
   #10   x1=1; x2=0; x3=1; x4=0;
   #10   x1=1; x2=0; x3=1; x4=1;
   #10   x1=1; x2=1; x3=0; x4=0;
   #10   x1=1; x2=1; x3=0; x4=1;
   #10   x1=1; x2=1; x3=1; x4=0;
   #10   x1=1; x2=1; x3=1; x4=1;

   #10   $stop;
end


//instantiate the module into the test bench
log_eqn_pos1 inst1 (
   .x1(x1),
   .x2(x2),
   .x3(x3),
   .x4(x4),
   .z1(z1)
   );

endmodule
```

```
x1x2x3x4 = 0000, z1 = 0
x1x2x3x4 = 0001, z1 = 0
x1x2x3x4 = 0010, z1 = 0
x1x2x3x4 = 0011, z1 = 1
x1x2x3x4 = 0100, z1 = 0
x1x2x3x4 = 0101, z1 = 1
x1x2x3x4 = 0110, z1 = 0
x1x2x3x4 = 0111, z1 = 1
x1x2x3x4 = 1000, z1 = 1
x1x2x3x4 = 1001, z1 = 1
x1x2x3x4 = 1010, z1 = 1
x1x2x3x4 = 1011, z1 = 1
x1x2x3x4 = 1100, z1 = 0
x1x2x3x4 = 1101, z1 = 1
x1x2x3x4 = 1110, z1 = 0
x1x2x3x4 = 1111, z1 = 1
```

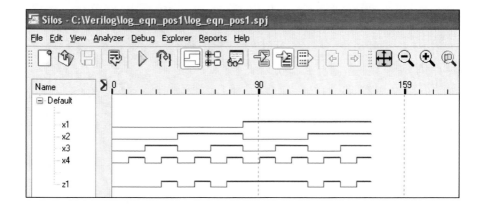

2.15   Minimize the following equation, then implement the result using structural modeling: $z_1 = x_1'x_2(x_3'x_4' + x_3'x_4) + x_1x_2(x_3'x_4' + x_3'x_4) + x_1x_2'x_3'x_4$

x_1x_2 x_3x_4	0 0	0 1	1 1	1 0
0 0	0 ⁰	0 ¹	0 ³	0 ²
0 1	1 ⁴	1 ⁵	0 ⁷	0 ⁶
1 1	1 ¹²	1 ¹³	0 ¹⁵	0 ¹⁴
1 0	0 ⁸	1 ⁹	0 ¹¹	0 ¹⁰

z_1

$z_1 = x_2x_3' + x_1x_3'x_4$

```verilog
//structural sum of products
module log_eqn_sop3 (x1, x2, x3, x4, z1);

input x1, x2, x3, x4;
output z1;
wire net1, net2, net3;

and2_df inst1 (    //instantiate the and gate
   .x1(x2),
   .x2(~x3),
   .z1(net1)
   );

and3_df inst2 (    //instantiate the and gate
   .x1(x1),
   .x2(~x3),
   .x3(x4),
   .z1(net2)
   );

or2_df inst3 (     //instantiate the or gate
   .x1(net1),
   .x2(net2),
   .z1(net3)
   );
assign z1 = net3;
endmodule
```

```verilog
//structural sum of products test bench
module log_eqn_sop3_tb;

reg x1, x2, x3, x4;
wire z1;

//apply input vectors
initial
begin : apply_stimulus
   reg [4:0] invect;
   for (invect = 0; invect < 16; invect = invect + 1)
   begin
      {x1, x2, x3, x4} = invect [4:0];
      #10 $display ("{x1x2x3x4} = %b, z1 = %b",
                    {x1, x2, x3, x4}, z1);
   end
end
//continued on next page
```

```
//instantiate the module into the test bench
log_eqn_sop3 inst1 (
   .x1(x1),
   .x2(x2),
   .x3(x3),
   .x4(x4),
   .z1(z1)
   );

endmodule
```

```
{x1x2x3x4} = 0000, z1 = 0
{x1x2x3x4} = 0001, z1 = 0
{x1x2x3x4} = 0010, z1 = 0
{x1x2x3x4} = 0011, z1 = 0
{x1x2x3x4} = 0100, z1 = 1
{x1x2x3x4} = 0101, z1 = 1
{x1x2x3x4} = 0110, z1 = 0
{x1x2x3x4} = 0111, z1 = 0
{x1x2x3x4} = 1000, z1 = 0
{x1x2x3x4} = 1001, z1 = 1
{x1x2x3x4} = 1010, z1 = 0
{x1x2x3x4} = 1011, z1 = 0
{x1x2x3x4} = 1100, z1 = 1
{x1x2x3x4} = 1101, z1 = 1
{x1x2x3x4} = 1110, z1 = 0
{x1x2x3x4} = 1111, z1 = 0
```

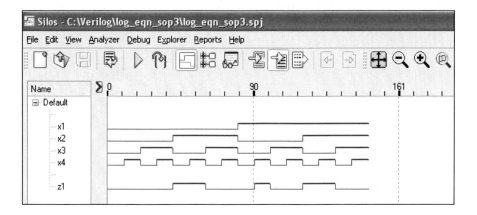

Chapter 3   Language Elements

3.1   Write the following decimal numbers as 16-bit binary numbers in 2s complement representation:

(a) 2170_{10} $= 0000\_1000\_0111\_1010_2$

The binary weight assigned to the 4-bit segments is:

$2^{12} = 4096$	$2^8 = 256$	$2^4 = 16$	$2^0 = 1$
0000	1000	0111	1010

(b) -858_{10} $= 1111\_1100\_1010\_0110_2$
(c) 32767_{10} $= 0111\_1111\_1111\_1111_2$
(d) -32768_{10} $= 1000\_0000\_0000\_0000_2$
(e) -1_{10} $= 1111\_1111\_1111\_1111_2$

3.4   Obtain the module, test bench, binary outputs, and waveforms for the equations shown below. Use built-in primitives and declare any internal wires.

$$z_1 = (x_1 \oplus x_2)x_3'$$
$$z_2 = (x_1 \oplus x_2)' \oplus x_3$$

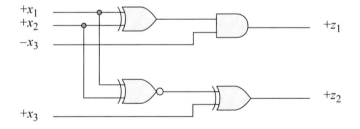

```
//logic equation using exclusive-OR
module log_eqn_xor1 (x1, x2, x3, z1, z2);

input x1, x2, x3;
output z1, z2;

wire net1, net2, net3, net4;

//continued on next page
```

```verilog
//instantiate the built-in primitives
xor (net1, x1, x2);
and (net3, net1, ~x3);
xnor (net2, x1, x2);
xor (net4, net2, x3);

assign z1 = net3;
assign z2 = net4;

endmodule
```

```verilog
//logic equation test bench
module log_eqn_xor1_tb;

reg x1, x2, x3;
wire z1, z2;

//display variables
initial
$monitor ("x1x2x3 = %b, z1 = %b, z2 = %b",
                {x1, x2, x3}, z1, z2);

//apply stimulus
initial
begin
      #0     x1=1'b0; x2=1'b0; x3=1'b0;
      #10    x1=1'b0; x2=1'b0; x3=1'b1;
      #10    x1=1'b0; x2=1'b1; x3=1'b0;
      #10    x1=1'b0; x2=1'b1; x3=1'b1;
      #10    x1=1'b1; x2=1'b0; x3=1'b0;
      #10    x1=1'b1; x2=1'b0; x3=1'b1;
      #10    x1=1'b1; x2=1'b1; x3=1'b0;
      #10    x1=1'b1; x2=1'b1; x3=1'b1;

      #10    $stop;
end

//continued on next page
```

```
//instantiate the module into the test bench
log_eqn_xor1 inst1 (
        .x1(x1),
        .x2(x2),
        .x3(x3),
        .z1(z1),
        .z2(z2)
        );

endmodule
```

x1x2x3 = 000, z1 = 0, z2 = 1	x1x2x3 = 100, z1 = 1, z2 = 0
x1x2x3 = 001, z1 = 0, z2 = 0	x1x2x3 = 101, z1 = 0, z2 = 1
x1x2x3 = 010, z1 = 1, z2 = 0	x1x2x3 = 110, z1 = 0, z2 = 1
x1x2x3 = 011, z1 = 0, z2 = 1	x1x2x3 = 111, z1 = 0, z2 = 0

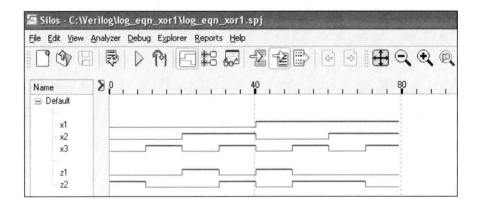

3.8   Which *one* of the following statements is true?

(a)   Vectors and arrays are the same.
False. A vector is a single element that is n bits wide.

(b)   Every element in an array can be assigned using one statement.
False. Not possible.

(c)   A RAM can be modeled by declaring an array of registers.
True. We define the RAM to have a word length equal to the register length and a size equal to the number of elements in the array.

Chapter 4   Expressions

4.1   Design a 4-bit adder using behavioral modeling with a parameter statement and blocking assignment. Design a test bench and obtain the outputs.

```
//dataflow for a 4-bit adder
module adder_behav_param (a, b, cin, sum);


parameter width = 4;


//list inputs and outputs
input [width - 1:0] a, b;          //a[3:0], b[3:0]
input cin;                         //cin is a scalar
output [width:0] sum;              //sum is 5 bits
                                   //to include cout


//define signals as wire for dataflow
wire [3:0] a, b;
wire cin;
reg [width:0] sum;                 //sum is used in an
                                   //always statement


//implement the 4-bit adder using nonblocking
//assignment and concatenate cout and sum
always @ (a or b or cin)
begin
    sum <= a + b + cin;
end

endmodule
```

```
//adder_behav_param test bench
module adder_behav_param_tb;

parameter width = 4;
reg [width-1:0] a, b;
reg cin;

wire [width:0] sum;

//display variables
initial
$monitor ("a b cin = %b_%b_%b, sum = %b",
          a, b, cin, sum);

//apply input vectors
initial
begin
    #0    a = 4'b0011;
          b = 4'b0100;
          cin = 1'b0;

    #5    a = 4'b1100;
          b = 4'b0011;
          cin = 1'b0;

    #5    a = 4'b0111;
          b = 4'b0110;
          cin = 1'b1;

    #5    a = 4'b1001;
          b = 4'b0111;
          cin = 1'b1;

    #5    a = 4'b1101;
          b = 4'b0111;
          cin = 1'b1;

    #5    a = 4'b1111;
          b = 4'b0110;
          cin = 1'b1;
    #5    $stop;
end
//continued on next page
```

```
//instantiate the module into the test bench
adder_behav_param inst1 (
        .a(a),
        .b(b),
        .cin(cin),
        .sum(sum)
        );

endmodule
```

```
a b cin = 0011_0100_0, sum = 00111
a b cin = 1100_0011_0, sum = 01111
a b cin = 0111_0110_1, sum = 01110
a b cin = 1001_0111_1, sum = 10001
a b cin = 1101_0111_1, sum = 10101
a b cin = 1111_0110_1, sum = 10110
```

4.4   Design a four-function arithmetic and logic unit for the three *logical* operations: logical AND, logical OR, and logical negation. There will be three 4-bit operands: *a*, *b*, and *c* and one result **reg** variable. Generate the test bench and obtain the outputs for the following operations:

AND operation $= (a \,\&\&\, b) \,\&\&\, c$
OR operation $= (a \,||\, c) \,||\, b$
AND OR operation $= (b \,\&\&\, c) \,||\, a$
NOT operation $= !((b \,\&\&\, c) \,||\, a)$

```
//demonstrate logical operations
module log_ops2 (a, b, c, opcode, rslt);

input [3:0] a, b, c;
input [1:0] opcode;
output rslt;

reg rslt;

//continued on next page
```

```
parameter              andop      = 2'b00,
                       orop       = 2'b01,
                       andorop    = 2'b10,
                       notop      = 2'b11;


always @ (a or b or c or opcode)
begin
      case (opcode)
            andop:      rslt = (a && b) && c;
            orop:       rslt = (a || c) || b;
            andorop:    rslt = (b && c) || a;
            notop:      rslt = !((b && c) || a);
      endcase
end

endmodule
```

```
//test bench for logical operators
module log_ops2_tb;

reg [3:0] a, b, c;
reg [1:0] opcode;
wire rslt;

initial
$monitor ("a=%b, b=%b, c=%b, opcode=%b, rslt=%b",
                        a, b, c, opcode, rslt);

//apply input vectors
initial
begin
      #0    a = 4'b0110;
            b = 4'b1100;
            c = 4'b0111;
            opcode = 2'b00;


      #5    a = 4'b0101;
            b = 4'b0000;
            c = 4'b1010;
            opcode = 2'b00;

//continued on next page
```

```
        #5      a = 4'b0000;
                b = 4'b1001;
                c = 4'b0000;
                opcode = 2'b01;

        #5      a = 4'b0000;
                b = 4'b0000;
                c = 4'b1111;
                opcode = 2'b10;

        #5      a = 4'b1100;
                b = 4'b1001;
                c = 4'b0101;
                opcode = 2'b11;

        #5      $stop;
end

//instantiate the module into the test bench
log_ops2 inst1 (
        .a(a),
        .b(b),
        .c(c),
        .opcode(opcode),
        .rslt(rslt)
        );
endmodule
```

```
a = 0110, b = 1100, c = 0111, opcode = 00, rslt = 1
a = 0101, b = 0000, c = 1010, opcode = 00, rslt = 0
a = 0000, b = 1001, c = 0000, opcode = 01, rslt = 1
a = 0000, b = 0000, c = 1111, opcode = 10, rslt = 0
a = 1100, b = 1001, c = 0101, opcode = 11, rslt = 0
```

4.8   Design a Verilog module that adds two 8-bit operands, then shifts the sum left four bit positions into a *rslt1* **reg** variable. Then shift the original sum right four bit positions into a *rslt2* **reg** variable. Generate a test bench and obtain the outputs.

```
//add shift operations
module add_shift (a, b, sum, rslt1, rslt2);


input [7:0] a, b;
output [7:0] sum;
output [15:0] rslt1, rslt2;


wire [7:0] a, b;
reg [7:0] sum;
reg [15:0] rslt1, rslt2;


always @ (a or b)
begin
     sum = a + b;
     rslt1 = sum << 4;
     rslt2 = sum >> 4;
end
endmodule
```

```
//add shift test bench
module add_shift_tb;


reg [7:0] a, b;
wire [7:0] sum;
wire [15:0] rslt1, rslt2;


initial              //display variables
$monitor ("a=%b, b=%b, sum=%b, rslt1=%b, rslt2=%b",
                a, b, sum, rslt1, rslt2);


initial              //apply input vectors
begin
     #0    a = 8'b0101_0101;
           b = 8'b0101_0101;

     #10   a = 8'b0000_1100;
           b = 8'b0000_0100;

     #10   a = 8'b1111_0000;
           b = 8'b0000_1111;

     #10   $stop;
end                  //continued on next page
```

```
//instantiate the module into the test bench
add_shift inst1 (
      .a(a),
      .b(b),
      .sum(sum),
      .rslt1(rslt1),
      .rslt2(rslt2)
      );

endmodule
```

```
a = 01010101, b = 01010101, sum = 10101010,
rslt1 = 0000101010100000, rslt2 = 0000000000001010

a = 00001100, b = 00000100, sum = 00010000,
rslt1 = 0000000100000000, rslt2 = 0000000000000001

a = 11110000, b = 00001111, sum = 11111111,
rslt1 = 0000111111110000, rslt2 = 0000000000001111
```

Chapter 5   Gate Level Modeling

5.2   Design the circuit for the equation shown below using built-in primitives. Design the module and test bench, then obtain the outputs and waveforms.

$$z_1 = [x_1 x_2 + (x_1 \oplus x_2)]\, x_3$$

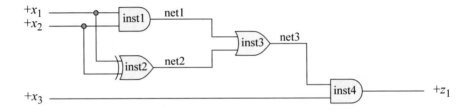

```
//xor product of sums
module xor_pos (x1, x2, x3, z1);

input x1, x2, x3;
output z1;

and inst1 (net1, x1, x2);
xor inst2 (net2, x1, x2);
or  inst3 (net3, net1, net2);
and inst4 (z1, net3, x3);
endmodule
```

```
//test bench for xor_pos
module xor_pos_tb;

reg x1, x2, x3;
wire z1;

initial
begin: apply_stimulus
reg [3:0] invect;
for (invect=0; invect<8; invect=invect+1)
   begin
      {x1, x2, x3} = invect [3:0];
      #10 $display ("{x1x2x3} = %b, z1 = %b",
                    {x1, x2, x3}, z1);
   end
end
//continued on next page
```

```
//instantiate the module into the test bench
xor_pos inst1 (
   .x1(x1),
   .x2(x2),
   .x3(x3),
   .z1(z1)
   );
endmodule
```

```
{x1x2x3} = 000, z1 = 0        {x1x2x3} = 100, z1 = 0
{x1x2x3} = 001, z1 = 0        {x1x2x3} = 101, z1 = 1
{x1x2x3} = 010, z1 = 0        {x1x2x3} = 110, z1 = 0
{x1x2x3} = 011, z1 = 1        {x1x2x3} = 111, z1 = 1
```

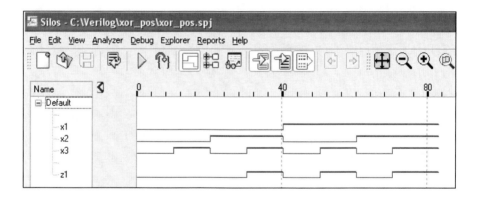

5.5   Design a logic circuit using built-in primitives that will convert a 4-bit binary
       code to a 4-bit excess-3 code. The excess-3 code is obtained by adding three
       to each binary number. Show only the low-order four bits of the excess-3
       code. Design the module and test bench, then obtain the outputs and wave-
       forms.

Binary				Excess-3			
x_1	x_2	x_3	x_4	z_1	z_2	z_3	z_4
0	0	0	0	0	0	1	1
0	0	0	1	0	1	0	0
0	0	1	0	0	1	0	1
0	0	1	1	0	1	1	0
0	1	0	0	0	1	1	1
0	1	0	1	1	0	0	0

Continued on next page

	Binary				Excess-3		
x_1	x_2	x_3	x_4	z_1	z_2	z_3	z_4
0	1	1	0	1	0	0	1
0	1	1	1	1	0	1	0
1	0	0	0	1	0	1	1
1	0	0	1	1	1	0	0
1	0	1	0	1	1	0	1
1	0	1	1	1	1	1	0
1	1	0	0	1	1	1	1
1	1	0	1	0	0	0	0
1	1	1	0	0	0	0	1
1	1	1	1	0	0	1	0

z_1

z_2

$$z_1 = x_1 x_2' + x_1 x_3' x_4' + x_1' x_2 x_4 + x_1' x_2 x_3$$
$$z_2 = x_2' x_3 + x_2' x_4 + x_2 x_3' x_4'$$
$$z_3 = x_3' x_4' + x_3 x_4$$
$$z_4 = x_4'$$

```
//binary to excess-3 code conversion
module bin_to_excess3 (x1, x2, x3, x4, z1, z2, z3, z4);
input x1, x2, x3, x4;
output z1, z2, z3, z4;

//generate output z1
and     inst1 (net1, x1, ~x2),
        inst2 (net2, x1, ~x3, ~x4),
        inst3 (net3, ~x1, x2, x4),
        inst4 (net4, ~x1, x2, x3);
or      inst5 (z1, net1, net2, net3, net4);

//continued on next page
```

```
//generate output z2
and    inst6 (net6, ~x2, x3),
       inst7 (net7, ~x2, x4),
       inst8 (net8, x2, ~x3, ~x4);
or     inst9 (z2, net6, net7, net8);

//generate output z3
xnor   inst10 (z3, x3, x4);

//generate output z4
buf    inst11 (z4, ~x4);

endmodule
```

```
x1x2x3x4} = 0000, {z1z2z3z4} = 0011
x1x2x3x4} = 0001, {z1z2z3z4} = 0100
x1x2x3x4} = 0010, {z1z2z3z4} = 0101
x1x2x3x4} = 0011, {z1z2z3z4} = 0110
x1x2x3x4} = 0100, {z1z2z3z4} = 0111
x1x2x3x4} = 0101, {z1z2z3z4} = 1000
x1x2x3x4} = 0110, {z1z2z3z4} = 1001
x1x2x3x4} = 0111, {z1z2z3z4} = 1010
x1x2x3x4} = 1000, {z1z2z3z4} = 1011
x1x2x3x4} = 1001, {z1z2z3z4} = 1100
x1x2x3x4} = 1010, {z1z2z3z4} = 1101
x1x2x3x4} = 1011, {z1z2z3z4} = 1110
x1x2x3x4} = 1100, {z1z2z3z4} = 1111
x1x2x3x4} = 1101, {z1z2z3z4} = 0000
x1x2x3x4} = 1110, {z1z2z3z4} = 0001
x1x2x3x4} = 1111, {z1z2z3z4} = 0010
```

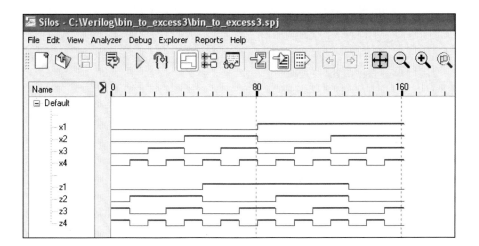

5.12   Design a logic circuit using built-in primitives that will generate an output if and only if a 4-bit binary input has a value greater than 12 or less than 3. Design the module and test bench, then obtain the outputs and waveforms.

x_1x_2 \\ x_3x_4	0 0	0 1	1 1	1 0
0 0	1	1	0	1
0 1	0	0	0	0
1 1	0	1	1	1
1 0	0	0	0	0

$$z_1$$

$$z_1 = x_1x_2x_4 + x_1x_2x_3 + x_1'x_2'x_3' + x_1'x_2'x_4'$$

```
//generate an output if input is
//greater than 12 or less than 3
module range (x1, x2, x3, x4, z1);

input x1, x2, x3, x4;
output z1;

and     inst1    (net1, x1, x2, x4);
and     inst2    (net2, x1, x2, x3);
and     inst3    (net3, ~x1, ~x2, ~x3);
and     inst4    (net4, ~x1, ~x2, ~x4);
or      inst5    (z1, net1, net2, net3, net4);
endmodule
```

```
//test bench for range module
module range_tb;

reg x1, x2, x3, x4;
wire z1;

//continued on next page
```

```
//apply input vectors
initial
begin: apply_stimulus
   reg [4:0] invect;
   for (invect=0; invect<16; invect=invect+1)
      begin
         {x1, x2, x3, x4} = invect [4:0];
         #10 $display ("x1x2x3x4 = %b, z1 = %b",
                       {x1, x2, x3, x4}, z1);
      end
end

//instantiate the module into the test bench
range inst1 (
   .x1(x1),
   .x2(x2),
   .x3(x3),
   .x4(x4),
   .z1(z1)
   );

endmodule
```

```
x1x2x3x4 = 0000, z1 = 1         x1x2x3x4 = 1000, z1 = 0
x1x2x3x4 = 0001, z1 = 1         x1x2x3x4 = 1001, z1 = 0
x1x2x3x4 = 0010, z1 = 1         x1x2x3x4 = 1010, z1 = 0
x1x2x3x4 = 0011, z1 = 0         x1x2x3x4 = 1011, z1 = 0
x1x2x3x4 = 0100, z1 = 0         x1x2x3x4 = 1100, z1 = 0
x1x2x3x4 = 0101, z1 = 0         x1x2x3x4 = 1101, z1 = 1
x1x2x3x4 = 0110, z1 = 0         x1x2x3x4 = 1110, z1 = 1
x1x2x3x4 = 0111, z1 = 0         x1x2x3x4 = 1111, z1 = 1
```

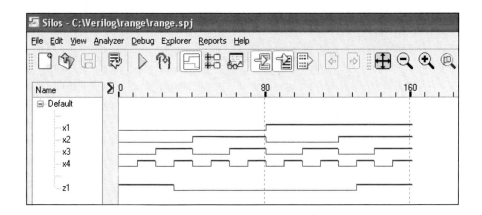

Chapter 6 · User-Defined Primitives

6.2   Design a built-in primitive for a 5-input majority circuit. Create a test bench and obtain the outputs for all combinations of the inputs.

```
//five-input majority circuit as a udp.
//Save as a .v file
primitive udp_maj5 (z1, x1, x2, x3, x4, x5);


output z1;
input x1, x2, x3, x4, x5;


table
//inputs are in same order as input list
// x1 x2 x3 x4 x5 :  z1;
    0  0  0  ?  ?  :  0;
    0  0  ?  0  ?  :  0;
    0  0  ?  ?  0  :  0;
    0  ?  0  0  ?  :  0;
    0  ?  ?  0  0  :  0;
    0  ?  0  ?  0  :  0;
    ?  0  0  0  ?  :  0;
    ?  ?  0  0  0  :  0;
    ?  0  0  ?  0  :  0;
    ?  0  ?  0  0  :  0;
    ?  ?  0  0  0  :  0;

    1  1  1  ?  ?  :  1;
    1  1  ?  1  ?  :  1;
    1  1  ?  ?  1  :  1;
    1  ?  1  1  ?  :  1;
    1  ?  ?  1  1  :  1;
    1  ?  1  ?  1  :  1;
    ?  1  1  1  ?  :  1;
    ?  ?  1  1  1  :  1;
    ?  1  1  ?  1  :  1;
    ?  1  ?  1  1  :  1;
    ?  ?  1  1  1  :  1;
endtable

endprimitive
```

```
//udp_maj5 test bench
module udp_maj5_tb;
reg x1, x2, x3, x4, x5;     //inputs are reg for tb
wire z1;                     //outputs are wire

//Declare a vector that has 1 more bit than the
//# of inputs.  This allows for the statement count
//to go 1 higher than the maximum count of the input
//combinations and prevents looping forever.
//If only 5 bits were used as the input vector,
//then the count would always be < 32 (the maximum
//count for 5 bits).

initial
begin: name //name required for this method
   reg [5:0] invect;
   for (invect = 0; invect < 32; invect = invect + 1)
      begin
         {x1, x2, x3, x4, x5} = invect [4:0];
         #10 $display ("x1x2x3x4x5 = %b%b%b%b%b, z1=%b"
                          x1, x2, x3, x4, x5, z1);
      end
end

//Instantiation must be done by position, not by name.

udp_maj5 inst1 (z1, x1, x2, x3, x4, x5);

endmodule
```

```
x1x2x3x4x5 = 00000, z1=0        x1x2x3x4x5 = 10000, z1=0
x1x2x3x4x5 = 00001, z1=0        x1x2x3x4x5 = 10001, z1=0
x1x2x3x4x5 = 00010, z1=0        x1x2x3x4x5 = 10010, z1=0
x1x2x3x4x5 = 00011, z1=0        x1x2x3x4x5 = 10011, z1=1
x1x2x3x4x5 = 00100, z1=0        x1x2x3x4x5 = 10100, z1=0
x1x2x3x4x5 = 00101, z1=0        x1x2x3x4x5 = 10101, z1=1
x1x2x3x4x5 = 00110, z1=0        x1x2x3x4x5 = 10110, z1=1
x1x2x3x4x5 = 00111, z1=1        x1x2x3x4x5 = 10111, z1=1
x1x2x3x4x5 = 01000, z1=0        x1x2x3x4x5 = 11000, z1=0
x1x2x3x4x5 = 01001, z1=0        x1x2x3x4x5 = 11001, z1=1
x1x2x3x4x5 = 01010, z1=0        x1x2x3x4x5 = 11010, z1=1
x1x2x3x4x5 = 01011, z1=1        x1x2x3x4x5 = 11011, z1=1
x1x2x3x4x5 = 01100, z1=0        x1x2x3x4x5 = 11100, z1=1
x1x2x3x4x5 = 01101, z1=1        x1x2x3x4x5 = 11101, z1=1
x1x2x3x4x5 = 01110, z1=1        x1x2x3x4x5 = 11110, z1=1
x1x2x3x4x5 = 01111, z1=1        x1x2x3x4x5 = 11111, z1=1
```

6.4   Design the logic circuit shown below using user-defined primitives (UDPs).
Obtain the design module, test bench, and outputs. Verify the outputs by obtaining the equation for z_1 directly from the logic diagram, then use Boolean
algebra to obtain the minimal sum-of-products expression.

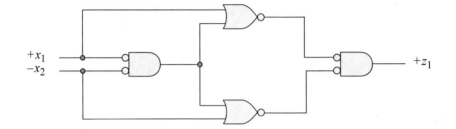

```
//2-input NOR gate as a user-defined primitive
primitive udp_nor2 (z1, x1, x2);
input x1, x2;
output z1;
//define state table
table
//inputs are in the same order as the input list
// x1 x2 :  z1;    comment is for readability
    0  0  :  1;
    0  1  :  0;
    1  0  :  0;
    1  1  :  0;
endtable

endprimitive
```

```
//module for logic diagram using NOR logic
//user-defined primitives
module log_diag_eqn5 (x1, x2, z1);

input x1, x2;
output z1;

//instantiate the udps
udp_nor2 inst1 (net1, x1, ~x2);
udp_nor2 inst2 (net2, x1, net1);
udp_nor2 inst3 (net3, net1, ~x2);
udp_nor2 inst4 (z1, net2, net3);

endmodule
```

```
//test bench for logic diagram equation 5
module log_diag_eqn5_tb;

reg x1, x2;
wire z1;

//display variables
initial
$monitor ("x1 x2 = %b %b, z1 = %b", x1, x2, z1);

//apply input vectors
initial
begin
    #0      x1 = 1'b0;x2 = 1'b0;
    #10     x1 = 1'b0;x2 = 1'b1;
    #10     x1 = 1'b1;x2 = 1'b0;
    #10     x1 = 1'b1;x2 = 1'b1;

    #10     $stop;
end

//instantiate the module into the test bench
log_diag_eqn5 inst1 (
    .x1(x1),
    .x2(x2),
    .z1(z1)
    );

endmodule
```

```
x1  x2 = 0 0,  z1 = 0
x1  x2 = 0 1,  z1 = 1
x1  x2 = 1 0,  z1 = 1
x1  x2 = 1 1,  z1 = 0
```

The equation from the logic diagram is:

$$
\begin{aligned}
z_1 &= [x_1 + (x_1'x_2)]\,[x_2' + (x_1'x_2)] \\
&= (x_1 + x_2)\,(x_2' + x_1') \\
&= (x_1 + x_2)x_2' + (x_1 + x_2)x_1' \\
&= x_1 x_2' + x_1'x_2 \\
&= x_1 \oplus x_2
\end{aligned}
$$

6.7   The logic block shown below generates a logic 1 output if the four inputs contain an odd number of 1s. Use only blocks of this type to design a 9-bit (eight data bits plus parity) odd parity checker. The output of the resulting circuit will be a logic 1 (indicating a parity error) if the parity of the nine bits is even. All inputs must be used. Use the least amount of logic.

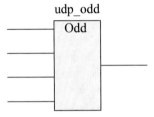

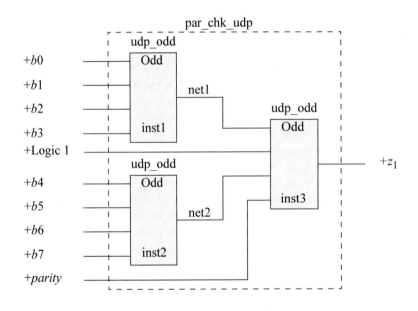

```
//circuit to generate a logic 1 output if
//there are an odd number of 1s on the inputs
primitive udp_odd (z1, x1, x2, x3, x4);

input x1, x2, x3, x4;
output z1;

//define state table
table
//inputs are in the same order as the input list
// x1 x2 x3 x4 :   z1;   comment is for readability
   0  0  0  0  :   0;
   0  0  0  1  :   1;
   0  0  1  0  :   1;
   0  0  1  1  :   0;
   0  1  0  0  :   1;
   0  1  0  1  :   0;
   0  1  1  0  :   0;
   0  1  1  1  :   1;
   1  0  0  0  :   1;
   1  0  0  1  :   0;
   1  0  1  0  :   0;
   1  0  1  1  :   1;
   1  1  0  0  :   0;
   1  1  0  1  :   1;
   1  1  1  0  :   1;
   1  1  1  1  :   0;
endtable

endprimitive
```

```
//a parity check circuit using udps
module par_chk_udp (b, par_in, par_err);

input [7:0] b;
input par_in;
output par_err;

//instantiate the udps
udp_odd inst1 (net1, b[7], b[6], b[5], b[4]);
udp_odd inst2 (net2, b[3], b[2], b[1], b[0]);
udp_odd inst3 (par_err, net1, 1'b1, net2, par_in);

endmodule
```

```
//test bench for 9-bit parity checker
module par_chk_udp_tb;


reg [7:0] b;
reg par_in;
wire par_err;


initial
$monitor ("b=%b, par_in=%b, par_err=%b",
           b, par_in, par_err);


//apply input vectors
initial
begin
   #0    b=8'b0000_0000;    par_in=1'b1;
   #10   b=8'b0000_0000;    par_in=1'b0;
   #10   b=8'b0000_1111;    par_in=1'b1;
   #10   b=8'b0101_0101;    par_in=1'b0;
   #10   b=8'b1010_1010;    par_in=1'b1;
   #10   b=8'b1111_0000;    par_in=1'b0;
   #10   b=8'b0000_1111;    par_in=1'b1;
   #10   b=8'b1111_1111;    par_in=1'b0;
   #10   b=8'b1111_1111;    par_in=1'b1;
   #10   $stop;
end


//instantiate the module into the test bench
par_chk_udp inst1 (
   .b(b),
   .par_in(par_in),
   .par_err(par_err)
   );
endmodule
```

```
b=00000000, par_in=1, par_err=0
b=00000000, par_in=0, par_err=1
b=00001111, par_in=1, par_err=0
b=01010101, par_in=0, par_err=1
b=10101010, par_in=1, par_err=0
b=11110000, par_in=0, par_err=1
b=00001111, par_in=1, par_err=0
b=11111111, par_in=0, par_err=1
b=11111111, par_in=1, par_err=0
```

6.10   Given the Karnaugh map shown below for the function z_1, obtain the minimum expression for z_1 in a sum-of-products form. Use NAND logic as UDPs. Obtain the design module, test bench, and waveforms.

$$z_1 = x_3'a + x_1'x_3' + x_2'x_3a' + x_1x_2'x_3$$

```
//a module using a MEV and NAND udps
module mev_udp (x1, x2, x3, a, z1);


input x1, x2, x3, a;
output z1;

//instantiate the udps
udp_nand2 inst1 (net1, ~x3, a);
udp_nand2 inst2 (net2, ~x1, ~x3);
udp_nand3 inst3 (net3, ~x2, x3, ~a);
udp_nand3 inst4 (net4, x1, ~x2, x3);

udp_nand4 inst5 (z1, net1, net2, net3, net4);
endmodule
```

```
//test bench for the map-entered variable udp
module mev_udp_tb;

reg x1, x2, x3, a;
wire z1;

//apply input vectors
initial
begin: apply_stimulus
   reg [4:0] invect;
   for (invect=0; invect<16; invect=invect+1)
      begin
         {x1, x2, x3, a} = invect [4:0];
         #10 $display ("x1 x2 x3 a = %b,  z1 = %b",
                        {x1, x2, x3, a}, z1);
      end
end
//continued on next page
```

```
//instantiate the module into the test bench
mev_udp inst1 (
    .x1(x1),
    .x2(x2),
    .x3(x3),
    .a(a),
    .z1(z1)
    );

endmodule
```

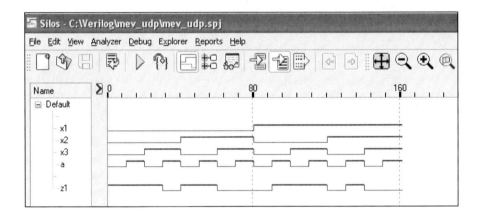

6.13   Design a synchronous modulo-16 counter using the positive-edge-triggered D
flip-flop of Section 6.3.2. There will be two scalar inputs: clk and rst_n; and
one vector output: $y[3:0]$. Obtain the design module, test bench, outputs, and
waveforms.

y_1	y_2	y_3	y_4
0	0	0	0
0	0	0	1
0	0	1	0
0	0	1	1
0	1	0	0
0	1	0	1
0	1	1	0
0	1	1	1

y_1	y_2	y_3	y_4
1	0	0	0
1	0	0	1
1	0	1	0
1	0	1	1
1	1	0	0
1	1	0	1
1	1	1	0
1	1	1	1

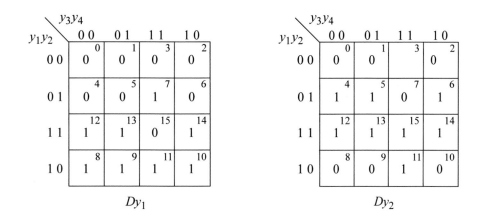

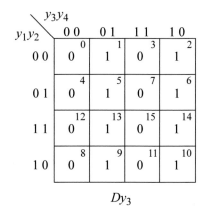

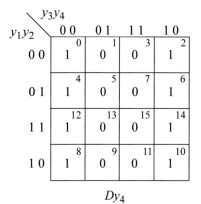

$$Dy_1 = y_1y_2' + y_1y_3' + y_1y_4' + y_1'y_2y_3y_4$$

$$Dy_2 = y_2y_3' + y_2y_4' + y_2'y_3y_4$$

$$Dy_3 = y_3'y_4 + y_3y_4' = y_3 \oplus y_4$$

$$Dy_4 = y_4'$$

```
//modulo-16 counter using a positive-edge-triggered
//D flip-flop user-defined primitive
module ctr_mod_16_udp (clk, rst_n, y1, y2, y3, y4);

input clk, rst_n;
output y1, y2, y3, y4;


//instantiate the udps for flip-flop y1
//flip-flop ports are: q, d, clk, rst_n
udp_and2      inst1  (net1, y1, ~y2);
udp_and2      inst2  (net2, y1, ~y3);
udp_and2      inst3  (net3, y1, ~y4);
udp_and4      inst4  (net4, ~y1, y2, y3, y4);
udp_or4       inst5  (net5, net1, net2, net3, net4);
udp_dff_edge1 inst6  (y1, net5, clk, rst_n);


//instantiate the udps for flip-flop y2
udp_and2      inst7  (net7, y2, ~y3);
udp_and2      inst8  (net8, y2, ~y4);
udp_and3      inst9  (net9, ~y2, y3, y4);
udp_or3       inst10 (net10, net7, net8, net9);
udp_dff_edge1 inst11 (y2, net10, clk, rst_n);


//instantiate the udps for flip-flop y3
udp_xor2      inst12 (net12, y3, y4);
udp_dff_edge1 inst13 (y3, net12, clk, rst_n);


//instantiate the udps for flip-flop y4
udp_dff_edge1 inst14 (y4, ~y4, clk, rst_n);

endmodule
```

```
//test bench for the modulo-16 counter
module ctr_mod_16_udp_tb;

reg clk, rst_n;
wire y1, y2, y3, y4;

//display variables
initial
$monitor ("{y1 y2 y3 y4} = %b", {y1, y2, y3, y4});

//continued on next page
```

```
//generate reset
initial
begin
    #0  rst_n = 1'b1;
    #2  rst_n = 1'b0;
    #5  rst_n = 1'b1;
end

//generate clock
initial
begin
    clk = 1'b0;
    forever
        #10clk = ~clk;
end

//determine length of simulation
initial
begin
    repeat (17) @ (posedge clk);
    $stop;
end

//instantiate the module into the test bench
ctr_mod_16_udp inst1 (
    .clk(clk),
    .rst_n(rst_n),
    .y1(y1),
    .y2(y2),
    .y3(y3),
    .y4(y4)
    );

endmodule
```

```
{y1 y2 y3 y4} = 0000        {y1 y2 y3 y4} = 1000
{y1 y2 y3 y4} = 0001        {y1 y2 y3 y4} = 1001
{y1 y2 y3 y4} = 0010        {y1 y2 y3 y4} = 1010
{y1 y2 y3 y4} = 0011        {y1 y2 y3 y4} = 1011
{y1 y2 y3 y4} = 0100        {y1 y2 y3 y4} = 1100
{y1 y2 y3 y4} = 0101        {y1 y2 y3 y4} = 1101
{y1 y2 y3 y4} = 0110        {y1 y2 y3 y4} = 1110
{y1 y2 y3 y4} = 0111        {y1 y2 y3 y4} = 1111
                            {y1 y2 y3 y4} = 0000
```

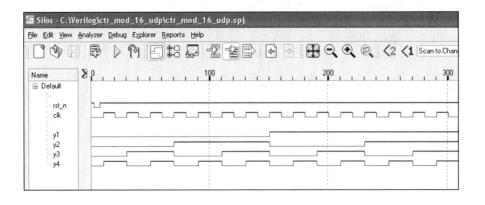

Chapter 7   Dataflow Modeling

7.1   Design the logic for a *D* flip-flop with an active-high reset using three *SR* latches. Then implement the design using dataflow modeling. Then use the *D* flip-flop to design a *T* flip-flop. Then use the *T* flip-flop to design modulo-10 4-bit ripple counter. Obtain the module, test bench, outputs, and waveforms.

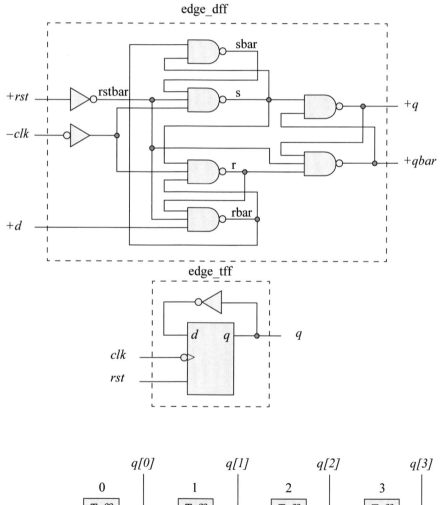

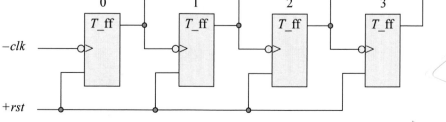

```verilog
//design a negedge-triggered d ff to be used later as a
//negedge-triggered t ff
module edge_dff (q, qbar, d, clk, rst);

//list i/o ports
input d, clk, rst;
output q, qbar;

wire q, qbar, d, clk, rst, rstbar;

//define internal wires
wire s, sbar, r, rbar, cbar;

//dataflow assign statements
assign rstbar = ~rst;

//an edge-triggered d ff is implemented with
//3 sr latches
//define input latches
assign sbar = ~(rbar & s),
       s    = ~(sbar & rstbar & ~clk),
       r    = ~(rbar & ~clk & s),
       rbar = ~(r & rstbar & d);

//define output latch
assign q    = ~(s & qbar),
       qbar = ~(q & r & rstbar);
endmodule
```

```verilog
//edge-T flip-flop using a D flip-flop.
//Toggles every clk cycle
module edge_tff (q, clk, rst);

input clk, rst;
output q;

wire q, clk, rst;

//instantiate the edge-triggered d ff
//the complement of output q is fed back
edge_dff inst1 (
    .q(q),
    .d(~q),
    .clk(clk),
    .rst(rst)
    );
endmodule
```

```
//design the ripple counter using the edge-triggered
//T flip-flop
module ctr_dff_tff (q, clk, rst);


//define i/o ports
input clk, rst;
output [3:0] q;


wire clk, rst;
wire [3:0] q;


//instantiate the T flip-flop
edge_tff inst1 (
    .q(q[0]),
    .clk(clk),
    .rst(rst)
    );


edge_tff inst2 (
    .q(q[1]),
    .clk(q[0]),
    .rst(rst)
    );


edge_tff inst3 (
    .q(q[2]),
    .clk(q[1]),
    .rst(rst)
    );


edge_tff inst4 (
    .q(q[3]),
    .clk(q[2]),
    .rst(rst)
    );

endmodule
```

```
//counter_dff_tff test bench
module ctr_dff_tff_tb;


reg clk, rst;//inputs are reg for tb
wire [3:0] q;//outputs are wire for tb


//display outputs at simulation times
initial
$monitor ($time, " count = %b, rst = %b", q, rst);


//define the reset signal
initial
begin
    #0     rst = 1'b1;     //reset initially
    #34    rst = 1'b0;     //at #34, deassert rst
    #350   rst = 1'b1;     //at #384, assert rst
    #50    rst = 1'b0;     //at #434, deassert rst
end


//define clk
initial
begin
    clk = 1'b0;
    forever #10 clk = ~clk;
end


//finish simulation at time 500
initial
begin
    #500 $finish;
end


//instantiate the counter into the test bench
ctr_dff_tff inst1 (
    .q(q),
    .clk(clk),
    .rst(rst)
    );

endmodule
```

```
0    count = 0000, rst = 1
34   count = 0000, rst = 0
40   count = 0001, rst = 0
60   count = 0010, rst = 0
80   count = 0011, rst = 0
100  count = 0100, rst = 0
120  count = 0101, rst = 0
140  count = 0110, rst = 0
160  count = 0111, rst = 0
180  count = 1000, rst = 0
200  count = 1001, rst = 0
220  count = 1010, rst = 0
240  count = 1011, rst = 0
260  count = 1100, rst = 0
280  count = 1101, rst = 0
300  count = 1110, rst = 0
320  count = 1111, rst = 0
340  count = 0000, rst = 0
360  count = 0001, rst = 0
380  count = 0010, rst = 0
384  count = 0000, rst = 1      //assert reset
434  count = 0000, rst = 0      //begin counting again
440  count = 0001, rst = 0
460  count = 0010, rst = 0
480  count = 0011, rst = 0
```

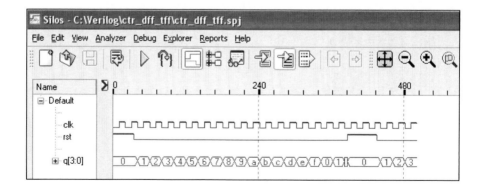

7.4   Design a full adder from two half adders using dataflow modeling. There are
       three scalar inputs: *a*, *b*, and *cin*; there are two scalar outputs: *sum* and *cout*.
       Obtain the design module, test bench, outputs, and waveforms for all combi-
       nations of the inputs.

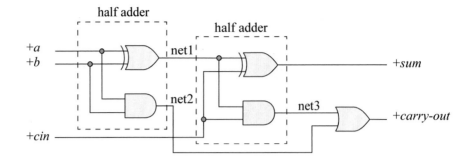

```
//dataflow full adder
module full_adder (a, b, cin, sum, cout);

input a, b, cin;
output sum, cout;

assign sum = (a ^ b) ^ cin;
assign cout = cin & (a ^ b) | (a & b);

endmodule
```

```
//full adder test bench
module full_adder_tb;

reg a, b, cin;
wire sum, cout;

//apply input vectors
initial
begin: apply_stimulus
   reg [3:0] invect;
      for (invect = 0; invect < 8; invect = invect + 1)
      begin
         {a, b, cin} = invect [3:0];
         #10 $display ("{abcin}=%b, sum=%b, cout=%b",
                         {a, b, cin}, sum, cout);
   end
end

//continued on next page
```

```
//instantiate the module into the test bench
full_adder inst1 (
    .a(a),
    .b(b),
    .cin(cin),
    .sum(sum),
    .cout(cout)
    );

endmodule
```

```
a b cin = 000, sum = 0, cout = 0
a b cin = 001, sum = 1, cout = 0
a b cin = 010, sum = 1, cout = 0
a b cin = 011, sum = 0, cout = 1
a b cin = 100, sum = 1, cout = 0
a b cin = 101, sum = 0, cout = 1
a b cin = 110, sum = 0, cout = 1
a b cin = 111, sum = 1, cout = 1
```

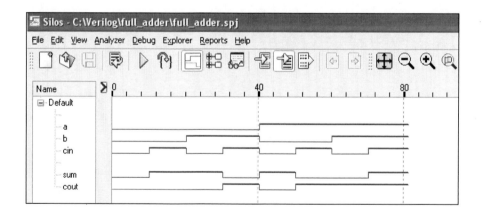

7.11   Use D flip-flops to design a counter that counts in the following sequence: $y_1y_2y_3y_4 = 0000, 1000, 0100, 0010, 0001, 1000, \ldots$, where y_4 is the low-order stage of the counter. Use the D flip-flop from Problem 7.1. The negative edge of the clock triggers the counter. The reset is active high. The Karnaugh maps are shown on the next page. Obtain the design module, test bench, outputs, and waveforms for the complete counting sequence. This counter is useful for obtaining four discrete time intervals. Alternatively, the counting sequence can be $y_1y_2y_3y_4 = 1000, 0100, 0010, 0001, 1000, \ldots$ if the D flip-flop has a *set* input. The counter would be initialized to $y_1y_2y_3y_4 = 1000$.

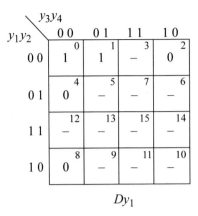

Dy_1

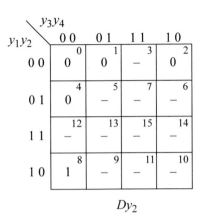

Dy_2

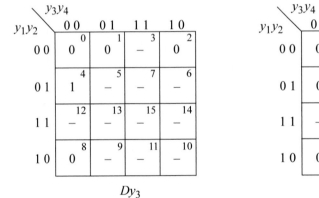

Dy_3

Dy_4

$Dy_1 = y_1'y_2'y_3'$
$Dy_2 = y_1$
$Dy_3 = y_2$
$Dy_4 = y_3$

```verilog
//counting sequence 0000, 1000, 0100, 0010, 0001, 1000
module ctr_seq1 (clk, rst, y);

input clk, rst;
output [1:4] y;

//instantiate the D flip-flop for y[1]
edge_dff inst1 (
   .q(y[1]),
   .d(~y[1] & ~y[2] & ~y[3]),
   .clk(clk),
   .rst(rst)
   );
//continued on next page
```

```verilog
//instantiate the D flip-flop for y[2]
edge_dff inst2 (
   .q(y[2]),
   .d(y[1]),
   .clk(clk),
   .rst(rst)
   );

//instantiate the D flip-flop for y[3]
edge_dff inst3 (
   .q(y[3]),
   .d(y[2]),
   .clk(clk),
   .rst(rst)
   );

//instantiate the D flip-flop for y[4]
edge_dff inst4 (
   .q(y[4]),
   .d(y[3]),
   .clk(clk),
   .rst(rst)
   );

endmodule
```

```verilog
//test bench for counting sequence 0000, 1000, 0100,
//0010, 0001, 1000
module ctr_seq1_tb;

reg clk, rst;
wire [1:4] y;

//display variables
initial
$monitor ("count = %b", y);

//define reset signal
initial
begin
   #0    rst = 1'b1;    //reset initially
   #5    rst = 1'b0;    //remove reset
end

//continued on next page
```

```
//define clock
initial
begin
   clk = 1'b0;
   forever
      #10 clk = ~clk;
end

//define length of simulation
initial
begin
   #120 $stop;
end

//instantiate the module into the test bench
ctr_seq1 inst1 (
   .clk(clk),
   .rst(rst),
   .y(y)
   );

endmodule
```

```
count = 0000
count = 1000
count = 0100
count = 0010
count = 0001
count = 1000
```

7.14   Design a Mealy pulse-mode asynchronous sequential machine that has two inputs x_1 and x_2 and one output z_1. For a Mealy machine, the outputs are a function of the present states and the present inputs. Output z_1 is asserted coincident with the x_2 pulse if the x_2 pulse is immediately preceded by a pair of x_1 pulses. Use SR latches and D flip-flops in a master-slave configuration. Use NAND logic for all gates and latches except for output z_1, which will be an AND gate.

Derive the state diagram, input maps, output map, and logic diagram. Then use continuous assignment to design the module. Obtain the test bench, outputs, and waveforms.

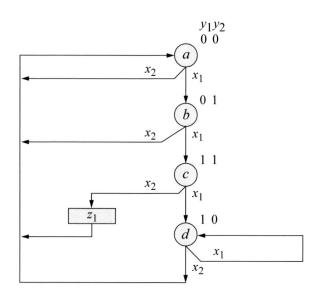

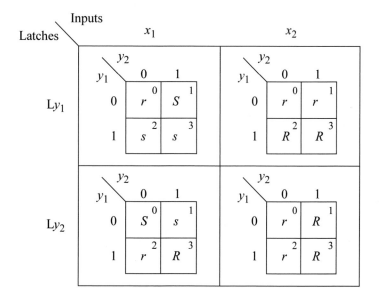

$$SLy_1 = y_2x_1 \qquad RLy_1 = x_2$$

$$SLy_2 = y_1'x_1 \qquad RLy_2 = y_1x_1 + x_2$$

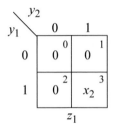

$$z_1 = y_1y_2x_2$$

Output z_1 is a function of the present state Ⓒ ($y_1y_2 = 11$) and the present input x_2.

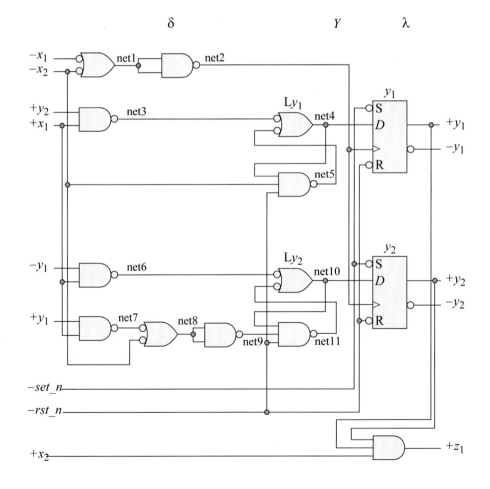

```verilog
//pulse-mode asynchronous sequential machine
module pm_asm4 (x1, x2, set_n, rst_n, y, y_n, z1);


input x1, x2, set_n, rst_n;
output [1:2] y, y_n;
output z1;


//define internal nets
wire net1, net2, net3, net4, net5, net6,
     net7, net8, net9, net10, net11;


//design for D flip-flop clock
assign    net1 = ~(~x1 & ~x2),
          net2 = ~(net1 & net1);


//design for latch Ly1
assign    net3 = ~(y[2] & x1),
          net4 = ~(net3 & net5),
          net5 = ~(net4 & ~x2 & rst_n);


//design for latch Ly2
assign    net6  = ~(~y[1] & x1),
          net7  = ~(y[1] & x1),
          net8  = ~(net7 & ~x2),
          net9  = ~(net8 & net8),
          net10 = ~(net6 & net11),
          net11 = ~(net10 & net9 & rst_n);


//instantiate the D flip-flop for y1
d_ff inst1 (
   .d(net4),
   .clk(net2),
   .q(y[1]),
   .q_n(y_n[1]),
   .set_n(set_n),
   .rst_n(rst_n)
   );

//continued on next page
```

```
//instantiate the D flip-flop for y2
d_ff inst2 (
   .d(net10),
   .clk(net2),
   .q(y[2]),
   .q_n(y_n[2]),
   .set_n(set_n),
   .rst_n(rst_n)
   );

//design for z1
assign z1  = (y[1] & y[2] & x2);
endmodule
```

```
//test bench for pulse-mode asynchronous sequential
//machine
module pm_asm4_tb;

reg x1, x2;
reg set_n, rst_n;
wire [1:2] y, y_n;
wire z1;

//display inputs and outputs
initial
$monitor ("rst_n = %b, x1x2 = %b, state = %b, z1 = %b",
           rst_n, {x1, x2}, y, z1);

//define input sequence
initial
begin
   #0    set_n = 1'b1;
         rst_n = 1'b0; //rst to state_a(00); no output
         x1    = 1'b0;
         x2    = 1'b0;

   #5    rst_n = 1'b1;

   #10   x1 = 1'b1;   //go to state_b(01) on posedge
                      //of 1st x1
   #10   x1 = 1'b0;   //no output

   #10   x1 = 1'b1;   //go to state_c(11) on posedge
                      //of 2nd x1
   #10   x1 = 1'b0;   //no output

//continued on next page
```

```
   #10    x2 = 1'b1;  //assert z1; go to state_a(00)
   #10    x2 = 1'b0;  //deassert output z1 on negedge
                      //of 1st x2

   #10    x1 = 1'b1;  //go to state_b(01)
   #10    x1 = 1'b0;  //no output

   #10    x1 = 1'b1;  //go to state_c(11)
   #10    x1 = 1'b0;  //no output

   #10    x1 = 1'b1;  //go to state_d(10)
   #10    x1 = 1'b0;  //no output

   #20    x2 = 1'b1;  //go to state_a(00) on posedge
                      //of 1st x2
   #10    x2 = 1'b0;  //no output

   #10    x1 = 1'b1;  //go to state_b(01) on posedge
                      //of 1st x1
   #10    x1 = 1'b0;  //no output

   #10    x1 = 1'b1;  //go to state_c(11) on posedge
                      //of 2nd x1
   #10    x1 = 1'b0;  //no output

   #10    x2 = 1'b1;  //assert z1; go to state_a(00)
   #10    x2 = 1'b0;  //deassert output z1 on negedge
                      //of 1st x2

   #30    $stop;
end


pm_asm4 inst1 (        //instantiate the module
                       //into the test bench
   .x1(x1),
   .x2(x2),
   .set_n(set_n),
   .rst_n(rst_n),
   .y(y),
   .y_n(y_n),
   .z1(z1)
   );

endmodule
```

```
rst_n = 0, x1x2 = 00, state = 00, z1 = 0
rst_n = 1, x1x2 = 00, state = 00, z1 = 0
rst_n = 1, x1x2 = 10, state = 00, z1 = 0
rst_n = 1, x1x2 = 00, state = 01, z1 = 0
rst_n = 1, x1x2 = 10, state = 01, z1 = 0
rst_n = 1, x1x2 = 00, state = 11, z1 = 0
rst_n = 1, x1x2 = 01, state = 11, z1 = 1
rst_n = 1, x1x2 = 00, state = 00, z1 = 0
rst_n = 1, x1x2 = 10, state = 00, z1 = 0
rst_n = 1, x1x2 = 00, state = 01, z1 = 0
rst_n = 1, x1x2 = 10, state = 01, z1 = 0
rst_n = 1, x1x2 = 00, state = 11, z1 = 0
rst_n = 1, x1x2 = 10, state = 11, z1 = 0
rst_n = 1, x1x2 = 00, state = 10, z1 = 0
rst_n = 1, x1x2 = 01, state = 10, z1 = 0
rst_n = 1, x1x2 = 00, state = 00, z1 = 0
rst_n = 1, x1x2 = 10, state = 00, z1 = 0
rst_n = 1, x1x2 = 00, state = 01, z1 = 0
rst_n = 1, x1x2 = 10, state = 01, z1 = 0
rst_n = 1, x1x2 = 00, state = 11, z1 = 0
rst_n = 1, x1x2 = 01, state = 11, z1 = 1
rst_n = 1, x1x2 = 00, state = 00, z1 = 0
```

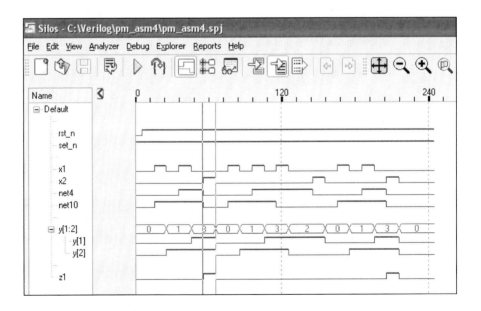

Chapter 8   Behavioral Modeling

8.2   Design a full adder using the **always** statement. The are three scalar inputs: the augend *a*, the addend *b*, and the carry-in *cin*. There are two outputs: *sum* and *cout*. Use blocking statements for *sum* and *cout* with a delay of five time units. Obtain the design module, test bench module, outputs, and waveforms. Test for all combinations of the inputs.

```
//behavioral full adder
module fulladder (a, b, cin, sum, cout);

input a, b, cin;
output sum, cout;
reg sum, cout;

//initialize sum and cout to avoid Xs until #5
initial
begin
   sum  = 1'b0;
   cout = 1'b0;
end

always @ (a or b or cin)
begin
   sum  = #5 (a ^ b ^ cin);
   cout = #5 ((a & b) | (a & cin) | (b & cin));
end
endmodule
```

```
//test bench for the behavioral full adder
module fulladder_tb;
reg a, b, cin;
wire sum, cout;

//apply input vectors and display variables
initial
begin: apply_stimulus
   reg [3:0] invect;
   for (invect=0; invect<8; invect=invect+1)
      begin
         {a, b, cin} = invect [2:0];
         #10 $display ("a b cin = %b, cout = %b, sum =
                        {a, b, cin}, cout, sum);
      end
end

//continued on next page
```

```
//instantiate the module into the test bench
fulladder inst1 (
    .a(a),
    .b(b),
    .cin(cin),
    .sum(sum),
    .cout(cout)
    );
endmodule
```

```
a b cin = 000, cout = 0, sum = 0
a b cin = 001, cout = 0, sum = 1
a b cin = 010, cout = 0, sum = 1
a b cin = 011, cout = 1, sum = 0
a b cin = 100, cout = 0, sum = 1
a b cin = 101, cout = 1, sum = 0
a b cin = 110, cout = 1, sum = 0
a b cin = 111, cout = 1, sum = 1
```

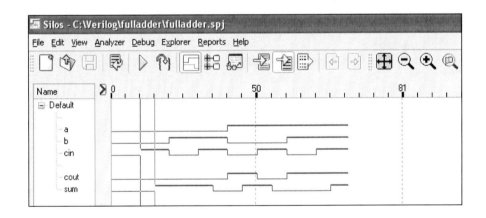

8.5   Design a 4-bit counter with a mode control line x_1 that operates according to the following specifications:

(a) If $x_1 = 0$, then the counter generates the sequence 0000, 1000, 1100, 1110, 1111, 0111, 0011, 0001, 0000, ...

(b) If $x_1 = 1$, then the counter generates the sequence 0000, 1000, 0001, 0100, 0010, 1100, 0011, 1110, 0111, 1111, 0000, ...

(c) Obtain the behavioral module using the **case** construct. Obtain the test bench, outputs, and waveforms.

```verilog
//behavioral 2-sequence counter
module ctr_2count_seq (clk, rst_n, x1, count);

input clk, rst_n, x1;
output [3:0] count;
wire clk, rst_n, x1;
reg [3:0] count, next_count;

always @ (posedge clk or negedge rst_n)
begin
   if (~rst_n)
      count = 4'b0000;
   else
      count = next_count;
end

always @ (count)
begin
   if (x1 == 0)
      case (count)
         4'b0000 : next_count = 4'b1000;
         4'b1000 : next_count = 4'b1100;
         4'b1100 : next_count = 4'b1110;
         4'b1110 : next_count = 4'b1111;
         4'b1111 : next_count = 4'b0111;
         4'b0111 : next_count = 4'b0011;
         4'b0011 : next_count = 4'b0001;
         4'b0001 : next_count = 4'b0000;
         default : next_count = 4'b0000;
      endcase
   else
      case (count)
         4'b0000 : next_count = 4'b1000;
         4'b1000 : next_count = 4'b0001;
         4'b0001 : next_count = 4'b0100;
         4'b0100 : next_count = 4'b0010;
         4'b0010 : next_count = 4'b1100;
         4'b1100 : next_count = 4'b0011;
         4'b0011 : next_count = 4'b1110;
         4'b1110 : next_count = 4'b0111;
         4'b0111 : next_count = 4'b1111;
         4'b1111 : next_count = 4'b0000;
         default : next_count = 4'b0000;
      endcase
end

endmodule
```

```verilog
//test bench for 2-sequence counter
module ctr_2count_seq_tb;


reg rst_n, clk, x1;        //inputs are reg for tb
wire [3:0] count;          //outputs are wire for tb


initial
$monitor ($time, "x1 = %b, count = %b", x1, count);


//define clk input
initial
begin
   clk = 1'b0;
   forever
      #10 clk = ~clk;
end


//define reset input
initial
begin
   #0     rst_n = 1'b0;
   #5     rst_n = 1'b1;
   #360   $stop;
end


//define x1 control input
initial
begin
   #0     x1 = 1'b0;
   #170   x1 = 1'b1;
end


//instantiate the module into the test bench
ctr_2count_seq inst1 (
   .clk(clk),
   .rst_n(rst_n),
   .x1(x1),
   .count(count)
   );


endmodule
```

0	x1 = 0, count = 0000	170	x1 = 1, count = 1000
10	x1 = 0, count = 1000	190	x1 = 1, count = 0001
30	x1 = 0, count = 1100	210	x1 = 1, count = 0100
50	x1 = 0, count = 1110	230	x1 = 1, count = 0010
70	x1 = 0, count = 1111	250	x1 = 1, count = 1100
90	x1 = 0, count = 0111	270	x1 = 1, count = 0011
110	x1 = 0, count = 0011	290	x1 = 1, count = 1110
130	x1 = 0, count = 0001	310	x1 = 1, count = 0111
150	x1 = 0, count = 0000	330	x1 = 1, count = 1111
		350	x1 = 1, count = 0000

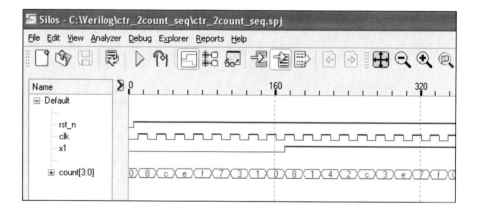

8.9   Design a modulo-11 counter using behavioral modeling with conditional statements and the **case** statement. Obtain the design module, test bench, outputs, and waveforms.

```verilog
//behavioral modulo-11 counter
module ctr_mod11a (clk, rst_n, count);

input clk, rst_n;
output [3:0] count;

reg [3:0] count, next_count;

always @ (posedge clk or negedge rst_n)
begin
   if (~rst_n)
      count = 4'b0000;
   else
      count = next_count;
end                      //continued on next page
```

```verilog
always @ (count)
begin
   case (count)
      4'b0000: next_count = 4'b0001;
      4'b0001: next_count = 4'b0010;
      4'b0010: next_count = 4'b0011;
      4'b0011: next_count = 4'b0100;
      4'b0100: next_count = 4'b0101;
      4'b0101: next_count = 4'b0110;
      4'b0110: next_count = 4'b0111;
      4'b0111: next_count = 4'b1000;
      4'b1000: next_count = 4'b1001;
      4'b1001: next_count = 4'b1010;
      4'b1010: next_count = 4'b0000;
      default: next_count = 4'b0000;
   endcase
end

endmodule
```

```verilog
//test bench for modulo-11 counter
module ctr_mod11a_tb;

reg clk, rst_n;
wire [3:0] count;

//display variables
initial
$monitor ("count = %b", count);

//define clock
initial
begin
   clk = 1'b0;
   forever
      #10   clk = ~clk;
end

//define reset and simulation duration
initial
begin
   #0    rst_n = 1'b0;
   #5    rst_n = 1'b1;
   #220  $stop;
end

//continued on next page
```

```
//instantiate the module into the test bench
ctr_mod11a inst1 (
    .clk(clk),
    .rst_n(rst_n),
    .count(count)
    );

endmodule
```

```
count = 0000
count = 0001
count = 0010
count = 0011
count = 0100
count = 0101
count = 0110
count = 0111
count = 1000
count = 1001
count = 1010
count = 0000
```

8.12   Design an asynchronous sequential machine using behavioral modeling. The machine has two inputs: x_1 and x_2 and one output z_1. Input x_1 will always be asserted whenever x_2 is asserted; that is, there will never be a situation where x_1 is deasserted and x_2 is asserted. Output z_1 is asserted coincident with every third x_2 pulse and remains active for the duration of x_2.

Obtain a representative timing diagram and a reduced primitive flow table. Assign state codes so that there are no transient states that could cause an erroneous output on z_1. Obtain the behavioral module, the test bench, and the waveforms showing a complete sequence to assert output z_1.

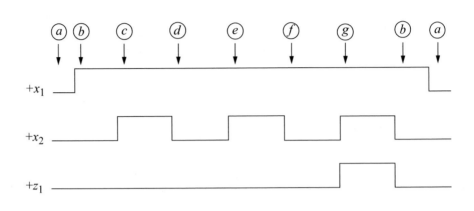

x_1x_2	00	01	11	10	z_1
ⓐ	–	–	b	0
a	–	c	ⓑ	0
–	–	ⓒ	d	0
a	–	e	ⓓ	0
–	–	ⓔ	f	0
a	–	g	ⓕ	0
–	–	ⓖ	b	1

ⓐ = 000, ⓑ = 001, ⓒ = 011, ⓓ = 010, ⓔ = 110, ⓕ = 111, ⓖ = 101

```verilog
//asynchronous sequential machine
module asm9 (rst_n, x1, x2, ye, z1);

input rst_n, x1, x2;
output [2:0] ye;
output z1;

wire rst_n, x1, x2;
reg [2:0] ye, next_state;
reg z1;

//assign state codes
parameter    state_a = 3'b000,
             state_b = 3'b001,
             state_c = 3'b011,
             state_d = 3'b010,
             state_e = 3'b110,
             state_f = 3'b111,
             state_g = 3'b101;

//latch next state
always @ (x1 or x2 or rst_n)
begin
   if (~rst_n)
      ye <= state_a;
   else
      ye <= next_state;
end

//define output
always @ (x1 or x2 or ye)
begin
   if (ye == state_a)
      z1 = 1'b0;

   if (ye == state_b)
      z1 = 1'b0;

   if (ye == state_c)
      z1 = 1'b0;

   if (ye == state_d)
      z1 = 1'b0;

   if (ye == state_e)
      z1 = 1'b0;

//continued on next page
```

```
   if (ye == state_f)
      z1 = 1'b0;

   if (ye == state_g)
      z1 = 1'b1;
end

//determine next state
always @ (x1 or x2)
begin
   case (ye)
      state_a:
         if ((x1==1'b1) && (x2==1'b0))
            next_state = state_b;
         else
            next_state = state_a;

      state_b:
         if ((x1==1'b1) && (x2==1'b1))
            next_state = state_c;
         else if ((x1==1'b0) && (x2==1'b0))
            next_state = state_a;
         else
            next_state = state_b;

      state_c:
         if ((x1==1'b1) && (x2==1'b0))
            next_state = state_d;
         else
            next_state = state_c;

      state_d:
         if ((x1==1'b1) && (x2==1'b1))
            next_state = state_e;
         else if ((x1==1'b0) && (x2==1'b0))
            next_state = state_a;
         else
            next_state = state_d;

      state_e:
         if ((x1==1'b1) && (x2==1'b0))
            next_state = state_f;
         else
            next_state = state_e;

//continued on next page
```

```
        state_f:
           if ((x1==1'b1) && (x2==1'b1))
              next_state = state_g;
           else if ((x1==1'b0) && (x2==1'b0))
              next_state = state_a;
           else
              next_state = state_f;

        state_g:
           if ((x1==1'b1) && (x2==1'b0))
              next_state = state_b;
           else
              next_state = state_g;

        default:
           next_state = state_a;
     endcase
end
endmodule
```

```
//test bench for asm9
module asm9_tb;

reg rst_n, x1, x2;
wire [2:0] ye;
wire z1;

//define state transitions
initial
begin
   #0     rst_n = 1'b0;
          x1 = 1'b0;
          x2 = 1'b0;
   #10    rst_n = 1'b1;

   #10    x1=1'b1; x2=1'b0;     //go to state_b
   #10    x1=1'b1; x2=1'b1;     //go to state_c
   #10    x1=1'b1; x2=1'b0;     //go to state_d
   #10    x1=1'b1; x2=1'b1;     //go to state_e
   #10    x1=1'b1; x2=1'b0;     //go to state_f
   #10    x1=1'b1; x2=1'b1;     //go to state_g;assert z1
   #10    x1=1'b1; x2=1'b0;     //go to state_b
   #10    x1=1'b0; x2=1'b0;     //go to state_a
   #10    $stop;
end

//continued on next page
```

```
//instantiate the module into the test bench
asm9 inst1 (
   .rst_n(rst_n),
   .x1(x1),
   .x2(x2),
   .ye(ye),
   .z1(z1)
   );

endmodule
```

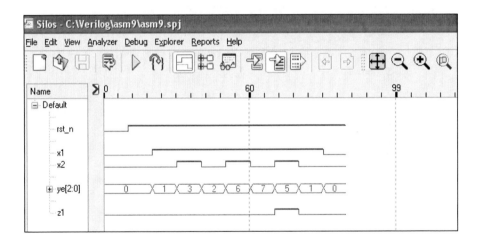

Chapter 9   Structural Modeling

9.3   Use structural modeling to design a 4-bit binary-to-excess-3 code converter, where the binary inputs are *b[3:0]* and the excess-3 outputs are *x[3:0]*. Design dataflow modules for the following functions: *and2_df*, *and3_df*, *xnor2_df*, *or3_df*, and *or4_df* that will be instantiated into the structural module. The equations for the excess-3 code are shown below. Obtain the structural module, test bench, and outputs.

x [3] = b[3] b[2]' + b[3] b[1]'b[0]' + b[3]' b[2] b[0] + b[3]' b[2] b[1]
x[2] = b[2]' b[1] + b[2]' b[0] + b[2] b[1]' b[0]'
x[1] = b[1]' b[0]' + b[1] b[0]
x[0] = b[0]'

```verilog
//structural binary-to-excess-3
module bin_to_excess3_struc (b, x);

input [3:0] b;
output [3:0] x;
wire [3:0] b;
wire net1, net2, net3, net4, net6, net7, net8;
wire [3:0] x;

//instantiate the logic for output x[3]
and2_df inst1 (
   .x1(b[3]),
   .x2(~b[2]),
   .z1(net1)
   );

and3_df inst2 (
   .x1(b[3]),
   .x2(~b[1]),
   .x3(~b[0]),
   .z1(net2)
   );

and3_df inst3 (
   .x1(~b[3]),
   .x2(b[2]),
   .x3(b[0]),
   .z1(net3)
   );

//continued on next page
```

```
and3_df inst4 (
   .x1(~b[3]),
   .x2(b[2]),
   .x3(b[1]),
   .z1(net4)
   );

or4_df inst5 (
   .x1(net1),
   .x2(net2),
   .x3(net3),
   .x4(net4),
   .z1(x[3])
   );

//instantiate the logic for output x[2]
and2_df inst6 (
   .x1(~b[2]),
   .x2(b[1]),
   .z1(net6)
   );

and2_df inst7 (
   .x1(~b[2]),
   .x2(b[0]),
   .z1(net7)
   );

and3_df inst8 (
   .x1(b[2]),
   .x2(~b[1]),
   .x3(~b[0]),
   .z1(net8)
   );

or3_df inst9 (
   .x1(net6),
   .x2(net7),
   .x3(net8),
   .z1(x[2])
   );

//continued on next page
```

```
//instantiate the logic for output x[1]
xnor2_df inst10 (
   .x1(b[1]),
   .x2(b[0]),
   .z1(x[1])
   );


//instantiate the logic for output x[0]
assign x[0] = ~b[0];

endmodule
```

```
//binary-to-excess3 test bench
module bin_to_excess3_struc_tb;


reg [3:0] b;
wire [3:0] x;


//apply stimulus
initial
begin : apply_stimulus
   reg [4:0] invect;
   for (invect = 0; invect < 16; invect = invect + 1)
      begin
         b = invect [4:0];
         #10    $display ("b = %b, x = %b", b, x);
      end
end


//instantiate the module into the test bench
bin_to_excess3_struc inst1 (
   .b(b),
   .x(x)
   );


endmodule
```

```
b = 0000, x = 0011
b = 0001, x = 0100
b = 0010, x = 0101
b = 0011, x = 0110
b = 0100, x = 0111
b = 0101, x = 1000
b = 0110, x = 1001
b = 0111, x = 1010
b = 1000, x = 1011
b = 1001, x = 1100
b = 1010, x = 1101
b = 1011, x = 1110
b = 1100, x = 1111
b = 1101, x = 0000
b = 1110, x = 0001
b = 1111, x = 0010
```

9.6   Use structural modeling to design a modulo-11 counter. Instantiate the following dataflow modules: *and2_df*, *and3_df*, *or2_df*, *or3_df*, and a behavioral module for a *D* flip-flop *d_ff*. Obtain the design module, the test bench, the outputs, and the waveforms. Display the simulation time.

```
//structural modulo_11 counter
module mod_11_ctr (clk, set_n, rst_n, y, y_n);

input clk, set_n, rst_n;
output [1:4] y, y_n;

wire clk, set_n, rst_n;
wire net1, net2, net3, net4, net5, net6, net7,
     net8, net9, net10, net11, net12, net13;
wire [1:4] y, y_n;

//instantiate gates to generate the input logic
//for flip-flop y[1]
and3_df inst1 (
    .x1(y[2]),
    .x2(y[3]),
    .x3(y[4]),
    .z1(net1)
    );

//continued on next page
```

```
and2_df inst2 (
   .x1(y[1]),
   .x2(~y[3]),
   .z1(net2)
   );

or2_df inst3 (
   .x1(net1),
   .x2(net2),
   .z1(net3)
   );

d_ff inst4 (
   .d(net3),
   .clk(clk),
   .q(y[1]),
   .q_n(y_n[1]),
   .set_n(set_n),
   .rst_n(rst_n)
   );

//instantiate gates to generate the input logic
//for flip-flop y[2]
and2_df inst5 (
   .x1(y[2]),
   .x2(~y[3]),
   .z1(net4)
   );

and2_df inst6 (
   .x1(y[2]),
   .x2(~y[4]),
   .z1(net5)
   );

and3_df inst7 (
   .x1(~y[2]),
   .x2(y[3]),
   .x3(y[4]),
   .z1(net6)
   );

or3_df inst8 (
   .x1(net4),
   .x2(net5),
   .x3(net6),
   .z1(net7)
   );                    //continued on next page
```

```
d_ff inst9 (
    .d(net7),
    .clk(clk),
    .q(y[2]),
    .q_n(y_n[2]),
    .set_n(set_n),
    .rst_n(rst_n)
    );

//instantiate gates to generate the input logic
//for flip-flop y[3]
and2_df inst10 (
    .x1(~y[3]),
    .x2(y[4]),
    .z1(net8)
    );

and3_df inst11 (
    .x1(~y[1]),
    .x2(y[3]),
    .x3(~y[4]),
    .z1(net9)
    );

or2_df inst12 (
    .x1(net8),
    .x2(net9),
    .z1(net10)
    );

d_ff inst13 (
    .d(net10),
    .clk(clk),
    .q(y[3]),
    .q_n(y_n[3]),
    .set_n(set_n),
    .rst_n(rst_n)
    );

//instantiate gates to generate the input logic
//for flip-flop y[4]
and2_df inst14 (
    .x1(~y[3]),
    .x2(~y[4]),
    .z1(net11)
    );

//continued on next page
```

```
and2_df inst15 (
   .x1(~y[1]),
   .x2(~y[4]),
   .z1(net12)
   );

or2_df inst16 (
   .x1(net11),
   .x2(net12),
   .z1(net13)
   );

d_ff inst17 (
   .d(net13),
   .clk(clk),
   .q(y[4]),
   .q_n(y_n[4]),
   .set_n(set_n),
   .rst_n(rst_n)
   );
endmodule
```

```
//test bench modulo_11 counter
module mod_11_ctr_tb;

reg clk, rst_n;
wire [1:4] y;

//display outputs at simulation time
initial
$monitor ($time, "Count = %b", y);

//define reset
initial
begin
   #0 rst_n = 1'b0;
   #5 rst_n = 1'b1;
end

//define clk
initial
begin
   clk = 1'b0;
   forever
      #10clk = ~clk;
end                        //continued on next page
```

```
//define length of simulation
initial
begin
   #300   $finish;
end

//instantiate the module into the test bench
mod_11_ctr inst1 (
   .clk(clk),
   .rst_n(rst_n),
   .y(y)
   );
endmodule
```

0	Count = 0000	150	Count = 1000
10	Count = 0001	170	Count = 1001
30	Count = 0010	190	Count = 1010
50	Count = 0011	210	Count = 0000
70	Count = 0100	230	Count = 0001
90	Count = 0101	250	Count = 0010
110	Count = 0110	270	Count = 0011
130	Count = 0111	290	Count = 0100

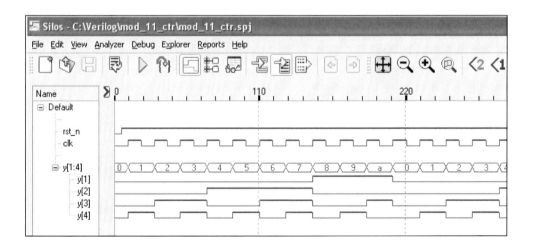

9.9   A serial data line x_1 consists of 3-bit words with one bit space between words as shown below:

$$x_1 = \begin{vmatrix} b_1 & b_2 & b_3 \end{vmatrix} \quad \begin{vmatrix} b_1 & b_2 & b_3 \end{vmatrix} \ldots$$

where $b_i = 0$ or 1. If there are exactly two 1s in any 3-bit word, then output z_1 is asserted during the last half of the clock cycle in the bit space between words. Since z_1 is asserted at time t_2 and deasserted at time t_3, there is no need to be concerned with output glitches because the machine will have stabilized before z_1 is asserted.

Derive the state diagram, Karnaugh maps for the flip-flops, and the input equations. Design the machine using structural modeling with D flip-flops and logic gates. Instantiate all logic gates as dataflow modules. Obtain the structural module, the test bench module, and the waveforms.

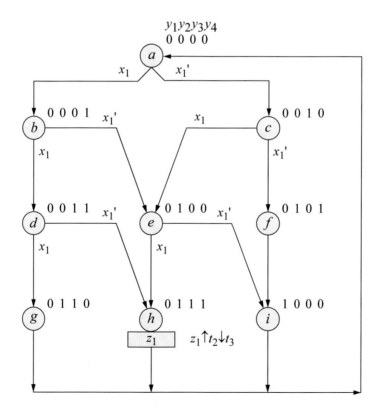

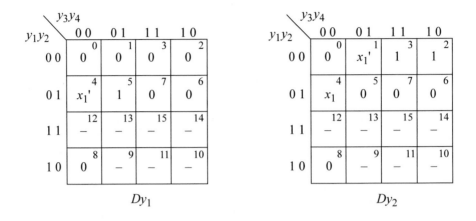

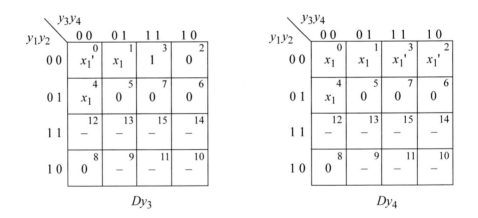

$$Dy_1 \quad = \quad y_2 y_3' x_1'$$
AND gate instantiation is *inst1*
AND gate output is *net1*

$$+ \, y_2 y_3' y_4$$
AND gate instantiation is *inst2*
AND gate output is *net2*

OR gate instantiation is *inst3*
OR gate output is *net3*

$Dy_2 \quad = \quad y_2 y_3' y_4' x_1$ AND gate instantiation is *inst5*
AND gate output is *net5*

$+ y_2' y_4 x_1'$ AND gate instantiation is *inst6*
AND gate output is *net6*

$+ y_2' y_3$ AND gate instantiation is *inst7*
AND gate output is *net7*

OR gate instantiation is *inst8*
OR gate output is *net8*

$Dy_3 \quad = \quad y_1' y_2' y_3' y_4' x_1'$ AND gate instantiation is *inst10*
AND gate output is *net10*

$+ y_2 y_3' y_4' x_1$ AND gate instantiation is *inst11*
AND gate output is *net11*

$+ y_2' y_4 x_1$ AND gate instantiation is *inst12*
AND gate output is *net12*

$+ y_2' y_3 y_4$ AND gate instantiation is *inst13*
AND gate output is *net13*

OR gate instantiation is *inst14*
OR gate output is *net14*

$Dy_4 \quad = \quad y_1' y_3' y_4' x_1$ AND gate instantiation is *inst16*
AND gate output is *net16*

$+ y_1' y_2' y_3' x_1$ AND gate instantiation is *inst17*
AND gate output is *net17*

$+ y_2' y_3 x_1'$ AND gate instantiation is *inst18*
AND gate output is *net18*

OR gate instantiation is *inst19*
OR gate output is *net19*

```
//structural synchronous machine to detect
//exactly two 1s in a 3-bit serial word
module moore_ssm6 (clk, set_n, rst_n, x1, y, z1);


input clk, set_n, rst_n, x1;
output [1:4] y;
output z1;


//define internal nets
wire   net1, net2, net3, net4, net5,
       net6, net7, net8, net9, net10,
       net11, net12, net13, net14, net15,
       net16, net17, net18, net19;


//design for flip-flop y[1]
and3_df inst1 (
    .x1(y[2]),
    .x2(~y[3]),
    .x3(~x1),
    .z1(net1)
    );

and3_df inst2 (
    .x1(y[2]),
    .x2(~y[3]),
    .x3(y[4]),
    .z1(net2)
    );

or2_df inst3 (
    .x1(net1),
    .x2(net2),
    .z1(net3)
    );

d_ff inst4 (
    .d(net3),
    .clk(clk),
    .q(y[1]),
    .set_n(set_n),
    .rst_n(rst_n)
    );

//continued on next page
```

```verilog
//design for flip-flop y[2]
and4_df inst5 (
    .x1(y[2]),
    .x2(~y[3]),
    .x3(~y[4]),
    .x4(x1),
    .z1(net5)
    );

and3_df inst6 (
    .x1(~y[2]),
    .x2(y[4]),
    .x3(~x1),
    .z1(net6)
    );

and2_df inst7 (
    .x1(~y[2]),
    .x2(y[3]),
    .z1(net7)
    );

or3_df inst8 (
    .x1(net5),
    .x2(net6),
    .x3(net7),
    .z1(net8)
    );

d_ff inst9 (
    .d(net8),
    .clk(clk),
    .q(y[2]),
    .set_n(set_n),
    .rst_n(rst_n)
    );

//design for flip-flop y[3]
and5_df inst10 (
    .x1(~y[1]),
    .x2(~y[2]),
    .x3(~y[3]),
    .x4(~y[4]),
    .x5(~x1),
    .z1(net10)
    );

//continued on next page
```

```
and4_df inst11 (
   .x1(y[2]),
   .x2(~y[3]),
   .x3(~y[4]),
   .x4(x1),
   .z1(net11)
   );

and3_df inst12 (
   .x1(~y[2]),
   .x2(y[4]),
   .x3(x1),
   .z1(net12)
   );

and3_df inst13 (
   .x1(~y[2]),
   .x2(y[3]),
   .x3(y[4]),
   .z1(net13)
   );

or4_df inst14 (
   .x1(net10),
   .x2(net11),
   .x3(net12),
   .x4(net13),
   .z1(net14)
   );

d_ff inst15 (
   .d(net14),
   .clk(clk),
   .q(y[3]),
   .set_n(set_n),
   .rst_n(rst_n)
   );

//design for flip-flop y[4]
and4_df inst16 (
   .x1(~y[1]),
   .x2(~y[3]),
   .x3(~y[4]),
   .x4(x1),
   .z1(net16)
   );

//continued on next page
```

```
and4_df inst17 (
    .x1(~y[1]),
    .x2(~y[2]),
    .x3(~y[3]),
    .x4(x1),
    .z1(net17)
    );

and3_df inst18 (
    .x1(~y[2]),
    .x2(y[3]),
    .x3(~x1),
    .z1(net18)
    );

or3_df inst19 (
    .x1(net16),
    .x2(net17),
    .x3(net18),
    .z1(net19)
    );

d_ff inst20 (
    .d(net19),
    .clk(clk),
    .q(y[4]),
    .set_n(set_n),
    .rst_n(rst_n)
    );

//design for output z1
and5_df inst21 (
    .x1(~y[1]),
    .x2(y[2]),
    .x3(y[3]),
    .x4(y[4]),
    .x5(~clk),
    .z1(z1)
    );

endmodule
```

```
//test bench for module moore_ssm6
module moore_ssm6_tb;

reg clk, set_n, rst_n, x1;
wire [1:4] y;
wire z1;

//define clock
initial
begin
   clk = 1'b0;
   forever
      #5 clk = ~clk;
end

//define input sequence
initial
begin
   #0 x1 = 1'b0;
      set_n = 1'b1;
      rst_n = 1'b0;                //initialize to state_a
   #3 rst_n = 1'b1;

//test for x1 = 00-
   @ (posedge clk)                         //go to state_c
   x1 = 1'b0;       @ (posedge clk)   //go to state_f
   x1 = $random;    @ (posedge clk)   //go to state_i
   x1 = $random;    @ (posedge clk)   //go to state_a

   x1 = 1'b0;       @ (posedge clk)   //go to state_c
   x1 = 1'b1;       @ (posedge clk)   //go to state_e
   x1 = 1'b0;       @ (posedge clk)   //go to state_i
   x1 = $random;    @ (posedge clk)   //go to state_a

//test for x1 = 011
   x1 = 1'b0;       @ (posedge clk)   //go to state_c
   x1 = 1'b1;       @ (posedge clk)   //go to state_e
   x1 = 1'b1;       @ (posedge clk)   //go to state_h;
                                      //assert z1
   x1 = $random;    @ (posedge clk)   //go to state_a

//test for x1 = 100
   x1 = 1'b1;       @ (posedge clk)   //go to state_b
   x1 = 1'b0;       @ (posedge clk)   //go to state_e
   x1 = 1'b0;       @ (posedge clk)   //go to state_i
   x1 = $random;    @ (posedge clk)   //go to state_a

//continued on next page
```

```
//test for x1 = 101
   x1 = 1'b1;      @ (posedge clk)      //go to state_b
   x1 = 1'b0;      @ (posedge clk)      //go to state_e
   x1 = 1'b1;      @ (posedge clk)      //go to state_h;
                                        //assert z1
   x1 = $random;   @ (posedge clk)      //go to state_a

//test for x1 = 110
   x1 = 1'b1;      @ (posedge clk)      //go to state_b
   x1 = 1'b1;      @ (posedge clk)      //go to state_d
   x1 = 1'b0;      @ (posedge clk)      //go to state_h;
                                        //assert z1
   x1 = $random;   @ (posedge clk)      //go to state_a

//test for x1 = 111
   x1 = 1'b1;      @ (posedge clk)      //go to state_b
   x1 = 1'b1;      @ (posedge clk)      //go to state_d
   x1 = 1'b1;      @ (posedge clk)      //go to state_g
   x1 = $random;   @ (posedge clk)      //go to state_a

   #20     $stop;
end

//instantiate the module into the test bench
moore_ssm6 inst1 (
   .clk(clk),
   .set_n(set_n),
   .rst_n(rst_n),
   .x1(x1),
   .y(y),
   .z1(z1)
   );
endmodule
```

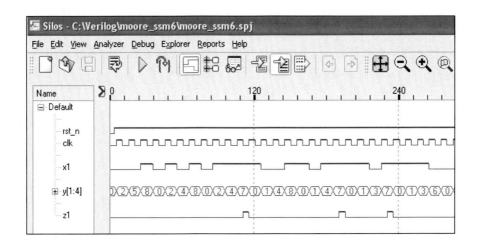

9.12   Design a Mealy pulse-mode asynchronous sequential machine that has three parallel inputs x_1, x_2, and x_3 and one output z_1 that is asserted coincident with x_3 whenever the sequence $x_1 x_2 x_3 = 100, 010, 001$ occurs. The storage elements will consist of SR latches and positive-edge-triggered D flip-flops. The latches are to be designed using NOR gates; therefore, the reset input will be a high logic level. The D flip-flops will be reset using a low logic level.

A representative timing diagram displaying valid input sequences and corresponding outputs is shown below together with the state diagram. State code assignment is arbitrary, because input pulses trigger all state transitions and the machine does not begin to sequence to the next state until the input pulse, which initiated the transition, has been deasserted. Thus, output z_1 will not glitch. In order to allow the machine to stabilize between input vectors, a vector of $x_1 x_2 x_3 = 000$ is inserted between each valid input.

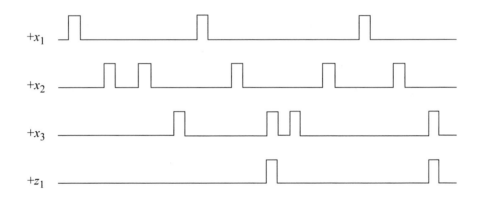

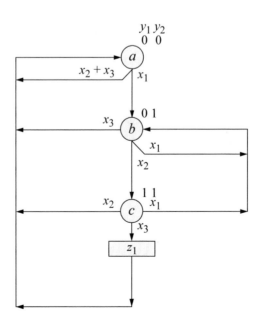

Obtain the structural module, the test bench for all state transitions, the outputs, and the waveforms.

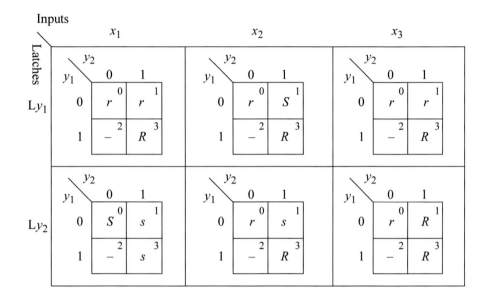

S indicates that the latch will be set.
s indicates that the latch will remain set.
R indicates that the latch will be reset.
r indicates that the latch will remain reset.

$$SLy_1 = y_1'y_2x_2$$

$$RLy_1 = x_1 + y_1x_2 + x_3$$

$$SLy_2 = x_1$$

$$RLy_2 = y_1x_2 + x_3$$

$$z_1 = y_1x_3$$

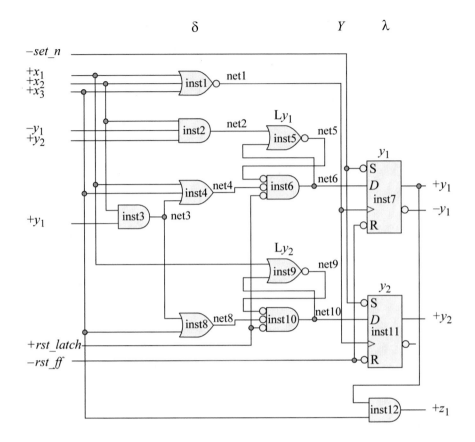

```
//structural module for a Mealy pulse-mode
//asynchronous sequential machine
module pm_asm6 (rst_latch, rst_ff, set_n, x1, x2, x3,
                y1, y2, z1);

input rst_latch, rst_ff, set_n, x1, x2, x3;
output y1, y2, z1;

//define internal nets
wire net1, net2, net3, net4, net5, net6, net8, net9,
     net10;

//design for clock input
   nor3_df inst1 (
   .x1(x1),
   .x2(x2),
   .x3(x3),
   .z1(net1)
   );

//continued on next page
```

```
//design for latch Ly1
and3_df inst2 (
    .x1(x2),
    .x2(~y1),
    .x3(y2),
    .z1(net2)
    );

and2_df inst3 (
    .x1(x2),
    .x2(y1),
    .z1(net3)
    );

or3_df inst4 (
    .x1(x1),
    .x2(x3),
    .x3(net3),
    .z1(net4)
    );

nor2_df inst5 (
    .x1(net2),
    .x2(net6),
    .z1(net5)
    );

nor3_df inst6 (
    .x1(net5),
    .x2(net4),
    .x3(rst_latch),
    .z1(net6)
    );

//design for D flip-flop y1
d_ff inst7 (
    .d(net6),
    .clk(net1),
    .q(y1),
    .set_n(set_n),
    .rst_n(rst_ff)
    );

//continued on next page
```

```
//design for latch Ly2
or2_df inst8 (
    .x1(net3),
    .x2(x3),
    .z1(net8)
    );

nor2_df inst9 (
    .x1(x1),
    .x2(net10),
    .z1(net9)
    );

nor3_df inst10 (
    .x1(net9),
    .x2(net8),
    .x3(rst_latch),
    .z1(net10)
    );

//design for D flip-flop y2
d_ff inst11 (
    .d(net10),
    .clk(net1),
    .q(y2),
    .set_n(set_n),
    .rst_n(rst_ff)
    );

//design for z1
and2_df inst12 (
    .x1(y1),
    .x2(x3),
    .z1(z1)
    );
endmodule
```

```
//test bench for Mealy pulse-mode machine
module pm_asm6_tb;

reg rst_latch, rst_ff, set_n, x1, x2, x3;
wire y1, y2, z1;

//display variables
initial
$monitor ("x1=%b, x2=%b, x3=%b, state=%b, z1=%b",
          x1, x2, x3, {y1, y2}, z1);    //next page
```

```
//apply stimulus
initial
begin
   #0    set_n=1'b1;
         rst_latch=1'b1;    //reset latches
         rst_ff=1'b0;       //reset flip-flops to
                            //state_a (00)
         x1=1'b0;
         x2=1'b0;
         x3=1'b0;
   #5    rst_latch=1'b0;    //remove reset from latches
         rst_ff=1'b1;       //remove reset from
                            //flip-flops

   #10   x1=1'b0; x2=1'b1; x3=1'b0; //go to state_a(00)
   #10   x1=1'b0; x2=1'b0; x3=1'b0;
   #10   x1=1'b0; x2=1'b0; x3=1'b1; //go to state_a(00)
   #10   x1=1'b0; x2=1'b0; x3=1'b0;
   #10   x1=1'b1; x2=1'b0; x3=1'b0; //go to state_b(01)
   #10   x1=1'b0; x2=1'b0; x3=1'b0;
   #10   x1=1'b0; x2=1'b0; x3=1'b1; //go to state_a(00)
   #10   x1=1'b0; x2=1'b0; x3=1'b0;
   #10   x1=1'b1; x2=1'b0; x3=1'b0; //go to state_b(01)
   #10   x1=1'b0; x2=1'b0; x3=1'b0;
   #10   x1=1'b0; x2=1'b1; x3=1'b0; //go to state_c(11)
   #10   x1=1'b0; x2=1'b0; x3=1'b0;
   #10   x1=1'b1; x2=1'b0; x3=1'b0; //go to state_b(01)
   #10   x1=1'b0; x2=1'b0; x3=1'b0;
   #10   x1=1'b0; x2=1'b1; x3=1'b0; //go to state_c(11)
   #10   x1=1'b0; x2=1'b0; x3=1'b0;
   #10   x1=1'b0; x2=1'b1; x3=1'b0; //go to state_a(00)
   #10   x1=1'b0; x2=1'b0; x3=1'b0;
   #10   x1=1'b1; x2=1'b0; x3=1'b0; //go to state_b(01)
   #10   x1=1'b0; x2=1'b0; x3=1'b0;
   #10   x1=1'b0; x2=1'b1; x3=1'b0; //go to state_c(11)

   #10   x1=1'b0; x2=1'b0; x3=1'b1; //go to state_a(00)
                                    //assert z1
   #10   x1=1'b0; x2=1'b0; x3=1'b0;
   #10   x1=1'b0; x2=1'b1; x3=1'b0; //go to state_a(00)

   #10   $stop;
end

//continued on next  page
```

```
//instantiate the module into the test bench
pm_asm6 inst1 (
   .rst_latch(rst_latch),
   .rst_ff(rst_ff),
   .set_n(set_n),
   .x1(x1),
   .x2(x2),
   .x3(x3),
   .y1(y1),
   .y2(y2),
   .z1(z1)
   );

endmodule
```

```
x1=0, x2=0, x3=0, state=00, z1=0
x1=0, x2=1, x3=0, state=00, z1=0
x1=0, x2=0, x3=0, state=00, z1=0
x1=0, x2=0, x3=1, state=00, z1=0
x1=0, x2=0, x3=0, state=00, z1=0
x1=1, x2=0, x3=0, state=00, z1=0
x1=0, x2=0, x3=0, state=01, z1=0
x1=0, x2=0, x3=1, state=01, z1=0
x1=0, x2=0, x3=0, state=00, z1=0
x1=1, x2=0, x3=0, state=00, z1=0
x1=0, x2=0, x3=0, state=01, z1=0
x1=0, x2=1, x3=0, state=01, z1=0
x1=0, x2=0, x3=0, state=11, z1=0
x1=1, x2=0, x3=0, state=11, z1=0
x1=0, x2=0, x3=0, state=01, z1=0
x1=0, x2=1, x3=0, state=01, z1=0
x1=0, x2=0, x3=0, state=11, z1=0
x1=0, x2=1, x3=0, state=11, z1=0
x1=0, x2=0, x3=0, state=00, z1=0
x1=1, x2=0, x3=0, state=00, z1=0
x1=0, x2=0, x3=0, state=01, z1=0
x1=0, x2=1, x3=0, state=01, z1=0
x1=0, x2=0, x3=0, state=11, z1=0
x1=0, x2=0, x3=1, state=11, z1=1
x1=0, x2=0, x3=0, state=00, z1=0
x1=0, x2=1, x3=0, state=00, z1=0
```

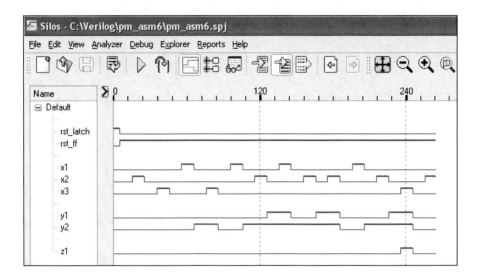

Chapter 10   Tasks and Functions

10.1   Design a module that contains a task to detect even or odd parity in a 16-bit register *reg_a*. The task returns a value of 0 if the parity is even and a value of 1 if the parity is odd. The task declaration returns the parity value to an integer called *parity*. Obtain the outputs from the module.

```
//module to illustrate a task to determine parity.
//0 indicates even parity, 1 indicates odd parity
module task_parity;

reg [15:0] reg_a;
integer parity;

initial
begin
    reg_a = 16'b1111_0000_1111_0000;    //even parity
  par (reg_a, parity);                //invoke the task

    reg_a = 16'b1111_0000_1111_0001;    //odd parity
  par (reg_a, parity);                //invoke the task

    reg_a = 16'b1111_1111_1111_1111;    //even parity
  par (reg_a, parity);                //invoke the task

    reg_a = 16'b0000_0000_0000_0001;    //odd parity
  par (reg_a, parity);                //invoke the task

    reg_a = 16'b1111_0010_1111_0000;    //odd parity
  par (reg_a, parity);                //invoke the task

    reg_a = 16'b1100_0000_1111_0001;    //odd parity
  par (reg_a, parity);                //invoke the task

end
task par;
   input [15:0] reg_a;
   output parity;

   begin
      parity = ^reg_a;
      $display ("parity = %b", parity);
   end
endtask

endmodule
```

```
parity = 0
parity = 1
parity = 0
parity = 1
parity = 1
parity = 1
```

10.5   Design a function for a half adder that adds operands *a* and *b* and produces a result *sum*. The **case** statement can be used in the function. Display the results.

```
//module for a half adder using a function
module fctn_half_add;

reg a, b;
reg [1:0] sum;

initial
begin
   sum = half_add (1'b0, 1'b0);
      $display ("a=0, b=0, cout, sum = %b", sum);

   sum = half_add (1'b0, 1'b1);
      $display ("a=0, b=1, cout, sum = %b", sum);

   sum = half_add (1'b1, 1'b0);
      $display ("a=1, b=0, cout, sum = %b", sum);

   sum = half_add (1'b1, 1'b1);
      $display ("a=1, b=1, cout, sum = %b", sum);
end

function [1:0] half_add;
input a, b;
reg [1:0] sum;
begin
   case ({a,b})
      2'b00:   sum = 2'b00;
      2'b01:   sum = 2'b01;
      2'b10:   sum = 2'b01;
      2'b11:   sum = 2'b10;
      default:sum = 2'bxx;
   endcase
      half_add = sum;
end
endfunction
endmodule
```

```
a=0, b=0, cout, sum = 00
a=0, b=1, cout, sum = 01
a=1, b=0, cout, sum = 01
a=1, b=1, cout, sum = 10
```

Chapter 11   Additional Design Examples

11.1   Design a behavioral module for an 8-bit serial-in, parallel-out shift register. Obtain the test bench, outputs, and waveforms for various input sequences.

```
//behavioral serial-in parallel-out shift register
module shift_reg_sipo (rst_n, clk, ser_in, shift_reg);

input rst_n, clk, ser_in;
output [7:0] shift_reg;

reg [7:0] shift_reg;

always @ (rst_n)
begin
   if (rst_n == 0)
      shift_reg <= 8'b0000_0000;
end

always @ (posedge clk)
begin
   shift_reg [7] <= ser_in;
   shift_reg [6] <= shift_reg [7];
   shift_reg [5] <= shift_reg [6];
   shift_reg [4] <= shift_reg [5];
   shift_reg [3] <= shift_reg [4];
   shift_reg [2] <= shift_reg [3];
   shift_reg [1] <= shift_reg [2];
   shift_reg [0] <= shift_reg [1];
end
endmodule
```

```
//test bench for serial-in parallel-out shift register
module shift_reg_sipo_tb;

reg rst_n, clk, ser_in;
wire [7:0] shift_reg;

initial                //define clock
begin
   clk = 1'b0;
   forever
      #10 clk = ~clk;
end
//continued on next page
```

```verilog
initial                    //display variables
$monitor ("ser_in = %b, shift_reg = %b",
          ser_in, shift_reg);


initial                    //apply inputs
begin
   #0     rst_n = 1'b0;
          ser_in = 1'b0;
   #5     rst_n = 1'b1;
          ser_in = 1'b1;


   #10    ser_in = 1'b1;
   #10    ser_in = 1'b0;
   #10    ser_in = 1'b1;
   #10    ser_in = 1'b0;
   #10    ser_in = 1'b1;
   #10    ser_in = 1'b1;
   #10    ser_in = 1'b0;
   #10    ser_in = 1'b1;
   #10    ser_in = 1'b1;
   #10    ser_in = 1'b1;
   #20    ser_in = 1'b0;
   #10    ser_in = 1'b1;


   #30    $stop;
end


//instantiate the module into the test bench
shift_reg_sipo inst1 (
   .rst_n(rst_n),
   .clk(clk),
   .ser_in(ser_in),
   .shift_reg(shift_reg)
   );

endmodule
```

```
ser_in = 0, shift_reg = 00000000
ser_in = 1, shift_reg = 00000000
ser_in = 1, shift_reg = 10000000
ser_in = 0, shift_reg = 10000000
ser_in = 0, shift_reg = 01000000
ser_in = 1, shift_reg = 01000000
ser_in = 0, shift_reg = 01000000
ser_in = 0, shift_reg = 00100000
ser_in = 1, shift_reg = 00100000
ser_in = 1, shift_reg = 10010000
ser_in = 0, shift_reg = 10010000
ser_in = 1, shift_reg = 10010000
ser_in = 1, shift_reg = 11001000
ser_in = 1, shift_reg = 11100100
ser_in = 0, shift_reg = 11100100
ser_in = 0, shift_reg = 01110010
ser_in = 1, shift_reg = 01110010
ser_in = 1, shift_reg = 10111001
```

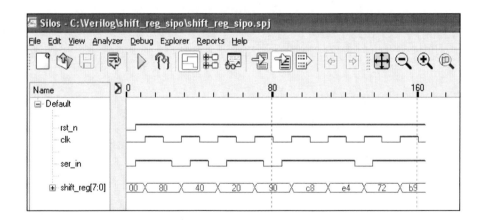

11.5   Design a 4-bit Gray code counter using structural modeling that incorporates a parity bit to maintain odd parity for the concatenated 4-bit Gray code and parity bit itself. Include error detection logic that generates an error output if the five bits have even parity. Use *JK* flip-flops and any additional dataflow logic primitives. Obtain the structural module, the test bench, the outputs, and the waveforms. After the counting sequence has been verified, the test bench will force a parity error to check the parity error detection logic.

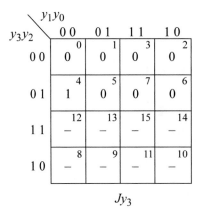

Jy_3

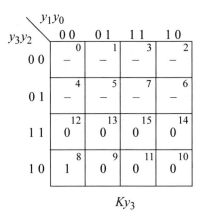

Ky_3

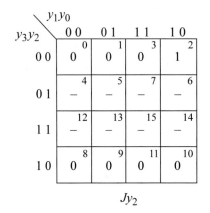

Jy_2

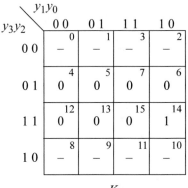

Ky_2

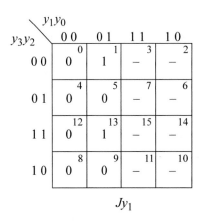

Jy_1

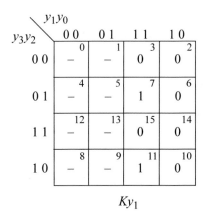

Ky_1

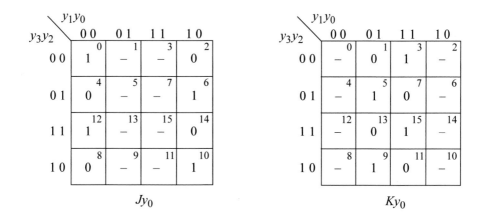

$$Jy_3 = y_2y_1'y_0'$$
$$Ky_3 = y_2'y_1'y_0'$$

$$Jy_2 = y_3'y_1y_0'$$
$$Ky_2 = y_3y_1y_0'$$

$$Jy_1 = y_3'y_2'y_0 + y_3y_2y_0$$
$$Ky_1 = y_3'y_2y_0 + y_3y_2'y_0$$

$$Jy_0 = y_3'y_2'y_1' + y_3'y_2y_1 + y_3y_2y_1' + y_3y_2'y_1$$
$$Ky_0 = y_3'y_2'y_1 + y_3'y_2y_1' + y_3y_2y_1 + y_3y_2'y_1'$$

```
//structural module for 4-bit Gray code counter
//with parity check
module ctr_gray_par_chk (sety3_n, set_n, rst_n, clk,
                         y, yp, par_err);

input sety3_n, set_n, rst_n, clk;
output [3:0] y;
output yp, par_err;

//define internal nets
wire  net1, net2, net4, net5, net7, net8,
      net9, net10, net12, net13;

//continued on next page
```

```
//design for flip-flop y[3]
and3_df inst1 (
    .x1(y[2]),
    .x2(~y[1]),
    .x3(~y[0]),
    .z1(net1)
    );

and3_df inst2 (
    .x1(~y[0]),
    .x2(~y[1]),
    .x3(~y[2]),
    .z1(net2)
    );

jkff_neg_clk inst3 (
    .clk(clk),
    .j(net1),
    .k(net2),
    .set_n(sety3_n),
    .rst_n(rst_n),
    .q(y[3])
    );

//design for flip-flop y[2]
and3_df inst4 (
    .x1(~y[3]),
    .x2(y[1]),
    .x3(~y[0]),
    .z1(net4)
    );

and3_df inst5 (
    .x1(y[1]),
    .x2(~y[0]),
    .x3(y[3]),
    .z1(net5)
    );

jkff_neg_clk inst6 (
    .clk(clk),
    .j(net4),
    .k(net5),
    .set_n(set_n),
    .rst_n(rst_n),
    .q(y[2])
    );

//continued on next page
```

```
//design for flip-flop y[1]
xnor2_df inst7 (
   .x1(y[3]),
   .x2(y[2]),
   .z1(net7)
   );

and2_df inst8 (
   .x1(net7),
   .x2(y[0]),
   .z1(net8)
   );

xor2_df inst9 (
   .x1(y[2]),
   .x2(y[3]),
   .z1(net9)
   );

and2_df inst10 (
   .x1(net9),
   .x2(y[0]),
   .z1(net10)
   );

jkff_neg_clk inst11 (
   .clk(clk),
   .j(net8),
   .k(net10),
   .set_n(set_n),
   .rst_n(rst_n),
   .q(y[1])
   );

//design for flip-flop y[0]
xnor3_df inst12 (
   .x1(y[3]),
   .x2(y[2]),
   .x3(y[1]),
   .z1(net12)
   );

xor3_df inst13 (
   .x1(y[1]),
   .x2(y[2]),
   .x3(y[3]),
   .z1(net13)
   );                    //continued on next page
```

```
jkff_neg_clk inst14 (
   .clk(clk),
   .j(net12),
   .k(net13),
   .set_n(set_n),
   .rst_n(rst_n),
   .q(y[0])
   );

//design for flip-flop yp
jkff_neg_clk inst15 (
   .clk(clk),
   .j(1'b1),
   .k(1'b1),
   .set_n(rst_n),
   .rst_n(1'b1),
   .q(yp)
   );

//design for parity error
xnor5_df inst16 (
   .x1(y[3]),
   .x2(y[2]),
   .x3(y[1]),
   .x4(y[0]),
   .x5(yp),
   .z1(par_err)
   );

endmodule
```

```
//test bench for the parity checked Gray code counter
module ctr_gray_par_chk_tb;

reg sety3_n, set_n, rst_n, clk;
wire [3:0] y;
wire yp, par_err;

//display variables
initial
$monitor ("count=%b, parity flag=%b, parity error=%b",
          y, yp, par_err);

//continued on next page
```

```verilog
//generate reset
initial
begin
   #0    rst_n = 1'b0;      //reset
         sety3_n = 1'b1;
         set_n = 1'b1;
   #5    rst_n = 1'b1;      //remove reset
   #345  sety3_n = 1'b0;
   #360  sety3_n = 1'b1;
end

//generate clock
initial
begin
   clk = 1'b0;
   forever
      #10clk = ~clk;
end

//determine length of simulation
initial
begin
   repeat (20) @ (negedge clk);
   $stop;
end

//instantiate the module into the test bench
ctr_gray_par_chk inst1 (
   .sety3_n(sety3_n),
   .set_n(set_n),
   .rst_n(rst_n),
   .clk(clk),
   .y(y),
   .yp(yp),
   .par_err(par_err)
   );
endmodule
```

```
count = 0000, parity flag = 1, parity error = 0
count = 0001, parity flag = 0, parity error = 0
count = 0011, parity flag = 1, parity error = 0
count = 0010, parity flag = 0, parity error = 0
count = 0110, parity flag = 1, parity error = 0
count = 0111, parity flag = 0, parity error = 0
count = 0101, parity flag = 1, parity error = 0
count = 0100, parity flag = 0, parity error = 0
//continued on next page
```

```
count = 1100, parity flag = 1, parity error = 0
count = 1101, parity flag = 0, parity error = 0
count = 1111, parity flag = 1, parity error = 0
count = 1110, parity flag = 0, parity error = 0
count = 1010, parity flag = 1, parity error = 0
count = 1011, parity flag = 0, parity error = 0
count = 1001, parity flag = 1, parity error = 0
count = 1000, parity flag = 0, parity error = 0
count = 0000, parity flag = 1, parity error = 0
count = 0001, parity flag = 0, parity error = 0
count = 1001, parity flag = 0, parity error = 1
count = 1000, parity flag = 1, parity error = 1
```

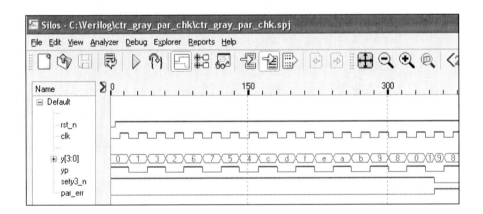

11.8   Design a Mealy synchronous sequential machine using structural modeling to detect a sequence of 0110 on a serial input line x_1. Overlapping sequences are valid. Output z_1 will be asserted during the last half of the clock cycle in the state in which the final 0 is detected. Obtain the state diagram, the Karnaugh maps for the input equations using D flip-flops, and the logic diagram.

Use dataflow modeling for any logic primitives that are required. Use behavioral modeling for the D flip-flop. Obtain the structural module, the test bench, the outputs, and the waveforms. In the test bench, check all possible paths in the state diagram for correct functional operation and include at least two valid overlapping sequences.

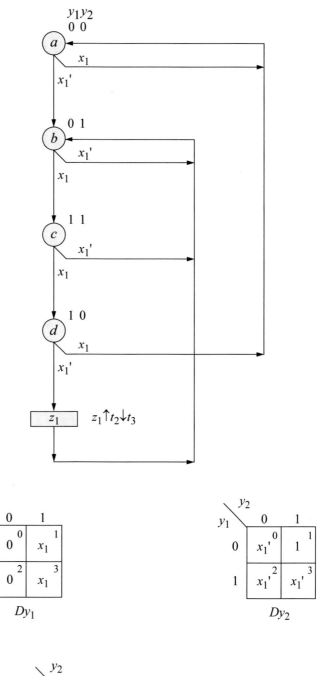

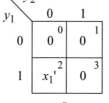

$$Dy_1 = y_2 x_1$$
$$Dy_2 = x_1' + y_1' y_2$$
$$z_1 = y_1 y_2' x_1'$$

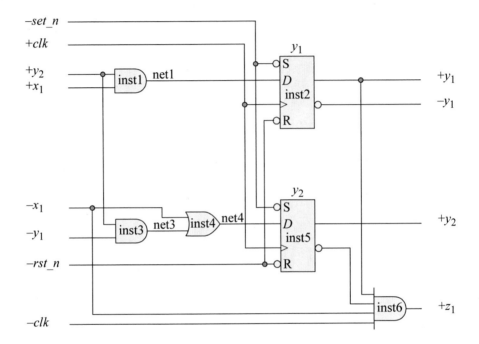

```
//structural mealy ssm to detect 0110
module mealy_ssm4 (set_n, rst_n, clk, x1, y, z1);

input set_n, rst_n, clk, x1;
output [1:2] y;
output z1;

//define internal nets
wire net1, net3, net4;

//design for flip-flop y[1]
and2_df inst1 (
    .x1(y[2]),
    .x2(x1),
    .z1(net1)
    );
//continued on next page
```

```
d_ff inst2 (
   .d(net1),
   .clk(clk),
   .q(y[1]),
   .set_n(set_n),
   .rst_n(rst_n)
   );


//design for flip-flop y[2]
and2_df inst3 (
   .x1(y[2]),
   .x2(~y[1]),
   .z1(net3)
   );


or2_df inst4 (
   .x1(~x1),
   .x2(net3),
   .z1(net4)
   );


d_ff inst5 (
   .d(net4),
   .clk(clk),
   .q(y[2]),
   .set_n(set_n),
   .rst_n(rst_n)
   );


//design for output z1
and4_df inst6 (
   .x1(y[1]),
   .x2(~y[2]),
   .x3(~x1),
   .x4(~clk),
   .z1(z1)
   );

endmodule
```

```verilog
//test bench for mealy ssm to detect 0110
module mealy_ssm4_tb;

reg set_n, rst_n, clk, x1;
wire [1:2] y;
wire z1;

//display variables
initial
$monitor ("y = %b, x1 = %b, z1 = %b", y, x1, z1);

//define clock
initial
begin
   clk = 1'b0;
   forever
      #10   clk = ~clk;
end

//define input sequence
initial
begin
   #0    x1 = 1'b0;
         set_n = 1'b1;
         rst_n = 1'b0;
   #5    rst_n = 1'b1;

   @ (posedge clk)
   //if x1 = 0 in state_a, go to state_b (01)

   x1 = 1'b1;  @ (posedge clk)
   //if x1 = 1 in state_b, go to state_c (11)

   x1 = 1'b1;  @ (posedge clk)
   //if x1 = 1 in state_c, go to state_d (10)

   x1 = 1'b0;  @ (posedge clk)
   //if x1 = 0 in state_d, assert z1 (t2 - t3)
   //go to state_b (01)

   x1 = 1'b0;  @ (posedge clk)
   //if x1 = 0 in state_b, go to state_b (01)

   x1 = 1'b1;  @ (posedge clk)
   //if x1 = 1 in state_b, go to state_c (11)


//continued on next page
```

```
    x1 = 1'b0;  @ (posedge clk)
    //if x1 = 1'b0 in state_c, go to state_b (01)

    x1 = 1'b1;  @ (posedge clk)
    //if x1 = 1'b1 state_b, go to state_c (110)

    x1 = 1'b1;  @ (posedge clk)
    //if x1 = 1'b1 in state_c, go to state_d (10)

    x1 = 1'b1;  @ (posedge clk)
    //if x1 = 1'b1 in state_d, go to state_a (00)

    x1 = 1'b0;  @ (posedge clk)
    //if x1 = 0 in state_a, go to state_b (01)

    x1 = 1'b1;  @ (posedge clk)
    //if x1 = 1 in state_b, go to state_c (11)

    x1 = 1'b1;  @ (posedge clk)
    //if x1 = 1 in state_c, go to state_d (10)

    x1 = 1'b0;  @ (posedge clk)
    //if x1 = 0 in state_d, assert z1 (t2 - t3)
    //go to state_b (01)

    x1 = 1'b1;  @ (posedge clk)
    //if x1 = 1'b1 in state_c, go to state_d (10)

    x1 = 1'b1;  @ (posedge clk)
    //if x1 = 1'b1 in state_d, go to state_a (00)

    x1 = 1'b0;  @ (posedge clk)
    //if x1 = 0 in state_a, go to state_b (01)

    #30     $stop;
end

//instantiate the module into the test bench
mealy_ssm4 inst1 (
    .set_n(set_n),
    .rst_n(rst_n),
    .clk(clk),
    .x1(x1),
    .y(y),
    .z1(z1)
    );

endmodule
```

```
y = 00, x1 = 0, z1 = 0          y = 00, x1 = 0, z1 = 0
y = 01, x1 = 1, z1 = 0          y = 01, x1 = 1, z1 = 0
y = 11, x1 = 1, z1 = 0          y = 11, x1 = 1, z1 = 0
y = 10, x1 = 0, z1 = 0          y = 10, x1 = 0, z1 = 0
y = 10, x1 = 0, z1 = 1          y = 10, x1 = 0, z1 = 1
y = 01, x1 = 0, z1 = 0          y = 01, x1 = 1, z1 = 0
y = 01, x1 = 1, z1 = 0          y = 11, x1 = 1, z1 = 0
y = 11, x1 = 0, z1 = 0          y = 10, x1 = 0, z1 = 0
y = 01, x1 = 1, z1 = 0          y = 10, x1 = 0, z1 = 1
y = 11, x1 = 1, z1 = 0          y = 01, x1 = 0, z1 = 0
y = 10, x1 = 1, z1 = 0
```

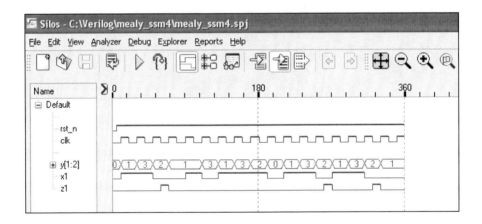

11.11 Using mixed-design modeling, design a 4-function arithmetic and logic unit to execute the following four operations: add, multiply, AND, and OR. The operands are eight bits defined as *a[7:0]* and *b[7:0]*. The result of all operations will be 16 bits *rslt[15:0]*. Determine the state of the following flags:

The parity flag *pf* = 1 if the result has an even number of 1s; otherwise, *pf* = 0.

The sign flag *sf* represents the state of the leftmost result bit.

The zero flag *zf* = 1 if the result of an operation is zero; otherwise, *zf* = 0.

Obtain the design module, the test bench module using several values for each of the four operations, the outputs, and the waveforms.

```verilog
//mixed-design modeling alu_4_functions
module alu_4fctn2 (a, b, opcode, rslt, pf, sf, zf);

input [7:0] a, b;
input [1:0] opcode;
output [15:0] rslt;
output pf, sf, zf;

reg [15:0] rslt;

// define internal operation results
reg [7:0] a_add_b;
reg [15:0] a_mul_b;
reg [7:0] a_and_b;
reg [7:0] a_or_b;

always @ (a or b)
begin
    a_add_b = a + b;
end

always @ (a or b)
begin
    a_mul_b = a * b;
end

always @ (a or b)
begin
    a_and_b = a & b;
end

always @ (a or b)
begin
    a_or_b = a | b;
end

assign pf = ~^rslt;
assign sf = rslt[15];
assign zf = (rslt == 16'h0000);

//define the opcodes
parameter add_op  = 2'b00;
parameter mul_op  = 2'b01;
parameter and_op  = 2'b10;
parameter or_op   = 2'b11;


//continued on next page
```

```verilog
always @ (opcode or a_add_b or a_mul_b or a_and_b
          or a_or_b)
begin
  case (opcode)
    add_op : rslt = {8'h00, a_add_b};
    mul_op : rslt = a_mul_b;
    and_op : rslt = {8'h00, a_and_b};
    or_op  : rslt = {8'h00, a_or_b};
    default: rslt = 16'hxxxx;
  endcase
end

endmodule
```

```verilog
//test bench for alu_4_functions
module alu_4fctn_tb;

reg [7:0] a, b;        //inputs are reg for tb
reg [1:0] opcode;
wire [15:0] rslt;      //outputs are wire for tb
wire pf, sf, zf;

initial
$monitor ("a=%h, b=%h, op=%h, rslt=%h, pfsfzf=%b",
          a, b, opcode, rslt, {pf,sf,zf});

//apply input vectors
initial
begin
//add op --------------------------------------------
  #0  a = 8'b0000_0001; b = 8'b0000_0010;
          opcode = 2'b00;
  #10 a = 8'b1100_0101; b = 8'b1011_0110;
          opcode = 2'b00;
  #10 a = 8'b0000_0101; b = 8'b1011_0110;
          opcode = 2'b00;
  #10 a = 8'b1000_1010; b = 8'b0011_1010;
          opcode = 2'b00;
  #10 a = 8'b1111_1111; b = 8'b0000_0001;
          opcode = 2'b00;
  #10 a = 8'b0000_0000; b = 8'b0000_0000;
          opcode = 2'b00;

//continued on next page
```

```
//mul_op -----------------------------------------------
   #10 a = 8'b0000_0100; b = 8'b0000_0011;
             opcode = 2'b01;
   #10 a = 8'b1000_0000; b = 8'b0000_0010;
             opcode = 2'b01;
   #10 a = 8'b1010_1010; b = 8'b0101_0101;
             opcode = 2'b01;
   #10 a = 8'b1111_1111; b = 8'b1111_1111;
             opcode = 2'b01;

//and op -----------------------------------------------
   #10 a = 8'b1111_0000; b = 8'b0000_1111;
             opcode = 2'b10;
   #10 a = 8'b1010_1010; b = 8'b1100_0011;
             opcode = 2'b10;
   #10 a = 8'b1111_1111; b = 8'b1111_0000;
             opcode = 2'b10;

//or op ------------------------------------------------
   #10 a = 8'b1111_0000; b = 8'b0000_1111;
             opcode = 2'b11;
   #10 a = 8'b1010_1010; b = 8'b1100_0011;
             opcode = 2'b11;
   #10 a = 8'b1111_1111; b = 8'b1111_0000;
             opcode = 2'b11;

   #10 $stop;

end


//instantiate the module into the test bench
alu_4fctn2 inst1 (
   .a(a),
   .b(b),
   .opcode(opcode),
   .rslt(rslt),
   .pf(pf),
   .sf(sf),
   .zf(zf)
   );

endmodule
```

```
a=01, b=02, op=0, rslt=0003, pfsfzf=100
a=c5, b=b6, op=0, rslt=007b, pfsfzf=100
a=05, b=b6, op=0, rslt=00bb, pfsfzf=100
a=8a, b=3a, op=0, rslt=00c4, pfsfzf=000
a=ff, b=01, op=0, rslt=0000, pfsfzf=101
a=00, b=00, op=0, rslt=0000, pfsfzf=101
a=04, b=03, op=1, rslt=000c, pfsfzf=100
a=80, b=02, op=1, rslt=0100, pfsfzf=000
a=aa, b=55, op=1, rslt=3872, pfsfzf=000
a=ff, b=ff, op=1, rslt=fe01, pfsfzf=110
a=f0, b=0f, op=2, rslt=0000, pfsfzf=101
a=aa, b=c3, op=2, rslt=0082, pfsfzf=100
a=ff, b=f0, op=2, rslt=00f0, pfsfzf=100
a=f0, b=0f, op=3, rslt=00ff, pfsfzf=100
a=aa, b=c3, op=3, rslt=00eb, pfsfzf=100
a=ff, b=f0, op=3, rslt=00ff, pfsfzf=100
```

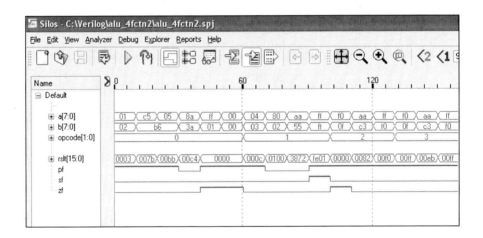

INDEX